Managing Forest Ecosystems

Volume 32

Aims & Scope

Well-managed forests and woodlands are a renewable resource, producing essential raw material with minimum waste and energy use. Rich in habitat and species diversity, forests may contribute to increased ecosystem stability. They can absorb the effects of unwanted deposition and other disturbances and protect neighbouring ecosystems by maintaining stable nutrient and energy cycles and by preventing soil degradation and erosion. They provide much-needed recreation and their continued existence contributes to stabilizing rural communities

Forests are managed for timber production and species, habitat and process conservation. A subtle shift from *multiple-use management* to *ecosystems management* is being observed and the new ecological perspective of *multi-functional forest management* is based on the principles of ecosystem diversity, stability and elasticity, and the dynamic equilibrium of primary and secondary production.

Making full use of new technology is one of the challenges facing forest management today. Resource information must be obtained with a limited budget. This requires better timing of resource assessment activities and improved use of multiple data sources. Sound ecosystems management, like any other management activity, relies on effective forecasting and operational control.

The aim of the book series ***Managing Forest Ecosystems*** is to present state-of-the-art research results relating to the practice of forest management. Contributions are solicited from prominent authors. Each reference book, monograph or proceedings volume will be focused to deal with a specific context. Typical issues of the series are: resource assessment techniques, evaluating sustainability for even-aged and uneven-aged forests, multi-objective management, predicting forest development, optimizing forest management, biodiversity management and monitoring, risk assessment and economic analysis.

More information about this series at http://www.springer.com/series/6247

Cathryn H. Greenberg • Beverly S. Collins
Editors

Natural Disturbances and Historic Range of Variation

Type, Frequency, Severity, and Post-disturbance Structure in Central Hardwood Forests USA

Springer

Editors
Cathryn H. Greenberg
USDA Forest Service, Southern Research Station, Bent Creek Experimental Forest
Asheville, NC, USA

Beverly S. Collins
Biology Department
Western Carolina University
Cullowhee, NC, USA

ISSN 1568-1319 ISSN 2352-3956 (electronic)
ISBN 978-3-319-21526-6 ISBN 978-3-319-21527-3 (eBook)
DOI 10.1007/978-3-319-21527-3

Library of Congress Control Number: 2015953434

Springer Cham Heidelberg New York Dordrecht London

Printed on acid-free paper

Springer International Publishing AG Switzerland is part of Springer Science+Business Media (www.springer.com)

Preface

This edited volume addresses the historic range of variation (HRV) in types, frequencies, severities, and scales of natural disturbances, and how they create heterogeneous structure within upland hardwood forests of Central Hardwood Region (CHR). The idea for this book was partially in response to a new (2012) forest planning rule which requires national forests to be managed to sustain 'ecological integrity' and within the 'natural range of variation' of natural disturbances and vegetation structure. This new mandate has brought to the forefront discussions of HRV (e.g., what is it?) and whether natural disturbance regimes should be the primary guide to forest management on national forests and other public lands. Natural resource professionals often seek 'reference conditions,' based on HRV, for defining forest management and restoration objectives. A large body of literature addresses changes in forest structure after natural disturbance, but most studies are limited to a specific site, disturbance event, forest type, or geographic area. Several literature reviews address a single natural disturbance type within a limited geographic area (often not the CHR), but do not address others or how their importance may differ among ecoregions. Synthesizing information on HRV of natural disturbance types, and their impacts on forest structure, has been identified as a top synthesis need.

Historically, as they are today, natural (non-anthropogenic) disturbances were integral to shaping central hardwood forests and essential in maintaining diverse biotic communities. In addition to a 'background' of canopy gaps created by single tree mortality, wind, fire, ice, drought, insect pests, oak decline, floods, and landslides recurringly or episodically killed or damaged trees, at scales ranging from scattered, to small or large groups of trees, and across small to large areas. Additionally, some animals, such as beavers, elks, bisons, and perhaps passenger pigeons, functioned as keystone species by affecting forest structure and thus habitat availability for other wildlife species. Prehistoric anthropogenic disturbances – fire and clearing in particular – also influenced forest structure and composition throughout much of the CHR and therefore the distribution of disturbance-dependent wildlife species. The spatial extent, frequencies, and severities differed among these natural disturbance types and created mosaics and gradients of structural conditions and canopy openness within stands and across the landscape.

A full-day symposium, organized by the editors, at the 2014 Association of Southeastern Biologists conference in Spartanburg, South Carolina, was the basis for this book. Our goal was to present original scientific research and knowledge synthesis covering major natural disturbance types, with a focus on forest structure and implications for forest management. Chapters were written by respected experts on each topic with the goal of providing current, organized, and readily accessible information for the conservation community, land managers, scientists, students and educators, and others interested in how natural disturbances historically influenced the structure and composition of central hardwood forests and what that means for forest management today.

Chapters in this volume address questions sparked by debated and sometimes controversial goals and 'reference conditions' in forest management and restoration, such as the following: What was the historic distribution, scale, and frequency of different natural disturbances? What is the gradient of patch sizes or level of tree mortality conditions created by these disturbances? How do gradual disturbances such as oak decline, occurring over a long period of time and across a broad landscape, differ in effects from discrete disturbances such as tornadoes? How does topography influence disturbance regimes or impacts? How do native biotic (insects or fungi, keystone wildlife species) and abiotic (precipitation, drought, temperature, wind, and soil) agents interact to alter disturbance outcomes? What was the diversity of age classes and gradient of forest structure created by natural disturbances alone? How might disturbance-adapted plants and animals have fared in the hypothetical historic absence of anthropogenic disturbances? How might climate change alter disturbance regimes and structure of upland hardwood forests in the future? And finally, should, and how, can land managers manage these forests within the HRV of natural disturbance frequencies, spatial extents, and gradient of conditions they create?

We sincerely thank all those who encouraged and aided in the development of this book. Each chapter was peer-reviewed by at least two outside experts and both coeditors, and we thank these colleagues for their useful suggestions: Chris Asaro, Robert Askins, Francis Ashland, Bart Cattanach, Steven Croy, Kim Daehyun, Dianne DeSteven, Chris Fettig, Mark Harmon, Matthew Heller, Louis Iverson, John Kabrick, Tara Keyser, Scott Lecce, William MacDonald, Henry McNab, Manfred Mielke, Billy Minser, Scott Pearson, Duke Rankin, Jim Rentch, John Stanturf, Scott Stoleson, Ben Tanner, and Thomas Wentworth. We also thank the Association of Southeastern Biologists for allowing us to host a conference symposium on this important topic, and the National Forests of North Carolina for assistance with travel costs for speakers. We especially thank each author for contributing, and for timely chapter revisions, which made this book possible.

Asheville, NC, USA Cathryn H. Greenberg
Cullowhee, NC, USA Beverly S. Collins

Contents

Contributors

Jennifer L. Aldred Department of Geography and Earth Sciences, University of North Carolina at Charlotte, Charlotte, NC, USA

Jeffery B. Cannon Department of Plant Biology, University of Georgia, Athens, GA, USA

William A. Carothers USDA Forest Service, Forest Health Protection, Asheville, NC, USA

Beverly S. Collins Department of Biology, Western Carolina University, Cullowhee, NC, USA

David R. Coyle D.B. Warnell School of Forestry and Natural Resources, University of Georgia, Athens, GA, USA

Virginia H. Dale Environmental Sciences Division, Oak Ridge National Laboratory, Oak Ridge, TN, USA

Kamal J.K. Gandhi D.B. Warnell School of Forestry and Natural Resources, University of Georgia, Athens, GA, USA

Christopher M. Godfrey Department of Atmospheric Sciences, University of North Carolina at Asheville, Asheville, NC, USA

Cathryn H. Greenberg USDA Forest Service, Southern Research Station, Bent Creek Experimental Forest, Asheville, NC, USA

Henri D. Grissino-Mayer Department of Geography, The University of Tennessee, Knoxville, TN, USA

Tristram C. Hales Hillslope Geomorphology, School of Earth and Ocean Sciences, Cardiff University, Cardiff, UK

Justin L. Hart Department of Geography and Program in Environmental Science, University of Alabama, Tuscaloosa, AL, USA

Daniel J. Hayes Environmental Sciences Division, Oak Ridge National Laboratory, Oak Ridge, TN, USA

M. Joseph Hughes Department of Earth and Environment, Boston University, Boston, MA, USA

Kier D. Klepzig USDA Forest Service, Southern Research Station, Asheville, NC, USA

Charles W. Lafon Department of Geography, Texas A&M University, College Station, TX, USA

David S. Leigh Department of Geography, The University of Georgia, Athens, GA, USA

W. Henry McNab USDA Forest Service, Southern Research Station, Bent Creek Experimental Forest, Asheville, NC, USA

Douglas K. Miller Atmospheric Sciences Department, University of North Carolina, Asheville, Asheville, NC, USA

Chelcy F. Miniat Coweeta Hydrologic Lab, Center for Forest Watershed Research, USDA Forest Service, Southern Research Station, Otto, NC, USA

Randall S. Morin USDA Forest Service, Northern Research Station, Forest Inventory and Analysis, Newtown Square, PA, USA

John T. Nowak USDA Forest Service, Forest Health Protection, Asheville, NC, USA

Steven W. Oak USDA Forest Service, Southern Region Forest Health Protection (retired), Asheville, NC, USA

Chris J. Peterson Department of Plant Biology, University of Georgia, Athens, GA, USA

Martin A. Spetich USDA Forest Service, Southern Research Station, Hot Springs, AR, USA

Julie P. Tuttle Curriculum for the Environment and Ecology, University of North Carolina, Chapel Hill, NC, USA

Gordon S. Warburton North Carolina Wildlife Resources Commission, Marion, NC, USA

Kendrick Weeks North Carolina Wildlife Resources Commission, Mills River, NC, USA

Gary R. Wein Highlands-Cashiers Land Trust, Highlands, NC, USA

Peter S. White Department of Biology, University of North Carolina, Chapel Hill, NC, USA

Peter A. Wilfahrt Curriculum for the Environment and Ecology, University of North Carolina Chapel Hill, Chapel Hill, NC, USA

Anne C. Witt Virginia Department of Mines, Minerals and Energy, Division of Geology and Mineral Resources, Charlottesville, VA, USA

Richard M. Wooten Geohazards and Engineering Geology, North Carolina Geological Survey, Swannanoa, NC, USA

Eric K. Zenner Department of Ecosystem Science and Management, The Pennsylvania State University, University Park, PA, USA

Chapter 1
Introduction to Natural Disturbances and Historic Range of Variation: Type, Frequency, Severity, and Post-disturbance Structure in Central Hardwood Forests

Cathryn H. Greenberg, Beverly S. Collins, W. Henry McNab, Douglas K. Miller, and Gary R. Wein

Abstract Throughout the history of upland hardwood forests of the Central Hardwood Region, natural disturbances have been integral to shaping forest structure and composition, and essential in maintaining diverse biotic communities. In this chapter, we introduce the geographic scope and dynamic history of climate, natural disturbances, and human influence on central hardwood forests. We briefly introduce biotic and abiotic agents of disturbance to provide a foundation for the book and further discussion of whether and how historic disturbance regimes should guide forest management within national forests and other public lands.

Keywords Central Hardwood Region • Upland hardwood forests • Disturbance agents • Forest management • Historic disturbance regime

C.H. Greenberg (✉) • W.H. McNab
USDA Forest Service, Southern Research Station, Bent Creek Experimental Forest, 1577 Brevard Road, Asheville, NC 28806, USA
e-mail: kgreenberg@fs.fed.us; hmcnab@fs.fed.us

B.S. Collins
Biology Department, Western Carolina University, NS 132, Cullowhee, NC 28734, USA
e-mail: collinsb@email.wcu.edu

D.K. Miller
Atmospheric Sciences Department, University of North Carolina, Asheville, One University Heights, Asheville, NC 28804, USA
e-mail: dmiller@unca.edu

G.R. Wein
Highlands-Cashiers Land Trust, PO Box 1703, Highlands, NC 28741, USA
e-mail: hitrust@earthlink.net

C.H. Greenberg, B.S. Collins (eds.), *Natural Disturbances and Historic Range of Variation*, Managing Forest Ecosystems 32,
DOI 10.1007/978-3-319-21527-3_1

1.1 Central Hardwood Forests: A Dynamic History of Climate, Natural Disturbances, and Human Influence

In upland hardwood forests of the Central Hardwood Region (CHR) (Fig. 1.1), natural disturbances, which we define here as non-anthropogenic disturbances, are integral to shaping forest structure and composition, and essential in maintaining diverse biotic communities. Wind, fire, ice, drought, insect pests, oak (*Quercus*) decline, floods, landslides, and beaver (*Castor canadensis*) damming are some recurring or episodic natural disturbances that can kill or damage trees at scales ranging from scattered to small, or large, groups of trees and across small to large areas (e.g., Fig. 1.2). The spatial extent, frequencies, and severities differ among these natural disturbance types and create mosaics and gradients of structural conditions and canopy openness within stands and across the landscape.

Land managers and planners have identified a need for additional information on natural disturbance types and the range of variation in their impacts on forest structure. A recent USDA Forest Service planning rule (USDA Forest Service 2012) requires that terrestrial and aquatic ecosystems and watersheds of national forests be managed to sustain or restore 'ecological integrity' and within the 'historic range

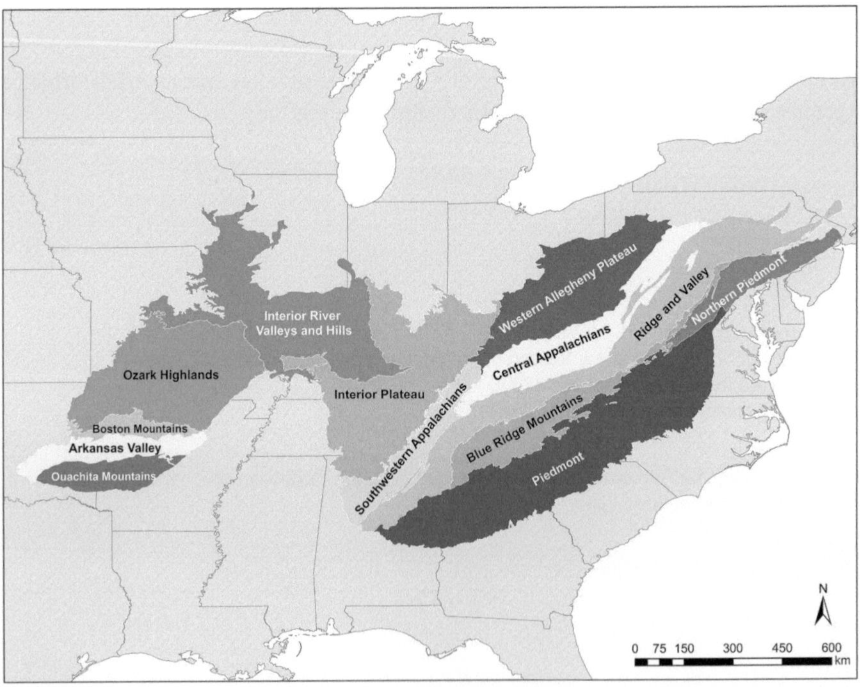

Fig. 1.1 The CHR encompasses 13 ecoregions, defined by similarity in climate, geology, physiography, soils, hydrology, land use, wildlife, and predominant types of disturbances (Omernik 1987)

Fig. 1.2 Examples of different natural disturbance types in central hardwood forests: (**a**) oak decline in the southern Appalachians (Photo courtesy of Steve Oak); (**b**) ice storm damage on the Daniel Boone National Forest, 2003 (Photo courtesy of USDA Forest Service State and Private Forestry); (**c**) springtime mountain wave, Great Smoky Mountains National Park (GSMNP; Photo

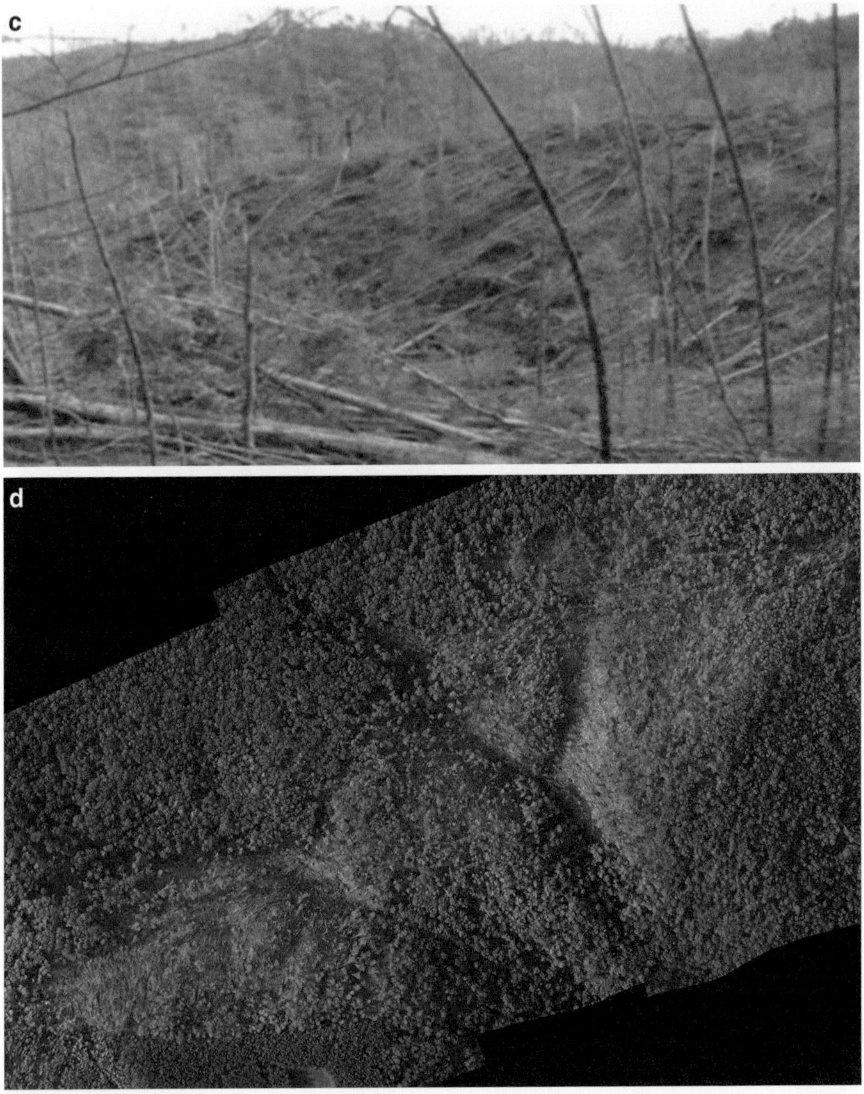

Fig. 1.2 (continued) courtesy of Jim Renfro); (**d**) tornado damage in Dade County, Georgia, 2011 (Photo by NOAA, National Ocean Service, April 2011 Tornado Response Imagery); (**e**) upper track of 2004 Peeks Creek debris flow, Macon County, North Carolina (Photo courtesy of Rick Wooten); (**f**) close up view of 2004 Peeks Creek debris flow, Macon County, North Carolina (Photo courtesy of Rick Wooten); (**g**) lightning-ignited Calderwood fire, ignited on private land, 17 August 2010, and spread into GSMNP (Photo courtesy of GSMNP staff); (**h**) flooding caused by beaver dam, Watauga County, North Carolina, July 2014 (Photo courtesy of Cathryn Greenberg)

Fig. 1.2 (continued)

Fig. 1.2 (continued)

of variation' of natural disturbances and vegetation composition and structure. This highlights the need to use the best available science to describe the historic types, frequencies, and severities of natural disturbances, and the scales and gradients of forest structural conditions they created.

Today's CHR forests are a legacy of natural and anthropogenic disturbances. Over the long-term, climate shifts through the Holocene have affected both forest types and (almost certainly) disturbance regimes. Pollen records indicate that just 18,000 years before present (BP), during the last glacial maximum, spruce (*Picea* spp.) and jack pine (*Pinus banksiana*) forests were widespread across much of this region, which is now occupied by hardwood forests (Delcourt and Delcourt 1981; Jackson et al. 2000). By around 14,000 years BP, a warming climate coincided with the arrival of humans (*Homo sapiens*) and the extinction of remaining megafauna (see Greenberg et al. Chap. 12), and thus began a slow transition to diverse deciduous forests; frequent burning by humans also facilitated the expansion of prairies, open woodlands, and savannas into eastern North America (Delcourt and Delcourt 1981, 2004). By 5,000–7,000 years ago, forest types familiar to us today were in place (Delcourt et al. 1993, 1998; Spetich et al. 2011). Climate fluctuations over the last several years include the warmer hypsithermal (7,500 to 5,000 years BP), followed by a cooling trend (5,000–120 years BP) excepting a warmer 'Medieval Climate Anomaly' (1,050–750 years BP) (Driese et al. 2008; Mann et al. 2009) and the 'Little Ice Age' (600 to 120 years BP) (Carroll et al. 2002). The confluence of changing climate and forest types with the increasing influence of humans over the last 3,000 years (Delcourt and Delcourt 2004) was certainly concurrent with shifting disturbance regimes and resulting forest composition and structure.

Although this book aims to characterize the range of variation in natural disturbances, the historic influence of humans on the landscape blurs separation of 'natural' and 'unnatural' disturbance effects. Current estimates of the human population in the eastern USA based on archaeological and historical information range between 0.5 and 2.6 million in 1500 AD (Milner and Chaplin 2010). Guyette et al. (2006a, b) found that human population density and culture were a major influence on historic fire frequencies; historically, human–caused fire was at least 200 times more frequent than lightning–caused fire in the eastern USA (see Spetich et al. 2011). Frequent burning and (or) abandonment of farmlands by Native Americans and later by European settlers likely created and maintained oak savannas, grassy balds, shortleaf- (*P. echninata*) and pitch pine (*P. rigida*)-dominated forests, and mixed pine-oak forests on sites that would otherwise, in many cases, be hardwood-dominated (Delcourt et al. 1998; Delcourt and Delcourt 2004). As a consequence of this conversion, effects of native southern pine beetle (*Dendroctus frontalis*) outbreaks on forest types that were heavily influenced by humans were historically greater than they might have been on landscapes where humans were absent. Similarly, the historic occurrence, distribution, and population sizes of some wildlife species, such as elk (*Cervus canadensis*), bobwhite quail (*Colinus virginianus*), or golden-winged warblers (*Vermivora chrysoptera*) that require different variants of open conditions, were likely predominantly a function of frequent burning by humans (see Greenberg et al. Chap. 12).

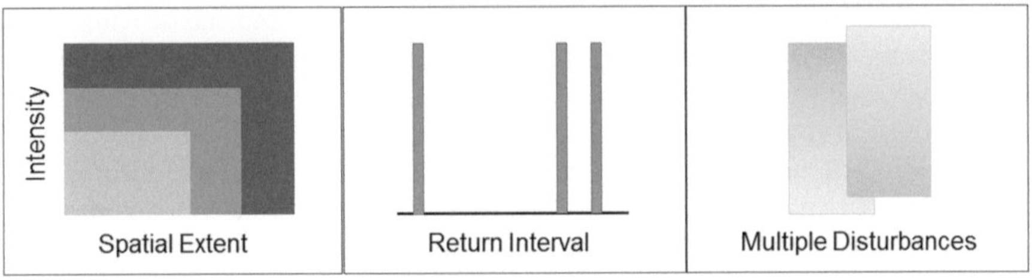

Fig. 1.3 Characteristics of disturbances over time and within a regional context

Anthropogenic disturbances have especially shaped CHR forests over the last 200 years. Key among these human impacts: (1) widespread, heavy logging in the early 1900s resulted in younger, relatively even-aged forests with smaller-diameter trees; (2) European settlement resulted in land being cleared for home sites and agriculture; (3) non-native invasive plants, pests, and pathogens affected forest composition, including a non-native 'blight' (*Cryphonectria parasitica*), introduced in 1904, which eliminated mature American chestnut (*Castanea dentata*) trees that historically composed up to 50 % of trees in eastern forests (Elliott and Swank 2008; Burke 2012); and (4) human population grew from an estimated fewer than 2.3 people per km^2 throughout most of the CHR in 1790 (US Census Bureau undated) to 29 – >77 people per km^2 today (worldatlas.com; 2012 census), leading to fragmented and more heavily used forests. Further, heavy use of fossil fuels by humans is accelerating climate change, which may already be affecting forests (McNab et al. 2014) and wildlife (Greenberg et al. 2014), or likely will in the near future (see Dale et al. Chap. 13). These anthropogenic impacts interact with natural disturbances and can enhance or diminish their effects throughout the CHR (see Dale et al. Chap. 13).

In addition to understanding the interactions with anthropogenic disturbances, assessing the range of variation of natural disturbances in forests requires moving beyond a single disturbance event at a specific site, to considering the entire disturbance regime; i.e., frequency, spatial extent, and multiple interacting types of disturbance, within the larger landscape, subregional, and regional context (Fig. 1.3). For example, severe wind events (see Peterson et al. Chap. 5) or ice storms (see Lafon et al. Chap. 7) may rarely reoccur in the same forest stand, but nonetheless occur with relatively higher frequency within a landscape or region (White et al. 2011). Similarly, forests often are subject to multiple, interacting biotic and abiotic disturbances. For example, drought may precipitate oak decline, particularly on dry, low-quality sites, and weakened oak trees may then be especially vulnerable to windthrow during a high wind event (Greenberg et al. 2011). Chapters in this volume assess the intensity, spatial extent, and return interval of both biotic (e.g., oak decline, Oak et al. Chap. 3; southern pine beetle, Nowak et al. Chap. 4; or keystone wildlife species and prehistoric humans, Greenberg et al. Chap. 12) and abiotic (e.g., wind, Peterson et al. Chap. 5; fire, Grissino-Mayer Chap. 6; ice storms, Lafon Chap. 7; flooding and erosion, Leigh Chap. 8, and landslides, Wooten et al. Chap. 9) natural disturbances at multiple scales to provide a broader picture of how the disturbance regime shapes forest structure or composition.

Reconstructing historic frequency, range of severity, and spatial extent of natural disturbances depends in part on availability of records and physical evidence. For example, comprehensive, systematic climate records are relatively recent (since 1895; National Oceanic and Atmospheric Administration (NOAA), National Climatic Data Center), and weather-severity rankings such as the Fujita scale of tornado severity are often based on the built environment (e.g., tornado damage to buildings), with less applicability to forests. Scientists in this book use a combination of intensive data such as stand or plot-level measurements of forest composition and structure, tree rings, soil charcoal, and sedimentology; extensive data such as southern pine beetle surveys; limited surveys such as disturbance damage for a salvage sale; anecdotal descriptions of specific disturbances (e.g., Abell 1934); and non-quantitative regional surveys (e.g., Ashe and Ayers 1901), or descriptions (Hursh and Haasis 1931) to capture the historic range of disturbances in central hardwood forests. In this chapter, we introduce the CHR and historic spatial and temporal range of variation of natural disturbances to provide a foundation for the book and further discussion of whether, and how, historic disturbance regimes should guide forest management within national forests and other public lands.

1.2 The Central Hardwood Region: Geographic Scope, Climate, and Forests

The CHR is characterized by a humid, temperate climate with evenly distributed annual precipitation and occasional soil moisture deficits during the middle to late summer growing season (Bailey 1995). This region of deciduous hardwood forests is centrally located in the eastern USA and is dominated by species of oaks and hickories (*Carya*) (Fig. 1.4). The predominantly oak-hickory forests of the CHR have long been recognized and described (Mattoon 1936), but their extent has been variously mapped (Mattoon 1937; Schnur 1937; Braun 1950; Society of American Foresters 1964; Eyre 1980; Merritt 1980; Dyer 2006), particularly in transition areas with adjoining forest regions. The region we define as the CHR generally follows the area delineated by Fralish (2003), with two exceptions. We include the hardwood-conifer forests of the Piedmont and the Ouachita Mountains, which had a high component of oaks and hickories before settlement (Nelson 1957; Foti and Glenn 1990) but now are largely dominated by pines; Fralish (2003) considers these areas as transition zones with the Southern Pine Region. The extent of the CHR in this book agrees well with 13 aggregated ecoregions identified by Omernik (1987) (Fig. 1.1), which are large ecological subdivisions that are relatively uniform in climate, geology, physiography, soils, hydrology, land use, wildlife, and predominant types of natural disturbances. However, other ecologists have used other terms to define regions of ecological similarity, as do some of the authors in this volume. For example, Braun (1950) referenced the area similar to the Blue Ridge Mountains ecoregion (Fig. 1.1) as the 'southern Appalachians,' whereas Ashe and Ayers (1901) defined the 'southern Appalachians' as (approximately) the group of seven ecoregions east of the Interior Plateau.

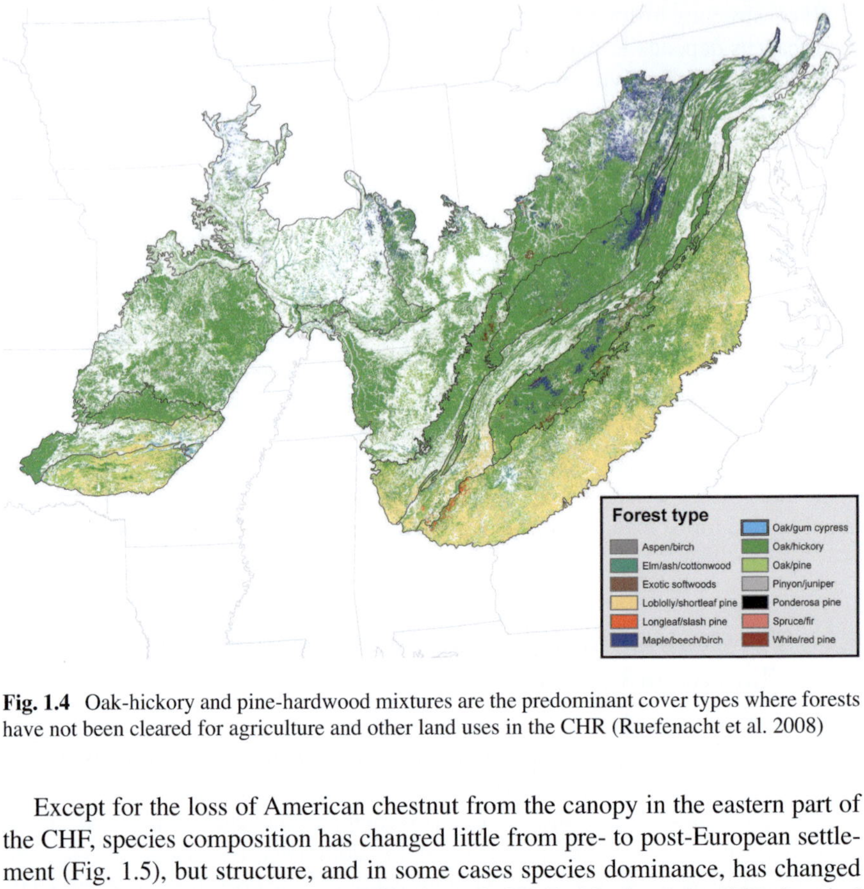

Fig. 1.4 Oak-hickory and pine-hardwood mixtures are the predominant cover types where forests have not been cleared for agriculture and other land uses in the CHR (Ruefenacht et al. 2008)

Except for the loss of American chestnut from the canopy in the eastern part of the CHF, species composition has changed little from pre- to post-European settlement (Fig. 1.5), but structure, and in some cases species dominance, has changed and fragmentation has increased (Riitters et al. 2002). Much of the CHR forested area has been selectively harvested several times, beginning with high-grading and fuelwood cutting associated with subsistence farming by pioneer European settlers during the late 1700s, and followed by several episodes of exploitation for commercial purposes (Schmidt and McWilliams 2003). The extent of this early human disturbance was described by Schnur (1937) who reported that most of the 100 mature, even-aged timber stands he sampled throughout the CHR in the early 1930s had originated from sprouts following harvest. At that time, more than half of the area of most states in the CHR was used for agriculture, farm size averaged about 30 ha, and only about 12 % of farmland was reported as non-pastured woodlands (US Census Bureau 1932).

Today, nearly 90 % of the CHR remains in private ownership with an average tract size of about 20 ha (Roberts et al. 1995). The largely private lands coupled with a climate favorable for agriculture has resulted in a mosaic of land uses and vegetative cover types throughout much of the region. Forested land ranges from 21 % in the Interior River Valleys and Hills ecoregion to 91 % in the Blue Ridge Mountains; overall, the vegetative cover on about 60 % of the CHR land area consists of trees (Fig. 1.6). Forests are less fragmented in the mountainous ecoregions, where steeper

Fig. 1.5 Oaks and hickories typically dominate the canopy of mature forests on middle and upper slopes in the CHR, such as this stand in the southwestern Appalachians ecoregion in Scott County, Tennessee (US Forest Service photo, F.E. Olmsted 1901; US Forest Service photograph collection #P9801. DH Ramsey Special Collections, UNC-Asheville)

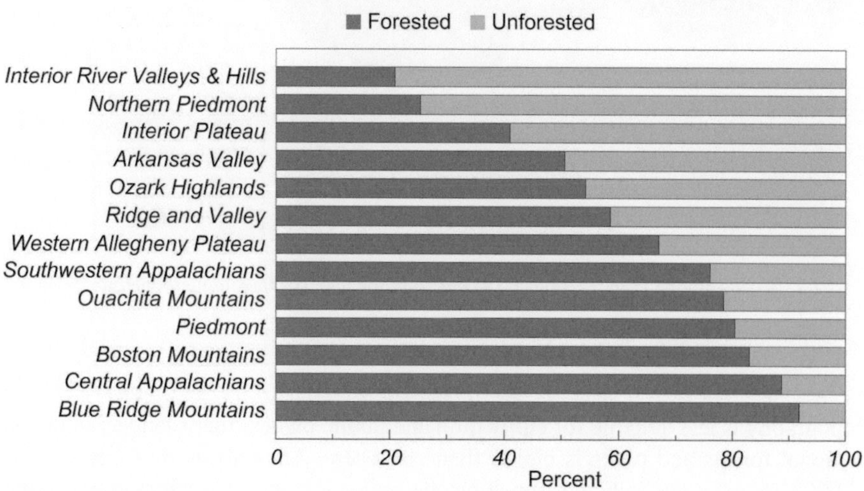

Fig. 1.6 Relative proportion (%) of land forested or unforested (e.g. water bodies, agricultural fields, built-up urban areas) by ecoregion in the CHR (Ruefenacht et al. 2008)

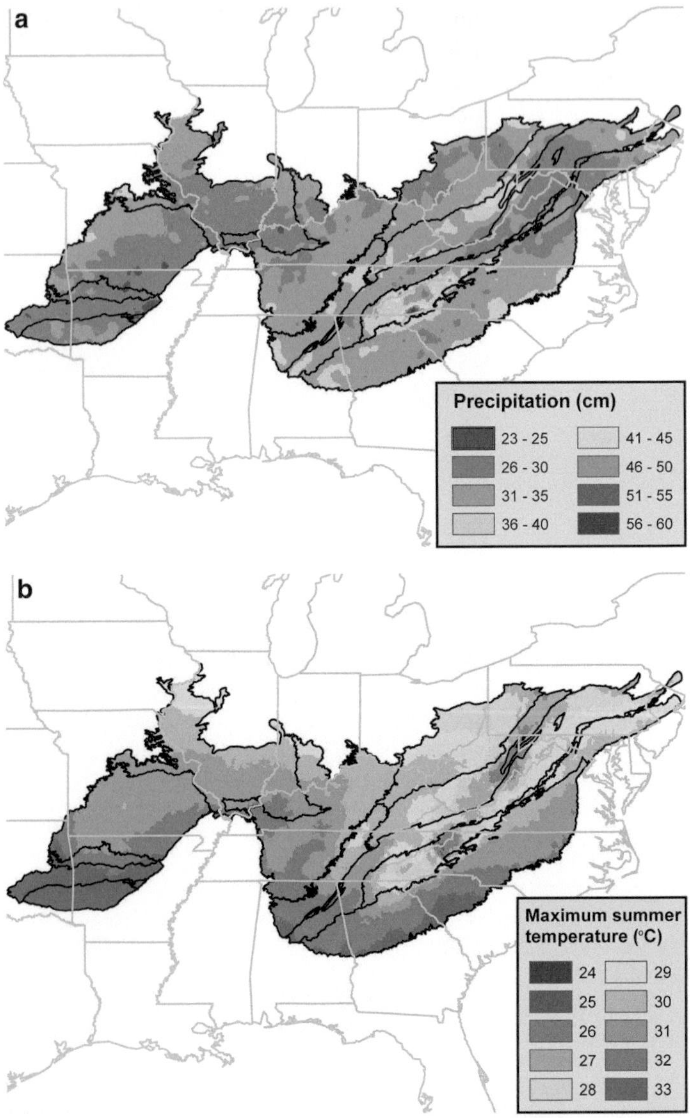

Fig. 1.7 Mean summer (June – August) (**a**) precipitation and; (**b**) maximum temperatures, in the CHR

topography is less suitable for cultivation and public ownership of land as state and national forests and parks is higher than elsewhere. An analysis by Gallant et al. (2000), for example, indicates forest patch size is more than five times larger in the mountainous Blue Ridge Mountains compared to the hilly topography of the adjacent Piedmont ecoregion. Forested areas are lowest in the Interior River Valleys and

Hills ecoregion, where the climate, and soils are well-suited for grain crops and in the Northern Piedmont ecoregion, where high population densities result in relatively greater land uses for urban purposes than elsewhere in the CHR.

Mean annual precipitation is variable throughout the CHR, ranging from 95–140 cm among ecoregions; higher amounts are associated with mountainous ecoregions. However, mean summer (June – August) precipitation, which influences forest species composition and productivity more than annual rainfall, is relatively uniform among ecoregions, ranging from 24–28 cm (Fig. 1.7a). Mean maximum summer temperature is relatively uniform throughout the CHR, except in the cooler eastern mountainous ecoregions (Fig. 1.7b). Except for small areas in the Interior Plateau and Interior River Valleys and Hills ecoregions, in southern Indiana and Illinois, the last period of Pleistocene glaciation did not extend into the CHR. Bedrock geology varies greatly in age and lithology throughout the CHR. Flat-lying to gently inclined sedimentary strata of sandstone, shale, limestone and dolomite underlie much of the CHR. Folded and faulted sedimentary strata are subordinate to the gently inclined sedimentary strata in the Southwestern and Central Appalachian ecoregions, but dominate the Ouachita Mountains and the Ridge and Valley ecoregions. Folded, faulted, and metamorphosed sedimentary and metamorphosed igneous bedrock underlies the greater part of Blue Ridge Mountains and Piedmont ecoregions, and consists mainly of varieties of slate, phyllite, schist, and gneiss. The Blue Ridge Mountains and Piedmont also include large, unmetamorphosed and weakly metamorphosed igneous intrusions. Soils in the CHR are mostly moderately productive Alfisols, formed from limestones. Ultisols are prevalent throughout but are typically lower in fertility and are more common in ecoregions underlain with highly weathered rocks of the Blue Ridge Mountains and Piedmont. Inceptisols also occur throughout, typically in concave areas of steep, hilly terrain. Table 1.1 summarizes environmental characteristics and common disturbances in each of the 13 ecoregions of the CHR.

Composition of the hardwood dominated forests of the CHR, consisting of more than 100 arborescent species, has long been described simply as oak-hickory or oak-chestnut (Pinchot and Ashe 1897; Holmes 1911). The more descriptive forest cover types identify 18 major communities, named for the one to three tree species that comprise the majority of canopy stocking (Table 1.2) (Eyre 1980). The upland group of cover types are dominated by one or more of nine species of oaks and generally occupy drier slopes and ridges; the other cover types, which form a group commonly occurring on bottomlands or moist slopes, usually include species of oak as minor stand components. One or more of six species of hickory (*Carya* spp.) are typical canopy associates with oaks in both upland and bottomland cover types. White oak (*Q. alba*) is the most extensive cover type, extending throughout the eastern USA and into Canada (Greeley and Ashe 1907); bear oak (*Q. ilicifolia*), is most restrictive, occurring in a patchy distribution primarily on xeric ridge tops in the Ridge and Valley, and barrens of other ecoregions in the northeastern area of the CHR (Little 1977). With only several exceptions (e.g., bur oak (*Q. macrocarpa*), chestnut oak (*Q. montana*)), the upland oak cover types occur throughout the CHR, but can vary in dominance in response to moisture gradients at both regional and

Table 1.1 Environmental characteristics within ecoregion subdivisions[a] of the CHR

Ecoregion area (km^2)[b]	Physiography & elevation range (m)	Geologic formations & soils	Mean annual Temp (°C)[c]	Mean annual Ppt (mm)[c]	Land cover & uses	Primary natural disturbances
Ouachita Mountains (26,896)	High hills, low mountains (88–820)	Sedimentary; Ultisols, Inceptisols	15–17	1,090–1,675	Mostly pine-oak forests; pasture and croplands	Ice storms, fire, tornados, landslides
Arkansas Valley (28,421)	Plains with hills (75–839)	Sandstone, shale; Ultisols, Inceptisols	15–17	1,040–1,575	Patchwork of agriculture, wood lands, built-up residential,	Fire, tornados, landslides
Boston Mountains (14,178)	Deeply dissected plateau (65–853)	Sandstone, shale; Ultisols, Inceptisols	13–14	1,118–1,372	Oak-hickory forests	Fire, tornados, landslides
Ozark Highlands (106,391)	Moderately dissected plateau; (80–560)	Limestone, sandstone, shale; Ultisols, Alfisols	12–15	965–1,244	About equal areas of forests and agriculture	Tornados, fire
Piedmont (166,117)	Moderately dissected plains (40–880)	Metamorphic schist, gneiss; Ultisols, Entisols	13–17	1,080–1,650	Mosaic of pine-oak forests, agriculture, pasture, large built-up urban areas	Ice storms, hurricane, occasional tornados, southern. pine beetle, fire
Northern Piedmont (30,459)	Low hills, plains (50–350)	Metamorphic gneiss-schist; Alfisols, Inceptisols	11–13	930–1,250	About 25 % urban and developed, forest and cropland about equal.	Ice storms, occasional remnants of hurricanes
Blue Ridge Mountains (46,595)	Hilly plateaus, mountains (250–2,040)	Metamorphic gneiss-schist; Inceptisols, Ultisols	7–14	1,100–2,500	Forests, urban in broad valleys	Ice and rain storms, occasional hurricane,landslides, fire
Ridge and Valley (115,483)	Low, parallel ridges and valleys (90–1,300)	Sedimentary, limestone, sandstone; Ultisols, Inceptisols	8–16	900–1,350	Woodland, cropland and pasture mosaic; about 60 % forested	Ice and rain storms, landslides,

Southwestern Appalachians (37,994)	Moderately dissected plateau (90–730)	Sedimentary, sandstone; Ultisols, Inceptisols	12–16	1,200–1,700	Forest and woodland; little agriculture	Rain storms, fire, landslides
Central Appalachians (62,050)	High, dissected plateau; (170–1,400)	Sedimentary, Sandstone, shale; Ultisols, Inceptisols	7–13	980–1,500	Mostly forests, pasture, surface coal mining	Ice storms, summer rain storms
Western Allegheny Plateau (81,440)	Dissected plateau, steep hills; (150–580)	Sedimentary formations; Alfisols, Incepsols, Ultisols	8–13	900–1,150	Forest and woodlands; extensive coal mining	Ice and rain storms, landslides, fire
Interior Plateau (123,523)	Plains, dissected plateaus, hills and ridges; (105–460)	Sedimentary formations, shale, limestone, siltstone; Ultisols, Alfisols	12–16	1,015–1,470	Forest, wood-lots, pasture, cropland, urban	Ice storms, fire, occasional tornados, severe drought
Interior River Valleys and Hills (120,405)	Wide, flat valleys, dissected glacial till plains (90–305)	Sedimentary, sandstone, limestone; Alfisols, Mollisols	10–14	860–1,320	Mostly cropland, pasture, scattered woodlands	Flooding in river bottoms, occasional tornados, severe drought

[a]See Fig. 1.1 for location of ecoregions
[b]Level III ecoregions (Omernik 1987)
[c]Range of mean annual temperature (Temp) or precipitation (Ppt) occurring within each ecoregion

Table 1.2 Distribution and ecological relationships of major forest cover types occurring in the CHR (Eyre 1980)

Forest cover type[a]	Geographical distribution[b]	Ecological relationships
Post oak – blackjack oak (*Q. stellata*, *Q. marilandica*)	Throughout all ecoregions of the CHR	Xeric sites – shallow clayey and deep sandy soils
Bur oak – (*Q. macrocarpa*)	Scattered occurrences; prairie-forest transitions of western ecoregions	Xeric to dry sites; prairie moist bottomlands
Chestnut oak (*Q. montana*)	Moderate to steep slopes of eastern ecoregions	Dry sites; ridge-tops and southerly slopes
White oak (*Q. alba*)	Throughout CHR, most common in Ozark Highlands, Interior Plateau, and Interior River Valleys and Hills ecoregions	Most common on dry-to moderately wet sites
Black oak (*Q. velutina*)	Common in Ozark Highlands, Interior Plateau, and Interior River Valleys and Hills ecoregions; scattered elsewhere	Generalist on many sites; occurrence largely result of disturbance
Northern red oak (*Q. rubra*)	Throughout; pure stands rare	Moderately moist to mesic sites
Yellow-poplar (*Liriodendron tulipifera*)	Large stands occur in ecoregions Northern Piedmont, Blue Ridge Mountains, Ridge and Valley ecoregions; patches elsewhere	Mesic coves and slopes, flood plains.
River birch – sycamore (*Betula nigra, Platanus americana*)	Throughout	Flood plains of large rivers, river bottoms,
Silver maple – American elm (*Acer saccharinum, Ulmus americana*)	Throughout	Mesic floodplains, old fields in bottomlands
Pin oak – sweetgum (*Q. palustris – Liquidambar styraciflua*)	Throughout, but particularly prevalent in Interior Plateau, Interior River Valleys and Hills ecoregions	Mesic sites of broad stream bottoms
Eastern red cedar (*Juniperus virginiana*)	Throughout; large stands uncommon except along prairie transition	Dry, shallow soils; rock outcrops

[a]Minor forest cover types include bear oak (*Q. ilicifolia*), black locust (*Robinia pseudoacacia*), yellow-poplar (*L. tulipifera*)/eastern hemlock (*Tsuga canadensis*), yellow-poplar (*L. tulipifera*)/white oak (*Q. alba*)/northern red oak (*Q. rubra*), sassafras (*Sassafras albidum*) – persimmon (*Diospyros virginiana*), pitch pine (*P. rigida*)
[b]See Fig. 1.1 for location of ecoregions

landscape scales. The widely occurring and xerophytic post oak (*Q. stellata*), for example, increases as a stand component in the drier western ecoregions and also responds to soil moisture variation in relation to topography at the landscape scale (Stahle and Hehr 1984; see Wilfahrt et al. Chap. 11). The association of vegetation with environmental gradients is most apparent at landscape scales within mountainous ecoregions where moisture and temperature gradients are formed from interactions of topography with elevation.

Elevation is a strong environmental factor in the eastern portion of the CHR that affects cover types in the mountainous Blue Ridge Mountains and Central Appalachians ecoregions, and somewhat in the Southwestern Appalachians ecoregion. In the high mountains of the Blue Ridge Mountains ecoregion, elevation and topography form temperature and moisture environments manifested by distinctive forest communities (Pinchot and Ashe 1897). Historically, a canopy mixture of pines and oaks was typical of the hotter and drier hilly topography of the low elevation intermountain valleys, referred to as plateau forests because of their similarity in soils and species with the adjacent Piedmont ecoregion. Forests of middle elevations, from 900 to 1,500 m, consisted primarily of oaks, hickories and patches of conifers on dry slopes and ridges (and American chestnut before 1920); yellow-poplar (*Liriodendron tulipifera*), northern red oak (*Q. rubra*) and other mesophytic species were common on moist slopes and in coves. The colder and wetter environments above 1,500 m in North Carolina results in a transition from oak-hickory forests to northern hardwoods and red spruce (*Picea rubens*). A similar transition from oak-hickory forests to northern hardwoods and spruce occurs above about 1,200 m elevation on the Dolly Sods plateau and near the summit of Spruce Mountain in the Central Appalachians ecoregion of West Virginia. The current distribution of red spruce stands in the Blue Ridge Mountains ecoregion is better related to disturbances associated with timber harvest during the early 1900s than with environmental factors (Korstian 1937; Pielke 1981). In the dissected plateau and low mountains of the Southwestern Appalachians ecoregion, where elevations reach only 730 m, moisture gradients associated with relief are the primary factor affecting distribution of the oak-hickory and mixed mesophytic upland forest vegetation.

In the western mountainous ecoregions of the CHR (i.e., Ouachita Mountains, Arkansas Valley, and Boston Mountains), the highest elevation occurs at Mt. Magazine (840 m), one of several flat-topped, mesa-like ridges in the Arkansas Valley ecoregion. This ecoregion, an 80-km wide synclinal trough with diverse landforms including plains, hills, ridges, mountains and narrow floodplains along the Arkansas River, is situated between the Ouachita Mountains and the Boston Mountains ecoregions. In the Ouachita Mountains to the south the highest peak is Rich Mountain (817 m). There, temperature-moisture gradients associated with aspect results in a greater proportion of conifers (e.g., shortleaf pine) on the warmer and drier southern slopes compared to the slightly more mesic northern slopes dominated by oaks and hickories (Pell 1983). The Boston Mountains ecoregion is a deeply dissected former tableland, with highest elevations of about 650 m. Forests there are dominated by white, northern red, and black oaks (*Q. velutina*) on well drained slopes, although pines are prominent on southern slopes and mesophytic hardwoods are common on moist slopes and in ravines. In comparison with vegetation, however, the distribution of aquatic insect fauna has a stronger alliance with elevation in the mountainous landscapes of the Arkansas ecoregions (Bowles and Mathis 1989).

The landscape-scale distribution of forest cover types in the seven non-mountainous ecoregions is primarily associated with topographic moisture gradients, although disturbance is an important factor for many of these, including the upland oak types. The pitch pine cover type can regenerate on some very dry sites, but usually follows fire elsewhere. The sassafras (*Sassafras albidum*) – persimmon (*Diospyros virginiana*) type is typically found only on dry abandoned fields with red cedar (*Juniperus virginiana*) and pines. Likewise, the widely distributed and successional black locust (*Robinia pseudoacacia*) type has been described as primarily 'man-made.' Except for the relatively long-lived nature of a few cover types, primarily those with species in the white oak group (e.g., white oak, chestnut oak, post oak) and perhaps yellow-poplar, most are transitory and dependent on periodic disturbances to maintain their presence on the forested landscape (Eyre 1980).

Species composition of upland oak-hickory stands in the CHR before the current policy of wildfire suppression (beginning in the 1930s) may be inferred from stand inventory data presented by Schnur (1937) (Table 1.3). He found a relationship of species with site quality; either white oak or black oak occurred in more than 90 % of all stands, and one of several species of hickory was present in 70 % of stands. Red maple (*Acer rubrum*) occurred in 52 % of stands across all site qualities, and yellow-poplar occurred in 20 % of stands, mostly on high-quality sites. Shortleaf, Virginia (*P. virginiana*), or pitch pine was present in over half of stands sampled. Altogether, Schnur found 15 species of oak, five species of hickory, and approximately 50 other associated tree species, including American chestnut, which by that time had been removed as a dominant canopy species in forests of eastern ecoregions by mortality from the introduced blight (Frothingham and Stuart 1931).

Oak-dominated forests are the most common upland type throughout the CHR (Table 1.4). Conifers are present in relatively high proportions in forests of the Ouachita Mountains and Piedmont ecoregions and elsewhere on low quality sites and topographic positions that are more vulnerable to summer drought, and where lightning-ignited fires may have historically been most common. Historically, some central hardwood forests now dominated by oak and hickory had a heavy overstory component of shortleaf pine and pitch pine, especially on dry sites where frequent burning by Native American and (later) European settlers provided seedbed and light conditions suitable for their regeneration. Additionally, nearly pure second-growth yellow pine forests occurred in abandoned clearings, covering large areas (Ashe and Ayers 1901; Mattoon 1915; Ashe 1922; Balch 1928; Delcourt and Delcourt 2004).

Across the CHR, both broadly and locally, an elevation gradient from pine, through dry site oaks and mesic oaks, to mesic hardwoods, including northern hardwoods, is associated with a decrease in 'ground based' natural and anthropogenic disturbances such as fire and grazing, and an increase in 'canopy' disturbances such as wind and ice (White et al. 2011). White et al. (2011) compared the relative frequency of disturbances among ecoregions of the CHR. Their analysis revealed that fire frequency, including both natural and anthropogenic sources, is higher in the Ouachita Mountains, Arkansas Valley, Boston Mountains, Ozark Highlands, and Interior Plateau ecoregions while remnant hurricanes and landslides are more common disturbances in the Blue Ridge Mountains, Ridge and Valley, Southwestern

Table 1.3 Percent occurrence and basal area of arborescent species ≥2.5 cm dbh on sample plots at nearly 100 locations in the CHR soon after initiation of a national policy of wildland fire control (Schnur 1937)

Species	Basal area (%)	Occurrence (%)
White oak (*Querus alba*)	28.23	95.3
Black oak (*Q. velutina*)	19.11	91.6
Scarlet oak (*Q. coccinea*)	17.08	79.7
Hickory (*Carya* spp)[c]	2.69	76.0
Chestnut oak (*Q. montana*)	13.73	63.4
Red maple (*Acer rubrum*)	1.31	52.2
Northern red oak (*Q. rubra*)	4.65	52.0
Black gum (*Nyssa sylvatica*)	0.48	37.6
Understory group[d]	0.57	30.2
American chestnut (*Castanea dentata*)	1.84	24.5
Yellow-poplar (*Liriodendron tulipifera*)	0.84	20.3
Black and red ash (*Fraxinus nigra, F. pennsylvanica*)	0.62	21.0
Black locust (*Robinia pseudoacacia*)	0.44	15.1
Virginia pine (*Pinus virginiana*)	1.86	14.1
Post oak (*Q. stellata*)	0.90	13.4
Sweetgum (*Liquidambar styraciflua*)	0.54	5.2
Pitch pine (*P. rigida*)	0.43	5.0
Shortleaf pine (*P. echinata*)	0.47	4.2
Southern red oak (*Q. falcata*)	0.60	4.2
Bottomland oaks[b]	0.05	3.0
Miscellaneous oaks[a]	0.14	2.0
Pin oak (*Q. palustris*)	0.40	1.7
Blackjack oak (*Q. marilandica*)	0.14	1.2

[a]Hills (*Q. ellipsoidalis*), bear (*Q. ilicifolia*), dwarf chinquapin (*Q. prinoides*)
[b]Swamp white (*Q. bicolor*), willow (*Q. phellos*), shingle (*Q. imbricaria*)
[c]Bitternut (*C. cordiformis*), shagbark (*C. ovata*)
[d]Hophornbeam (*Ostrya virginiana*) blue beech (*Carpinus caroliniana*), persimmon (*Diospyros virginiana*), sourwood (*Oxydendrum arboreum*), American holly (*Ilex americana*), sassafras (*Sassafras albidum*)

and Central Appalachians, and Western Allegheny Plateau ecoregions, and ice storms are common in the northernmost ecoregions. Below, we discuss the primary natural disturbances and how they affect the structure and composition of forests within the within the CHR.

1.3 Agents of Natural Disturbance: Biological Process and Living Organisms

The ongoing, continuous process of decline, death, or breakage of forest trees is a 'background' disturbance that creates canopy openings and promotes fine-scale structural heterogeneity and diversity in central hardwood forests. Current gap

Table 1.4 Forest composition by species type[a] and ecoregion in the CHR

Ecoregion	Misc. conifer forests (%)[b]	Oak-conifer forests (%)[c]	Upland oak-hickory forests (%)[d]	Misc. hardwood forests (%)[e]
Ouachita Mountains	51.3	20.0	26.3	2.4
Arkansas Valley	19.7	9.0	63.9	7.4
Boston Mountains	6.7	2.8	90.5	0.0
Ozark Highlands	2.0	2.9	94.1	1.0
Piedmont	52.3	10.1	36.2	1.4
Northern Piedmont	6.6	3.6	83.3	6.5
Blue Ridge Mountains	4.4	7.9	82.2	5.5
Ridge and Valley	12.0	3.1	76.7	8.2
Southwestern Appalachians	23.0	4.9	71.3	0.8
Central Appalachians	1.1	0.3	83.8	14.8
Western Allegheny Plateau	1.2	1.4	80.9	16.5
Interior Plateau	2.6	8.2	82.3	6.9
Interior River Valleys and Hills	0.2	0.8	78.6	20.4

[a]Species types are combinations of USDA Forest Service, Forest Inventory and Analysis (FIA) forest groups shown on Fig. 1.4

[b]Miscellaneous conifer forests may consist of species groups including White/red/jack pine; Spruce/fir; Longleaf/slash pine; Loblolly/shortleaf pine; Pinyon/juniper; Ponderosa pine; and Exotic softwoods

[c]Oak-conifer forests consist of species in the oak/pine group

[d]Upland oak-hickory forests consist of species in the oak/hickory group and includes mesophytic species in a forest type commonly referred to as 'cove hardwoods'

[e]Miscellaneous hardwood forests may consist of species groups including Oak/gum/cypress; Elm/ash/cottonwood; Maple/beech/birch; and Aspen/birch

formation rates are about 1–2 % per year, which may be lower than historical rates and linked to changes in forest regeneration patterns. Gaps vary in size, shape, and rate of closure, all of which combine to produce a fine-scale patchwork of different tree species, ages, diameters, heights, crown spreads, and growth rates. In this volume, Hart (Chap. 2) discusses how fine-scale gap dynamics have historically affected structural and compositional heterogeneity in central hardwood forests.

Tree mortality is periodically accelerated by drought, native insects, or pathogens, resulting in pulses of more, variably-sized 'gaps' and potentially affecting forest composition. For example, oak decline, which can be induced by drought in combination with other factors such as low site quality, shoe string fungus (*Armillaria mellea*), or insects such as red oak borer (*Enaphalodes rufulus*), causes progressive crown dieback and eventual death of physiologically mature oaks in the red oak group (Erythrobalanus species) (Starkey et al. 2004). Oak decline has been documented across all surveyed central hardwood forest ecoregions (Oak et al. Chap. 3; Table 3.5) (Starkey and Oak 1989), and can impact a few trees in forests with diverse species composition and age structure, to several thousand hectares in landscapes dominated by Erythrobalanus species, especially in the western ecoregions. Southern pine beetle outbreaks in the CHR, also induced by drought, occur at 7 – 25 year intervals and accelerate gap dynamics by killing pines (Nowak et al. Chap. 4; Table 4.1). Historically, these cyclical outbreaks likely created a continuum of perforated forest structure, where pines were killed in mixed pine-oak stands, to large patches of dead trees where pines were dominant, such as on ridge tops dominated by Table Mountain (*P. pungens*) and pitch pine forests. Over the past 80 years, southern pine beetle outbreaks, in combination with decades of suppressing fire (most anthropogenically-ignited), have promoted a gradual shift in the composition of pine or mixed pine-oak-hickory forests (historically influenced by human activities such as burning, grazing, and field abandonment) to predominantly oak-hickory forests. In this volume, Oak et al. (Chap. 3) discuss historic spatial and temporal variability in the occurrence of oak decline, and the probability and severity of oak decline events using models based on regional long-term monitoring surveys, as part of the Forest Vegetation Simulator. Nowak et al. (Chap. 4) discuss the southern pine beetle as an agent of natural disturbance that periodically killed pines in small- to large patches, creating incremental, long-lasting structural heterogeneity across central hardwood forest landscapes.

Historically, some wildlife species also functioned as agents of disturbance. Grazing by elk and bison that once commonly occurred within the CHR likely helped to maintain grass-dominated open woodlands or prairies that probably originated from, and were also maintained by, frequent anthropogenic burning in many cases. Beavers, historically abundant within the CHR, also functioned as 'ecosystem engineers' by damming streams and creating large wetlands and 'beaver meadows,' providing habitat for amphibians, aquatic reptiles, and birds and mammals associated with wetlands, edge, and moist, grassy habitats (Askins 2000; Greenberg et al. Chap. 12). In this volume, Greenberg et al. (Chap. 12) discuss how some animal species were themselves important agents of natural disturbance that helped to create and maintain heterogeneous habitats within the CHR.

1.4 Agents of Natural Disturbance: Abiotic Process and Severe Weather

Across the CHR, many natural disturbances are directly or indirectly due to severe weather. Individual weather events, such as wind, rain, and ice storms, are influenced primarily by season, latitude, altitude, and proximity to large water bodies. In the eastern USA, most transitory weather events, which last only one or two days, are strongly influenced by the Gulf of Mexico, Atlantic Ocean, and Appalachian Mountains, in addition to the global weather patterns that support their development and decay. Extreme or repeated events can lead to landslides or local flooding. Persistent weather events, such as drought, can last months to years and owe their existence primarily to a stagnant regional weather pattern. The frequency of both persistent (drought) (Table 1.5, Fig. 1.8) and transitory (precipitation and wind) severe weather events (e.g., Peterson et al. Chap. 5) vary over the CHR.

Table 1.5 Total number of years (1895–2013) with moderate (PDSI[1] ≤2) or severe (PDSI ≤4) growing season droughts[a,b] and number of multi-year droughts in the 13 ecoregions[c] of the CHR

Ecoregion	Ecoregion area (km^2)	Moderate drought (years)	Severe drought (years)	Total drought (years)	Total multi-year (>2 years) droughts
Ouachita Mountains	26,896	15	2	17	2 (2-years)
Arkansas Valley	28,421	14	0	14	0
Boston Mountains	14,178	14	2	14	1 (2-year)
Ozark Highlands	106,391	14	3	17	1 (5-year)
Piedmont	166,177	17	2	19	2 (3-year) + 1 (2-year)
Northern Piedmont	30,459	12	2	14	1 (2-year) + 1 (4-year)
Blue Ridge Mountains	46,595	9	2	11	2 (2-year)
Ridge and Valley	115,483	10	2	12	1 (2-year)
Southwestern Appalachians	37,994	15	3	18	3 (3-year) + 1 (4-year)
Central Appalachians	62,050	19	2	21	2 (2-year) + 1 (3-year)
Western Allegheny Plateau	81,440	15	1	16	3 (2-year) + 1 (3-year)
Interior Plateau	123,523	20	4	24	4 (2-year) + 1 (3-year)
Interior River Valleys & Hills	120,405	15	7	22	5 (2-year) + 1 (3 year)

[a]Based on Palmer's Drought Severity Index (PDSI). Source of PDSI data: http://www.ncdc.noaa.gov/temp-and-precip/drought/historical-palmers.php; Website maintained by Department of Commerce, National Oceanic and Atmospheric Administration, National Environmental Satellite Data and Information Service, National Climatic Data Center, 151 Patton Avenue, Asheville, NC 28801. Accessed on March 21, 2014, 12:18 PM; data range 1895–2013

[b]PDSI (1895–2013) defined based on average summer (June, July, August) PDSI from one climate division per ecoregion. Moderate is PDSI −2.0 to −3.99, and severe is PDSI ≤ −4.0. Data columns here include total number of years (whether single- or multiple successive years) with drought, and number of multiple, successive-year droughts

[c]Frequencies in this table are not adjusted for per unit area of each respective ecoregion (Fig. 1.1). Comparisons of frequencies among ecoregions should first adjust for different ecoregion sizes

Persistent severe weather events, such as drought, occur with stagnant regional weather patterns, which are almost always associated with a stalled high pressure system at low- and mid-levels in the atmosphere. When stagnant high pressure occurs during the warm season, it often is accompanied by anomalously warm temperatures, which accelerate drying at the ground. Drought is also self-perpetuating, in that evaporation and evapotranspiration slow or cease as the soil and vegetation become increasingly dry, causing the atmosphere to become drier. Relief from drought only comes when the regional weather pattern shifts to bring humid air and precipitation sufficient to saturate the ground. Often, an active tropical cyclone season will provide enough humid air and accompanying precipitation to 'break' a drought, allowing water tables to begin to return to normal levels. Although the CHR usually receives enough precipitation to balance evapotranspiration, drought occurs sporadically throughout the region (Fig. 1.8) and transient summer drought is common on sites with sandy or thin soil. Severe growing season drought func-

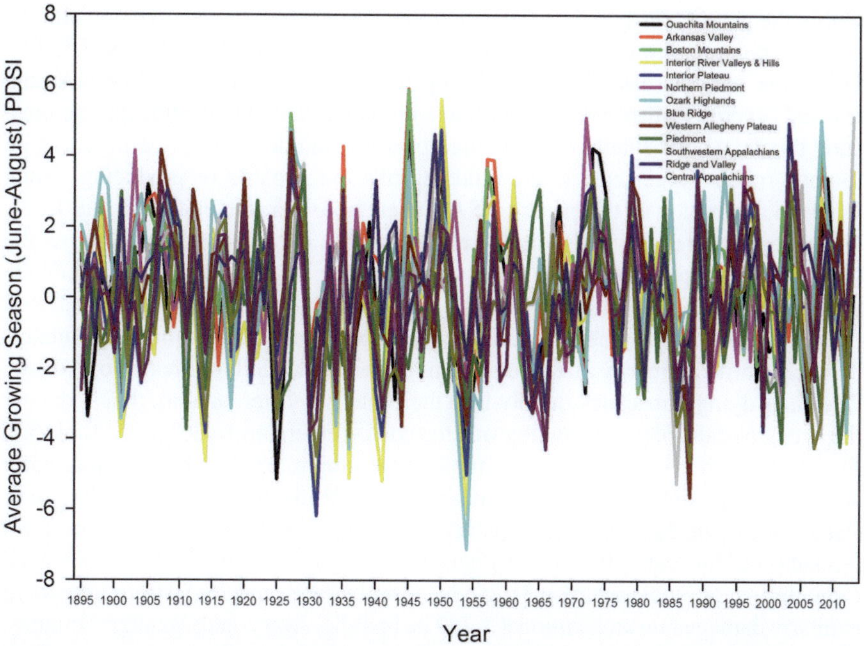

Fig. 1.8 Annual growing season (June-August) Palmer's Drought Severity Index (PDSI) in ecoregions of the CHR. PDSI ≤ 2 is considered moderate, and PDSI ≤ 4 is considered severe

tions as a natural disturbance both directly, by killing trees, and indirectly, by increasing vulnerability of weakened trees to pathogens and pests (such as oak decline or attack by southern pine beetles), windthrow, or breakage from subsequent severe weather events. Rare climatic events such as intense drought can initiate a pulse of recruitment following heavy tree mortality at the CHR scale, with effects on forest composition, age class, and structure lasting for centuries (Pederson et al. 2014). Droughts also increase the threat of wildfire.

Transitory severe weather events range in scale from convective storms (e.g., thunderstorms and squall lines) that last up to six hours, to tropical storms (e.g., dying hurricanes) that last a day or two, to large-scale mid-latitude storms that last three to five days. On the small end of the spatial spectrum, convective storms affect tens of km, while large-scale mid-latitude storms cover thousands of km. The speed of motion of these storms also varies, from stalled (zero km per hour) to 121 km per hour. At all scales, transitory storms are capable of producing damaging precipitation and wind.

Wind is produced in transitory convective weather events such as thunderstorms and squall lines when a cold downdraft of air encounters the ground and is forced to move horizontally. The winds at the edge of the downdraft air, called the 'gust front,' sometimes produce sufficient damage to be mistaken for a tornado. Destructive downdrafts are more common outside of the Appalachian Mountains, as storms are often disrupted as they cross the mountain chain. Air circulations within the parent convective storm can produce tornadoes, which, like downdrafts, are more common away from the mountains (Peterson et al. Chap. 5; Figs. 5.3, 5.4). Damaging wind is also produced during transitory large-scale mid-latitude and tropical severe weather events when there are large gradients in atmospheric pressure; that is, when centers of strong high pressure and deep low pressure are close to each other. Large-scale mid-latitude storms are capable of producing strong winds in any region of the eastern USA. When there is a stable layer at mid-levels of the atmosphere, an intense pressure gradient can drive strong winds across the spine of the Appalachian Mountains after passage of a large-scale mid-latitude storm toward the northeast. These winds drive downslope windstorms, usually in the cool season, along the eastward-facing slopes of the Appalachian Mountains. Tropical storms primarily affect ecoregions closest to the Gulf of Mexico or Atlantic Ocean, as dying hurricanes rapidly lose their intensity after landfall, thus reducing the strength of the pressure gradient and the winds (Peterson et al. Chap. 5, Table 5.1, Fig. 5.2). High winds can create single- or multiple-tree windthrows that form small- to large gaps, or widespread, nearly complete destruction across large areas. For example, in the southern Appalachians, reports of timberland damage from remnants of Hurricane Hugo (1989) indicate that up to 25 % of trees were damaged (windthrown or blown tops) on 31,255 ha of timberland, and 25–50 % of trees were similarly damaged on an additional 1,769 ha in Ashe, Avery, and Watauga Counties, North Carolina (Dogett 1993). Further, heavy precipitation often accompanies high wind, saturating the soil and increasing the vulnerability of shallow-rooted trees to windthrow (Coutts et al. 1999). In this volume, Peterson et al. (Chap. 5) examine patterns and consequences of wind disturbance in central hardwood forests.

Damaging precipitation involves either (1) flooding, where runoff associated with precipitation exceeds the transport capacity of channels (see Leigh et al. Chap. 8); (2) mudslides and/or debris flows caused by high pore water pressures in steep, shallow soils (see Wooten et al. Chap. 9) or; (3) receiving precipitation in a form (e.g., ice, snow, or hail) that excessive accumulation can cause the loss of limbs or entire sections of trees (see Lafon Chap. 7). Excessive rainfall can occur in the eastern USA at any time of year due to stalled or slow moving convective storms, weather fronts, tropical storms, or large-scale mid-latitude storms. Excessive precipitation can also occur in the central hardwood ecoregions located on the windward side of the Appalachian Mountains when humid air masses are forced upward, either triggering a convective storm or enhancing pre-existing large-scale upward motion to produce a localized region of heavy precipitation. Excessive rainfall can result in floods, and, in hilly or mountainous regions (e.g., Blue Ridge Mountains, Ridge and Valley, and Southwestern Appalachians ecoregions), landslides and debris flows (Tao and Barros 2014). For example, heavy rainfall from tropical and extra-tropical cyclones triggered hundreds to thousands of debris flows on steep slopes in 1916 (North Carolina), 1940 (North Carolina), (1969 (Virginia, West Virginia), 1977 (North Carolina), 1985 (Virginia, West Virginia) and 2004 (North Carolina); rainfall from other more localized storms in these and other years triggered tens to hundreds of debris flows. Debris flows can carry whole trees, rocks, and soil (Band et al. 2012; Tao and Barros 2014), altering forest structure and hydrology by creating large canopy gaps and disrupting aquatic ecosystems. Forest cover is an important stabilizing factor through precipitation interception, evapotranspiration, and root reinforcement. Landslides can increase recruitment of early successional tree species, leading to persisting patches of early successional vegetation in older forests (Seiwa et al. 2013). In this volume, Wooten et al. (Chap. 9) examine the historical frequency and spatial patterning of landslide events, and their influence on forest structure, and Leigh (Chap. 8) examines paleoenvironmental floods and fires in the Blue Ridge Mountains.

Excessive accumulation of frozen precipitation can occur during the warm (hail) and cool (ice and snow) seasons. Excessive hail accumulation is most likely in central hardwood ecoregions located away from the coastlines of the Gulf of Mexico and Atlantic Ocean, and west of the Appalachian Mountains (Ouachita and Boston Mountains, Arkansas Valley, Ozark Highland, Western Allegheny and Interior Plateau, Interior River Valleys and Hills ecoregions), where mid-level dry layers that promote hail production are more likely. Ice storms can occur in any region of the eastern USA when the cold or warm front of a transiting large-scale mid-latitude storm moves in the vicinity during the cool season. Another scenario that commonly contributes to excessive ice accumulation occurs along the eastward-facing slopes of the Appalachian Mountains (Piedmont, Northern Piedmont, Blue Ridge Mountains, and Ridge and Valley ecoregions). Surface high pressure located over the northeastern USA in the winter and early springtime often forces a dome of cold air against the slopes of the eastward-facing Appalachian Mountains, sometimes called 'cold air damming' or a 'wedge,' and in the presence of a precipitating mid-latitude storm, can result in excessive ice accumulation. Ice storms also can be restricted to a 'belt' within an altitudinal range that extends for several km (e.g., Ashe 1918), and forests on exposed slopes are most often affected. Ice storms can affect large areas. For example, a 1987 ice storm damaged trees across an estimated 12,140 ha, and 60 % of the 482 km of trails were blocked by fallen trees in the Pisgah District of the Pisgah National Forest in the southern Appalachians (Osborne 1987); heavy ice storm damage covered an area approximately 270 km long and up to 170 km wide in eastern Arkansas through northern Mississippi, and in 2000 two ice storms damaged approximately 40 % of the 7.4 million ha of forest in Arkansas (see Bragg et al. 2003). In general, ice storms can create fine-scale to intermediate structural heterogeneity in forests, depending on the severity of damage to trees, the number of trees, and the area affected. In this volume, Lafon (Chap. 7) examines the historical frequency and range of damage by ice storms in the CHR.

Lightning-ignited fires are relatively uncommon in central hardwood forests (Table 1.6) (Mitchener and Parker 2005; Guyette et al. 2006a; Cohen et al. 2007). When they do occur, they are most frequent along ridgetops, or dry south-facing slopes where microclimates are more conducive to burning (Delcourt and Delcourt 1997; Harmon 1982). Not coincidentally, fire-adapted plant species such as Table Mountain pine, pitch pine, and others are also found primarily in these dry, low quality, more fire-prone sites (see Chaps. 6, 10, 11). Fire frequency is strongly associated with human populations, giving further indication that human ignitions were likely the source of most fires in central hardwood forests (Guyette et al. 2006a, b; Greenberg et al. Chap. 12). Fire reconstruction studies using soil charcoal or tree ring analyses indicate that fire frequencies vary among ecoregions, landscapes, and

Table 1.6 Mean ± SE annual number of reported[a] human-caused and lightning-ignited wildfires (1970–2013; n = 44 years)[b] per 2,000 km² in National Forests of the CHR

National forest	Area[c] (km²)	Ecoregion(s)[d]	Mean ± SE (range) annual number of human-caused wildfires per 2,000 km²	Mean ± SE (range) annual number of lightning-ignited wildfires per 2,000 km²
Bankhead NF	733.4	Southwestern Appalachians	52.1 ± 6.2 (8.2–199.1)	1.1 ± 0.4 (0.0–13.6)
Chatahoochee NF	3,036.2	Blue Ridge Mountains & Ridge and Valley	51.7 ± 5.6 (14.5–171.3)	2.5 ± 0.4 (0.0–13.2)
Cherokee NF	2,658.4	Blue Ridge Mountains	73.2 ± 7.8 (13.5–227.2)	3.8 ± 0.8 (0.0–27.8)
Daniel Boone NF	2,269.6	Central Appalachians	93.7 ± 8.0 (26.4–268.8)	1.8 ± 0.4 (0.0–13.2)
George Washington NF	4,311.4	Ridge and Valley & Northern Piedmont	12.9 ± 1.3 (2.3–38.0)	2.9 ± 0.5 (0.0–11.1)
Hoosier NF[2]	820.8	Interior Plateau	60.4 ± 11.4 (7.3–180.3)	1.0 ± 0.5 (0.0–9.7)
Jefferson NF	2,928.0	Blue Ridge Mountains	16.1 ± 1.5 (3.4–44.4)	2.0 ± 0.3 (0.0–8.9)
Mark Twain NF	6,040.4	Ozark Highlands	65.8 ± 5.3 (14.6–207.3)	1.1 ± 0.2 (0.0–6.6)
Monongahela NF	3,728.6	Central Appalachians	4.8 ± 0.5 (0.0–17.2)	0.4 ± 0.1 (0.0–2.1)
Nantahala NF	2,150.2	Blue Ridge Mountains	46.0 ± 3.4 (10.2–103.2)	2.7 ± 0.5 (0.0–14.9)
Oconee NF	472.5	Piedmont	34.1 ± 5.0 (0.0–152.4)	4.2 ± 1.2 (0.0–42.3)
Ouachita NF	7,236.8	Ouachita Mountains & Arkansas Valley	25.4 ± 1.8 (6.6–60.8)	5.9 ± 0.7 (0.8–18.5)
Ozark NF	4,609.7	Boston Mountains	28.9 ± 2.1 (7.8–66.8)	2.5 ± 0.4 (0.0–15.6)
Pisgah NF	2,076.4	Blue Ridge Mountains	52.7 ± 10.3 (9.6–392.0)	4.2 ± 0.7 (0.0–17.3)
Shawnee NF	1,076.7	Interior River Valleys and Hills	58.0 ± 6.6 (3.7–219.2)	0.5 ± 0.2 (0.0–5.6)
Sumter NF	1,501.0	Piedmont & Blue Ridge Mountains	27.6 + 2.7 (6.7–98.6)	3.1 ± 0.6 (0.0–17.3)
Talladega NF	1,597.4	Piedmont	69.7 ± 5.6 (20.0–180.3)	7.8 ± 1.1 (0.0–30.0)
Uwharrie NF	205.0	Piedmont	77.6 ± 6.7 (0.0–195.2)	1.3 ± 0.6 (0.0–19.5)
Wayne NF[2]	971.9	Western Allegheny Plateau	90 ± 15.2 (12.3–240.8)	0.3 ± 0.2 (0.0–4.1)

[a]Source: Fire Family Plus (database): (Main et al. 1990)
[b]Data for Hoosier NF and Wayne NF reported 1994–2013 (n = 20 years)
[c]Source: USDA Forest Service (2013) Land areas of the National Forest system. USDA For Serv FS-383, Washington DC
[d]See Fig. 1.1; the majority of National Forest area is within the ecoregion listed first in table

even specific sites (Hart and Buchanan 2012). In this book, Leigh (Chap. 8) uses river valley sediments and soil charcoal to examine the paleoenvironmental range of variation in floods and fires, and Grissino-Mayer (Chap. 6) uses fire reconstruction data to examine historical fire frequencies in yellow pine and mixed pine-hardwood forests of the southern Appalachians, and how fire (whether lightning- or human-caused) helped to shape the composition and distribution of these forest types across the landscape.

1.5 Disturbance and Diversity: Adaptation and Distribution Across a Dynamic Landscape

Historically, background (gap dynamics) and pulses of natural disturbances through time have been integral in shaping structurally diverse forests and maintaining a diversity of flora and fauna. Anthropogenic disturbances, such as clearing and frequent burning by Native Americans, and later by European settlers, were also essential in maintaining diverse biotic communities. The occurrence and persistence of disturbance-adapted species of flora and fauna provides clues that disturbances historically created structural conditions with sufficient spatial and temporal frequency to sustain them, although their distributions and populations likely ebbed and flowed over time, and across landscapes. In plants, traits such as shade intolerance (e.g., yellow-poplar and yellow birch (*Betula alleghaniensis*)) or serotiny (e.g., Table Mountain pine), are clues that open forest conditions were continuously available historically, in quantities adequate to sustain them. Similarly, the association between fire-adapted plant species and fire-prone topographic positions suggest that fire was sufficiently common on those landscapes to sustain fire-adapted traits. Narrow habitat requirements by some animal species, such grassy sites with absent canopy (e.g., Eastern meadowlark (*Sturnella magna*) at lower elevations, or golden-winged warblers at higher elevations), or forests with open-canopy conditions and snag availability (e.g., Eastern bluebirds (*Sialia sialis*)), are clues that such sites were available historically. In this volume, Wilfahrt et al. (Chap. 11) explore the relationship between disturbance-adapted traits in tree species, productivity, and disturbance. Tuttle et al. (Chap. 10) use 1930s vegetation surveys and recent data to compare historic and current variation in forest structure and composition caused by natural and anthropogenic disturbances in the Great Smoky Mountains National Park. Greenberg et al. (Chap. 12) discuss how natural disturbances might have influenced the historic distribution patterns and relative abundance of disturbance-adapted wildlife species in the CHR, and suggest that humans functioned as a keystone species by creating suitable habitats in sufficient quantities and varieties to enable the persistence of a wide range of disturbance-adapted species.

1.6 Managing Central Hardwood Forests of Today and Tomorrow Within the Historic Range of Variation in Natural Disturbances

Today's central hardwood forests are different from those of yesterday. In general, forests are younger, reduced in area, more fragmented, isolated by surrounding urban development, and, in many cases, different in species dominance from the forests of 200 years ago. The historic structure and function of forests is further threatened by the introduction of non-native pests and pathogens, some of which have decimated entire tree species such as the American chestnut and, currently, eastern hemlock (*Tsuga canadensis*). Additionally, climate change could alter the severity and frequency of natural disturbances, such as severe storms, fires, droughts, floods, and insect outbreaks. Managing today's forests within the historic range of variation in natural disturbances may not be possible given their altered current condition and possible future condition in a changing environment. Further, managing solely for HRV in natural (non-anthropogenic) disturbances may not be adequate for attaining important objectives, such as sufficient habitat with specific and differing structural conditions (e.g., oak woodland or prairie) required by many disturbance-adapted plant and animal species. An important step in forest planning will be clear and careful definition of management objectives and desired reference conditions, which may or may not be entirely 'natural,' particularly if human activities that historically created different types of disturbed habitats used by disturbance-adapted species are not considered natural. In this volume, Dale et al. (Chap. 13) discuss climate change and the future of natural disturbance regimes, and Zenner (Chap. 14) concludes with a discussion of whether today's forests can, or should, be managed within the historic range of variation in natural disturbance, and silvicultural tools for attaining specific management objectives.

Acknowledgments We acknowledge the contributions of Katherine Reidy (UNCA student) for developing a conceptual draft map of hurricane tracks in the CHR and Ida Evretjarn (Forest Service volunteer) for production of Fig. 1.1 and development of draft versions of other figures. Scott Goodrick contributed to accessing data for Table 1.6. Tara Keyser contributed to accessing data for Table 1.5 and Fig. 1.8, and provided useful insights through discussion.

References

Abell, C. A. (1934). Influence of glaze storms upon hardwood forests in the southern Appalachians. *Journal of Forestry, 32*, 35–37.

Ashe, W. W. (1918). Note on the preceding. *Monthly Weather Review, 46*, 374.

Ashe, W. W. (1922). *Forest types of the Appalachian and White Mountains*. Chapel Hill: Elisha Mitchell Scientific Society.

Ashe, W. W., & Ayers, H. B. (1901). Forests and forest conditions in the southern Appalachians. In *A report of the Secretary of Agriculture in relation to the forests, rivers, and mountains of the southern Appalachian Region*. Washington, DC: US Gov Printing Office.

Askins, R. A. (2000). *Restoring North America's birds*. New Haven: Yale University Press.

Bailey, R. G. (1995). *Description of the ecoregions of the United States* (Misc Pub No 1391 2nd ed.). Washington, DC: USDA Forest Service.

Balch, R. E. (1928). *The influence of the southern pine beetle on forest composition in western North Carolina*. Ithaca: New York State College of Forestry.

Band, L. E., Hwang, T. T., Hales, T. C., Vose, J., & Ford, C. (2012). Ecosystem processes at the watershed scale: Mapping and modeling ecohydrological controls of landslides. *Geomorphology, 137*, 159–167.

Bowles, D. E., & Mathis, M. L. (1989). Caddisflies (Insecta: Trichoptera) of mountainous regions in Arkansas, with new state records for the order. *Journal of the Kansas. Entomological Society, 62*, 234–244.

Bragg, D. C., Shelton, M. G., & Zeide, B. (2003). Impacts and management implications of ice storms on forests in the southern United States. *Forest Ecology and Management, 186*, 99–123.

Braun, E. L. (1950). *Deciduous forests of eastern North America*. New Jersey: The Blackburn Press.

Burke, K. L. (2012). Niche contraction of American chestnut in response to chestnut blight. *Canadian Journal of Forest Research, 42*, 614–620.

Carroll, W. D., Kapeluck, P. R., Harper, R. A., & Van Lear, D. H. (2002). Gen Tech Rep SRS-53. In D. N. Wear & J. G. Greis (Eds.), *Southern Forest Resource Assessment* (Gen Tech Rep SRS-53). Asheville: USDA Forest Service Southern Research Station.

Cohen, D., Dellinger, B., Klein, R., & Buchanan, B. (2007). Patterns in lightning-caused fires at Great Smoky Mountains National Park. *Fire Ecology Special Issue, 3*, 68–82.

Coutts, M. P., Nielsen, C. C. N., & Nicoll, B. C. (1999). The development of symmetry, rigidity and anchorage in the structural root system of conifers. *Plant and Soil, 217*, 1–15.

Delcourt, P. A., & Delcourt, H. R. (1981). Vegetation maps for eastern North America: 40,000 yr BP to the present. In R. C. Romans (Ed.), *Geobot* (II). New York: Plenum Press.

Delcourt, P. A., & Delcourt, H. R. (1997). The influence of prehistoric human set fires on oak-chestnut forests in the southern Appalachians. *Castanea, 63*, 337–345.

Delcourt, P. A., & Delcourt, H. R. (2004). *Prehistoric Native Americans and ecological change: Human ecosystems in eastern North America since the Pleistocene*. Cambridge: Cambridge University Press.

Delcourt, P. A., Delcourt, H. R., Morse, D. F., & Morse, P. A. (1993). History, evolution, and organization of vegetation and human culture. In W. H. Martin, S. G. Boyce, & A. C. Echternact (Eds.), *Biodiversity of the southeastern United States: Lowland terrestrial communities*. New York: Wiley.

Delcourt, P. A., Delcourt, H. R., Ison, C. R., Sharp, W. E., & Gremillion, K. G. (1998). Prehistoric human use of fire, the eastern agricultural complex, and Appalachian oak-chestnut forests: Paleoecology of Cliff Palace pond, Kentucky. *American Antiquity, 63*, 263–278.

Dogett, C. A. (1993). A method to assess large-scale forest damage: A case study. *Southern Journal of Applied Forestry, 17*, 197–199.

Driese, S. G., Zheng-Hua, L., & McKay, L. D. (2008). Evidence for multiple, episodic, mid-Holocene Hypsithermal recorded in two soil profiles along an alluvial floodplain catena, southeastern Tennessee, USA. *Quaternary Research, 69*, 276–291.

Dyer, J. M. (2006). Revisiting the deciduous forests of eastern North America. *Bioscience, 56*, 341–352.

Elliott, K., & Swank, W. (2008). Long-term changes in forest composition and diversity following early logging (1919–1923) and the decline of American chestnut (*Castanea dentata*). *Plant Ecology, 197*, 155–172.

Eyre, F. H. (Ed.). (1980). *Forest cover types of the United States and Canada*. Washington, DC: Soc Am For.

Foti, T. L., & Glenn, S. M. (1990) The Ouachita mountain landscape at the time of settlement. In *Proceedings conference restoring old growth forests in the interior highlands of Arkansas and Oklahoma*, 19–20 Sept 1990, Morilton.

Fralish, J. S. (2003). The central hardwood forest: Its boundaries and physiographic provinces. In J. W. Van Sambeek, J. O. Dawson, F. Ponder Jr., E. F. Lowenstein, & J. S. Fralish (Eds.), *Proceedings thirteenth central hardwood for conference, April 1–3 2002, Urbana IL* (Gen Tech Rep NC-234). St Paul: USDA Forest Service North Central Station.
Frothingham, E. H., & Stuart, R. Y. (1931). *Timber growing and logging practice in the southern Appalachian region* (Tech Bull No 250). Washington, DC: USDA Forest Service.
Gallant, A. L., Loveland, T. R., Sohl, T. L., & Napton, D. E. (2000). Using an ecoregion framework to analyze land-cover and land-use dynamics. *Environmental Management, 34*(Supp 1), S89–S110.
Greeley, W. B., & Ashe, W. W. (1907). *White oak in the southern Appalachians* (Circular 105). Washington, DC: USDA Forest Service.
Greenberg, C. H., Keyser, T. L., & Speer, J. H. (2011). Temporal patterns of oak mortality in a southern Appalachian forest. *Natural Areas Journal, 31*, 131–137.
Greenberg, C. H., Perry, R. W., Franzreb, K. E., Loeb, S. C., Saenz, D., Rudolph, D. C., Winters, E., Fucik, E. M., Kwiatkowski, M. A., Parresol, B. R., Austin, J. D., & Tanner, G. W. (2014). Climate change and wildlife in the southern United States: Potential effects and management options. In J. M. Vose & K. E. Klepzig (Eds.), *Climate change adaptation and mitigation management options: A guide for natural resource managers in southern forest ecosystems* (pp. 379–420). Boca Raton: CRC Press.
Guyette, R. P., Spetich, M. A., & Stambaugh, M. C. (2006a). Historic fire regime dynamics and forcing factors in the Boston mountains, Arkansas, USA. *Forest Ecology and Management, 234*, 293–303.
Guyette, R. P., Dey, D. C., Stambaugh, M. C., & Muzika, R. (2006b). Fire scars reveal variability and dynamics of eastern fire regimes. In M. B. Dickinson (Ed.), *Proc Fire in eastern oak forests: Delivering science to land managers* (Gen Tech Rep NRS-p-1). Northern Research Station Newtown Square: USDA Forest Service.
Harmon, M. (1982). Fire history of the westernmost portion of Great Smoky Mountains National Park. *Bull Torrey Bot Club, 109*, 74–79.
Hart, J. L., & Buchanan, M. L. (2012). History of fire in eastern oak forests and implications for restoration. In *Proceedings fourth fire in eastern oak for conf* (Gen Tech Rep NRS-P-102). Newtown Square: USDA Forest Service Northern Research Station.
Holmes, J. S. (1911). *Forest conditions in western North Carolina* (Bull No 23). Raleigh: The North Carolina Geol and Econ Surv.
Hursh, C. R., & Haasis, F. W. (1931). Effects of a 1925 summer drought on southern Appalachian hardwoods. *Ecology, 12*, 380–386.
Jackson, S. T., Webb, R. S., Anderson, K. H., Overpeck, J. T., Webb, T., III, Williams, J. W., & Hansen, B. C. S. (2000). Vegetation and environment in Eastern North America during the Last Glacial Maximum. *Quaternary Science Reviews, 19*, 489–508.
Korstian, C. F. (1937). Perpetuation of spruce on cut-over and burned lands in the higher southern Appalachian Mountains. *Ecological Monographs, 7*, 125–167.
Little, E. L. (1977). *Atlas of United States trees* (Minor eastern hardwoods, Misc Pub 1342 4th ed.). Washington, DC: USDA Forest Service.
Main, W. A., Paananen, D. M., & Burgan, R. E. (1990). *Fire Family Plus* (USDA For Serv Gen Tech Rep NC-138). St. Paul: North Central Forest Experiment Station.
Mann, M. E., Zhihua, Z., Rutherford, S., Bradley, R. S., Hughes, M. K., Shindell, D., Ammann, C., Faluvegi, G., & Ni, F. (2009). Global signatures and dynamical origins of the Little Ice Age and Medieval Climate Anomaly. *Science, 326*, 1256–1260.
Mattoon, W. R. (1915). *Life history of the shortleaf pine*. Washington, DC: Bull 244, US Department of Agriculture.
Mattoon, W. R. (1936). *Forest trees and forest regions of the United States* (Misc Pub 217). Washington, DC: Misc Pub 217, US Department of Agriculture.
Mattoon, W. R. (1937). *Forest regions of the United States (map)*. Washington, DC: USDA Forest Service.

McNab, W. H., Spetich, M. A., Perry, R. W., Haywood, J. D., Laird, S. G., Clark, S. L., Hart, J. L., Torreano, S. J., & Buchanan, M. L. (2014). Climate-induced migration of native tree populations and consequences for forest composition. In J. M. Vose & K. D. Klepzig (Eds.), *Climate change adaptation and mitigation management options: A guide for natural resource managers in southern forest ecosystems* (pp. 307–378). Boca Raton: CRC Press.

Merritt, C. (1980). The Central Region. In J. W. Barnett (Ed.), *Regional silviculture of the United States*. New York: John Wiley & Sons.

Milner, G. R., & Chaplin, G. (2010). Eastern North American population at CA AD 1500. *American Antiquity, 75*, 707–726.

Mitchener, L. J., & Parker, A. J. (2005). Climate, lightning, and wildfire in the National Forests of the southeastern United States: 1989–1998. *Physical Geography, 26*, 147–162.

Nelson, T. C. (1957). The original forests of the Georgia Piedmont. *Ecology, 38*, 390–397.

Omernik, J. M. (1987). Ecoregions of the conterminous United States. *Annals Association of American Geographers, 77*, 118–125.

Osborne, C. (1987). Pisgah National Forest heavily damaged. The Asheville Citizen, 13 Apr 1987.

Pederson, N., Dyer, J. M., McEwan, R. W., Hessi, A. E., Mock, C. J., Orwig, D. A., Rieder, H. E., & Cook, B. I. (2014). The legacy of episodic climatic events in shaping temperate, broadleaf forests. *Ecological Monographs, 84*, 599–620.

Pell, B. (1983). The natural divisions of Arkansas: A revised classification and description. *Natural Areas Journal, 3*, 12–23.

Pielke, R. A. (1981). The distribution of spruce in west-central Virginia before lumbering. *Castanea, 46*, 201–216.

Pinchot, G., & Ashe, W. W. (1897). Timber trees and forests of North Carolina. NC Geol Surv Bull No. 6, Winston.

Riitters, K. H., Wickham, J. D., O'Neill, R. V., Jones, K. B., Smith, E. R., Coulston, J. W., Wade, T. G., & Smith, J. H. (2002). Fragmentation of continental United States forests. *Ecosystems, 5*, 815–822.

Roberts, S. D., Dunning, J. B., & Miller, B. K. (1995). *Management of biological diversity in the central hardwood region* (Nat Res Prof Ser FNR-147). West Lafayette: Perdue University.

Ruefenacht, B., Finco, M. V., Nelson, M. D., Czaplewski, R., Helmer, E. H., Blackard, J. A., Holden, G. R., Lister, A. J., Salajanu, D., Weyermann, D., & Winterberger, K. (2008). Conterminous US and Alaska forest type mapping using forest inventory and analysis data. *Photogrammetric Engineering and Remote Sensing, 74*, 1379–1388.

Schmidt, T. L., & McWilliams, W. H. (2003). Shifts and future trends in the forest resources of the Central Hardwood Region. In J. W. VanSambeek, J. W. Dawson, F. Ponder Jr., E. F. Loewenstein, & J. S. Fralish (Eds.), *Proceedings thirteenth central hardwood for conference, Apr 1–3 2002, Urbana, Ill* (Gen Tech Rep NC-234). St. Paul: USDA Forest Service North Central Station.

Schnur, G. L. (1937). *Yield, stand, and volume tables for even-aged upland oak forests*. Washington, DC: U.S. Department of Agriculture.

Seiwa, K., Miwa, Y., Akasaka, S., Kanno, H., Tomita, M., Saitoh, T., Ueno, N., Kimura, M., Hasegawa, Y., Konno, M., & Masaka, K. (2013). Landslide-facilitated species diversity in a beech-dominant forest. *Ecological Restoration, 28*, 29–41.

Society of American Foresters. (1964). *Forest cover types of North America*. Washington, DC: Society of American Foresters.

Spetich, M. A., Perry, R. W., Harper, C. A., & Clark, S. L. (2011). Fire in eastern hardwood forests through 14,000 years. In C. H. Greenberg, B. S. Collins, & F. R. Thompson III (Eds.), *Sustaining young forest communities: Ecology and management of early successional habitats in the Central Hardwood Region* (pp. 41–58). New York: Springer.

Stahle, D. W., & Hehr, J. G. (1984). Dendroclimatic relationships of post oak across a precipitation gradient in the south central United States. *Annual Association of American Geographers, 74*, 561–573.

Starkey, D. A., & Oak, S. W. (1989). Site factors and stand conditions associated with oak decline in southern upland hardwood forests. In G. Rink & C. A. Budelsky (Eds.), *Gen Tech Rep NC-132*. St Paul: USDA Forest Service North Central Station.

Starkey, D. A., Oliveria, F., Mangini, A., & Mielke, M. (2004). Oak decline and red oak borer in the interior highlands of Arkansas and Missouri: natural phenomena, severe occurrences. In *Upland oak ecology symposium: History, current conditions, and sustainability* (pp. 217–222). Asheville: USDA Forest Service Southern Research Station.

Tao, J. J., & Barros, A. P. (2014). Coupled prediction of flood response and debris flow initiation during warm- and cold-season events in the southern Appalachians, USA. *Hydrology and Earth System Sciences, 18*(1), 367–388.

US Census Bureau. (1932). *Fifteenth census of the United States: 1930*, vol II, Report on agricultural lands. USDC Bureau of the Census, Washington, DC. www.agcensus.usda.gov/Publications/Historical_Publications. Accessed 17 Sept 2014.

US Census Bureau. (Undated). Map of the population of the United States compiled from returns of population at the first census, 1790. Washington, DC: USDI. www.census.gov/history/img/1790. Accessed 17 Sept 2014.

USDA Forest Service. (2012). National forest system land management planning, final rule (36 CFR Part 219). *Federal register, 77*(68), 21161–21276.

White, P. S., Collins, B., & Wein, G. (2011). Natural disturbances and early successional habitats. In C. H. Greenberg, B. S. Collins, & F. R. Thompson III (Eds.), *Sustaining young forest communities: Ecology and management of early successional habitats in the Central Hardwood Region, USA* (pp. 27–40). New York: Springer.

Worldatlas. (2012). US state population densities based on 2012 census. http://www.worldatlas.com/aatlas/populations/usadensityh.htm. Accessed 8 May 2014.

Chapter 2
Gap-Scale Disturbances in Central Hardwood Forests with Implications for Management

Justin L. Hart

Abstract All forest ecosystems are subject to canopy disturbance events that influence species composition and stand structure, and drive patterns of succession and stand development. Disturbances may be categorized by a variety of characteristics, but they are most often classified along a gradient according to their spatial extent, magnitude, and frequency. This gradient spans from broad-scale, stand-replacing events where most of the overstory is removed to fine-scale events which result from the removal of a single canopy individual or a small cluster of trees. The disturbance regimes of most stands in the Central Hardwood Region are characterized by fine-scale events. At the stand scale, these localized and asynchronous events can create a patchwork mosaic of microsites comprised of different tree species, ages, diameters, heights, crown spreads, and growth rates. Through the modification of fine-scale biophysical conditions, these localized canopy disturbances promote heterogeneity and biodiversity in forest ecosystems. Forest management based on natural disturbance processes should consider elements of the gap-scale disturbance regime, such as frequency, size, shape, and closure mechanisms, and the historical range of variation associated with these characteristics. Silvicultural prescriptions are available for gap-based management designed to promote oak regeneration and mimic natural disturbance processes.

Keywords Canopy gap • Disturbance • Silviculture • Succession • Stand development

2.1 Introduction

All forests are subject to disturbance events which influence forest composition and structure and thus, drive successional and developmental pathways. Forest disturbances may be categorized by a variety of characteristics, but are most often

J.L. Hart (✉)
Department of Geography and Program in Environmental Science,
University of Alabama, Tuscaloosa, AL 35487, USA
e-mail: hart013@ua.edu

C.H. Greenberg, B.S. Collins (eds.), *Natural Disturbances and Historic Range of Variation*, Managing Forest Ecosystems 32,
DOI 10.1007/978-3-319-21527-3_2

Fig. 2.1 Disturbance classification gradient based on spatial extent and magnitude of a discrete disturbance event (Photographs by author)

classified according to spatial extent, magnitude, and frequency, and occur along a gradient that spans from fine-scale events that result from the loss of a single canopy tree or a small cluster of individuals to broad-scale, stand-replacing events when most of the overstory is removed (Fig. 2.1; Oliver and Larson 1996). The disturbance regimes of most hardwood stands in the Central Hardwood Region (CHR) (Greenberg et al. Chap. 1, Fig. 1.1) are characterized by fine-scale events (Lorimer 1980; Barden 1981; Runkle 1981, 1982, 1996, 2000; Cho and Boerner 1991). At the stand-scale, these localized canopy disturbances create a patchwork mosaic of microsites which may be comprised of different tree species, ages, diameters, heights, and crown spreads (Runkle 1981, 1985; Canham and Marks 1985; Phillips and Shure 1990). By modifying fine-scale biophysical conditions, these gap-scale canopy disturbances may increase heterogeneity and biodiversity in forest ecosystems (Putz 1983; Abe et al. 1995).

Canopy gaps are typically defined as visible void spaces in the main forest canopy within which gap stems are shorter than a specified threshold (e.g., 20 m) or shorter than a percentage of the canopy trees surrounding the void (e.g., <75 % canopy height), and a gapmaker tree or trees are present (Yamamoto 2000; Richards and Hart 2011). The area directly beneath the canopy void is typically considered to be the canopy gap (sometimes called the true gap, light gap, or observed gap). However, because this void area changes with time since gap formation, and insolation and other abiotic variables are not limited to the area directly underneath the canopy void, canopy gaps can also be described as the total terrestrial area below the gap and extending to the bases of the canopy trees surrounding the gap. This is termed the expanded or extended gap (Fig. 2.2; Runkle 1981). Canopy gaps may be characterized by a range of different physical parameters, but the metrics most commonly measured are canopy gap formation mechanisms, gap formation frequency, gap size and shape distributions, fraction of land in gaps, gap closure mechanisms, and advanced reproduction within gaps. In this chapter, I attempted to synthesize information on these characteristics from studies conducted throughout the CHR and provide recommendations for management based on natural gap-scale disturbance processes.

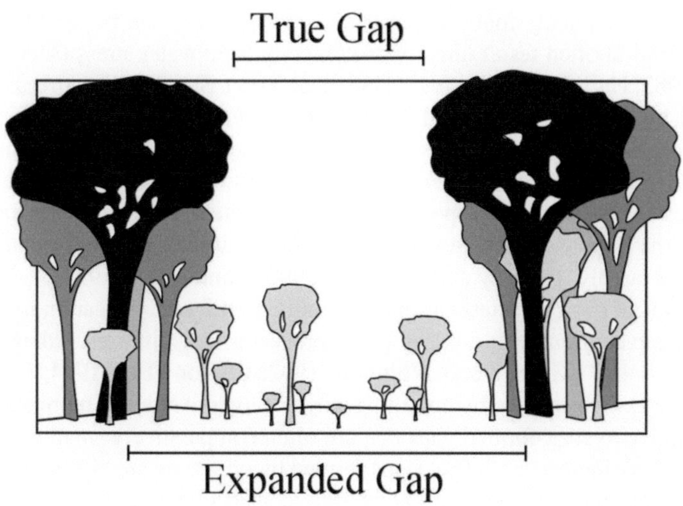

Fig. 2.2 The true gap (also called observed or light gap) and expanded gap (also called extended gap) environments. The *black* trees represent canopy dominants along the periphery of the gap

2.2 Gap Formation Mechanisms

Canopy gaps occur on sites where partial or total death of one or a small cluster of canopy individuals has occurred (Watt 1947; Runkle 1981, 1985). As such, canopy gaps may be caused by a variety of disturbance agents such as strong winds and insect outbreaks. Individual canopy gaps are formed by trees that have been uprooted (root network uplifted), stems that have been snapped (bole broken below the crown), or snags (standing dead trees with crowns mostly intact; Putz 1983; Clinton et al. 1993; Yamamoto 2000; Richards and Hart 2011).

Tree uprooting typically results from strong winds and may be related to soil depth where trees are not firmly anchored (Schaetzl et al. 1989), soil saturation which reduces soil cohesion and shear strength (Beatty and Stone 1986; Schaetzl et al. 1989), or biotic factors such as infection by the fungus *Armillaria mellea*, which weakens root systems (Williams et al. 1986). Uprooting may also be caused by crown asymmetry, which occurs as gap neighbors benefit from adjacent canopy gaps and undergo lateral branch growth to fill the void from the side, unequal crown growth of trees at different elevations on steep slopes, and/or twig loss and abrasion from wind (Young and Hubbell 1991; Rentch et al. 2010; Peterson et al. Chap. 5). In mesic hardwood stands of the CHR, wind-induced mortality is the predominant gap formation mechanism (Barden 1979; Runkle 1981, 1982, 1996). Snapped stem-formed gaps may be caused by strong wind events when the stem fails above ground level (Quine and Gardiner 2007) or by trees that first formed as snags that subsequently snapped. Snag-formed gaps are common in upland stands of the CHR (Clinton et al. 1993, 1994; Richards and Hart 2011). In this region, snag-formed gaps are most typically found on xeric sites where water can be limiting and are hypothesized to often result directly or indirectly from water stress (McComb and Muller 1983; Hart and Kupfer 2011; Hart et al. 2012). Standing dead trees are often removed by mild to severe wind events and this process complicates formation-specific gap investigations because the category likely represents a combination of gaps that formed directly by stem snapping and those that first formed as snags. Snags that are eventually snapped likely create distinct microenvironmental conditions and the response of residual trees may differ between these gaps and those that are formed rapidly (Krasney and Whitmore 1992; Clinton et al. 1994).

Biophysical gap characteristics may be influenced by the formation mechanism and thus, gap environments and gap-phase processes may differ according to mode of tree mortality (Krasney and Whitmore 1992; Clinton et al. 1994; Himes and Rentch 2013). Gaps caused by uprooting may be larger in size and more elliptical in shape relative to snag-formed gaps that are often comparatively small and circular (Hart and Grissino-Mayer 2009). Gaps formed by uprooting are also more likely to involve two or more canopy trees compared to snag-formed gaps as the fall of an uprooted gapmaker has a greater probability of removing neighboring individuals (Yamamoto and Nishimura 1999). Snag gaps are typically smaller in size and they release growing space more gradually as they shed limbs rather than instantaneously. I note however, that once the snag is barren of leaves the majority of the growing space has been released as woody material restricts a relatively small proportion of insolation. Additionally, composition of gaps created by uprooting events may differ significantly from that in gaps created by snags or snapped stems (Hart and Kupfer 2011). Such patterns may be attributed to the physical alteration of the gap environment by the uprooting process as intra-gap heterogeneity caused by uprooting canopy trees (e.g., pit and mound topography) has been shown to be an important determinant of species composition in gaps (Hutnik 1952; Putz 1983; Runkle 1985). However, gap formation is coupled with other factors. For example, soils, slope aspect, and steepness influence the gap formation mechanism and the gap formation mechanism in turn influences gap size. Hart and Kupfer (2011) found that snag-formed gaps were most common on south-facing slopes on soils with low moisture holding capacity on the Cumberland Plateau in Tennessee. Species composition of these gaps differed from that of gaps caused by tree uprooting; however, tree uprooting was largely restricted to north-facing slopes on soils with higher moisture holding capacities. Thus, although forest community response to a disturbance event is constrained by the physical environment, the disturbance regime itself may also be strongly influenced by the physical setting.

2.3 Gap Formation Frequency

Disturbance regimes and forest response to discrete events vary by forest developmental stage attributed largely to differences in species composition, stand structure, and tree age distributions (Table 2.1). Exogenous disturbance events are stochastic

Table 2.1 Typical canopy gap characteristics in relatively young and old central hardwood forest stands

Characteristic	Young	Old
Gap frequency	High	Low
Gap size	Small	Large
Gap duration	Short	Long
Gap fraction	–	–
Gap closure	Lateral crown expansion	Sub-canopy height growth

and not related to stand age, but responses of residual trees to these discrete events may differ (e.g., some old trees may be less likely to respond to increased resources and growing space compared to younger, more vigorous individuals).

Although gap formation frequency varies through time (because of the stochastic nature of exogenous disturbance) and by stage of development (because of the factors listed above), estimates on the rate of canopy gap formation have been developed. For hardwood stands in the CHR, the canopy gap formation rate is typically 0.5–2 % per year (Runkle 1982, 1985). Based on these values, the average time between natural canopy disturbances for a given site is 50–200 years (Runkle 1985). However, some recent research indicates that exogenous canopy gap formation rates may have declined over the past 400 years in white oak (*Quercus alba*) stands of the eastern USA (Buchanan and Hart 2012). The authors speculated this pattern may have been attributed to changes in drought frequency and intensity, changes is anthropogenic land-use patterns, and the extinction of Passenger Pigeon (*Ectopistes migratorious*) (Greenberg et al. Chap. 12). Thus, the estimated gap formation rate (i.e., background mortality) of 0.5–2 % per year may be slightly lower than that of centuries prior because of changes in the frequency of exogenous disturbance events. I stress that this pattern has not been found in all canopy disturbance reconstruction studies in the region. For example, Rentch et al. (2003) did not find differences in canopy gap formation frequency in five old-growth oak stands across the 1700s, 1800s, or 1900s. More research on potential changes in canopy gap formation in central hardwood forests is warranted.

The proportion of stands in the eastern USA in the complex stage of development at European settlement was estimated to be much higher than at present (Whitney 1994; Lorimer 2001). In complex stage stands canopy tree heights and crown volumes are more highly variable, which creates more complex canopy topography (Oliver and Larson 1996). Stands with old trees and with more complex canopy topography are more likely to experience damage (i.e., localized tree mortality) from strong wind events (Runkle 1985; Foster 1988; Quine and Gardiner 2007). Thus, a single storm event may influence stands differently across stages of development, with stands in the complex stage being the most sensitive to wind-induced damage. Although the frequency and intensity of severe wind events in the eastern USA may not have changed over the past few centuries, the conversion of stands throughout the CHR to more simple structures may have reduced the frequency of localized canopy disturbances in these systems, which may in turn have influenced regeneration patterns.

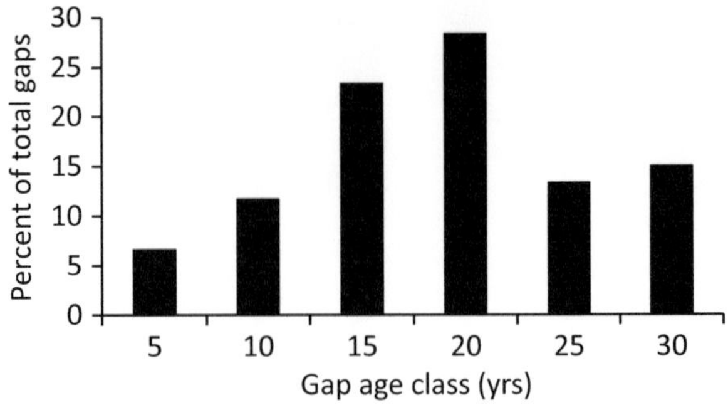

Fig. 2.3 Distribution of 60 canopy gaps by five-year age class bins in upland hardwood stands on the Bankhead National Forest, Alabama (Adapted from Richards and Hart 2011)

If the gap closure rate approximates the gap formation rate in a stand, gap fraction is considered to be in dynamic equilibrium (Himes and Rentch 2013). If gap formation and gap closure are in equilibrium, the gap age distribution should reveal a high frequency of young gaps and the number of gaps should decline with increased age. However, superimposed over the natural background mortality rate are exogenous disturbances; events which may create a high frequency of localized, gap-scale disturbances throughout a stand. These punctuated events may cause the gap age distribution to become irregular (Fig. 2.3).

The most commonly used disturbance classification terminology is based on spatial extent and magnitude of damage caused by a discrete event (Oliver and Larson 1996). However, for some events it may be the timing of formation that dictates the disturbance classification (e.g., gap-scale v. intermediate-scale disturbances). For example, strong wind events may remove trees singularly or in small groups (i.e., create localized canopy disturbances) throughout a stand. The size of individual canopy disturbances may be of the scale that constitutes a gap, but if the gaps were created across a broad area of the stand the disturbance may in fact have removed enough basal area to be considered of the intermediate scale. In such instances it would be gap formation rate that determines the disturbance classification. Disturbance history reconstructions using tree-ring records and forest inventory data have been conducted in some hardwood stands of the CHR. The common convention in these studies is to classify gap-scale disturbances that removed trees from at least 25 % of the stand as stand-wide events. The return interval of these stand-wide events was typically 20–40 years (Nowacki and Abrams 1997; Ruffner and Abrams 1998; Hart and Grissino-Mayer 2008; Hart et al. 2012). At the stand scale, these disturbance events may have removed enough basal area to be considered intermediate-scale disturbances, but the mortality was localized (i.e., gap scale) throughout the stands.

2.4 Canopy Gap Sizes and Shapes

Canopy gap size is highly variable and is influenced by factors such as the number of trees removed, the height and crown volume of removed trees, and the gap formation mechanism. The range of gap sizes reported from hardwood stands in the CHR ranges from 5 m^2, in gaps that are just about to close, to greater than 4,000 m^2 (Barden 1980; Clinton and Baker 2000; Hart and Grissino-Mayer 2009). Some authors have suggested that canopy disturbances exceed the size of canopy gaps (i.e., they are not gap scale, but intermediate- or stand-scale events) if the canopy void space exceeds 1,000 m^2 (Yamamoto 2000). Nonetheless, gap-scale disturbance studies from the central hardwood forests have typically found true canopy gaps to range from 30–140 m^2 and expanded canopy gaps to typically range from 200–500 m^2 (Barden 1980, 1981; Runkle 1981, 1982, 1990; Runkle and Yetter 1987; Clinton et al. 1993, 1994; Hart and Grissino-Mayer 2009; Richards and Hart 2011; Himes and Rentch 2013). Gap size may also be expressed in relation to adjacent canopy tree height by comparing gap diameter to mean canopy height (D:H). Such comparisons in central hardwood forests have found that the D:H of most gaps is <1.0 (Runkle 1985; Richards and Hart 2011).

Canopy gap shapes typically range from circular to elliptical, but gap shape varies and the patterns can be blocky rather than elliptical (Lima 2005). However, the majority of gap-based research in central hardwood forests has noted elliptical shapes (Runkle 1982, 1992; Clinton et al. 1993; Hart and Grissino-Mayer 2009, Richards and Hart 2011). Gap shapes may be quantified by calculating the gap length (longest axis of the gap) to gap width (longest distance perpendicular to length) and analyzing the length-to-width ratio (L:W; Hart and Grissino-Mayer 2009; Rentch et al. 2010; Richards and Hart 2011). The gap formation mechanism may be a strong determinant of gap shape. Snag-formed gaps tend be more circular in shape and have L:W of about 1.0. In contrast, gaps formed by uprooted and snapped stems tend to be more ellipsoidal with L:W of >2.0 (Hart and Grissino-Mayer 2009).

2.5 Canopy Gap Fraction

Canopy gap fraction is the percent of a stand that is within a true or expanded canopy gap. Gap fraction in central hardwood forests for true gaps typically ranges from 3–25 % and for expanded gaps often ranges from 8–30 % (Romme and Martin 1982; Runkle 1982; Runkle 1985; Keller and Hix 1999; Busing 2005; Hart and Grissino-Mayer 2009; Himes and Rentch 2013; Weber 2014). Although gap frequency and size vary by stage of stand development, gap fraction may be rather similar. Hart and Grissino-Mayer (2009) found gap fraction in upland oak stands on the Cumberland Plateau at age 80 years approximated the values reported from much older stands throughout the region. Thus, the percentage of land area in gap environments in relatively young and old stands was similar, but the distribution of the gap area was quite different (Table 2.1). Young stands are characterized by a high frequency of small gaps and older stands are characterized by a relatively small number of large gaps.

2.6 Gap Closure and Structural Development

Hardwood stands during the stem exclusion and understory reinitiation phases of development are characterized by high densities of relatively small individuals of similar age (Oliver and Larson, 1996). Intense competition for resources and self-thinning in developing stands result in a high frequency of localized canopy disturbances (Clebsch and Busing 1989; Hart and Grissino-Mayer, 2009). In such systems, when a canopy individual dies residual neighboring stems are able to quickly capture the released growing space and close the canopy void via lateral crown extension, thereby altering tree size, tree architecture, and stand structure (Hart and Grissino-Mayer 2008). Conversely, during the complex stage of development stands contain fewer individuals and have reduced competition and mortality rates resulting in a reduced frequency of endogenous canopy disturbance events (Zeide 2005). Most canopy trees in complex stage hardwood stands have comparatively large crowns and when one of these individuals is removed from the canopy, a relatively large void is created and peripheral trees are often incapable of closing the gaps through lateral crown extension (Tyrell and Crow 1994; Yamamoto 2000). These larger gaps should require relatively long periods to close because of their size, which increases the probability of a new individual recruiting to a dominant or codominant position through subcanopy ascension (Runkle 1985; Rentch et al. 2003; Webster and Lorimer 2005; Zeide 2010). These comparatively large gaps in old stands may also allow for the establishment of new individuals and may therefore promote multi-aged stands. In the absence of exogenous disturbance events, structural changes with maturity are driven by these localized canopy disturbances (Johnson et al. 2009). Indeed, it is gap-scale disturbance processes that create the complexity that defines old-growth structure in hardwood systems (Oliver and Larson 1996; Frelich 2002; Richards and Hart 2011).

2.7 Gap-Phase Succession

Forest community responses to gap-scale disturbances are influenced by a range of gap characteristics such as size (Runkle and Yetter 1987), age (Brokaw 1985), formation frequency (Canham 1989), formation mechanism (Putz 1983; Clinton et al. 1993), distance from edge (Kupfer et al. 1997), topographic position (Clinton et al. 1994; Abe et al. 1995) and orientation (Poulson and Platt 1988) among others. Throughout the CHR, light is commonly the most limiting factor (Oliver and Larson 1996) and gap characteristics are important because of their direct influence on understory light regimes (Canham et al. 1990). Gap-scale disturbances influence all forest strata, but the biophysical changes caused by local canopy removal are typically most evident in the regeneration layer (Brokaw and Busing 2000; Yamamoto 2000). Responses in this stratum are important because saplings represent the pool of species likely to recruit to larger size classes, so sapling composition in gaps is often an important determinant of future canopy composition in stands with disturbance regimes dominated by gap-scale processes (Wilder et al. 1999; Taylor and Lorimer 2003).

Gaps are stochastic and favor species in the understory or midstory of the gap environment. Without competition from shade-tolerant mesophytes, oaks have the physiological capability for long-term survival beneath an oak-dominated canopy, and may be considered a gap-phase genus (Orwig and Abrams 1995; Abrams 1996; Rentch et al. 2003) as these trees are able to persist in low light conditions and maintain the ability to respond to increased resources associated with canopy disturbance in their vicinity. During the twentieth century, the understory strata of oak-dominated stands across a variety of site types came to support a high density of shade-tolerant individuals, principally red maple (*Acer rubrum*) and sugar maple (*Acer saccharum*) (i.e., the oak-to-maple transition; McEwan et al. 2011). The driver of this widespread understory composition shift may vary according to site, but was likely a function of climate change, herbivore population density fluctuation, loss of American chestnut (*Castanea dentata*) and Passenger Pigeon (*Ectopistes migratorious*) (Greenberg et al. Chap. 12), changes in land-use patterns, and modification of the fire regime (Lorimer 1993; Abrams 2003; Nowacki and Abrams 2008; McEwan et al. 2011; Grissino-Mayer Chap. 6). When a canopy gap forms in a stand exhibiting the oak-to-maple transition, the probability of a shade-tolerant individual being in the gap environment is greater than the likelihood of a shade-intolerant or moderately-tolerant species being in the gap. Furthermore, small canopy gaps typically close quickly by lateral crown expansion and may not permit enough time for even fast growing shade-intolerant species to colonize the gap environment and then ascend to the canopy prior to gap closure. For these reasons, gap-scale disturbances typically favor shade-tolerant species. As canopy oaks die in oak-dominated stands, the gaps formed provide a mechanism for shade-tolerant stems that are abundant in the understory to recruit to larger size classes and to ascend to canopy positions. This gap-scale process is driving the observed successional replacement of oak by maple throughout central hardwood forests.

2.8 Gap-Based Management

In recent decades, there has been a fundamental philosophical change in the management of forest resources. Increasingly, managers are utilizing approaches that emulate natural ecological processes including natural disturbance regimes (Franklin and Johnson 2012; Hanson et al. 2012; Zenner Chap. 14). This approach emphasizes creating structures and community assemblages through silviculture that are similar to those that were historically produced by natural disturbance processes (Seymour and Hunter 1999). Wind is the most common and arguably the most influential canopy disturbance agent in hardwood forests of the eastern USA (Runkle 1996). The goal of natural disturbance-based management is not to mimic the actual disturbance event (i.e., trees are not typically felled by winching to emulate the effects of strong winds), but rather to use the effects of such events (e.g., the altered light regime) as models for individual and cumulative silvicultural treatments with the goal of minimizing the structural, compositional, and functional disparities between managed and unmanaged stands. The rationale for such an approach is that emulation of natural events restores and/or maintains resilience to a range of environmental challenges, critical ecosystem functions, and native forest biodiversity (Long 2009). The success of this management approach requires clear and tangible guidelines that are based on quantitative data from stands that are situated in similar biophysical settings and are therefore appropriate analogues (Seymour et al. 2002; Franklin et al. 2007).

Uneven-aged management approaches have commonly relied upon single tree and group selection systems (Nyland 2002). Removing trees singly or in small groups will certainly emulate the sizes of naturally formed canopy gaps. However, through much of the CHR, markets have historically not supported such an approach. As a consequence, operators are often not experienced with single tree and group selection harvests. In addition, single tree and small group selections typically promote shade-tolerant species, and with few exceptions, this approach has not worked to maintain strong components of moderately-tolerant taxa such as oak and hickory (*Carya* spp.) in the CHR (Nyland 2002; Johnson et al. 2009).

A gap-based silvicultural approach that utilizes even-aged treatments applied in groups rather than evenly throughout an entire stand may be sufficient to maintain dominance of moderately-tolerant taxa such as oak and hickory, be economically viable throughout central hardwood forests, and fall within the historical range of variation of the disturbance regime. Oak stands in this region are most often managed with an even-aged approach; by modifying the size and spatial arrangement of even-aged treatments, managers may be able to fulfill multiple management objectives: commodity production, oak maintenance, and entries that fall within the historical range of variation.

I emphasize that there is no specific gap size that is guaranteed to promote oak recruitment (Lhotka 2013). Rather, gap size should be viewed based largely on the D:H. Oak reproduction will usually grow well if stems receive 20–50 % of full sunlight, which is typically met in gaps with a D:H of about 1.0 (Marquis 1965). Gaps of this size may be slightly larger than most naturally formed single tree fall gaps of central hardwood forests (Runkle 1985), but may still fall within the historical range of variation and may approximate the size of natural multi-tree fall events. In addition, such approaches should be placed strategically around existing oak advanced reproduction. The return interval of stand-wide, canopy disturbance events reported from hardwood stands through the CHR is 20–40 years. This return interval may be used to help guide the timing between entries in a single stand. Subsequent entries could create new group selection openings or could expand the size of the gaps created prior (i.e., femelschlag; Lhotka and Stringer 2013). I also note that the canopy gap formation rate of 0.5–2 % per year reported from the region may be somewhat low compared to centuries prior (Buchanan and Hart 2012) although more data are needed to verify this claim. Nonetheless, larger or more frequent harvest-created gaps may not be outside the historical range of variation in the disturbance regime.

For those who wish to adopt a management approach that emulates natural disturbance processes, it is important to recognize that many of the contemporary oak-dominated stands that occur throughout the CHR of the eastern USA were not the result of natural disturbance events (Cowell 1998, Foster et al. 2002). Managers that wish to maintain oak dominance and adhere to a natural disturbance-based management approach will likely need to make a compromise between a silvicultural system designed to emulate natural disturbances and one designed to maintain desired species assemblages. Oak regeneration failure has been reported widely across all but the most xeric site conditions throughout the CHR (Abrams 1992, Lorimer 1993, Nowacki and Abrams 2008, McEwan et al. 2011). Although variability exists at the species-level, oak are generally considered only moderately tolerant of shade, and canopy disturbance events that increase insolation in the understory are required for regeneration (Dey 2002). These canopy disturbances must be sufficiently large to provide adequate light levels for small oak, but not so large that they allow for the establishment of shade-intolerant species that can outcompete oak in high light environments (Runkle 1985, Grayson et al. 2012). Thus, gap opening size and the density and size of oak and its competition are critical factors to be considered in developing a silvicultural prescription to maintain oak dominance. In stands with a significant component of shade-tolerant mesophytes in the understory and where the management objective is to maintain oak, entries designed to release advanced oak reproduction should be implemented in conjunction with competition reduction measures such as fire or herbicide application (Loftis 1990, Schweitzer and Dey 2011, Hutchinson et al. 2012, Brose et al. 2013). Oak seedlings are often abundant in successional stands with abundant shade tolerant stems in the understory, but most of these oak seedlings will not recruit to sapling or small tree size classes and may be considered ephemeral (i.e., the oak bottleneck). Without competition reduction measures, advanced oak reproduction will be sparse, and oak reproduction should be in place before overstory trees are removed (Johnson et al. 2009). Although competition removal may fall outside the historical range of variation, these actions may be essential to maintain oak dominance in stands with strong maple or other shade tolerant components. Competition reduction is not without its problems. For example, many managers have found that prescribed fire is not an effective control measure for a prolific sprouting species like red maple and herbicide can be cost prohibitive.

Acknowledgments I would like to thank Merrit Cowden, Lauren Cox, Tom Weber, James Rentch, Beverly Collins, and Cathryn Greenberg for their many helpful comments on earlier drafts.

References

Abe, S., Masaki, T., & Nakashizuka, T. (1995). Factors influencing sapling composition in canopy gaps of a temperate deciduous forest. *Vegetatio, 120*, 21–32.

Abrams, M. D. (1992). Fire and the development of oak. *Bioscience, 42*, 346–353.

Abrams, M. D. (1996). Distribution, historical development and ecophysiological attributes of oak species in the eastern United States. *Annales des Sciences Forestieres 53*, 487–512.

Barden, L. S. (1979). Tree replacement in small canopy gaps of a *Tsuga canadensis* forest in the southern Appalachian Highlands, Tennessee. *Oecologia, 44*, 141–142.

Barden, L. S. (1980). Tree replacement in a cove hardwood forest of the southern Appalachian Highlands. *Oikos, 35*, 16–19.

Barden, L. S. (1981). Forest development in canopy gaps of a diverse hardwood forest of the southern Appalachian mountains. *Oikos, 37*, 205–209.

Beatty, S. W., & Stone, E. L. (1986). The variety of soil microsites created by tree falls. *Canadian Journal of Forest Research, 16*, 539–548.

Brokaw, N. V. L. (1985). Gap-phase regeneration in a tropical forest. *Ecology, 66*, 682–687.

Brokaw, N. V. L. & Busing, R. T. (2000). Niche versus chance and tree diversity in forest gaps. *Trends in Ecology and Evolution 15*, 183–188.

Brose, P. H., Dey, D. C., Phillips, R. J., & Waldrop, T. A. (2013). A meta-analysis of the fire-oak hypothesis: Does prescribed burning promote oak reproduction in eastern north America? *Forest Science, 59*, 322–334.

Buchanan, M. L., & Hart, J. L. (2012). Canopy disturbance history of old-growth *Quercus alba* sites in the eastern United States: Examination of long-term trends and broad-scale patterns. *Forest Ecology and Management, 267*, 28–39.

Busing, R. T. (2005). Tree mortality, canopy turnover, and woody detritus in old cove forests of the southern Appalachians. *Ecology, 86*, 73–84.

Canham, C. D. (1989). Different responses to gaps among shade-tolerant tree species. *Ecology, 70*, 548–550.

Canham, C. D., & Marks, P. L. (1985). The response of woody plants to disturbance: Patterns of establishment and growth. In S. T. A. Pickett & P. S. White (Eds.), *The ecology of natural disturbance and patch dynamics* (pp. 197–216). San Diego: Academic.

Canham, C. D., Denslow, J. S., Platt, W. J., Runkle, J. R., Spies, T. A., & White, P. S. (1990). Light regimes beneath closed canopies and tree-fall gaps in temperate and tropical forests. *Canadian Journal of Forest Research, 20*, 620–631.

Cho, D., & Boerner, R. E. J. (1991). Canopy disturbance patterns and regeneration of *Quercus* species in two old-growth forests. *Vegetatio, 93*, 9–18.

Clebsch, E. E. C., & Busing, R. T. (1989). Secondary succession, gap dynamics, and community structure in a Southern Appalachian Highlands cove forest. *Ecology, 70*, 728–735.

Clinton, B. D., & Baker, C. R. (2000). Catastrophic windthrow in the southern Appalachians: Charcteristics of pits and mounds and initial vegetation responses. *Forest Ecology and Management, 126*, 51–60.

Clinton, B. D., Boring, L. R., & Swank, W. T. (1993). Canopy gap characteristics and drought influence in oak forests of the Coweeta Basin. *Ecology, 74*, 1551–1558.

Clinton, B. D., Boring, L. R., & Swank, W. T. (1994). Regeneration patterns in canopy gaps of mixed-oak forests of the Southern Appalachian Highlands: Influences of topographic position and evergreen understory. *American Midland Naturalist, 132*, 208–319.

Cowell, C. M. (1998). Historical change in vegetation and disturbance on the Georgia Piedmont. *American Midland Naturalist, 140*, 78–89.

Dey, D. (2002). The ecological basis for oak silviculture in eastern North America. In W. J. McShea & W. M. Healy (Eds.), *Oak forest ecosystems*. Baltimore: Johns Hopkins University Press.

Foster, D. R. (1988). Species and stand responses to catastrophic wind in central New England, USA. *Journal of Ecology, 76*, 135–151.

Foster, D. R., Clayden, S., Orwig, D. A., Hall, B., & Barry, S. (2002). Oak, chestnut and fire: Climatic and cultural controls of long-term forest dynamics in New England USA. *Journal of Biogeography, 29*, 1359–1379.

Franklin, J. F., & Johnson, K. N. (2012). A Restoration framework for federal forests in the Pacific Northwest. *Journal of Forestry, 110*, 429–439.

Franklin, J. F., Mitchell, R. J., & Palik, B. J. (2007). *Natural disturbance and stand development principles of ecological forestry* (General technical report NRS 19). Newton Square: USDA Forest Service, Northern Research Station.

Frelich, L. E. (2002). *Forest dynamics and disturbance regimes: Studies from temperate evergreen deciduous forests*. Cambridge: Cambridge University Press.

Grayson, S. F., Buckley, D. S., Henning, J. G., Schweitzer, C. J., Gottschalk, K. W., & Loftis, D. L. (2012). Understory light regimes following silvicultural treatments in central hardwood forests of Kentucky, USA. *Forest Ecology and Management, 279*, 66–76.

Hanson, J. J., Lorimer, C. G., Halpin, C. R., & Palik, B. J. (2012). Ecological forestry in an uneven-aged, late-successional forest: Simulated effects of contrasting treatments on structure and yield. *Forest Ecology and Management, 270*, 94–107.

Hart, J. L., & Grissino-Mayer, H. D. (2008). Vegetation patterns and dendroecology of a mixed hardwood forest on the Cumberland Plateau: Implications for stand development. *Forest Ecology and Management, 255*, 1960–1975.

Hart, J. L., & Grissino-Mayer, H. D. (2009). Gap-scale disturbance processes in secondary hardwood stands on the Cumberland Plateau, Tennessee, USA. *Plant Ecology, 201*, 131–146.

Hart, J. L., & Kupfer, J. A. (2011). Sapling richness and composition in canopy gaps of a southern Appalachian mixed *Quercus* forest. *The Journal of the Torrey Botanical Society, 138*, 207–219.

Hart, J. L., Clark, S. L., Torreano, S. J., & Buchanan, M. L. (2012). Composition, structure, and dendroecology of an old-growth *Quercus* forest on the tablelands of the Cumberland Plateau, USA. *Forest Ecology and Management, 266*, 11–24.

Himes, J. M., & Rentch, J. S. (2013). Canopy gap dynamics in a second-growth Appalachian hardwood forest in West Virginia. *Castanea, 78*, 171–184.

Hutchinson, T. F., Boerner, R. E. J., Sutherland, S., Sutherland, E. K., Ortt, M., & Iverson, L. R. (2012). Prescribed fire effects on the herbaceous layer of mixed-oak forests. *Canadian Journal of Forest Research, 35*, 877–890.

Hutnik, R. J. (1952). Reproduction on windfalls in a northern hardwood stand. *Journal of Forestry, 50*, 693–694.

Johnson, P. S., Shifley, S. R., & Rogers, R. (2009). *The ecology and silviculture of oaks* (2nd ed.). Cambridge: CAB International.

Keller, J. A., & Hix, D. M. (1999). Canopy gap fraction and origin in second-growth forests of Washington County, Ohio. *Castanea, 64*, 252–258.

Krasney, M. E., & Whitmore, M. C. (1992). Gradual and sudden forest canopy gaps in Allegheny northern hardwood forests. *Canadian Journal of Forest Research, 22*, 139–143.

Kupfer, J. A., Runkle, J. R., & Malanson, G. P. (1997). Factors influencing species composition in canopy gaps: The importance of edge proximity in Hueston Woods, Ohio. *The Professional Geographer, 49*, 165–178.

Lhotka, J. M. (2013). Effect of gap size on mid-rotation stand structure and species composition in a naturally regenerated mixed broadleaf forest. *New Forests, 44*, 311–325.

Lhotka, J. M., & Stringer, J. W. (2013). Forest edge effects on *Quercus* reproduction within naturally regenerated mixed broadleaf stands. *Canadian Journal of Forest Research, 43*, 911–918.

Lima, R. A. F. (2005). Gap size measurement: The proposal of a new field method. *Forest Ecology and Management, 214*, 413–419.

Loftis, D. L. (1990). A shelterwood method for regenerating red oak in the southern Appalachians. *Forest Science, 36*, 917–929.

Long, J. N. (2009). Emulating natural disturbance regimes as a basis for forest management: A North American view. *Forest Ecology and Management, 257*, 1868–1873.

Lorimer, C. G. (1980). Age structure and distribution history of a Southern Appalachian Highlands virgin forest. *Ecology, 61*, 1169–1184.

Lorimer, C. G. (1993). *Causes of the oak regeneration problem* (General technical report SE 84). Asheville: USDA Forest Service Southeastern Forest Experiment Station.

Lorimer, C. G. (2001). Historical and ecological roles of disturbance in eastern North American forests: 9,000 years of change. *Wildlife Society Bulletin, 29*, 425–439.

Marquis, D. A. (1965). *Controlling light in small clearcuttings* (US Forest Service research paper NE 38). Newtown Square: USDA Forest Service Northeastern Forest Experiment Station.

McComb, W. C., & Muller, R. N. (1983). Snag densities in old-growth and second-growth Appalachian forests. *Journal of Wildlife Management, 47*, 376–382.

McEwan, R. W., Dyer, J. M., & Pederson, N. (2011). Multiple interacting ecosystem drivers: Toward an encompassing hypothesis of oak forest dynamics across eastern North America. *Ecography, 34*, 244–256.

Nowacki, G. J., & Abrams, M. D. (1997). Radial-growth averaging criteria for reconstructing disturbance histories from presettlement-origin oaks. *Ecological Monographs, 67*, 225–249.

Nowacki, G. J., & Abrams, M. D. (2008). The demise of fire and "mesophication" of forests in the eastern United States. *BioScience, 58*, 123–138.

Nyland, R. D. (2002). *Silviculture: Concepts and applications*. New York: McGraw-Hill.

Oliver, C. D., & Larson, B. C. (1996). *Forest stand dynamics* (Updateth ed.). New York: McGraw-Hill Book Company.

Orwig, D. A. & Abrams, M. D. (1995). Dendroecological and ecophysiological analysis of gap environments in mixed-oak understories of northern Virginia. *Functional Ecology 9*, 799–806.

Phillips, D. L., & Shure, D. J. (1990). Patch-size effects on early succession in southern Appalachian forests. *Ecology, 71*, 204–212.

Poulson, T. L., & Platt, W. J. (1988). Light regeneration niches. *Bulletin of the Ecological Society of America, 69*, 264.

Putz, F. E. (1983). Treefall pits and mounds, buried seeds, and the importance of soil disturbance to pioneer trees on Barro Colorado Island, Panama. *Ecology, 64*, 1069–1074.

Quine, C. P., & Gardiner, B. A. (2007). Understanding how the interaction of wind and trees results in wind-throw, stem break-age, and canopy gap formation. In *Plant disturbance ecology: The process and the response*. San Diego: Academic.

Rentch, J. S., Fajvan, M. A., & Hicks, R. R., Jr. (2003). Spatial and temporal disturbance characteristics of oak-dominated old-growth stands in the Central Hardwood Forest Region. *Forest Science, 49*, 778–789.

Rentch, J. S., Schuler, T. M., Nowacki, G. W., Beane, N. R., & Ford, W. M. (2010). Canopy gap dynamics of second-growth red spruce-northern hardwood stands in West Virginia. *Forest Ecology and Management, 260*, 1921–1929.

Richards, J. D., & Hart, J. L. (2011). Canopy gap dynamics and development patterns in secondary *Quercus* stands on the Cumberland Plateau, Alabama, USA. *Forest Ecology and Management, 262*, 2229–2239.

Romme, W. H., & Martin, W. H. (1982). Natural disturbance by tree-falls in old-growth mixed mesophytic forest: Lilley Cornett Woods, Kentucky. In R. N. Muller (Ed.), *Proc fourth central hardwood for conf.* Lexington: Univ Kentucky.

Ruffner, C. M., & Abrams, M. D. (1998). Relating land-use history and climate to the dendroecology of a 326-year-old *Quercus prinus* talus slope forest. *Canadian Journal of Forest Research, 28*, 347–358.

Runkle, J. R. (1981). Gap regeneration in some old-growth forests of the eastern United States. *Ecology, 62*, 1041–1051.

Runkle, J. R. (1982). Patterns of disturbance in some old-growth mesic forests of the eastern North America. *Ecology, 63*, 1533–1546.

Runkle, J. R. (1985). Disturbance regimes in temperate forests. In S. T. A. Pickett & P. S. White (Eds.), *The ecology of natural disturbance and patch dynamics*. San Diego: Academic.

Runkle, J. R. (1990). Gap dynamics in an Ohio *Acer-Fagus* forest and speculations on the geography of disturbance. *Canadian Journal of Forest Research, 20*, 632–641.

Runkle, J. R. (1992). Guidelines and sample protocol for sampling forest gaps. USDA, Forest Service, Portland.

Runkle, J. R. (1996). Central mesophytic forests. In *Eastern old-growth forests: Prospects for rediscovery and recovery*. Washington, DC: Island Press.

Runkle, J. R. (2000). Canopy tree turnover in old-growth mesic forests of eastern North America. *Ecology, 81*, 554–567.

Runkle, J. R., & Yetter, T. C. (1987). Treefalls revisited: Gap dynamics in the Southern Appalachian Highlands. *Ecology, 68*, 417–424.

Schaetzl, R. J., Johnson, D. L., Burns, S. F., & Small, T. W. (1989). Tree uprooting: Review of terminology, process, and environmental implications. *Canadian Journal of Forest Research, 19*, 1–11.

Schweitzer, C. J., & Dey, D. C. (2011). Forest structure, composition, and tree diversity response to a gradient of regeneration harvests in the mid-Cumberland Plateau escarpment region, USA. *Forest Ecology and Management, 262*, 1729–1741.

Seymour, R. S., & Hunter, M. L., Jr. (1999). Principles of ecological forestry. In M. L. Hunter Jr. (Ed.), *Managing biodiversity in forest ecosystems* (pp. 22–61). Cambridge: Cambridge University Press.

Seymour, R. S., White, A. S., & Demaynadier, P. G. (2002). Natural disturbance regimes in northeastern North America—evaluating silvicultural systems using natural scales and frequencies. *Forest Ecology and Management, 155*, 357–367.

Taylor, S. O., & Lorimer, C. G. (2003). Loss of oak dominance in dry-mesic deciduous forests predicted by gap capture methods. *Plant Ecology, 167*, 71–88.

Tyrell, L. E., & Crow, T. R. (1994). Structural characteristics of old-growth hemlock-hardwood forests in relation to age. *Ecology, 75*, 370–386.

Watt, A. S. (1947). Pattern and process in the plant community. *Journal of Ecology, 35*, 1–22.

Weber, T. A. (2014). The influence of gap-scale disturbance on development and succession in a Cumberland Plateau *Quercus-Pinus* forest. Thesis, University of Alabama, Tuscaloosa.

Webster, C. R., & Lorimer, C. G. (2005). Minimum opening sizes for canopy recruitment of mid-tolerant tree species: A retrospective approach. *Ecological Applications, 15*, 1245–1262.

Whitney, G. G. (1994). *From coastal wilderness to fruited plain: A history of environmental change in temperate North America from 1500 to the present*. Cambridge: Cambridge University Press.

Wilder, C. M., Holtzclaw, F. W., Jr., & Clebsch, E. E. C. (1999). Succession, sapling density and growth in canopy gaps along a topographic gradient in a second growth east Tennessee forest. *American Midland Naturalist, 142*, 201–212.

Yamamoto, S. (2000). Forest gap dynamics and tree regeneration. *Journal of Forest Research, 5*, 223–229.

Yamamoto, S., & Nishimura, N. (1999). Canopy gap formation and replacement pattern of major tree species among developmental stages of beech (*Fagus crenata*) stands, Japan. *Plant Ecology, 140*, 167–176.

Young, T. P., & Hubbell, S. P. (1991). Crown asymmetry, treefalls and repeat disturbance of broad-leaved forest gaps. *Ecology, 72*, 1464–1471.

Zeide, B. (2005). How to measure stand density. *Trees, 19*, 1–14.

Zeide, B. (2010). Comparison of self-thinning models: An exercise in reasoning. *Trees, 24*, 1117–1126.

Chapter 3
Oak Decline in Central Hardwood Forests: Frequency, Spatial Extent, and Scale

Steven W. Oak, Martin A. Spetich, and Randall S. Morin

Abstract Oak decline is a widely distributed disease that results from an interacting set of factors in the Central Hardwood Region. Episodes of decline have been reported since before the turn of the twentieth century and from every state in the region. It is a stress-mediated disease that results from the interactions of physiologically mature trees, abiotic and biotic stressors that alter carbohydrate physiology, and opportunistic fungal pathogens and inner bark-feeding insects. Symptoms include reduced radial growth and slow, progressive crown dieback. Decline occurs over several years or decades, ending in death of vulnerable trees. Patterns of oak decline vary from a few trees in stands with diverse species composition and age structure, to areas covering several thousand ha in landscapes with more uniform composition of susceptible, physiologically mature red oak group species. Prolonged periods of drought that occur in combination with repeated spring defoliations by leaf-feeding insects exacerbate decline. Past disturbances have shaped current forest species composition and age structure, favoring physiologically mature stands with a large oak component, and are thus inextricably linked to oak decline vulnerability. Noteworthy examples are the functional extirpation of the American chestnut by the non-indigenous chestnut blight pathogen, combined with changing disturbance patterns, including fire suppression and reduced harvesting, during the early twentieth century. Data from extensive regional surveys have been used to develop models predicting the probability and impacts of oak decline events as part of the Forest Vegetation Simulator.

S.W. Oak (✉)
USDA Forest Service, Southern Region Forest Health Protection (retired),
347 Ridge View Drive, Asheville, NC 28803, USA
e-mail: swoclo@charter.net

M.A. Spetich
USDA Forest Service, Southern Research Station, 100 Reserve Street,
Hot Springs, AR 71902, USA
e-mail: mspetich@fs.fed.us

R.S. Morin
USDA Forest Service, Northern Research Station, Forest Inventory and Analysis,
11 Campus Blvd., Ste. 200, Newtown Square, PA 19073, USA
e-mail: rmorin@fs.fed.us

C.H. Greenberg, B.S. Collins (eds.), *Natural Disturbances and Historic Range of Variation*, Managing Forest Ecosystems 32,
DOI 10.1007/978-3-319-21527-3_3

Keywords Crown dieback • Oak mortality • Predisposing factors • Oak decline survey • Oak decline event monitor

3.1 Historical Context for Oaks in Central Hardwood Forests

Oak (*Quercus*) forest types currently dominate the Central Hardwood Region (CHR), and oak decline is a widely distributed change agent altering species composition and forest structure throughout the region. The disease has regulated oak populations since oak species and interacting antagonists first appeared in the CHR, and that role continues to the present. Impacts of oak decline would have varied through time, as oak composition has shifted over the past several millennia with changing climate, weather, and the extent, frequency, and intensity of fire (e.g., Delcourt and Delcourt 2004). Human interventions and preferences have intentionally or unintentionally shaped forest composition and structure, either favoring or disfavoring oak populations. The expression of these preferences began with the arrival of the first aboriginal people into the region, who undoubtedly used fire to create desirable forest structures around settlements and in hunting lands (see also Greenberg et al. Chap. 1). Although managing tree species composition may not have been a primary objective, the use of fire certainly favored species well-adapted to it such as oaks, American chestnut (*Castanea dentata*), and pines (*Pinus* spp.). The extent of such landscape management increased with the human population, and accelerated with the arrival of European migrants. The introduction of destructive non-indigenous pathogens and insects continues into present decades, stimulated by increased international movement of products and people. Even though the forces shaping CHR forests operated for millennia, the relevant historical context for oak decline as observed from the mid-twentieth to the present is relatively recent. It was triggered by biological, social, and political events in the early twentieth century which altered historic disturbance regimes.

Prior to the turn of the twentieth century, forests in many parts of the CHR were dominated in composition by American chestnut and subject to frequent disturbance by fire (sometimes by natural causes, but mostly by human ignitions; see Greenberg et al. Chap. 1, Table 1.6). In 1880, 98.6 % of fires in the CHR were human caused with the top three causes attributed to land clearing, hunters and locomotives (Spetich et al. 2011). Logging to supply fuel and building materials to a pre-, and later, emerging- industrial society also was an important and widespread disturbance factor in the late nineteenth and early twentieth centuries.

Two events then occurred that had far-reaching consequences on forest landscapes. First, the chestnut blight was discovered in New York City in 1904. The cause of the disease was a non-indigenous fungal pathogen, most likely introduced with Asian chestnut varieties imported by the nursery trade to many locations in the eastern USA as early as 1876. The native American chestnut had no inherent resistance to the pathogen, and within a few decades of the discovery of the disease in New York City, the species was functionally extirpated throughout its range. Since then, it has survived only as root sprouts before once again becoming blighted and killed in a cycle that is repeated to the present day.

The second event was the 1911 enactment of the Weeks Act authorizing acquisition of land for national forests to protect headwaters of navigable streams. It was drafted in part as a response to a disastrous fire season the previous year that included 'The Great Fire of 1910' that burned over 808,000 ha and resulted in the deaths of 87 people in northeast Washington, northern Idaho, and western Montana. Though it occurred in vastly different ecosystems than those in the CHR, this catastrophe crystallized national policy and prompted the USDA Forest Service to make fire suppression a primary mission everywhere (see also Zenner Chap. 14). Further, the Weeks Act incorporated provisions for the development of federal-state cooperative fire control programs.

Prior to the Weeks Act, oaks dominated the CHR due to traits that made them more resilient to fire. The species builds large, belowground reserves of carbohydrates in root systems protected from fire. When fire killed aboveground shoots of small trees, these belowground carbohydrate reserves allowed new shoots to resprout rapidly. Large oaks also have relatively thick bark that helps reduce damage from ground-level fires. Thus, oaks had a competitive advantage over other tree species less well adapted to fire. In 1924 the Clark-McNary Act was passed, expanding the Weeks Act. Among other provisions, this Act encouraged states to form their own forestry agencies and further advanced fire suppression programs. These agencies and the laws and practices they spawned, combined with rapidly increasing efficiencies in agricultural production in the early twentieth century, gradually began to transform forests of the CHR. Open woodland conditions that had been maintained by fire, grazing and harvesting for millennia gradually became more closed as cohorts of oaks already established in the understory grew into dominant and codominant crown positions in the absence of frequent fire and other disturbance (Abrams 1992). Concurrent with these ecological changes, the USA population was growing and society was shifting from agrarian and rural, with a resource utilization ethic, to industrial and urban, with an emerging conservation ethic. By the late twentieth century, oak cohorts 80–100 years old dominated the CHR, especially on publicly owned lands not subject to development and urbanization. Table 3.1 summarizes general social and forest dynamics attributes before and after the turn of the twentieth century that help explain oak density and oak decline patterns of the more recent past.

Oak density for contemporary forests of CHR ecoregions resulting from these historic influences was determined using plots in the USDA Forest Service Forest Inventory and Analysis (FIA) Eastwide data base (Hansen et al. 1992). Data collected during the 1980s and 1990s were extracted from 26,662 plots in 20 states (Table 3.2). Overall, one third of the basal area (BA) in the CHR was comprised of oak species, with the highest average oak density values occurring in the Ozark Highlands, Boston Mountains, Arkansas Valley, and Ridge and Valley ecoregions. The Ozark Highlands had, by far, the highest oak density of all ecoregions (63.5 %)

Table 3.1 Prevalent historic attributes of CHR forests

Attribute	Pre-twentieth century	Twentieth century to present
Social/cultural	Small agrarian population	Large urbanized population
	Resource utilization/exploitation perspective	Resource conservation/protection perspective
Disturbance	Frequent fire of mostly anthropogenic origin	Near-complete fire suppression
	Frequent logging/utilization	Limited logging
Forest composition	American chestnut regionally abundant	Mixed oak predominant
	Vigorously sprouting woody species favored (e.g. oaks)	More shade tolerant woody species favored (e.g. red maple)
Forest structure	Diverse herbaceous understory; woody understory persists as sprouts	Dense woody understory
	Widely spaced, large diameter overstory	Dense, smaller diameter overstory
	More complex age structure	Aging oak cohorts 80–100 years old

Table 3.2 Mean proportion of BA in oak species and number of FIA plots in each oak BA proportion category by CHR ecoregion

Ecoregion	Mean % oak BA	Percentage oak BA category				
		0	1–25	26–50	51–75	>75
Arkansas Valley	44.0	53	103	102	110	102
Blue Ridge Mountains	31.8	305	351	418	345	182
Boston Mountains	48.3	24	59	111	130	136
Central Appalachians	28.6	475	716	477	375	253
Interior Plateau	25.3	597	755	551	415	346
Interior River Valleys and Hills	31.4	505	439	391	374	320
Northern Piedmont	33.4	113	102	75	62	72
Ouachita Mountains	31.6	192	230	243	134	78
Ozark Highlands	63.5	114	242	478	907	2,121
Piedmont	21.8	2,031	1,429	1,048	667	355
Ridge and Valley	37.1	463	630	617	604	618
Southwestern Appalachians	31.7	143	247	309	237	106
Western Allegheny Plateau	25.0	715	698	455	329	268
Entire Central Hardwood Region	33.2	5,730	6,001	5,275	4,689	4,957

and the highest proportion of plots in the densest category (55 % of all plots with >75 % oak BA). The ecoregion with the next highest oak density was the Boston Mountains – 48.3 % mean oak BA and 30 % of all plots with >75 % oak BA. The only ecoregion with mean percent oak density below 25 % was the Piedmont at 21.8 % (see Greenberg et al. Chap. 1, Fig. 1.1 for ecoregions map).

The geographic distribution of the proportion of oak BA on forested lands was mapped by interpolating plot values (Fig. 3.1). Plots in the >75 % oak BA category were particularly concentrated in the Ozark Highlands of Missouri. In addition, oak BA exceeded 50 % in parts of the Arkansas Valley, and Boston and Ouachita Mountains as well as the Western Highland Rim area of the Interior Plateau in Tennessee and the northern half of the Ridge and Valley in Virginia, West Virginia and western Maryland. Non-forest land was most noticeable in the Northern Piedmont and Interior River Valleys and Hills, though a few pockets of high oak BA forest were detected in the latter ecoregion. Details of the methodology employed to create the interpolated BA surface displayed in Fig. 3.1 are available in Morin et al. (2005).

There is no doubt that contemporary forest composition and structure are products of prevalent disturbances over at least the past two centuries. Choosing the reference condition upon which to base management practices compatible with the historic range of variation of natural disturbances presents a dilemma for land managers concerning the role of humans in forest ecosystems. The range of natural disturbance and the shape of forest composition and structure prior to human habitation of the CHR can only be generally inferred and ignores the reality of the past five to ten millennia (see Greenberg et al. Chap. 12). Alternatively, selecting a dis-

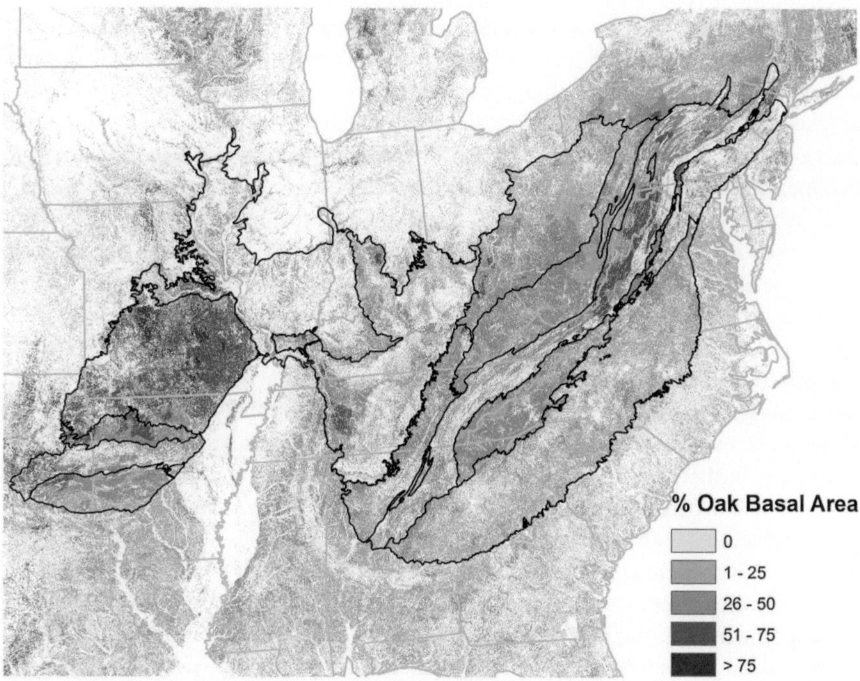

Fig. 3.1 The geographic distribution of the proportion (%) of oak basal area on forested lands in the eastern USA. The map was created by interpolating the values from 93,611 forest inventory plots in 37 states (Morin et al. 2005). The data were extracted from the USDA Forest Service Forest FIA Eastwide data base, which consists of data collected during the 1980s and 1990s (Hansen et al. 1992) (Map credit to Randall S. Morin)

turbance regime that includes human interventions along a spectrum from early aboriginal to contemporary times is a philosophical and arbitrary process. In any case, CHR forests cannot be termed 'natural' until the full suite of forest plants (including American chestnut) and animals present before the arrival of humans in the CHR are restored and functioning as part of ecosystems (see Zenner Chap. 14).

3.2 Oak Decline Etiology and Symptoms

Oak decline is a stress-mediated disease that results from the interactions of three groups of factors first described by Sinclair (1965), and later elaborated by Manion (1991). The individual factors in each group that combine to result in a specific oak decline episode can vary widely. The first group includes long-term predisposing factors that act to reduce the resilience of healthy trees to stress or attack by pathogens and insects. Among these are edaphic conditions such as soil depth and texture; topographic factors such as slope and aspect; and physiological maturity (distinguished from chronological age). Hyink and Zedaker (1987) characterized the concept of physiological age as having greater biological significance than chronological age. They described advanced physiological age (senescence) as the progression toward critical levels of physiological relationships such as water transport and translocation efficiencies, hormone balances, and the balance between photosynthesis and respiration. When critical levels are exceeded, internal resources are unavailable for effective tree response to stressors thereby predisposing them to decline. Oak et al. (1991) created an index of physiological age using site index (SI) and chronological age that was useful in oak decline risk rating (Oak et al. 1996).

The second group is comprised of inciting factors associated with the initiation of decline and the earliest, non-specific symptoms: depletion of stored food reserves, reduced growth, and dieback. Factors in this group include prolonged drought or spring defoliation caused by some insect species or late frost. The third group is comprised of contributing factors. These are biotic agents and often are implicated as the cause of mortality but in fact, are opportunists normally incapable of killing vigorous trees. However, they are well adapted to exploit predisposed trees that have been further weakened by the inciting factors. The most commonly cited fungal parasites involved in oak decline mortality are *Armillaria mellea* (though complex interactions with other *Armillaria* species have been described by Bruhn et al. 2000) and *Biscogniauxia atropunctata* (cause of hypoxylon canker of oaks). Both are widely distributed facultative parasites in natural ecosystems. *Armillaria mellea* is common in soil, decaying roots, and dead wood; *Biscogniauxia atropunctata* resides as an endophyte in stems and branches decaying sapwood. They become more aggressive pathogens when conditions are appropriate for pathogenesis and cause root disease and stem cankers. The most commonly cited insect pest is the two-lined chestnut borer (*Agrilus bilineatus*) which creates meandering galleries in the inner bark of weakened trees (Wargo et al. 1983).

The earliest visible aboveground symptom of oak decline is dieback of the live crowns of trees in upper canopy positions beginning with the outer twigs and branches. This can sometimes occur during the growing season, leaving dead foliage attached. More commonly it occurs during the dormant period with affected limbs failing to refoliate the following spring. Relatively slow, progressive dieback downward and inward, involving larger limbs occurs over years or even decades and is a distinctive symptom of oak decline. The death of branches in the crown results in the production of sprouts along the larger limbs and main stem. Eventually, severely declined trees die. Species in the red oak group (e.g., black oak (*Q. velutina*), scarlet (*Q. coccinea*), northern red oak (*Q. rubra*), southern red oak (*Q. falcata*), and blackjack oak (*Q. marilandica*)) are more susceptible to decline-induced mortality than are species in the white oak group including white oak (*Q. alba*), chestnut oak (*Q. montana*), and post oak (*Q. stellata*).

Crown dieback reflects root disease progression belowground where armillaria root disease is an important contributing factor. Carbohydrate chemistry is altered in roots of trees stressed by drought and defoliation and is accompanied by decreased levels of starch and increased levels of simple sugars (Parker 1970; Wargo 1972, 1977). Growth of *A. mellea* is stimulated by these changes and becomes more aggressive, attacking more of the stressed tree's root system. The crown must die back to accommodate the impaired root system. Long-term monitoring of symptomatic trees has shown that dieback (and presumably root disease) may abate 10 years after the return of good growing conditions so long as it has not progressed beyond about one-third of the live crown volume (Oak, unpubl. data). Moisture stress also is important in stimulating *Biscogniauxia atropunctata* to transform from a sapwood endophyte to a more aggressive cankering pathogen and sapwood rotter (Bassett and Fenn 1984).

3.3 Distribution

Millers et al. (1989) reviewed the literature of forest tree declines and reported 57 episodes in the eastern USA between 1856 and 1986 where oak mortality was higher than expected in areas covering at least 400 ha. Details of survey methodology and data collection often were not included, and authors of cited reports usually attributed the mortality to one or two causes without naming oak decline specifically. This was true even after the first elucidation of decline etiology and symptoms was published (Sinclair 1965). Despite this lack of specificity, the summarized causes included multiple interactions consistent with the published disease etiology. The first systematic regional surveys of oak decline using consistent data collection protocols were conducted mostly in the South in the mid-1980s using ground and aerial photo methodologies (Starkey et al. 1989, 2000; Oak et al. 1990), and continuous forest inventory plot networks (Oak et al. 1991, 2004). Combining these sources reveals that every state in the CHR has experienced oak decline damage and mortality in at least one decade dating back to earliest reports in the mid-nineteenth century (Table 3.3). Notable concentrations of oak decline episodes were evident in Appalachian and Ozark Mountain states (Arkansas, Georgia, Missouri, North

Table 3.3 Oak decline and mortality reports for states within the CHR by decade

	Pre-1900	1900s									2000s	
State[a]		00	10	20	30	40	50	60	70	80	90	00	10–13
AL									■	■			
AR							■		■	■	■	■	
DE							■			■			
GA	■		■	■			■	■	■	■	■		
IA									■	■			
IN									■				
IL									■	■			
KY									■	■			
MD							■			■			
MO									■	■		■	
NC	■		■	■		■	■	■	■	■	■		
NJ							■	■	■				
NY							■	■	■	■			
OH									■				
OK									■				
PA					■	■	■	■	■	■			
SC										■			
TN			■	■		■				■	■		
VA	■		■	■		■	■	■	■	■	■		
WV	■						■		■	■			

Adapted from Millers et al. (1989), Starkey et al. (1989, 2000), Oak et al. (2004)
[a]Most states are only partially within the CHR

Carolina, Pennsylvania, Tennessee, Virginia, and West Virginia) from the 1950s through the 1990s. Intensified survey activities may have contributed to part of the increase in reported oak decline distribution, incidence, and severity over this period. The USDA Forest Service formed the Division of Forest Pest Control in 1956 (later variously named Forest Insect and Disease Control, Forest Insect and Disease Management, Forest Pest Management, and Forest Health Protection) to survey and interpret forest health conditions on federal lands. State forest health programs servicing state and private forest landowners began in the 1960s under cost sharing agreements with the USDA Forest Service State and Private Forestry and have flourished in the decades since.

3.4 Oak Decline Patterns at Various Spatial Scales

Oak decline patterns have been described and analyzed across a range of spatial scales with varying methodologies tailored to survey objectives. Reports commonly included estimates of areal extent (patch size), incidence and severity of symptoms, and (less frequently) description of effects on growth, species composition, and forest structure. Patch size, incidence and severity, and effects will be detailed within this section.

Initial surveys were at a local scale and limited to the characterization of decline-affected stands (Millers et al. 1989; Starkey et al. 1989; Law and Gott 1987). Random surveys representing all health classes were conducted later and broadened the scope to establish disease incidence and impacts in a landscape context. These surveys used large format aerial photography (Oak et al. 1990) or aerial sketch mapping (Starkey et al. 2000) and were supported by ground sampling for validation. Description of stand and site features from the ground validation was used later for risk rating and effects modeling (Oak et al. 1996). Regional scale analyses were conducted using data collected from risk-based polygon sampling (Guldin et al. 2006) and by continuous forest inventory plot networks (Oak et al. 1991, 2004; Fan et al. 2012).

3.4.1 Patch Size

The pattern of oak decline on the landscape varies widely with tree species composition, age structure, and mortality incidence. Small patches consisting of scattered individuals or small groups of oaks occur in landscapes where age structure, tree species composition, and correlated site conditions are relatively diverse. In contrast, patches encompassing several thousand ha can occur where species composition and site conditions are relatively less diverse. Such large areas have developed on landscapes in the Blue Ridge Mountains and Ridge and Valley ecoregions of western Virginia, and more recently in the Ozark Highlands, Boston Mountains, and Ouachita Mountains ecoregions of Missouri, Arkansas, and Oklahoma. Landscapes in these provinces are dominated by cohorts of physiologically mature trees in species of the red oak subgenus growing on sites of average to lower productivity and are subject to periodic drought (Greenberg et al. Chap. 1, Fig. 1.8). Since the mid-1980s in the east, recurrent defoliation by the non-indigenous gypsy moth has also been an important inciting factor. During the 1990s in the west, unprecedented outbreaks of the indigenous red oak borer (*Enaphalodes rufulus*) contributed.

Aerial survey methods supplemented with ground truth assessments provide the perspective for estimating patch size that is lacking in ground-based surveys alone. Oak decline and mortality were evaluated on two national forests in the Ridge and Valley and Blue Ridge Mountains ecoregions of Virginia by Rauschenberger and Ciesla (1966) using aerial sketch mapping of about 70 % of the forest land inside the forest boundaries, supplemented with ground survey. Areas delineated with at least 5 % mortality totaled approximately 42,016 ha on the George Washington National Forest in northwestern Virginia. Patch size ranged from 61 to 2,424 ha. Mortality was less prevalent and patches were smaller overall on the Jefferson National Forest in southwestern Virginia (range 113–485 ha; mean = 297). This survey predated by 20 years the widespread infestation of Virginia forests by the gypsy moth (*Lymantria dispar*). Outbreaks of this insect usually are of longer duration, the intensity of defoliation greater and return interval shorter than for native defoliators. These dynamics often incite very severe decline episodes with catastrophic levels of mortality. While patch size estimates are lacking for post-gypsy moth decline episodes in Virginia, they are likely substantially larger.

Law and Gott (1987) interpreted large-scale color infrared aerial photos acquired over the Mark Twain National Forest, Missouri within the Ozark Highlands ecoregion and found mortality areas ranged from <0.5 to 28 ha (mean = 4 ha). The decline episode that prompted this assessment followed prolonged drought and several other predisposing and inciting conditions. However, about a decade later a much more severe and widespread episode occurred, accompanied by an unprecedented outbreak of the red oak borer. As was the case for gypsy moth-associated decline events in Virginia, the size of mortality patches were not measured, but probably increased significantly over earlier estimates.

3.4.2 Incidence and Severity

Starkey et al. (1989) surveyed 38 decline-affected stands from Virginia to Georgia and west to Arkansas and Missouri. All were on public lands with most located in National Forests. Oaks dominated the composition, with 50 % in red oak group species, 31 % in white oak group species, and 7 % hickory (*Carya*) species. Diagnosis of decline and decline mortality was confined to dominant and codominant trees with progressive dieback symptoms. Dieback and mortality among trees of all species in intermediate and suppressed crown positions was attributed to suppression and not to decline. Decline was observed in 80 % of dominant and codominant trees of all species. Hickories were the only non-oak species exhibiting appreciable symptom incidence. Advanced decline (>33 % live crown loss) was present in 20 % of all trees, and 17 % were dead with decline symptoms. Red oak group species were more prone to decline-associated mortality compared with white oak group species (24 % vs. 8 %, respectively). Among red oak group species, black oak was most vulnerable to oak decline mortality (34 %) followed by scarlet oak (23 %). The mortality incidence among hickory species was comparable to that recorded among white oak trees (12 %).

In addition to mortality impacts, Starkey et al. (1989) also analyzed the impact of oak decline on tree growth. Radial growth for 77 pairs of declined and healthy red oak trees was compared by in stands located in North Carolina, Tennessee, and Arkansas. Overall, declined trees grew 17 % more slowly than healthy trees for the last 45 years of the growth history, and 27 % more slowly for the last 20 years.

Tainter et al. (1990) further elaborated the predisposing effects of drought that were evident for several decades after the cessation of stress on oak populations of different physiological ages. They suggested that severe drought over several years in the early 1950s altered oak populations resulting in two health classes. One class had diminished resilience to subsequent droughts and eventually died while the other survived and recovered at least some of its former growth rate. Dwyer et al. (1995) observed similar drought dynamics in different age classes of black and scarlet oaks in Missouri dating back to stress events up to 45 years earlier.

Reports of increased oak mortality in the Ozark Highlands ecoregion on the Mark Twain National Forest, Missouri prompted surveys of declined areas on the Fristoe Unit in 1982 (Law and Gott 1987). Mortality areas were detected by interpretation of large-scale aerial photographs on 15.6 % of the 2,384 ha surveyed. Ground validation surveys confirmed that 53 % of the volume in scarlet oak was dead or dying, along with 35 % of black oak and 26 % of northern red oak. The 71–80 year age class was the most severely affected with 42 % of stand volume dead or dying.

Aerial sketch mapping followed by ground validation surveys were initiated in 1999 after concentrations of oak decline damage were reported on the Pleasant Hill Ranger District, Ozark National Forest, Arkansas in the Boston Mountains ecoregion. Moderate-to-severe damage was detected on approximately 17,372 ha (16 %) of the Ranger District. In the severe damage stratum, 24 % of the BA was declined or dead (Starkey et al. 2000).

Oak decline in a larger, landscape, context was evaluated in surveys of three national forest ranger districts: the Lee Ranger District on the George Washington National Forest in the Ridge and Valley ecoregion in Virginia; the Wayah Ranger District on the Nantahala National Forest in the Blue Ridge Mountains ecoregion in North Carolina; and the Buffalo Ranger District on the Ozark National Forest in the Boston Mountains ecoregion in Arkansas (Oak et al. 1990). These areas represented much of the diversity in climate, physiography, soils, and hardwood tree species composition where oak decline had been a recurring problem (Millers et al. 1989). A two-stage sampling design was used. Large-scale aerial photo samples were interpreted and stratified by tree size and damage class, with the results validated by ground plot sub-sampling. The survey yielded decline damage area and damage severity estimates. The Lee Ranger District had the highest incidence of decline (56 % of hardwood forest type). The Wayah Ranger District had intermediate incidence (35 % of hardwood forest type), whereas 28 % was affected on the Buffalo Ranger District. Within damaged strata, incidence of mortality plus advanced decline ranged from 10 % to 16 % of dominant and codominant trees compared with 1–3 % in undamaged strata.

Guldin et al. (2006) evaluated oak decline on 181 plots systematically distributed across the Interior Highlands of Arkansas, Oklahoma, and Missouri, and found about 12 % of the area in the most heavily damaged class (>6.67 m^2/ha of BA unhealthy, a category which included dead trees and those displaying at least 34 % crown dieback). Mean stem density over the entire surveyed area was 95.3 trees/ha, of which 12.9 trees/ha were unhealthy (13.4 %). The percentage of mean total BA in an unhealthy condition was slightly higher (14.5 %). This damage was concentrated in the red oak group. Thirty percent of the BA in species of this group was unhealthy compared with about 9 % of species in the white oak group.

Other regional oak decline assessments were possible using large-scale continuous forest inventory plot networks. A 1986 inventory in Virginia afforded the opportunity to evaluate oak decline in the northern Piedmont and western mountains survey units (Oak et al. 1991) that lie within the Northern Piedmont, Blue Ridge Mountains, and Ridge and Valley ecoregions. Virginia had experienced chronic and severe oak decline since the earliest reports of the disease (Millers et al. 1989) and was in the midst of a widely reported severe decline episode during the inventory year. Decline occurred on an estimated 444,400 ha of oak forest (16.4 %) with the northern mountain unit (Blue Ridge Mountains and Ridge and Valley ecoregions) sustaining the highest incidence (29.7 %). Estimated annual mortality was greatest for counties with concentrations of decline. Shenandoah County in the Ridge and Valley ecoregion sustained average losses of 1.74 m^3 per ha per year from 1977 to 1986 which represented the highest rate in the assessment area. Average annual mortality in decline-affected plots overall was 1.84 m^3 per ha compared with 1.03 m^3 per ha in unaffected plots.

Stand and site factors associated with oak decline incidence (vulnerability) and severity (risk, as measured by volume losses when decline did occur) were examined for potential use in predicting oak decline. Factors showing promise included tree species composition, site quality, stand age, SI to age ratio, physiography, and stand density. The relationships between individual factors and oak decline vulnerability and risk were complex. For example, less productive sites were more vulnerable to oak decline but 33 % of the total affected area and 36 % of the oak mortality still occurred on sites with higher productivity (SI$\geq$21 m). Chestnut oak forest types were the most vulnerable to oak decline, but risk was highest in oak-hickory forest types.

Oak et al. (2004) used FIA inventory data collected by the USDA Forest Service Southern and Southeastern Forest Experiment Stations to conduct an analysis of geographic and temporal decline trends in 12 southern states over two survey periods, 1984–89 and 1990–97. Data were originally interpreted by state, but were partitioned by CHR ecoregions within the inventoried area for this discussion. The Northeastern Forest Experiment Station FIA unit used different damage coding methods during these periods which precluded analysis in the northern part of the CHR (Fig. 3.2).

Just under half of the total CHR area, 47 million ha, was included in the oak decline analyses (Table 3.4). Ecoregions poorly or not represented were the Interior River Valleys and Hills (0 % inventoried), Western Allegheny Plateau (0 %), Central Appalachians (11 %), Ozark Highlands (24 %), northern Piedmont (25 %) and Interior Plateau (40 %). Among these, the Central Appalachians and Western Allegheny Plateau have high relative oak density (Table 3.2; Fig. 3.1) and have experienced recurrent, and sometimes severe, oak decline episodes in historical accounts (Millers et al. 1989).

Inventories conducted during the 1980s detected forests vulnerable to oak decline on about 9.8 million ha in CHR ecoregions, of which about 10.3 % were affected

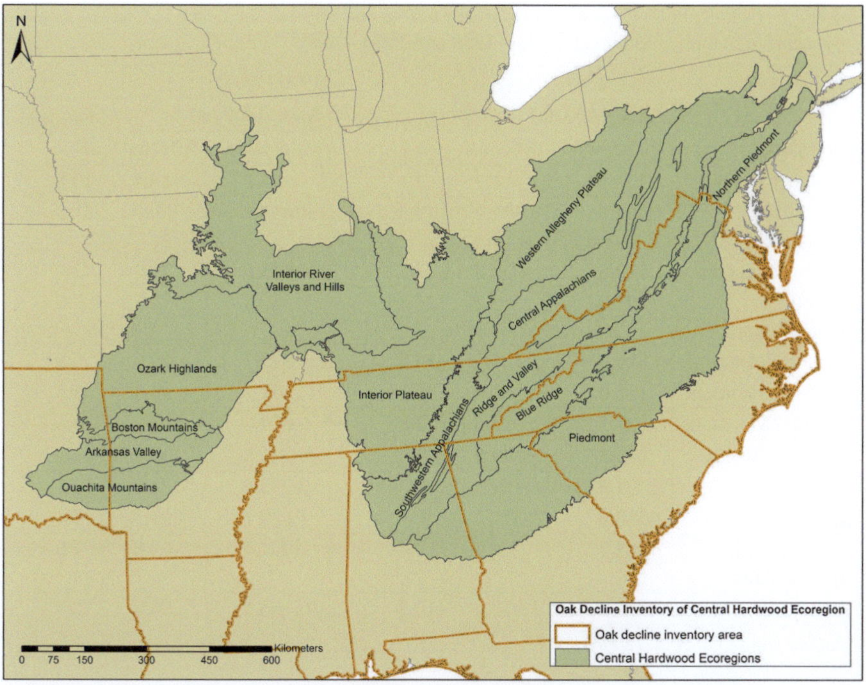

Fig. 3.2 CHR ecoregions and area included in USDA Forest Service FIA oak decline assessments conducted between 1984 and 1997 (Oak et al. 2004) (Map credit to Ida Evretjarn)

Table 3.4 Area of CHR ecoregions inventoried for oak decline by USDA Forest Service Southeastern and Southern Research Station FIA work units, 1984–1997

Ecoregion	Area (1000 ha) Total	Inventoried	Percent
Arkansas Valley	2,842.1	2,842.1	100
Blue Ridge Mountains	4,659.5	4,506.5	96.7
Boston Mountains	1,417.8	1,417.8	100
Central Appalachians	6,205.0	674.6	10.9
Interior Plateau	12,352.3	4,951.8	40.1
Interior River Valleys and Hills	12,040.5	0	0.0
Northern Piedmont	3,045.9	748.0	24.6
Ouachita Mountains	2,689.6	2,689.6	100
Ozark Highlands	10,639.1	2,597.2	24.4
Piedmont	16,611.7	16,611.7	100
Ridge and Valley	11,548.3	6,511.3	56.4
Southwestern Appalachians	3,799.4	3,272.4	86.1
Western Allegheny Plateau	8,144.0	0	0.0
Total	95,995.2	46,823.1	48.8

Table 3.5 Oak decline vulnerable area, affected area, and affected incidence in CHR ecoregions in successive FIA inventory cycles 1984–1989 and 1990–1997

	1984–1989			1990–1997		
	Area (1000 ha)		Incidence	Area (1000 ha)		Incidence
Ecoregion	Vulnerable	Affected	(%)	Vulnerable	Affected	(%)
Arkansas Valley	474.1	14.7	3.1	438.0	46.7	10.7
Blue Ridge Mountains	1,497.9	191.9	12.8	1,454.1	309.7	21.3
Boston Mountains	512.8	36.8	7.2	693.6	79.1	11.4
Central Appalachians	227.0	50.2	22.1	232.5	18.1	7.8
Interior Plateau	818.0	126.5	15.5	875.7	59.6	6.8
Northern Piedmont	135.0	32.6	24.1	131.1	33.0	25.2
Ouachita Mountains	621.3	17.6	2.8	670.0	44.2	6.6
Ozark Highlands	695.3	49.0	7.0	811.2	107.8	13.3
Piedmont	2,437.6	139.3	5.7	2,233.9	224.8	10.1
Ridge and Valley	1,588.9	319.1	20.1	1,629.1	299.3	18.4
Southwestern Appalachians	788.9	36.0	4.6	997.4	69.6	7.0
Total	9,797.0	1,013.6	10.3	10,166.6	1,291.9	12.7

Adapted from Oak et al. (2004)

(1.0 million ha; Table 3.5). Ecoregions with incidence greater than the overall mean included the Blue Ridge Mountains, Central Appalachians, Northern Piedmont, and Ridge and Valley. However, relatively small portions of the Central Appalachians and Northern Piedmont were inventoried, yielding small sample sizes (around 100 vulnerable plots each). Therefore, confidence in the oak decline incidence estimates for these ecoregions overall is low compared to incidence estimates for ecoregions receiving more intensive inventory. The Ridge and Valley ecoregion had over 20.1 % oak decline incidence based on about 1.6 million acres of vulnerable forest (806 plots). Western ecoregions of the CHR (Arkansas Valley, Boston Mountains, Ouachita Mountains, and Ozark Highlands) had among the lowest oak decline incidences (2.8–7.2 % individually; 5.1 % combined).

The geographic distribution of plots vulnerable to oak decline during the 1980s inventories (Fig. 3.3a) generally reflected oak density displayed in Fig. 3.1, with high oak concentrations throughout the Blue Ridge Mountains, Ridge and Valley, Ozark Highlands, Boston Mountains and Ouachita Mountains, and in portions of the Southwestern Appalachians and Interior Plateau. High density of plots vulnerable to oak decline did not, however, translate directly to high density of affected plots. Instead, these were concentrated in the northern Ridge and Valley in Virginia; the southern Blue Ridge Mountains in North Carolina; and on the Western Highland Rim area of the Interior Plateau in Tennessee (Fig. 3.3b).

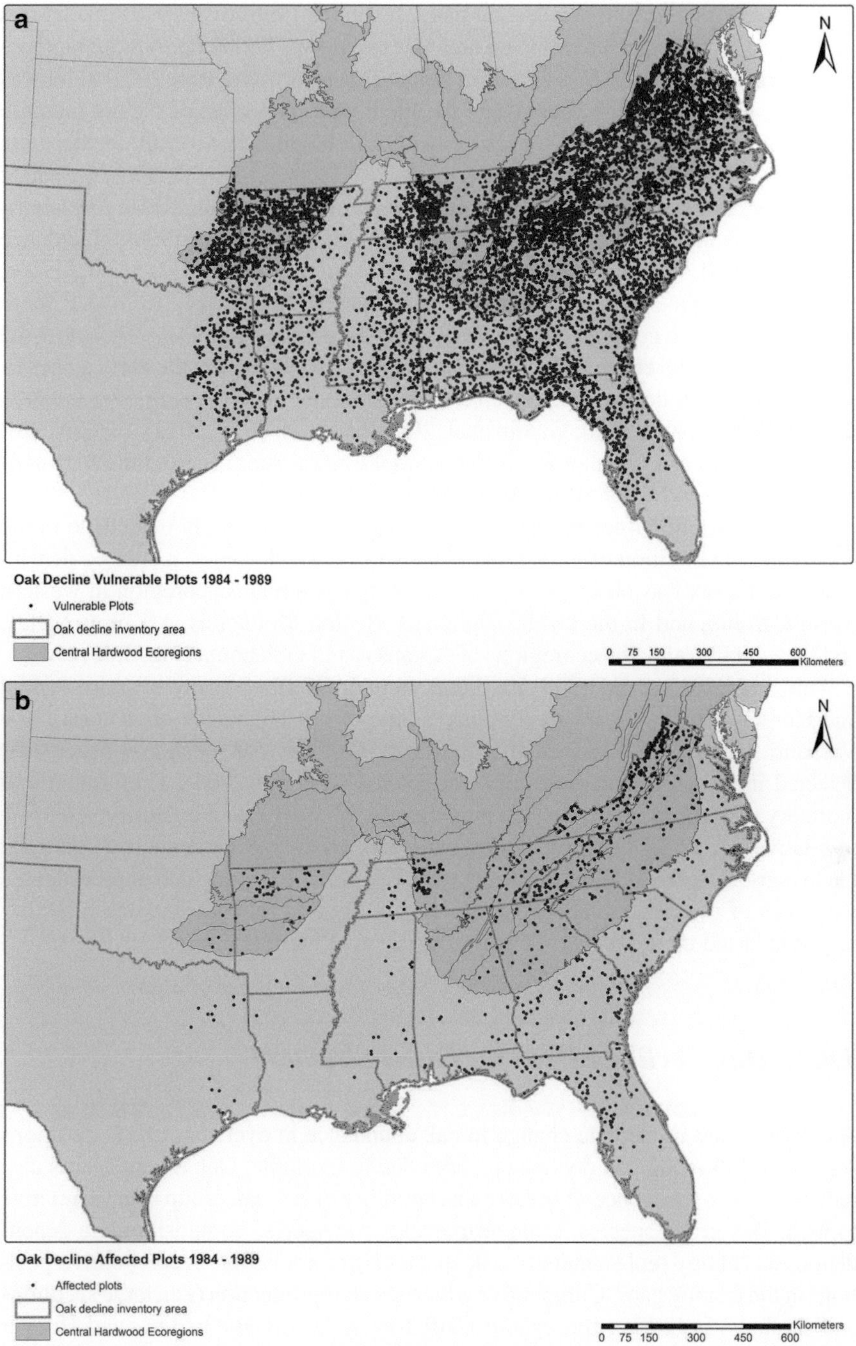

Fig. 3.3 Geographic distribution of USDA Forest Service FIA plots inventoried between 1984 and 1989 (**a**) vulnerable to and; (**b**) affected by oak decline within CHR ecoregions (Adapted from Oak et al. (2004); map credit to Ida Evretjarn)

The area vulnerable to oak decline increased in the 1990s over 1980s inventories by nearly 370,000 ha while affected area increased by 278,000 ha. A large increase in vulnerable area in the Southwestern Appalachians of more than 200,000 ha was offset by an equally large decrease in the Piedmont, with most of the net increase accounted for in the Boston Mountains and Ozark Highlands. Overall, incidence in inventoried CHR ecoregions increased to 12.7 % (Table 3.5). Incidence in the Ridge and Valley remained high while it increased markedly in the Blue Ridge Mountains (from 12.8 % incidence in the 1980s inventories to 21.3 % in the 1990s). Incidence nearly doubled in the westernmost ecoregions of the CHR (Ozark Highlands, Boston Mountains, Arkansas Valley, and Ouachita Mountains) from 5.1 % to 10.6 %, though the combined mean for these ecoregions was still slightly below the 1990s mean for the entire CHR. These inventories detected only the early stages of a very severe oak decline episode that would continue and intensify over the next decade (Starkey et al. 2000; Guldin et al. 2006; Fan et al. 2008, 2012). FIA inventories in Arkansas and Oklahoma which encompass the western ecoregions were conducted in 1995 and 1999, respectively (Hansen et al. 1992).

The increase in vulnerable area noted in Table 3.5 was barely detectable in the geographic distribution of vulnerable plots (Fig. 3.4a). However, increased density of affected plots was observed in the Blue Ridge Mountains ecoregion in western North Carolina and in the Ozark Highlands, Boston Mountains, Arkansas Valley, and Ouachita Mountains ecoregions of Arkansas and Oklahoma (Fig. 3.4b).

Fan et al. (2012) used 1999–2010 data from 6,997 FIA plots to examine spatial and temporal trends of oak decline across the Ozark Highlands of Arkansas and Missouri. This period marked the culmination of the oak decline episode first detected in the preceding inventory evaluated Oak et al. (2004). They found that mortality of red oak group species increased by 11 % of relative density and 15 % of relative BA while mortality among white oak group species remained comparable to non-oak species. Drought events were key inciting factors with unprecedented outbreaks of the red oak borer serving as contributing factors. The oak mortality response lasted up to 10 years after the cessation of inciting drought.

3.4.3 *Oak Decline Effects on Forest Structure*

An obvious and immediate change in oak abundance in overstory crown positions was noted following oak decline episodes due to mortality. Oak diversity was also reduced as a consequence of greater susceptibility of red oak group species relative to white oak group species. Long term changes in species composition are dependent upon canopy replacement of oak species by reproduction in competitive positions in the canopy gaps. Competitive advance oak reproduction (i.e., large seedlings and saplings) is lacking throughout CHR forests (Loftis 1983; Beck and Hooper 1986), as are disturbance regimes necessary for development and subsequent recruitment into the forest overstory (McEwan et al. 2011). As a result, oaks are

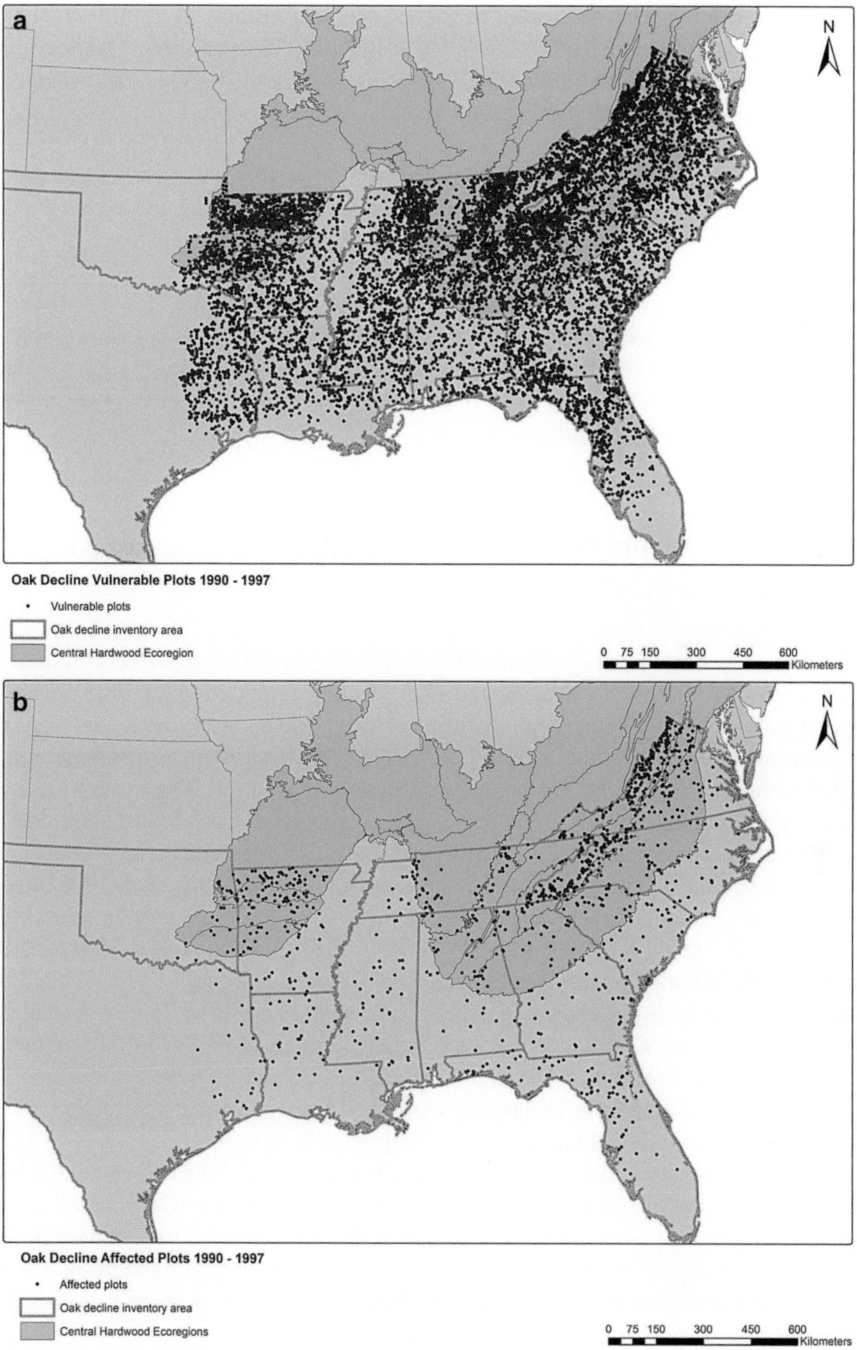

Fig. 3.4 Geographic distribution of USDA Forest Service FIA plots inventoried between 1990 and 1997 (**a**) vulnerable to and; (**b**) affected by oak decline within CHR ecoregions (Adapted from Oak et al. (2004); map credit to Ida Evretjarn)

already decreasing relative to other hardwood species (Abrams 1992; Aldrich et al. 2005; McGee and Hooper 1970; Loftis 1983; Beck and Hooper 1986). Oak recruitment into canopy positions after silvicultural disturbances is widely acknowledged to be problematic on more productive sites (McGee and Hooper 1970; Loftis 1983; Beck and Hooper 1986) but not on less productive sites (Roach and Gingrich 1968; Sander and Clark 1971). However, oak regeneration performance has been studied only in the context of silvicultural disturbances (e.g., clearcutting and shelterwood cutting, with or without treatment of competing vegetation). Whether these site productivity relationships and oak reproduction performance will hold following oak decline mortality, with or without silvicultural interventions, is unresolved. However, the relationship between the current importance values of oaks and maples and a regeneration index presented by McEwan et al. (2011) strongly suggests the oak composition will continue to decrease over a wide range of sites throughout the CHR under prevailing disturbance regimes.

3.5 Modeling and Managing Oak Decline Using the Forest Vegetation Simulator

The probability and severity of oak decline events in the CHR and effects on forest dynamics can be simulated using the Oak Decline Event Monitor (http://www.fs.fed.us/foresthealth/technology/od_rating.shtml), which was developed using data from extensive regional surveys of affected and healthy areas (Starkey et al. 1989; Oak et al. 1990; Oak and Croll 1995; Starkey et al. 2000). The Event Monitor runs within the structure of the Forest Vegetation Simulator (FVS), an individual-tree, distance-independent, growth and yield model (Dixon 2002). The probability of an oak decline event is computed from stand and site data (Oak et al. 1996), and mortality is scheduled according to the stand risk rating, with greater mortality scheduled for stands with elevated risk. Variants of FVS and the Event Monitor are available for the southern and central states sub-regions within the CHR. Table 3.6 displays output from simulation of a mixed oak stand in the Blue Ridge Mountains ecoregion of North Carolina using the Southern Variant of FVS. In this example, oak decline risk was high in the first simulation cycle and remained so through 7 cycles (35 years) Based on probability computed from stand and site factors, oak decline events were scheduled at the end of the second and seventh cycles (+10 and +35 years). Total stand BA was reduced by 8.3 m^2 per ha after the first event and by 6.4 m^2 per ha after the second. The effect on species composition was a depletion of the oak component from 84 % of stocking at the beginning of the simulation to 23 % after 10 cycles (50 years). The simulated changes in overall stand density and oak composition after the eighth cycle resulted in a reduction of decline risk to the low category.

FVS and the Oak Decline Event Monitor can be used to evaluate the potential of management actions for reducing oak decline risk and for mitigating changes

Table 3.6 Output selected from a 50-year simulation of a mixed oak stand in the Blue Ridge Mountains ecoregion, North Carolina. The probability of an oak decline event in a subject stand is calculated using a logistic regression (Oak et al. 1996) from stand and site factors collected during standard inventories. An oak decline event is scheduled (value of 1 in this table) when the calculated probability exceeds a generated random number between 0.00 and 1.00. The severity of the event is determined by oak decline risk classification (Oak and Croll 1995) and mortality is imposed based on expectations synthesized from numerous published local and regional oak decline assessments. Different mortality rates are imposed on red oak group species (highest mortality rate), white oak group species (intermediate mortality rate), and hickory species (lowest mortality rate). After imposition of mortality, stand growth is simulated using the appropriate regional FVS variant for the selected time interval (the Southern Variant at 5 year intervals for this case), and a new oak decline probability computed from the new stand attributes. Results were converted to metric units. In this simulation, oak decline events were scheduled at 10 and 35 years. The consequence of these events resulted in a reduction of oak BA from 18.86 m^2/ha (84 % of total stand BA) to 5.06 m^2/ha (23 % of total stand BA) after 10 simulation intervals (50 years). Oak decline risk was reduced to the low category after the +35 year oak decline event due a reduction of oak density resulting from cumulative mortality

			BA (m^2/ha)		
Time (years)	Risk	Decline event	Total	Oak	% Oak
+5	High	0	22.54	18.86	84
+10	High	1	23.91	19.78	83
+15	High	0	15.64	11.04	71
+20	High	0	17.02	11.50	68
+25	High	0	18.17	12.19	67
+30	High	0	20.01	12.88	64
+35	High	1	22.08	13.34	60
+40	Low	0	15.64	4.60	29
+45	Low	0	20.01	4.83	24
+50	Low	0	22.08	5.06	23

deemed detrimental for various desired future stand compositions and structures. Managers may choose to change outcomes by altering susceptibility (risk or probability of an oak decline event) or vulnerability (severity of damage should a decline event occur). Susceptibility is influenced by changing species composition while vulnerability is reduced by improving overall stand vigor through removal of trees likely to die in such a decline event. Fan et al. (2008) examined more than 4,000 randomly selected trees in the Ozark Highlands during an oak decline event from 2002 to 2006. They found that oak mortality was mainly related to crown width and amount of crown dieback and produced models useful for marking trees for thinning or harvest. Though intended for stands threatened by defoliation caused by gypsy moth, many prescriptions described by Gottschalk (1993) are useful for managing stands susceptible or vulnerable to oak decline. Spring defoliation is a major inciting factor in oak decline etiology, and the outbreak dynamics of this non-indigenous insect compared with native defoliators (outbreaks of longer duration, shorter return interval, and with more complete defoliation) have often resulted in catastrophic mortality from oak decline.

3.6 Summary

Oak decline has been recorded throughout the CHR since the 1800s. Affected contiguous areas may range from a few to thousands of hectares and severity also can be highly variable. Severe and recurrent damage has been reported in the Northern Piedmont, Blue Ridge Mountains, and Ridge and Valley ecoregions in the eastern CHR, and in the Ozark Highlands, Boston Mountains, and Ouachita Mountains ecoregions in the western CHR. Advanced physiologic age and drought are common interacting factors throughout, with catastrophic mortality occurring where gypsy moth defoliation has occurred in the eastern CHR, and more recently in concert with outbreaks of a contributing factor, red oak borer, in the western CHR.

Visible symptomology includes slow, progressive crown dieback from the top down and from the outside inward of trees in upper canopy positions followed by tree mortality and typically occurs over many years or decades. This crown dieback is an indicator of the progression of root disease belowground. These symptoms are the result of a complex of many interacting factors. Etiology includes issues that stress healthy trees and are classified into three general groups: predisposing, long-term factors that reduce tree resilience, inciting factors that add further stresses to trees invoking the decline event, and contributing factors that take advantage of stressed trees but by themselves do not invoke oak decline. There can be considerable variation in the combination of these factors from one decline event or area to another.

Composition and structure of CHR forests have been altered through the influence of human activity, forest management policies and introduced pathogens. Two influences are particularly notable. The loss of American chestnut by way of an introduced pathogen, and the reduced incidence of fire used as a cultural practice for millennia by people inhabiting the CHR have had far-reaching impacts on these forests. Changes include less complex age structure, more shade tolerant fire sensitive woody species, and a dense, relatively small-diameter overstory. Thus, what was once an ecosystem dominated by American chestnut and mediated by fire of mostly anthropogenic origin has been radically transformed. Land managers are presented with a dilemma concerning the role of humans in forest ecosystems when choosing the reference condition upon which to base management practices compatible with the 'historic range of variability' in natural disturbances, particularly concerning fire, non-indigenous plants, pathogens, and insects, and extirpation of native flora and fauna. The range of natural disturbance effects on the shape of forest composition and structure prior to human habitation of the CHR can only be generally inferred and ignores the reality of the past five to ten millennia (see Greenberg et al. Chap. 12). Alternatively, selecting a disturbance regime that includes human interventions along a spectrum from early aboriginal to contemporary times is a philosophical and arbitrary process. In any case, CHR forests cannot be termed 'natural' until the full suite of forest plants (including American chestnut) and animals (see Greenberg et al. Chap. 12) present before the arrival of humans in the CHR are restored and functioning as part of ecosystems.

Systematic surveys of oak decline began in the mid-1980s and were invaluable in elucidating its spatial and temporal context. In the southern half of the CHR, the oak decline-affected area was shown to encompass over a million hectares from 1984–1987 and 1.3 million hectares from 1990–1997. Survey methodologies were refined and standardized during this time, resulting in a more comprehensive understanding of decline events, revealing factors that may help in the prediction of future decline.

The most significant impact of oak decline is due to tree mortality resulting in major changes in forest structure and composition with red oak group species typically having much higher mortality rates than species of the white oak group. Drought has been a major inciting factor in oak decline across the CHR and the impact of the cumulative influence of drought on oak decline related mortality may last for decades.

Acknowledgments We recognize with gratitude the career contributions of Dale A. Starkey in helping to increase understanding of oak decline in the southeastern USA. The assistance of Anthony Elledge and Edwin Yockey in recovery of FIA oak decline assessment data is gratefully acknowledged. We thank Ida Evretjarn for GIS analysis of these data and production of map products.

References

Abrams, M. D. (1992). Fire and the development of oak forests. *Bioscience, 42*, 346–353.

Aldrich, P. R., Parker, G. R., Romero-Severson, J., & Michler, C. H. (2005). Confirmation of oak recruitment failure in Indiana old-growth forest: 75 years of data. *Forest Science, 51*, 406–416.

Bassett, E. N., & Fenn, P. (1984). Latent colonization and pathogenicity of *Hypoxylon atropunctatum* on oaks. *Plant Disease, 68*, 317–319.

Beck, D. E., & Hooper, R. M. (1986). Development of a southern Appalachian hardwood stand after clearcutting. *Southern Journal of Applied Forestry, 10*, 168–172.

Bruhn, J. N., Wetteroff, J. J., Mihail, J. D., Kabrick, J. M., & Pickens, J. B. (2000). Distribution of *Armillaria* species in upland Ozark mountain forests with respect to site, overstory species composition and oak decline. *European Journal of Forest Pathology, 30*, 43–60.

Delcourt, P. A., & Delcourt, H. R. (2004). *Prehistoric native Americans and ecological change: Human ecosystems in eastern North America since the Pleistocene*. Cambridge, MA: Cambridge University Press.

Dixon, G. W. (2002, modified frequently). *Essential FVS: A user's guide to the Forest Vegetation Simulator* (Intern Rep). Fort Collins: USDA Forest Service Forest Management Service Center.

Dwyer, J. P., Cutter, B. E., & Wetteroff, J. J. (1995). A dendrochronological study of black and scarlet oak decline in the Missouri Ozarks. *Forest Ecology and Management, 75*, 69–75.

Fan, Z., Kabrick, J. M., Spetich, M. A., Shifley, S. R., & Jensen, R. G. (2008). Oak mortality associated with crown dieback and oak borer attack in the Ozark Highlands. *Forest Ecology and Management, 255*, 2297–2305.

Fan, Z., Fan, X., Crosby, M. K., Moser, W. K., He, H. S., Spetich, M. A., & Shifley, S. R. (2012). Spatio-temporal trends of oak decline and mortality under periodic regional drought in the Ozark Highlands of Arkansas and Missouri. *Forests, 3*, 614–631.

Gottschalk, K. W. (1993). *Silvicultural guidelines for forest stands threatened by gypsy moth* (Gen. Tech Rep NE-171). Radnor: USDA Forest Service Northeastern Forest Experiment Station.

Guldin, J. M., Poole, E. A., Heitzman, E., Kabrick, J. M., & Muzika, R. (2006). Ground truth assessments of forests affected by oak decline and red oak borer in the Interior Highlands of Arkansas, Oklahoma, and Missouri: Preliminary results from overstory analysis. In K. F. Connor (Ed.), *Proceedings of 13th biennial southern silvicultural research conference* (General technical report SRS-92). Asheville: USDA Forest Service Southern Research Station.

Hansen, M. H., Frieswyk, T., Glover, J. F., & Kelly, J. E. (1992). *The Eastwide forest inventory data base: Users manual* (Gen Tech Rep NC-151). St. Paul: USDA Forest Service North Central Forest Experiment Station.

Hyink, D. M., & Zedaker, S. M. (1987). Stand dynamics and the evaluation of forest decline. *Tree Physiology, 3*, 17–26.

Law, J. R., & Gott, J. D. (1987). Oak mortality in the Missouri Ozarks. In R. L. Hay, F. W. Woods, & H. DeSelm (Eds.), *Proceedings of 6th central hardwood for conference* (pp. 427–436). Knoxville: University of Tennessee.

Loftis, D. L. (1983). Regenerating southern Appalachian mixed hardwood stands with the shelterwood method. *Southern Journal of Applied Forestry, 7*, 212–217.

Manion, P. D. (1991). *Tree disease concepts* (2nd ed., pp. 328–348). Englewood Cliffs: Prentice-Hall.

McEwan, R. W., Dyer, J. M., & Pederson, N. (2011). Multiple interacting ecosystems drivers: Toward an encompassing hypothesis of oak forest dynamics across eastern North America. *Ecography, 34*, 244–256.

McGee, C. W., & Hooper, R. M. (1970). *Regeneration after clearcutting in the southern Appalachians* (Res Pap SE-70). Asheville: USDA Forest Service Southeastern Forest Experiment Station.

Millers, I., Shriner, D. S., & Rizzo, D. (1989). *History of hardwood decline in the eastern United States* (Gen Tech Rep NE-126). Broomall: USDA Forest Service Northeastern Forest Experiment Station.

Morin, R. S., Liebhold, A. M., Luzader, E. R., Lister, A. J., Gottschalk, K. W., & Twardus, D. B. (2005). *Mapping host-species abundance of three major exotic pests* (Res Pap NE-726). Newtown Square: USDA Forest Service Northeastern Research Station.

Oak, S. W., & Croll, P. M. (1995). *Evaluation of oak decline risk rating using the CISC database on the Cherokee National Forest, Tennessee* (Rep No 95-1-22). Asheville: USDA Forest Service Southern Region State and Private Forestry Forest Health Protection.

Oak, S. W., Starkey, D. A., & Ishikawa, P. (1990). Application of color infrared aerial photography for detecting oak decline damage and change in southern forests. In J. D. Greer (Ed.), *Protect natural resources with remote sensing, proceedings of 3rd for service remote sensing application conference, April 9–13 1990* (pp. 95–107). Tucson: The American Society for Photogrammetry and Remote Sensing, Bethesda.

Oak, S. W., Huber, C. M., & Sheffield, R. M. (1991). *Incidence and impact of oak decline in western Virginia, 1986* (Resource Bull SE-123). Asheville: USDA Forest Service Southeastern Forest Experiment Station.

Oak, S., Tainter, F., Williams, J., & Starkey, D. (1996). Oak decline risk rating for the southeastern United States. *Annales des Sciences Forestières, 53*, 721–730.

Oak, S. W., Steinman, J. R., Starkey, D. A., & Yockey, E. K. (2004). Assessing oak decline incidence and distribution in the southern US using forest inventory and analysis data. In M. A. Spetich (Ed.), *Upland oak ecol sympos. History, current conditions, and sustainability* (Gen Tech Rep SRS-73, pp. 236–242). Asheville: USDA Forest Service Southern Research Station.

Parker, J. (1970). *Effects of defoliation and drought on root food reserves in sugar maple seedlings* (Res Pap NE-169). Upper Darby: USDA Forest Service Northeastern Forest Experiment Station.

Rauschenberger, J. L., & Ciesla, W. M. (1966). *Evaluation of oak mortality on the George Washington and Jefferson National Forests, Virginia* (Rep No 66-1-31). Asheville: USDA Forest Service Southern Region State and Private Forestry Forest Insect and Disease Control.

Roach, B. A., & Gingrich, S. F. (1968). *Even-aged silviculture for upland central hardwoods* (Agriculture handbook, Vol. 355). Washington, DC: USDA Forest Service.

Sander, I. L., & Clark, F. B. (1971). *Reproduction of upland hardwood forests in the Central United States*. Washington, DC: USDA Forest Service.

Sinclair, W. A. (1965). Comparisons of recent declines of white ash, oaks, and sugar maple in northeastern woodlands. *Cornell Plantations, 20*, 62–67.

Spetich, M. A., Perry, R. W., Harper, C. A., & Clark, S. L. (2011). Fire in eastern hardwood forests through 14,000 years. In C. H. Greenberg, B. S. Collins, & F. R. Thompson III (Eds.), *Sustaining young forest communities: Ecology and management of early successional habitats in the central hardwood region* (pp. 41–58). New York: Springer.

Starkey, D. A., Oak, S. W., Ryan, G. W., Tainter, F. H., Redmond, C., & Brown, H. D. (1989). *Evaluation of oak decline areas in the South* (Prot Rep R8 PR-17). Atlanta: USDA Forest Service Southern Region.

Starkey, D. A., Mangini, A., Oliveria, F. L., Clarke, S., Bruce, B., Kertz, R. C., & Menard, R. (2000). *Forest health evaluation of oak mortality and decline on the Ozark National Forest, 1999* (Rep No 2000-02-02). Alexandria: USDA Forest Service Southern Region State and Private Forestry Forest Health Protection.

Tainter, F. H., Retzlaff, W. A., Starkey, D. A., & Oak, S. W. (1990). Decline of radial growth in red oaks is associated with short-term changes in climate. *European Journal of Forest Pathology, 20*, 95–105.

Wargo, P. M. (1972). Defoliation-induced chemical changes in sugar maple roots stimulate growth of *Armillaria mellea*. *Phytopathology, 62*, 1278–1283.

Wargo, P. M. (1977). *Armillaria mellea* and *Agrilus bilineatus* and mortality of defoliated oak trees. *Forest Science, 23*, 485–492.

Wargo, P. M., Houston, D. R., & LaMadeleine, L. A. (1983). *Oak decline*. Washington, DC: USDA Forest Service Forest Insect and Disease Leaflet 165.

Chapter 4
Southern Pine Beetles in Central Hardwood Forests: Frequency, Spatial Extent, and Changes to Forest Structure

John T. Nowak, Kier D. Klepzig, David R. Coyle, William A. Carothers, and Kamal J.K. Gandhi

Abstract The southern pine beetle (SPB) is a major disturbance in pine forests throughout the range of southern yellow pines, and is a significant influence on forests throughout several Central Hardwood Region (CHR) ecoregions. At endemic levels, SPB colonizes individual stressed or lightning-struck trees, acting as a natural thinning agent. During outbreaks, tree mortality from SPB may impact CHR forests by indirectly converting stands to other species types, or changing the stand age and structure. Southern pine beetle can also create disturbance in stands by causing mortality in large clusters of pine trees or by hastening the succession from pine-hardwood forests to late-successional forest by killing single or groups of overstory pine trees. Populations are cyclical and have traditionally impacted CHR forests every 7–25 years, depending on location. The most significant outbreaks in this region in the past 65 years occurred 1974–1976 and 1997–2003, with the most recent impacting more than 405,000 ha, and caused an estimated economic loss of more than $1 billion across six states. In this chapter we examine the spatial extent and frequency of SPB outbreaks in the CHR. We also discuss the severity of disturbance caused by SPB to forests in this region over the past 65 years; how this

J.T. Nowak (✉) • W.A. Carothers
USDA Forest Service, Forest Health Protection,
200 W.T. Weaver Blvd, Asheville, NC 28804, USA
e-mail: jnowak@fs.fed.us; wcarothers@fs.fed.us

K.D. Klepzig
USDA Forest Service, Southern Research Station,
200 W.T. Weaver Blvd, Asheville, NC 28804, USA
e-mail: kklepzig@fs.fed.us

D.R. Coyle • K.J.K. Gandhi
D.B. Warnell School of Forestry and Natural Resources, University of Georgia,
180 East Green Street, Athens, GA 30602, USA
e-mail: drcoyle@uga.edu; kjgandhi@uga.edu

C.H. Greenberg, B.S. Collins (eds.), *Natural Disturbances and Historic Range of Variation*, Managing Forest Ecosystems 32,
DOI 10.1007/978-3-319-21527-3_4

disturbance has altered the resultant forests; and the potential impacts of changes in climate and anthropogenic effects on preventing infestations and reducing levels of tree mortality attributed to SPB.

Keywords Southern pine beetle • Table mountain pine • Restoration • Southern pine beetle outbreak history • Shortleaf pine

4.1 Introduction

Southern pine beetle (*Dendroctonus frontalis*; SPB) (Fig. 4.1a) is a native bark beetle species (Coulson and Klepzig 2011) that has long been a significant disturbance factor across the Central Hardwood Region (CHR) and all of the southern USA. Trees most commonly used as hosts are loblolly pine (*Pinus taeda*) and shortleaf pine (*P. echinata*), but this insect can also colonize and kill all of the southern yellow pines (Fig. 4.1b). This insect has also been known to kill eastern white pine (*P. strobus*), Table Mountain pine (*P. pungens*), pitch pine (*P. rigida*), Virginia pine (*P. virginiana*), red spruce (*Picea rubens*), and others during significant SPB outbreaks (Chamberlin 1939). This insect can be found from New Jersey south to central Florida and west to eastern Texas. There are also disjunct populations in Arizona and Central America (Thatcher and Barry 1982; Clarke and Nowak 2009). Significant economic losses (an average of $43 million per year; Pye et al. 2011) and ecological impacts, including changes in the physical environment, hydrology, wildlife, forest structure, and more (Tchakerian and Coulson 2011), can occur during SPB outbreaks. In this chapter we examine the spatial extent, frequency, and severity of disturbance that this insect has caused to forests in the CHR over the past several decades and how this disturbance has altered these forests. We will also discuss silvicultural techniques (e.g., thinning to reduce basal area) used to prevent and mitigate the impacts of this insect (Fig. 4.1c).

4.2 SPB Biology

Southern pine beetle is a small bark beetle, ranging from 2.2 to 4 mm long (Chamberlin 1939; Coulson and Witter 1984). This tree-killing species relies on mass attack to overcome tree defenses. An individual SPB eating (or tunneling) its way into a host tree is initially met with an exudation of oleoresin. If enough resin flows from the initial attack wound, the attacking beetle may be ‘pitched out,’ or sealed in a globule of crystallized pine resin. If, however, the attacking female is able to continue tunneling to cambium, she releases an aggregation pheromone which attracts other beetles (including males which mate with the attacking females in ‘nuptial chambers’). In this way, a mass attack of a single tree is orchestrated through chemical communication (Klepzig and Hofstetter 2011). While the exact

Fig. 4.1 (**a**) Adult southern pine beetle (Photo by Erich Vallery USDA Forest Service, www.bugwood.org); (**b**) shortleaf pine forest heavily impacted by southern pine beetle; and (**c**) and low hazard stand conditions – basal area lower than 21 m^2 per ha and open understory stand conditions

role of the fungi they carry continues to be debated, maternal females next tunnel through the phloem, laying eggs and inoculating the phloem and outer xylem with at least three different species of fungi. One of these fungi (a bluestain fungus, *Ophiostoma minus*) may aid the beetle in killing the tree, and the other two fungi are nutritional mutualists without which the hatching larvae cannot complete development (Klepzig and Hofstetter 2011).

Southern pine beetles also carry a number of other microbes and mites (Hofstetter 2011) and are prey for several associated predators and parasitoids (Reeve 2011). As a group, SPB generally first colonize single or small groups of two or three trees that have been struck by lightning (Lorio 1986) or weakened by some other factor. These small infestations may grow, but may not grow, into larger infestations or 'spots.' Most spots range in size from about 4–50 trees attacked, or about 14 trees attacked in an average spot (Ayres et al. 2011) (Fig. 4.1b). However, individual pines within hardwood stands can still be attacked during outbreak periods. Under most circumstance, tree crowns begin to fade to yellow within 2–4 weeks of attack, indicating the trees were successfully colonized and have died. Tree crowns then turn red and bark begins to slough. By this point SPB have completed their life cycle, emerged, and will have likely attacked adjacent trees. SPB can complete its life cycle within 30 days during summer months and have four to eight overlapping generations per year, depending on local climatic conditions (Hopkins 1909; Coulson and Witter 1984; Hain et al. 2011).

4.2.1 Susceptible Forest Conditions

Southern pine beetles prefer pine and pine-hardwood forests that are overstocked and more than 12–15 years old, although, pine stands with a greater hardwood component are considered to be less susceptible to SPB (Fettig et al. 2007). Natural and plantation forests that have high basal areas of greater than 28 m^2 per ha are considered highly susceptible to SPB. These higher basal area forest conditions could have been created by planting too high density if the stand was artificially regenerated or by the natural succession process with hardwoods growing up under the pine overstory. The standard recommendation for decreasing a forest's susceptibility to SPB is to reduce basal area to 18 m^2 per ha (Fettig et al. 2007; Nowak et al. 2008) (Fig. 4.1c). Trees in overstocked stands are generally less vigorous due to competition for light, water, and nutrients and therefore are considered to have less adequate defense (e.g., resin flow) than trees growing with less competition. The greater spacing between tree also lowers the effectiveness of SPB's pheromone communications system (Gara and Coster 1968; Fettig et al. 2007) due to more airflow and direct sunlight, which can make the pheromone plume more diffuse (Thistle et al. 2004) and less likely to be found by colonizing beetles. Individual tree vigor and more open stand structure are the two main reasons that thinning stands to reduce basal area is recommended to reduce levels of tree mortality attributed to SPB (Nebeker and Hodges 1983; Fettig et al. 2007; Nowak et al. 2008). Prescribed burning in both pine and pine-hardwood stands might also be a tool to create a more open stand and reduce understory competition (Nowak et al. 2008), which can significantly lower a forest's susceptibility to SPB. Thinning from below to retain an overstory of pine and pine-hardwood trees and maintaining low understory competition through chemical treatment or prescribed fire is the most effective way to protect a forest from SPB spot initiation and spot expansion.

4.3 Historical Outbreaks

The SPB has been documented as a pest since the 1750s and was formally described in 1868 by Zimmermann (1868). In the early 1900s more research was conducted on the biology, behavior, and impact of this insect (Hopkins 1909, 1911; Balch 1928; St. George and Beal 1929; St. George 1930) due to an increase in the understanding of its economic impact. Southern pine beetle populations are considered pulse eruptive (Berryman 1986) with periods of low endemic populations and periods of high epidemic population levels. Return intervals for epidemic populations vary within the CHR. Outbreaks occur every 7–12 years in the Piedmont and every 20–25 years in the more mountainous northern parts of the CHR, including the Blue Ridge Mountains and Southwestern Appalachians. At low populations, SPB are generally confined to weak and dying trees, particularly trees struck by lightning, but they can kill more vigorous trees during outbreaks. SPB outbreak periods usually last 1–3 years and populations fluctuate due to a variety of abiotic and biotic factors, such as climate, natural enemies, and host condition and abundance (Birt 2011).

Several SPB outbreaks have occurred in the CHR as noted in the literature since the 1850s. It should be noted that for this discussion on outbreak history we did not include SPB outbreak occurrences in the Piedmont because significant literature on that exists (Coulson and Klepzig 2011) and SPB activity in the Piedmont is more similar to the Coastal Plain than other regions of the CHR. Balch (1928) reported several instances of SPB outbreaks and severe pine mortality in the CHR, including 1890–1893 in Virginia and North Carolina, 1902–1905 in western North Carolina, and 1910–1915 in the Blue Ridge Mountains of Alabama, Georgia, North Carolina, South Carolina, Tennessee, and Virginia. In more recent times and since better records have been maintained, SPB has occurred in much of the CHR every decade since 1950, with the most significant outbreaks in the CHR occurring in 1974–1975 and 1997–2003 (Table 4.1). There has been a noticeable absence of even slight to moderate SPB activity in the CHR since 2002–2003. Concentrating on the mountainous regions of the CHR (Blue Ridge Mountains, Central Appalachians, Ridge and Valley, and Southwestern Appalachians), most of the SPB activity during the past 60 years occurred in the Blue Ridge Mountains and the Southwestern Appalachians in Alabama. There has been limited SPB activity in the Central Appalachians and Ridge and Valley only during the most severe outbreaks in the region (Table 4.1).

The most recent outbreak in the CHR occurred on state land, private farms, industry land, national forests, and national parks from 1997 to 2003 (Nowak et al. 2008), and was considered one of the most impactful SPB outbreaks of all time. This is outbreak was similar in scope and intensity to the 1974–1976 outbreak, and likely earlier outbreaks described by Balch (1928), which also impacted several ecoregions and states in the CHR (Table 4.1). However, record keeping for the past outbreaks was not as thorough as was done for the 1997–2003 outbreak. We will use the 1997–2003 outbreak as an example of the type of ecological and economic impact

Table 4.1 SPB activity in the CHR (excluding the Piedmont) 1950–2014

Ecoregion	State	No. years	Outbreak years	No. spots	Est. no. trees killed
Blue Ridge Mountains	GA	24 years	1959	Significant activity reported, but no quantification of activity	
			1968	4,539	29,276
			1974	2,204	102,051
			1988	330	176,760
			1995	264	202,658
			2000–2001	1,100	252,800
	NC	22 years	1953–1957	General reference to outbreak, but no quantification	
			1974–1975	713	58,589
			1990	427	114,265
			2000–2001	1,450	122,450
	SC	23 years	1968–1969	2,393	2,572 trees/ 405 ha surveyed
			1971–1976	4,337	10,750 trees/ 405 ha surveyed
	TN	14 years	1969–1972	5,236	177,304
			1975–1977	786	179,951
			2000–2001	982	150,554
Central Appalachians	KY	1 year	1997	32	3,130
	VA	3 years	1975–1976	101	3,682
Ridge and Valley	AL	4 years	1964	320	5,542
	GA	11 years	1987–1988	407	38,691
			2000	500	349,800
	VA	8 years	1975–1976	1,657	200,979
			1994	229	54,044
Southwestern Appalachians	AL	25 years	1953–1956	15,000+	No estimates of trees available
			1968–1969	3,861	16,343
			1996	1,151	94,382
			1999–2000	1,043	135,380
	KY	4 years	1975–1976	451	8,081
			2000–2001	4,200	500,000+

Data collected from USDA Forest Service Forest Health Protection Asheville Field Office reports
No. Years = number of years with any SPB activity; outbreak years = years of an outbreak with at least 1 spot per 405 ha host; No. Spots = number of SPB spots during outbreak years

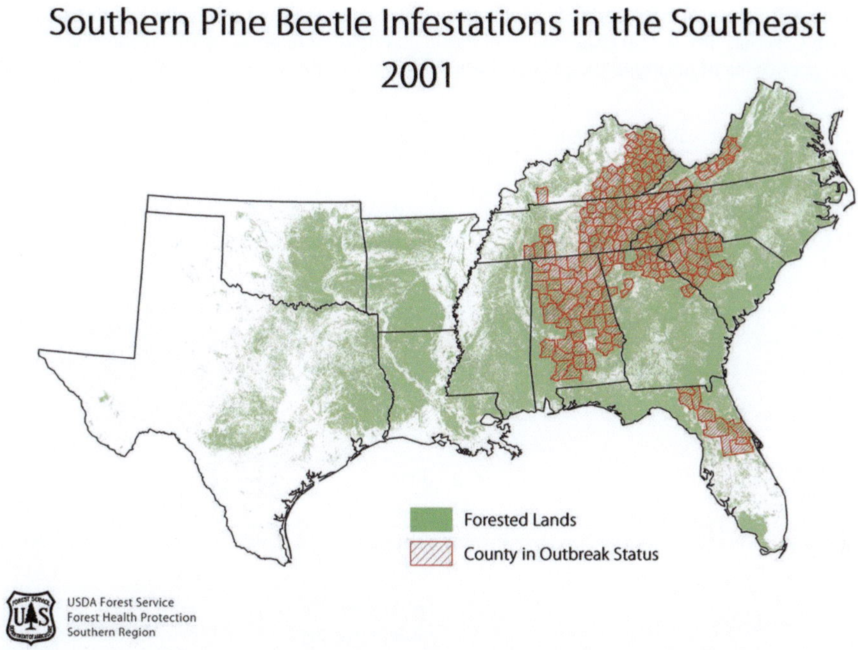

Fig. 4.2 Counties impacted by the 1997–2003 SPB outbreak

during severe SPB outbreaks in the CHR. The outbreak caused more than $1 billion in losses and impacted more than 405,000 ha across six states: Alabama, Georgia, Kentucky, North Carolina, South Carolina, and Tennessee (Fig. 4.2). This outbreak affected multiple ecoregions in the CHR, including the Piedmont, Blue Ridge Mountains, Ridge and Valley, Southwestern Appalachians, Central Appalachians, and Interior Plateau (for ecoregion map, see Greenberg et al. Chap. 1, Fig. 1.1).

Outbreak status is defined as one SPB spot per 405 ha of host type (Price et al. 1998) with high outbreak status at three spots per 405 ha of host type. During the 1997–2003 outbreak, multiple counties in multiple states exceed 20 SPB spots per 405 ha of host type. This level was rarely seen prior to this outbreak and is considered unprecedented although good records do not exist prior to 1990 (Pye et al. 2008). In 2001, 187 counties were in outbreak status and 126 exceeded the high outbreak level (Table 4.2). At the peak of the infestation in 2001, there were approximately 57,000 spots reported in 310 counties.

4.3.1 Need for Restoration

Widespread SPB outbreaks, like the 1974–1976 and 1997–2003 outbreaks, alter forest structure, species composition, and function of pine and pine-hardwood forests. Table 4.3 shows the number of individual spots across the region. Each of

Table 4.2 Extent of SPB outbreak in six states (Alabama, Georgia, Kentucky, North Carolina, South Carolina, Tennessee) (Adopted from Pye et al. (2008)). Outbreak status is defined as 1 SPB spot per 405 ha of host type (Price et al. 1998) with high outbreak status at 3 spots per 405 ha of host type

Year	Outbreak (# of counties)	High outbreak (# of counties)	Dollar amount
1997	7	0	$8,615,789
1998	16	1	$18,407,434
1999	41	17	$78,268,052
2000	150	69	$362,574,736
2001	187	126	$233,890,952
2002	147	96	$328,136,210
2003	30	7	$9,707,703

Table 4.3 Total number of SPB spots in six states at the peak of the 1997–2003 outbreak

State	2000	2001
Alabama	26,407	11,945
Georgia	2,582	2,604
Kentucky	1,664	3,513
North Carolina	2,219	3,860
South Carolina	12,996	22,149
Tennessee	9,352	12,766
Totals	58,839	60,628

these spots would have altered stand structure and species composition and all of these stands were in need of restoration after the outbreak in order to return the forest to previous conditions of pine dominated forests. Southern pine beetle can create disturbance in stands by causing mortality in large clusters of pine trees or by hastening the succession from pine-hardwood forests to late-successional forest by killing single or groups of overstory pine trees. Economic impacts are the easiest to quantify because the trees have an easily identifiable value (value per metric ton, population suppression and stand restoration costs, and reduction in market value due to sudden increase in wood supply).

Ecological impacts of SPB are more difficult to quantify, but are no less important. Negative ecological impacts can include loss of habitat for wildlife species that require pine forests (see Greenberg et al. Chap. 12). One particularly significant impact concerns the red-cockaded woodpecker (*Picoides borealis*; RCW) in Kentucky where, due to the 1997–2003 SPB outbreak, there are currently no known individuals living in the state. Before the SPB outbreak in 2001, RCW occupied several locations on the Daniel Boone National Forest in multiple counties– Laurel, McCreary, Pulaski, and Whitley counties. Because of widespread habitat destruction by the SPB, all of the remaining RCW were moved to South Carolina in 2001 by the USDA Forest Service (AWAKE 2014). Other ecological impacts may not be considered negative. As mentioned earlier, SPB is a native species that acts as a natural landscape scale disturbance factor, much like fire. SPB disturbance can include changes in nutrient cycling, hastening of forest succession, and changes in species composition (Tchakerian and Coulson 2011). SPB outbreaks create single and multi-tree gaps and stand level disturbance.

Pine forests in the CHR are also impaired by years of wildfire suppression and significant reductions in human-caused fires (see also Greenberg et al. Chap. 1; Grissino-Mayer Chap. 6; Greenberg et al. Chap. 12) and the forest communities that returned naturally after the 2001 SPB outbreak are different than previous conditions. Pine is not coming back in these areas without intervention (Elliott et al. 2012) because pine are early successional, shade-intolerant species that usually require bare mineral soil and abundant light in order to regenerate. Forest management intervention is needed in order to restore the unique pine and mixed pine-hardwood communities in the CHR (Elliott et al. 2012; Xi et al. 2012). There is also a desire by the USDA Forest Service and several partners to restore fire-adapted communities and increase heterogeneity on the landscape (CFLRP 2012) because, with the lack of landscape level fires and significant SPB outbreaks, there has been a trend towards pine overstory mortality and a lack of oak and pine regeneration (Elliott et al. 1999). Forest restoration efforts have focused on shortleaf and Table Mountain pine stands (Elliott et al. 2012).

Because of the significant economic and ecological impacts of the 1997–2003 SPB outbreak, there was a strong desire by the Southern Group of State Foresters and the USDA Forest Service (Southern Research Station and Forest Health Protection) to restore some of the pine and pine-hardwood forests impacted by SPB (Nowak et al. 2008). This outbreak spurred Congressional approval and appropriation for the SPB Initiative and the SPB Prevention and Restoration Program (Nowak et al. 2008; USDA Forest Service 2014). A central focus of the initiative and program was to provide the necessary funding to restore these pine communities and to enhance our knowledge of the impact on these communities and the most efficacious restoration practices. Additionally, a main objective of the SPB Prevention and Restoration Program is to create more open stands consisting of the appropriate tree species for the site, particularly shortleaf pine, and shortleaf pine-hardwood stands. These treatments include forest thinning, prescribed burning and restoring native pine forests. These activities have multiple benefits beyond protecting forests from SPB, such as increasing stand resiliency to changes in climate and pressure from invasive species, improving wildlife habitat and reducing fire risk. This program has treated nearly 500,000 ha, mostly through landowner cost-share programs that have directly benefitted more than 15,000 landowners. Approximately 25 % of the on-the-ground accomplishments and funding have gone towards restoring pine forests (Nowak et al. 2008; USDA Forest Service 2014).

In order to examine the efficacy of the SPB Prevention and Restoration Program treatments on reducing stand susceptibility to future outbreaks of SPB, plots were established in stands that had received either precommercial thinning or prescribed burns under the SPB Prevention and Restoration Program in the Piedmont ecoregion of the CHR in Georgia and South Carolina. Three plots were established in each stand, each separated by at least 50 m. Measurements included live crown

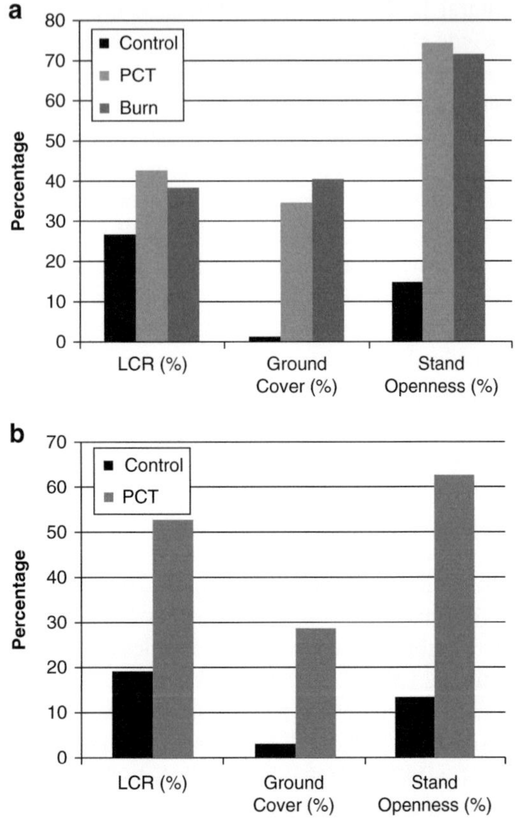

Fig. 4.3 Study in Piedmont ecoregion of (**a**) South Carolina; and (**b**) Georgia, evaluating SPB prevention and restoration treatments on percent live crown ratio, ground cover, and stand openness

ratio, percent ground cover, stem openness at 2 m (estimated percentage of area without stems in view) (Fig. 4.3), and stem diameter at breast height (Figs. 4.4 and 4.5). Overall, treatments by the SPB Prevention and Restoration Program had a very positive effect on stand characteristics, and may greatly reduce stand susceptibility to future SPB outbreaks. Thinning loblolly pine stands in South Carolina reduced stem density by over 390 %, but increased average stem diameter by nearly 40 %. Burning has been shown to lower a stands susceptibility to SPB (Nowak et al. in press) perhaps by creating a more open understory and promoting air movement (Fig. 4.6) (Thistle et al. 2004; Fettig et al. 2007). In the study examining the SPB Prevention and Restoration plots in both Georgia and South Carolina, burned and precommercially thinned treatments had higher live crown ratios, stand openness (Fig. 4.3), and stem diameter (Fig. 4.4).

Fig. 4.4 Study in Piedmont ecoregion of South Carolina and Georgia evaluating SPB prevention and restoration treatments on stem diameter

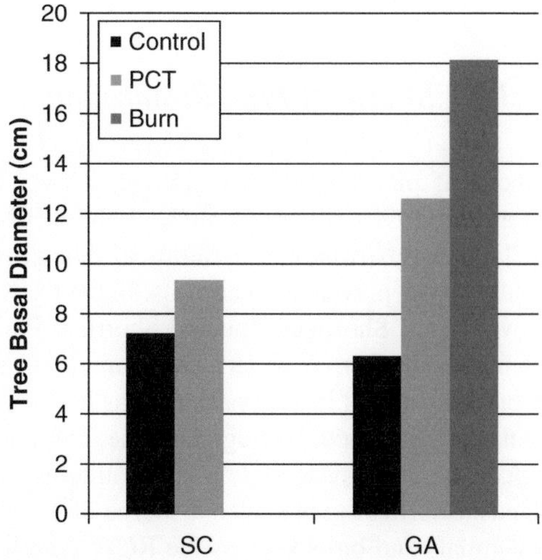

Fig. 4.5 (**a**) Unthinned pine stand with significant understory competition; and (**b**) thinned and burned stand showing greater openness and less competition

Fig. 4.6 (**a**) Mature restored (thinned and burned) shortleaf pine – chalky bluestem forest on the Ouachita National Forest; and (**b**) planted shortleaf pine in stand impacted by SPB (Both projects are partially funded by the SPB Prevention and Restoration Program)

4.3.2 Shortleaf Pine Restoration

Shortleaf pine has a natural range of over 688,000 km^2 in more than 22 states (Lawson 1990) covering much of the CHR and having a similar range as SPB (Clarke and Nowak 2009). Shortleaf pine grows best on a variety of soils, but especially on deep, well-drained soils. In the CHR, shortleaf pine is associated with two cover types, Shortleaf Pine and Shortleaf Pine-Oak (Lawson 1990). Shortleaf pine is important from a variety of commercial, ecological, and wildlife perspectives. Commercially, it is used primarily as lumber, plywood, logs, other structural material, and pulpwood. Ecologically, the diversity and integrity of shortleaf pine stands and related ecosystems has led to interest in restoration efforts on many public lands. Shortleaf pine is also a source of food, shelter, and nesting habitat for small mammals and birds such as the RCW. Shortleaf pine stands grown on appropriate sites with low basal area and open stand conditions would have low susceptibility to SPB (Fig. 4.6a).

Shortleaf pine can be regenerated through natural seeding if there is bare mineral soil (Lawson 1986) or by planting seedlings (Fig. 4.6b) (Barnett et al. 1986). Elliott et al. (2012) found that in shortleaf pine stands killed by SPB, burning the sites increased oak seedling density, but without shortleaf pine in the overstory there was no shortleaf pine regeneration. It is recommended that these sites be burned prior to planting seedlings with a follow-up herbicide release in order to achieve the most successful regeneration (Cassidy 2005; Nowak et al. 2008).

4.3.3 Table Mountain Pine Restoration

Table Mountain pine is endemic to the southern Appalachians, and is limited to xeric, rocky sites at high elevations from Georgia to Pennsylvania (Della-Bianca 1990). This species is often associated with red maple (*Acer rubrum*), black gum (*Nyssa sylvatica*), pitch pine, scarlet oak (*Quercus coccinea*), and chestnut oak (*Q. montana*) (Della-Bianca 1990). Table Mountain pine is not known as a significant timber species, but it is an important species from an aesthetic, erosion control and wildlife perspective (Zobel 1969; Della-Bianca 1990).

The 1997–2003 SPB outbreak and changes in fire frequency have severely impacted mature seed-bearing Table Mountain pine. Table Mountain pine need stand and site disturbance, light, and heat for successful regeneration (Della-Bianca 1990). In undisturbed conditions, succession trends toward oaks and red maple (Welch et al. 2000). Table Mountain pine has a bimodal age distribution with most of the trees in large tree age classes and a smaller percentage in the seedling age class (Williams and Johnson 1990). The authors concluded that without a change in the disturbance regime, such as reintroduction of fire, Table Mountain pine would be difficult to maintain because of the lack of younger age classes. Another study found that the SPB outbreak from 1997–2003 removed up to 30 % of the Table Mountain pine basal area in southwestern Virginia (Lafon and Kutac 2003). These authors also found that due to fire suppression, and reductions in the number of human caused fires (see also Chaps. 1, 6, 12), hardwoods had become more prevalent in the Table Mountain pine stands, and without the reintroduction of landscape-level fire Table Mountain pine would be unlikely to return from disturbances such as SPB.

4.4 Effects of Future Change on SPB

The SPB has a wide host range, high genetic plasticity, and an ability to sustain epidemics even in nontraditional hosts. As such, its geographic range is mostly constrained by host availability and climatic conditions (Ungerer et al. 1999). Certainly host material is abundant further north and would not limit the spread of this insect further into the CHR. While extremely hot temperatures can kill SPB (Wagner et al. 1984), an increase in minimum winter temperature of just a few degrees could result in a substantial increase in the geographic range of the SPB (Ungerer et al. 1999; Tran et al. 2007; Hain et al. 2011). However, while warmer winters in the next 50 years might result in increased numbers of SPB generations (Duerr and Mistretta 2013), they could also disrupt the natural timing of adult emergence and new infestations in the spring (Olatinwo et al. 2014). Widely used future climate projection scenarios agree that such a change to warmer temperatures is probable (Wear and Greis 2013; Dale et al. Chap. 13). Therefore it may be prudent to include SPB management considerations into future forest plans even in the northern portions of the CHR (Duerr and Mistretta 2013; Olatinwo et al. 2014). These management strategies have been discussed above, but include appropriate planting and forest composition, as well as preventative thinning and direct suppression of new infestations.

Other anthropogenic factors may influence SPB success in the CHR. Forest fragmentation (or parcelization) may shift age distributions to younger stands, and increase thinning (assuming a more active management approach – perhaps including thinning – in residential areas to meet goals of fire safe landscapes, aesthetic values and selection for bigger, healthy yard trees). This would likely result in lower SPB numbers and less tree mortality (Olatinwo et al. 2014). Conversely, weather extremes (flooding and drought, windstorms) could stress trees, predisposing them to SPB attack. Regardless, maintaining healthy, vigorous forests via appropriate spacing, stand density, and thinning will continue to be the best methods for mitigating SPB impacts to pine forests and ecosystems.

4.5 Summary

SPB is a significant, but natural disturbance factor in the CHR. It has been reported in this region since the nineteenth century, with some of the most significant outbreaks occurring between 1974 and 1976 and from 1997 through 2003. The most recent outbreak impacted pine and pine-hardwood forests across six states in the CHR, leaving behind groups of standing dead trees created by more than 100,000 SPB spots across 405,000 ha. SPB activity in the mountainous areas of the CHR has been conspicuously absent since 2003. We do predict that SPB will return to the CHR within the next 15 years. Without forest management intervention, these forests will likely not return to pine. It is the desire of the Southern Group of State Foresters and USDA Forest Service for many of these stands in southern ecoregions to return to fire adapted forests such as shortleaf- and Table Mountain pine-dominated forests. These pine forests are valuable to the region for the obvious economic reasons, but also for several ecological and aesthetic reasons. The SPB Prevention and Restoration Program was developed in 2003 in order to restore resilient pine stands in areas that had been impacted and to create stands that would be less susceptible to future outbreaks.

References

AWAKE. (2014). *Red-cockaded woodpecker on all wild about Kentucky's environment.* http://www.kentuckyawake.org/Red-Cockaded_Woodpecker. Accessed 30 Jul 2014.

Ayres, M. P., Martinson, S. J., & Friedenberg, N. A. (2011). Southern pine beetle ecology: Populations within stands. In R. N. Coulson & K. D. Klepzig (Eds.), *Southern pine beetle II* (General technical report SRS 140). Asheville: USDA Forest Service Southern Research Station.

Balch, R. E. (1928). *The influence of the southern pine beetle on forest composition in western North Carolina.* Thesis, New York State College of Forestry, Ithaca

Barnett, J. P., Brissette, J. C., & Carlson, W. C. (1986). Artificial regeneration of shortleaf pine. In P. A. Murphy (Ed.), *Proceedings, symposium on the shortleaf pine ecosystem, March 31–April 2 1986.* Monticello: Arkansas Cooperative Extension Service.

Berryman, A. A. (1986). *Forest insects: Principles and practice of population management.* New York: Plenum Press.

Birt, A. (2011). Regional population dynamics. In R. N. Coulson & K. D. Klepzig (Eds.), *Southern pine beetle II* (General technical report SRS 140). Asheville: USDA Forest Service Southern Research Station.

Cassidy, D. (2005). *A southern pine management guide for Tennessee landowners* (PB1751). Knoxville: University of Tennessee.

CFLRP. (2012). *The Grandfather restoration project: A collaborative forest landscape restoration program.* http://www.fs.fed.us/restoration/documents/cflrp/2012Workplans/GrandfatherRestorationWorkPlan2012.pdf. Accessed 25 Sept 2014.

Chamberlin, W. J. (1939). *The bark and timber beetles of North America: North of Mexico.* Corvallis: Oregon State College Cooperative Association.

Clarke, S. R., & Nowak, J. T. (2009). *Southern pine beetle* (Forest insect and disease leaflet 49). Portland: USDA Forest Service Forest Health Protection.

Coulson, R. N., & Klepzig, K. D. (Eds.). (2011). *Southern pine beetle II* (General technical report SRS 140). Asheville: USDA Forest Service Southern Research Station.

Coulson, R. N., & Witter, J. A. (1984). *Forest entomology: Ecology and management.* New York: John Wiley and Sons.

Della-Bianca, L. (1990). Table Mountain pine. In R. M. Burns & B. H. Honkala (Eds.), *Silvics of North America: Vol 1 conifers* (USDA Agriculture handbook 654). Washington, DC: USDA.

Duerr, D. A., & Mistretta, P. A. (2013). Invasive pests: Insects and diseases. In D. N. Wear & J. G. Greis (Eds.), *The southern forest futures project: Technical report* (General technical report SRS 178). Asheville: USDA Forest Service Southern Research Station.

Elliott, K. J., Hendrick, R. L., Major, A. E., Vose, J. M., & Swank, W. T. (1999). Vegetation dynamics after a prescribed fire in the southern Appalachians. *Forest Ecology and Management, 114*, 199–213.

Elliott, K. J., Vose, J. M., Knoepp, J. D., & Clinton, B. D. (2012). Restoration of shortleaf pine (*Pinus echinata*)-hardwood ecosystems severely impacted by the southern pine beetle (*Dendroctonus frontalis*). *Forest Ecology and Management, 274*, 181–200.

Fettig, C. J., Klepzig, K. D., Billings, R. D., Munson, A. S., Nebeker, T. E., Negron, J., & Nowak, J. T. (2007). The effectiveness of vegetation management practices for prevention and control of bark beetle infestations in coniferous forests of the western and southern United States. *Forest Ecology and Management, 238*, 24–53.

Gara, R. I., & Coster, J. E. (1968). Studies on the attack behavior of the southern pine beetle III. Sequence of tree infestation within stands. *Contributions from Boyce Thompson Institute, 24*, 77–86.

Hain, F. P., Duehl, A. J., Gardner, M. J., & Payne, T. L. (2011). Natural history of the southern pine beetle. In R. N. Coulson & K. D. Klepzig (Eds.), *Southern pine beetle II* (General technical report SRS 140). Asheville: USDA Forest Service Southern Research Station.

Hofstetter, R. W. (2011). Mutualists and phoronts of the southern pine beetle. In R. N. Coulson & K. D. Klepzig (Eds.), *Southern pine beetle II* (General technical report SRS 140). Asheville: USDA Forest Service Southern Research Station.

Hopkins, A. D. (1909). *Practical information on the scolytid beetles of North American forests I. Bark beetles of the genus Dendroctonus* (Bull 83 part I). Washington, DC: USDA Bureau of Entomology.

Hopkins, A. D. (1911). *The dying of pines in the southern states: Cause, extent and remedy* (Farmers' bulletin 476). Washington, DC: USDA.

Klepzig, K. D., & Hofstetter, R. W. (2011). From attack to emergence: Interactions between southern pine beetle, mites, microbes and trees. In R. N. Coulson & K. D. Klepzig (Eds.), *Southern pine beetle II* (General technical report SRS 140). Asheville: USDA Forest Service Southern Research Station.

Lafon, C. W., & Kutac, M. J. (2003). Effects of ice storms, southern pine beetle infestations and fire on Table Mountain pine forests of southwestern Virginia. *Physical Geography, 24*, 502–519.

Lawson, E. R. (1986). Natural regeneration of shortleaf pine. In P. A. Murphy (Ed.), *Proceedings symposium on the shortleaf pine ecosystem, March 31–April 2 1986*. Monticello: Arkansas Cooperative Extension Service.

Lawson, E. R. (1990). Shortleaf pine. In R. M. Burns & B. H. Honkala (Eds.), *Silvics of North America: Vol 1 conifers* (USDA Agriculture handbook 654). Washington, DC: USDA.

Lorio, P. L., Jr. (1986). Growth-differentiation balance: A basis for understanding southern pine beetle-tree interactions. *Forest Ecology and Management, 14*, 259–273.

Nebeker, T. E., & Hodges, J. D. (1983). Influence of forestry practices on host-susceptibility to bark beetles. *Journal of Applied Entomology, 96*, 194–208.

Nowak, J. T., Asaro, C., Klepzig, K., & Billings, R. (2008). The southern pine beetle prevention initiative: Working for healthier forests. *Journal of Forestry, 106*, 261–267.

Nowak et al. (in press). Southern pine beetle infestations in relation to forest stand conditions, previous thinning and prescribed burning: Evaluation of the Southern Pine Beetle Prevention Program. Vol. 113.

Olatinwo, R., Guo, Q., Fei, S., Otrosina, W. J., Klepzig, K. D., Streett, D. A., Laird, S., McNab, H., & Liu, Y. (2014). Climate-induced vulnerability to biological threats in the southern United States. In J. R. Vose & K. D. Klepzig (Eds.), *Climate change adaptation and mitigation management options: A guide for natural resource managers in southern forest ecosystems*. Boca Raton: CRC Press.

Price, T. S., Doggett, C., Pye, J. M., & Smith, B. (1998). *A history of southern pine beetle outbreaks in the southeastern United States*. Macon: Georgia Forestry Commission.

Pye, J. M., Price, T. S., Clarke, S. R., & Huggett, R. J. (2008). *A history of southern pine beetle outbreaks in the southeastern United States through 2004*. http://www.srs.fs.usda.gov/econ/data/spb/index.htm. Accessed 28 Jul 2014.

Pye, J. M., Holmes, T. P., Prestemon, J. P., & Wear, D. N. (2011). Economic impacts of the southern pine beetle. In R. N. Coulson & K. D. Klepzig (Eds.), *Southern pine beetle II* (General technical report SRS 140). Asheville: USDA Forest Service Southern Research Station.

Reeve, J. D. (2011). Predators of the southern pine beetle. In R. N. Coulson & K. D. Klepzig (Eds.), *Southern pine beetle II* (General technical report SRS 140). Asheville: USDA Forest Service Southern Research Station.

Saint George, R. A. (1930). Drought-affected and injured trees attractive to bark beetles. *Journal of Economic Entomology, 23*, 825–828.

Saint George, R. A., & Beal, J. A. (1929). *The southern pine beetle: A serious enemy of pine in the South* (Farmers' bulletin 1586). Washington, DC: US Department of Agriculture.

Tchakerian, M. D., & Coulson, R. N. (2011). Ecological impacts of southern pine beetle. In R. N. Coulson & K. D. Klepzig (Eds.), *Southern pine beetle II* (General technical report SRS 140). Asheville: USDA Forest Service Southern Research Station.

Thatcher, R. C., & Barry, P. J. (1982). *Southern pine beetle* (Forest insect and disease leaflet 49). Washington, DC: USDA Forest Service.

Thistle, H. W., Peterson, H., Allwine, G., Lamb, B., Strand, T., Holsten, E. H., & Shea, P. J. (2004). Surrogate pheromone plumes in three forest trunk spaces: Composite statistics and case studies. *Forest Science, 50*, 610–625.

Tran, J. K., Ylioja, T., Billings, R., Régnière, J., & Ayres, M. P. (2007). Impact of minimum winter temperatures on the population dynamics of *Dendroctonus frontalis* (Coleoptera: Scolytinae). *Ecological Applications, 17*, 882–899.

Ungerer, M. J., Ayers, M. P., & Lombardero, M. J. (1999). Climate and the northern distribution limits of *Dendroctonus frontalis* Zimmermann (Coleoptera: Scolytidae). *Journal of Biogeography, 26*, 1133–1145.

USDA Forest Service. (2014). *Southern pine beetle prevention program*. www.fs.fed.us/r8/fhp. Accessed 30 Jul 2014.

Wagner, T. L., Gagne, J. A., Sharpe, P. J. H., & Coulson, R. N. (1984). Effects of constant temperature on longevity of adult southern pine beetles, *Dendroctonus frontalis* Coleoptera: Scolytidae. *Environmental Entomology, 13*, 1125–1130.

Wear, D. N., & Greis, J. G. (2013). *The southern forest futures project: Technical report* (General technical report SRS 178). Asheville: USDA Forest Service, Southern Research Station.

Welch, N. T., Waldrop, T. A., & Buckner, E. R. (2000). Response of southern Appalachian Table Mountain pine (*Pinus pungens*) and pitch pine (*P. Rigida*) stands to prescribed burning. *Forest Ecology and Management, 136*, 185–197.

Williams, C. E., & Johnson, W. C. (1990). Age structure and the maintenance of *Pinus pungens* in pine-oak forests of southwestern Virginia. *The American Midland Naturalist, 124*, 130–141.

Xi, W., Waldron, J. D., Cairns, D. M., Lafon, C. W., Birt, A. G., Tchakerian, M. D., Klepzig, K. D., & Coulson, R. N. (2012). Restoration of southern pine forests after the southern pine beetle. In J., Stanturf & others (Eds.), *A goal-oriented approach to forest landscape level restoration*, Dordrecht: Springer Science+Business Media.

Zimmermann, C. (1868). Synopsis of the Scolytidae of America north of Mexico, with notes and an appendix by JL Leconte. *American Entomological Society Transactions, 2*, 149.

Zobel, D. (1969). Factors affecting the distribution of *Pinus pungens*, an Appalachian endemic. *Ecological Monographs, 39*, 304–333.

Chapter 5
First Steps Toward Defining the Wind Disturbance Regime in Central Hardwoods Forests

Chris J. Peterson, Jeffery B. Cannon, and Christopher M. Godfrey

Abstract Wind disturbance is one of the most prevalent natural disturbances in the Central Hardwoods Region (CHR). All ecoregions within the CHR are subject to a greater or lesser degree to tornado, derecho or thunderstorm wind damage, with an east-to-west increase in the importance of tornadoes and derechos. At the regional scale, hurricanes decrease in importance with distance from the Atlantic and Gulf coasts. The disturbed patch sizes created by these various storms include occasional very large (e.g., >25 ha) patches, but the great majority are a few ha or less, perhaps differing from common visual impressions. Hurricane and derecho disturbance patterns appear to be more predictable in relation to topographic features, whereas tornado damage is much more stochastic. All wind disturbance types cause greater damage to larger trees, and most studies reveal interspecific differences in levels of wind damage, although such patterns are not always consistent among studies or locations. Wind disturbance commonly advances succession in second-growth forest but may set succession back in primary forests. The greatest research needs are landscape-scale patterns of damage; relationships of damage to topography and soils; clarifying the tree characteristics (e.g., architecture, wood strength, rooting depth) that underlie interspecific differences in vulnerability; and documenting ecosystem effects of wind disturbance.

Keywords Wind • Blowdown • Tornado • Hurricane • Disturbance regime

---

C.J. Peterson (✉) • J.B. Cannon
Department of Plant Biology, University of Georgia,
2502 Miller Plant Science Building, Athens, GA 30602, USA
e-mail: chris@plantbio.uga.edu; jbcannon@plantbio.uga.edu

C.M. Godfrey
Department of Atmospheric Sciences, University of North Carolina at Asheville,
231 Robinson Hall, Asheville, NC 28804, USA
e-mail: cgodfrey@unca.edu

C.H. Greenberg, B.S. Collins (eds.), *Natural Disturbances and Historic Range of Variation*, Managing Forest Ecosystems 32,
DOI 10.1007/978-3-319-21527-3_5

5.1 Introduction

Wind may be the most widespread agent of natural disturbance in the Central Hardwoods Region (CHR), as well as across temperate forests of eastern North America (White 1979; Everham and Brokaw 1996). This pervasiveness may increase or decrease over time in a nonstationary climate, though just how climate changes may alter storm size, frequency, or severity is a topic of active discussion (Trapp et al. 2007; Grinsted et al. 2013; Dale et al. Chap. 14). Nevertheless, the recent shift towards forest management modeled on natural disturbance regimes requires quantitative information on characteristics of the wind disturbance regime (e.g., Perera et al. 2004).

In this chapter, we seek to present a synthesis of wind disturbance characteristics in the CHR (Fig. 5.1; see Greenberg et al. Chap. 1, Fig. 1.1 for CHR ecoregions map), as well as the immediate impacts of these disturbances on forest structure and composition. Due to space constraints, we are not examining the longer-term patterns of regrowth and recovery following wind disturbance (e.g., Everham and Brokaw 1996; Webb 1999; Busing et al. 2009). Xi and Peet (2011) propose that the temporal scale of research be categorized as immediate (a few months to 1 year), short-term (several months to several years) and long-term (a decade to centuries); within this rubric, we will focus mostly on immediate effects, with just a few comments on longer-term effects in light of possible effects of storms on forest succession. We consider forest damage from hurricanes, tornadoes, derechos, and other types of wind event.

We should begin with a note on the organization of this chapter. The meteorology and climatology, as well as broad-scale patterns in wind event frequency and severity are distinctive for the different types of wind storms, so we organize around types of storm for these topics. In contrast, where we discuss damage patterns to forests at the stand and individual scale, we organize around forest characteristics or response variables, because at the finer scales the differences among types of storm diminish while the differences among response variables remain substantial. Further, we will attempt to highlight CHR forests to the degree possible, but will of necessity also include research from outside the CHR (Fig. 5.1). This is necessary because of the limited number of studies within the CHR, in comparison to many studies in northern hardwoods and sub-boreal forest types to the north, and coastal plain and other forests to the south. As with any active area of research, we will note areas in which research, and thus understanding, is scant.

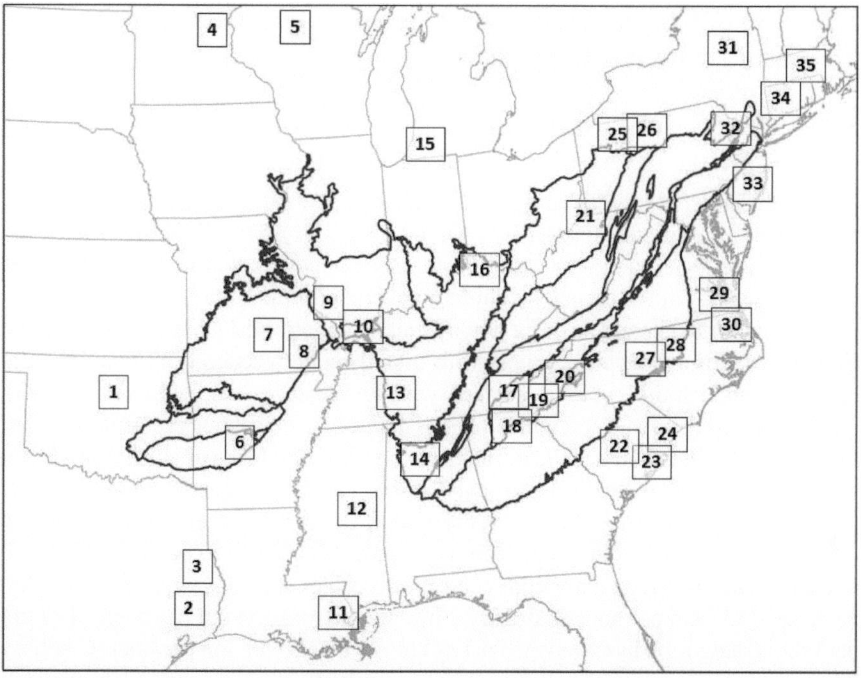

Fig. 5.1 Geographic distribution of wind disturbance studies in and near the CHR; ecoregions delineated with *dark black* lines (see Greenberg et al. Chap. 1, Fig 1.1 for ecoregion names). Numbers denote study locations, roughly from west to east, and correspond with citation number. Some citations (e.g., Peterson 2007) are listed more than once because the study reports findings from multiple locations. *1* – Myster and Malahy 2010; *2* – Glitzenstein and Harcombe 1988; *3* – Harcombe et al. 2009; *4* – Palmer et al. 2000; Arévalo et al. 2000; Allen et al. 2012; *5* – Dunn et al. 1983; Lang et al. 2009; *6* – Phillips et al. 2008; *7* – Rebertus and Meier 2001; *8* – Peterson and Rebertus 1997; *9* – Holzmueller et al. 2012; *10* – Nelson et al. 2008; *11* – Chapman et al. 2008; *12* – Kupfer et al. 2008; *13* – Peterson and Leach 2008a, b; *14* – Cowden et al. 2014; *15* – Brewer and Merritt 1978; *16* – Held and Winstead 1976; Held and Bryant 1989; Held et al. 1998; *17* – Cannon et al. unpubl; *18* – Cannon et al. unpubl; Sobhani et al. 2014; *19* – Elliott et al. 2002; *20* – Greenberg and McNab 1998; McNab et al. 2004; *21* – Rentch 2010; *22* – Putz and Sharitz 1991; Battaglia et al. 1999; *23* – Duever and McCollum 1993; *24* – Hook et al. 1991; Gresham et al. 1991; *25* – Peterson et al. 1990; Peterson and Pickett 1991; Peterson 2000; *26* – Evans et al. 2007; Peterson et al. 2013; *27* – Busing et al. 2009; *28* – Xi et al. 2008a, b; *29* – Prengaman et al. 2008; Kribel and Ware 2014; *30* – Laing et al. 2011; *31* – Robinson and Zappieri 1999; Canham et al. 2001; *32* – Peterson 2007; Sobhani et al. 2014; *33* – Matlack et al. 1993; *34* – Peterson 2007; *35* – Foster 1988; Foster and Boose 1992

5.2 Storm Characteristics and Geographic Distribution; Landscape-Scale Patterns

5.2.1 *Hurricane Disturbance*

Hurricanes are certainly the largest and probably best-studied agent of forest wind disturbance (Foster and Boose 1995; Xi and Peet 2011). The terms 'hurricane' and 'typhoon' are local names for strong tropical cyclones; they occur over all of the world's oceans, but are especially common over the tropical north Atlantic. A tropical cyclone is the generic term for a non-frontal synoptic scale low-pressure system over tropical or sub-tropical waters with organized convection (i.e., thunderstorm activity) and definite cyclonic surface wind circulation. Tropical cyclones with maximum sustained surface winds of less than 63 km per hour are usually called 'tropical depressions.' Once the tropical cyclone reaches winds of at least 63 km per hour they are typically called a 'tropical storm' and are assigned a name. If winds reach 119 km per hour, then they are called (in North America) a 'hurricane' (Landsea 2014). In the USA, hurricanes are ranked according to the Saffir-Simpson intensity scale, based on maximum wind speeds: Category 1 has winds from 119 km per hour – 153 km per hour; Category 2 has winds from 154 km per hour – 177 km per hour; Category 3 has winds from 179 km per hour – 208 km per hour; Category 4 has winds from 209 km per hour – 156 km per hour; and Category 5 has winds in excess of 253 km per hour.

The physical structure of a hurricane consists of a central circular 'eye' typically several tens of km in diameter with relatively calm winds, clear skies and low pressure, immediately surrounded by the eyewall, which is the location of intense precipitation and the highest wind speeds. Further out, multiple arms of the vortex carry rain bands that can deposit months-worth of rain within a few hours. Although some hurricanes can be compact with heavy damage restricted to areas within 30–50 km of the path of the eye, some are much larger, with damage extending out to 300 km or more from the eye (Rauber et al. 2002). Due to their extensive size, damage from hurricanes can spread over an exceptionally large area: Hurricane Katrina (2005) damaged roughly 2 million ha of timberland (Stanturf et al. 2007). Earlier, Hurricane Hugo (1989) caused damage to 1,821,000 ha in South Carolina and 1,092,600 ha in North Carolina (Xi and Peet 2011).

Hurricane intensity may be evaluated on the basis of either low atmospheric pressure (measured in millibars), or maximum wind speeds. The most intense Atlantic basin hurricanes to strike the mainland USA have been Wilma (2005, 882 millibars), and Hurricanes Camille (1969) and Allen (1980), both of which produced wind gusts to 306 km per hour (Landsea 2014). There is apparently little or no correlation between hurricane size and intensity (Landsea 2014).

5.2.1.1 Formation and Movement

A hurricane is a vast rotating vortex that is powered by warm ocean surface waters; thus hurricanes rapidly lose intensity once they move over large land masses, a process that is further enhanced by friction with the rougher surface of land masses. The result is that such storms typically decrease to tropical storm and then to tropical depression intensity as they move further inland (Foster and Boose 1995).

Formation of a tropical cyclone requires several conditions: (a) warm ocean surface waters (at least 26.7 °C); (b) an atmosphere which cools fast enough with height such that it is potentially unstable to moist convection; (c) relatively moist layers near the mid-troposphere; and (d) low values (less than about 37 km per hour) of vertical wind shear. All this must take place at least 300 km from the equator in order for the Coriolis effect to influence the force balance of the wind. For Atlantic basin hurricanes, these conditions are usually met in one of three ways. Some tropical cyclones develop from convection within the Intertropical Convergence Zone (ITCZ), a region of persistent thunderstorm activity near the equator that corresponds with the convergence of the equatorward surface branch of the Hadley circulation in both the northern and southern hemispheres. Tropical cyclones may also form from stalled cold fronts that provide a lifting mechanism for convection in the oceans. By far the most common formation mechanism involves tropical easterly waves. These troughs of low pressure and associated regions of surface convergence result from previous convection over Africa and move westward along the easterly trade winds. Once convection initiates through any of these formation mechanisms, latent heat release from water vapor condensation in the atmosphere induces a low pressure center at the surface, which increases the surface convergence, enhancing lift, producing more thunderstorms, and so on in a positive feedback cycle. If a collection of thunderstorms does in fact organize into a tropical cyclone in this area, the system may last for several days or even 2–3 weeks as it moves westward and subsequently often northward.

Hurricanes that form over the subtropical north Atlantic ocean are initially pushed westward by the easterly trade winds, but as they near the Caribbean and/or North American continent, they eventually get pushed towards the northeast, causing many to have a parabolic overall track (Rauber et al. 2002). A weak northerly push occurs if the Bermuda high is strong and directs hurricanes into the Gulf of Mexico; a stronger push occurs if the Bermuda high is weak, and directs storms back into the central Atlantic. If the Bermuda high is intermediate in strength, the moderate push steers storms along the eastern seaboard of North America.

5.2.1.2 Climatology

Atlantic hurricanes form predominantly from July through October. The average number of hurricanes that make landfall on the USA coast follow a monthly pattern: June – 0.12; July – 0.15; August – 0.48; September – 0.67; October – 0.33; November – 0.03 (average number of hurricanes per year for the month). Their long-term trends in frequency, size and intensity have been the subject of much scrutiny, particularly because of interest in how these storm characteristics may change in the future. It is clear that hurricane occurrence fluctuates on decadal and perhaps centennial time scales (Zeng et al. 2009), with the period 1870–1900 being a time of somewhat greater hurricane activity along the eastern USA coast than most of the twentieth century.

On average, 6.2 systems per year reach hurricane force in the Atlantic basin, with a record high of 15 (2005) and low of 2 (2013). Not all of these storms make landfall, and some of those that do strike islands of the Caribbean or along the Central American coast. On average, 1.7 landfalling hurricanes strike the continental USA each year, with a high of seven in 1886 and lows of zero in numerous years (Landsea 2014). For the entire USA East and Gulf coasts, a total of 209 hurricanes have made landfall between 1901 and 2012; of these, 79 were Category 1, 49 were Category 2, 62 were Category 3, 16 were Category 4, and 3 were Category 5 (NOAA/NCDC undated a).

Different portions of the USA Atlantic and Gulf coasts have quite different risk of being hit by hurricanes. Along the southern Atlantic coast, risk is greatest along the coast of North Carolina (every 5–7 years), least on the Georgia and northern Florida coasts (every 10–13 years), and much higher again along the south (east) Florida coast (Fig. 5.2). In the Gulf, risk is greatest for sections of coastal Louisiana along the Mississippi River delta (once every 7–8 years, on average), but quickly decreases a short distance away. Coastal areas of Georgia, northern Florida and southern South Carolina have a comparatively lower frequency of hurricanes; this area comprises the South Atlantic Bight, which due to its concave shape and broad

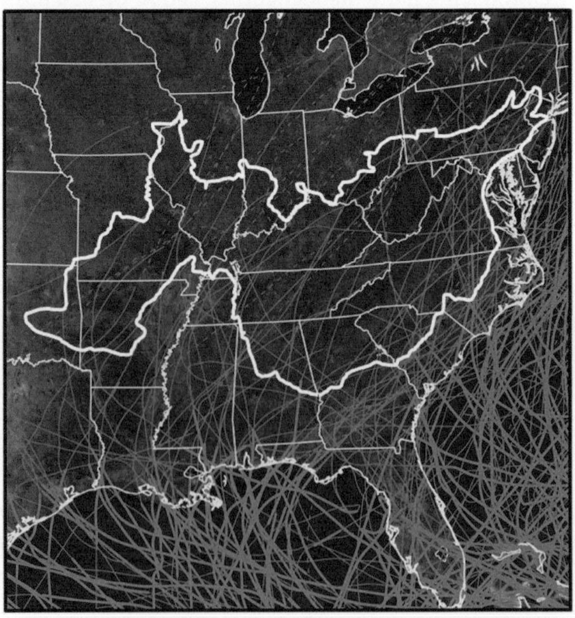

Fig. 5.2 All North Atlantic major Hurricanes (At least category 3 on the Saffir-Simpson hurricane scale), 1851–2013. Figure from NOAA Coastal Services Center 'Historical Hurricane Tracks' online tool. Heavy *white* line delineates boundary of CHR

sloping continental shelf, experiences far fewer landfalling hurricanes than areas just to the north or south.

The loss of storm intensity as a system moves over land has important implications for the geographic distribution of hurricane damage to forests (Xi et al. 2008a, b; Xi and Peet 2011), most notably that CHR forests will rarely experience a hurricane in its initial full intensity (see Fig. 5.2; dotted lines indicate storm tracks after intensity has dropped to tropical storm intensity). Maritime and lower Coastal Plain areas of the southeastern USA bear the full or substantial brunt of hurricanes, but by the time they reach the southern Appalachians, hurricane-force winds have usually diminished to tropical depression levels. In the study by Zeng et al. (2009) all areas with an expected hurricane return interval (hurricane-force winds) of <10 years were within 20 km of the coast, and all areas with an expected return interval of <100 years were within 50 km of the coast. Table 5.1 presents hurricane frequencies for several ecoregions within the CHR, broken down by storm intensity category. Several trends are immediately apparent. First, none of the CHR ecoregions experience Category 4 or Category 5 hurricane impacts, and the only ecoregion to experience Category 3 impacts (very rarely) is the Piedmont. However, even the Piedmont is seldom hit by Category 2 or Category 3 hurricanes (4 and 2 times, respectively, over the 172 years of the data archives). Moreover, Category 1 hurricanes have never been recorded in 8 of the 13 CHR ecoregions. Conversely, all ecoregions within the CHR do experience tropical storms to a greater or lesser degree; these are more often extreme precipitation than extreme wind events. Tropical storm frequencies are highest in the Blue Ridge Mountains, Northern Piedmont, and Boston

Table 5.1 Hurricane frequencies among ecoregions of the CHR, by intensity category. Frequencies are based on number of hurricane (eye) tracks that passed through or within 50 km of each ecoregion during the period 1842–2013

	Hurricane intensity category			
Ecoregion (area in km^2)	Trop. storm	Cat. 1	Cat. 2	Cat. 3
Ouachita Mountains (26,896)	1.96 (9)	0	0	0
Arkansas Valley (28,421)	1.85 (9)	0	0	0
Boston Mountains (14,178)	2.48 (6)	0	0	0
Ozark Highlands (106,391)	0.38 (7)	0	0	0
Piedmont (166,117)	1.68 (48)	0.35 (10)	0.14 (4)	0.07 (2)
Northern Piedmont (30,459)	2.50 (13)	0.57 (3)	0.19 (1)	0
Blue Ridge Mountains (46,595)	2.01 (16)	0.50 (4)	0.12 (1)	0
Ridge and Valley (115,483)	1.57 (31)	0.25 (5)	0	0
Southwestern Appalachians (37,994)	1.54 (10)	0.15 (1)	0	0
Central Appalachians (62,050)	0.47 (5)	0	0	0
Western Allegheny Plateau (81,440)	0.29 (4)	0	0	0
Interior Plateau (123,523)	0.52 (11)	0	0	0
Interior River Valleys (120,405)	0.15 (3)	0	0	0

Values are frequency per century per 10,000 km^2. Values in parentheses are raw (not adjusted for ecoregion area or unit time) number of hurricanes during the 172 years per ecoregion and category

Mountains ecoregions (Table 5.1), although the latter is probably an anomaly due to the very small area of the Boston Mountains ecoregion, compared to most others within the CHR.

5.2.1.3 Landscape-Scale Patterns

The landscape-scale distribution of hurricane damage to forests is best understood based on storm size coupled with the counter-clockwise rotating vortex storm structure. From these factors, several expectations can be derived, which have mostly been confirmed in empirical studies. First, because of the combination of both rotational and translational (forward movement) winds, storm intensity is greatest on the right side of a forward-moving hurricane; on the left side, the realized wind velocities are the difference between rotational and translational speed. Therefore, damage is usually substantially greater to the right of the path of the eye (Foster and Boose 1995; Dahal et al. 2014).

Second, damage severity is usually distributed in a 'nested' pattern, with most severe forest damage in the vicinity of the eyewall track, and lesser severities at increasing distances from the track (Stanturf et al. 2007; Xi et al. 2008b). This is driven by decreasing wind speeds at greater distances from the hurricane's eye. In their study of landscape- and regional-scale influences on forest damage from Hurricanes Hugo and Fran in North and South Carolina, Xi et al. (2008b) documented a significant decrease in damage with increasing distance from the eye.

Third, perhaps most prominent is that windward and leeward topographic positions are easily recognized given the size and location of a hurricane; forest damage is often greater on the windward positions than the leeward positions (Boose et al. 1994; Negron-Juarez et al. 2014). McNab et al. (2004) found that the remnants of Hurricane Opal created intermediate-sized gaps most frequently on southeasterly aspects at the Bent Creek Experimental Forest (BCEF) watershed in the southern Appalachians near Asheville, North Carolina. Xi et al. (2008b) also found that damage from Hurricane Fran at Duke Forest increased with exposure and slope, although the influence of aspect was not significant in that study, perhaps because the more modest terrain in the Piedmont diminished the dramatic differences in aspect that occur in the mountains. A corollary of this is that flat or gently rolling landscapes may have few sheltered, leeward positions, and consequently, forest damage may be much more widespread than in landscapes with mountains or other rugged topography that might provide some sheltered locations (Boose et al. 1994).

There are also elevational effects on forest damage in hurricanes, but they are complex. For example, Xi et al. (2008b) reported greatest damage from Hurricane Fran on ridgetops, but also found that valley bottoms sometimes were severely disturbed. McNab et al. (2004) found that localized gaps created by Hurricane Opal at the BCEF in western North Carolina were much more likely at lower elevations, with few damaged patches on ridges.

Disturbed patch size after hurricanes can be quite large in areas closer to the coast, but small blowdown patches are much more frequent. At Duke Forest

(36.017° N, 78.997° W) after Hurricane Fran, Carpino (1998) found that most patches of severe damage were ~0.2 ha in size, and the largest gaps reported by Xi et al. (2008a) were only 0.8 ha. McNab et al. (2004) documented gaps created by the remnants of Hurricane Opal with a mean size of 0.7 ha, ranging from 0.1 to 3.9 ha, at the BCEF in western North Carolina. At the small end of the size scale, scattered treefalls created by hurricanes are very similar to small gaps created by other processes (Hart Chap. 2). Nevertheless, even further inland the occasional large gap can be formed from hurricane remnants; Elliott et al. (2002) studied regeneration in a 10 ha hurricane-created gap at the Coweeta Hydrologic Laboratory in western North Carolina, 400 km from the coast.

More than a few km from the Atlantic or Gulf shoreline, most of the forest damage caused by hurricanes is directly caused by winds, although the sometimes extreme quantities of precipitation can saturate soils and cause trees to lose stability, thereby increasing the amount of damage caused by a given wind velocity. Whereas this seemingly straightforward pattern is often seen, it is not universal. Some hurricanes add relatively little precipitation prior to the impacts of the highest winds, leaving the soil less saturated and tree stability uncompromised (e.g., Hurricane Hugo 1989, in Xi et al. 2008a). In other cases, bottomland and swamp tree species (e.g., Putz and Sharitz 1991; Oswalt and Oswalt 2008; Harcombe et al. 2009) that under normal conditions grow in wet or saturated soils, may not be greatly weakened by the large amounts of precipitation in a hurricane, and therefore experience less damage.

5.2.2 *Tornado Disturbance*

Tornadoes are mesoscale convective phenomena that harbor the strongest known winds of all types of storms. Very powerful tornadoes have winds in excess of 375 km per hour, and individual suction vortices within a large tornado may exceed even these velocities. Countering these extreme wind velocities, individual tornadoes are far smaller than other types of wind event, with damage swaths ranging from 50 m to 4 km wide, and a few to several dozen km long; in rare cases, though, exceptionally long-lived tornadoes may have damage tracks that exceed 200 km in length (Rauber et al. 2002). Tornadoes occur on all continents, although North America is home to far greater frequencies than anywhere else on earth; the average annual number of tornadoes in the continental USA is 1,253 (NOAA/NCDC undated b), with most of those located in the eastern USA (Fig. 5.3). Tornadoes may occur singly, in small clusters, or in very large outbreaks; the infamous April 26–29, 2011 outbreak in the southeastern USA produced > 300 tornadoes, and claimed 322 lives.

In the USA, tornado intensity was classified from the 1970s through 2007 according to the Fujita scale (or F-scale). In this system, levels of damage to a variety of structures or objects were used to assign an F-scale value of 0 through 5 to any tornado, with 0 indicating the least damage and weakest tornadoes, and 5 indicating the most extreme damage and strongest tornadoes (Fujita 1971). In 2007, an

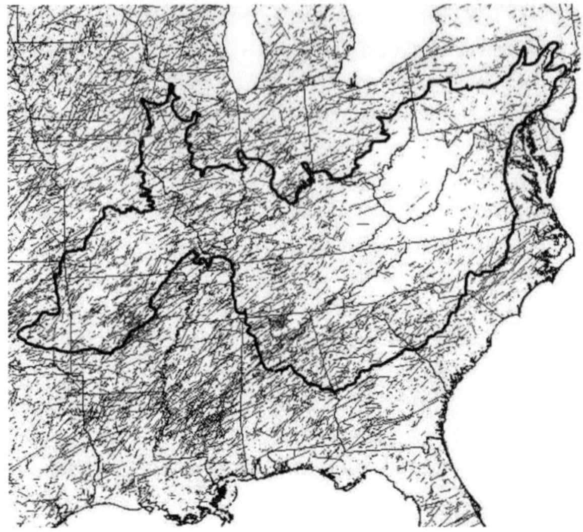

Fig. 5.3 All known tornado tracks in the eastern USA (1950–2013) with the CHR delineated in *dark* outline (Map from NOAA Storm Prediction Center's SVRGIS)

improved damage rating scale, the Enhanced Fujita scale or EF-scale, was put into practice across the United States (WSEC 2006). The EF-scale modified the wind speed estimates used in the original F-scale, and included a broader range of damage indicators (the structures or objects whose damage is used to derive the EF-scale classification). Notably, the EF-scale includes two damage indicators (DI 27 and DI 28) based on trees, but very recently, several publications have begun to point out the limitations of the tree-related DIs (Frelich and Ostuno 2012; Edwards et al. 2013). For purposes of this review, it is important to note that an entire tornado track is given a single EF-scale rating based on the most severe damage recorded in that damage track (the maximum damage); thus, a single EF-scale rating does not necessarily characterize damage along the entire track.

5.2.2.1 Formation and Movement

Tornadoes can form in a variety of circumstances, but most – especially strong tornadoes – are associated with supercell thunderstorms. Supercell thunderstorms are thunderstorms with a rotating updraft and one or more downdrafts. These storms generally require an environment with strong vertical wind shear, plus the ingredients for general thunderstorm formation, which include sufficient low-level moisture, a conditionally unstable atmosphere, and an initial lifting mechanism. Individual tornadoes may descend from a small fraction of well-developed supercell storms. The mechanisms for tornadogenesis are varied and remain a subject of ongoing research. Davies-Jones et al. (2001) review a number of these formation mechanisms in detail.

Typically, vortex rotation in tornadoes is counter-clockwise, although up to 10 % rotate clockwise. The flow field will usually contain both a tangential and a radial component. In many cases, the inward-directed radial component contributes significantly to the overall wind speed within the tornadic vortex in order to balance the rapidly rising air near the core of the tornado (Wurman et al. 2013). Indeed, studies have hinted that in some tornadic systems, a substantial fraction of the forest damage is caused by this inflow rather than by the spinning vortex (e.g., Karstens et al. 2013). The rotational velocity waxes and wanes in stochastic fashion during the brief life of a tornado. Indeed, the short periods in which vortex rotation is slow and/or very compact may produce undamaged areas in many tornado paths, giving the appearance that the vortex has lifted off the ground. However, current thinking among tornado researchers is that such 'skipping' does not actually occur, and that many historical long-track tornadoes that appeared to have gaps were either several serial vortices, or a single one that was occasionally too small or weak to leave a damage signature.

Translational movement can proceed in any direction. However, because tornadoes follow the motion of parent thunderstorms, which typically follow the general west to east motion of the mid-latitude westerlies in North America, most (roughly 70 %) tornado track movement is typically toward the east, northeast, or east-northeast. Forward speed can range from stationary to 100 km per hour (Suckling and Ashley 2006).

The small size of tornado vortices at ground level, combined with their incredible wind velocities, often results in very steep spatial gradients in wind speed. This is frequently seen in post-storm damage surveys, either of forests or of built-up areas, where very heavy damage and very light or no damage may be separated by only 10 or 20 m (Foster et al. 1998). Indeed, aerial damage surveys of residential neighborhoods sometimes reveal large single-family homes that are devastated on one side but entirely intact on the other side (Peterson and Godfrey 2014).

Somewhat distinct from other types of tornadoes are those that form within the spiral rainbands of landfalling hurricanes and are often called 'hurricane-induced tornadoes' (Agee and Hendricks 2011). These phenomena are concentrated within 400–500 km of the Atlantic or Gulf coasts. Hurricane-induced tornadoes are usually small and weak, having path lengths and widths that are substantially less than lengths and widths for supercell tornadoes (Moore and Dixon 2011). Slightly more than 80 % of landfalling Gulf coast hurricanes produce hurricane-tornadoes, with a mean of 12.2 ± 18.8 (SD) per hurricane. Moore and Dixon (2011) reported 734 tornadoes produced by 50 hurricanes during the 55-year period from 1950–2005. Of these, 42 % were F0, 36 % were F1, 17 % were F2, 4 % were F3, and 0.4 % were F4 tornadoes.

5.2.2.2 Climatology

Systematic recording of standardized data on tornadoes by the National Weather Service (NWS) began in the 1950s, although anecdotal records of course go much further back. Even in the 75 years of systematic record-keeping changes in reporting, detection, classification, and prediction mean that use of the long-term records requires caveats. Nevertheless, the NWS tornado database does contain information on >40,000 tornadoes across the entire USA since 1950, and will therefore be our primary information source on climatology and geographic distribution.

Tornado climatology is well-established. Although tornadoes can and do occur in any month of the year in the southeastern USA, the peak period is April through July. However, the period of peak tornado activity varies geographically. Peak activity is earliest in the southeastern and southern plains states, and activity is progressively later further north. Thus the Gulf states have their most frequent tornadoes in March and April, whereas tornadoes in Minnesota and the Dakotas are most frequent in July. There is a secondary peak in the Gulf coast states in late fall. Diurnal distribution of tornadoes shows that they are concentrated in the late afternoon and early evening (Rauber et al. 2002).

Nationally, the majority of tornadoes are 'weak' (EF0 and EF1), a modest fraction are 'strong' (EF2 and EF3), and only a few are 'devastating' (EF4 and EF5). Within the CHR 4,358 tornadoes were recorded during 1993–2012 (Fig. 5.4), and were distributed among intensity classes as: EF0 – 42.1 %; EF1 – 38.2 %; EF2 – 14.4 %; EF3 – 4.4 %; EF4 – 0.8 %; and EF5 – 0.01 %. Thus for the CHR as a whole, an average of 218 tornadoes touch down annually.

Tornadoes have been documented in all states of the USA, but they are by far more common east of the Rocky Mountains (Fig. 5.3). Tornado occurrence is also markedly lower in the Appalachian sections of West Virginia and Virginia, the upper peninsula of Michigan, and northern New England. The greatest concentrations of tornadoes in historical times has been in two areas: the widely-known 'tornado alley' running from Oklahoma, through Kansas, Nebraska and Iowa, and

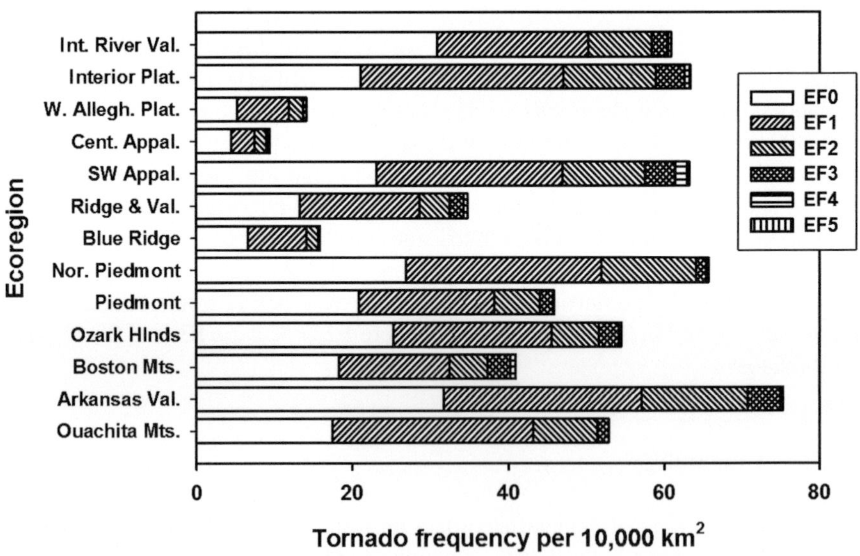

Fig. 5.4 Frequency of tornado touchdowns in CHR ecoregions by EF-scale rating (F-scale prior to 2007) (1993–2012)

the less-well-known 'Dixie alley,' consisting of Arkansas, northern Louisiana, Mississippi and Alabama.

The average numbers of tornadoes for the CHR as a whole is useful to compare to other forest types, but such numbers belie substantial variation among ecoregions within the CHR (Fig. 5.4). During the 20-year period of 1993–2012 (more accurate than the entire historical tornado record, because weather radars came into widespread use in the early 1990s and therefore probably revealed tornadoes that would have previously gone undetected or unreported) the highest tornado density was in the Arkansas Valley (75.3 per 10,000 km^2) and Northern Piedmont (65.7 per 10,000 km^2), and the lowest density was in the Central Appalachians (9.5 per 10,000 km^2) and Western Allegheny Plateau (14.1 per 10,000 km^2). Thus, different ecoregions within central hardwood forests experience approximately an 8-fold difference in frequency of tornado touchdown. Even these numbers do not fully convey risk of severe disturbance, which of course has the greatest ecological effects. Considering only strong and devastating tornadoes (EF2 and greater) for the 20 years of 1993–2012, the highest frequency is found in Arkansas Valley (18.29 per 10,000 km^2), the Interior Plateau (16.35 per 10,000 km^2), and Southwestern Appalachians (16.32 per 10,000 km^2). These three ecoregions are also those with the highest proportion of all tornadoes in the EF2 or greater category.

Distributions of tornado path lengths and widths were recently synthesized by Brooks (2004) for the period 1950–2001, and by Malamud and Turcotte (2012, length only) for the period 1982–2011, who found that both path length and path width increase substantially with tornado intensity, albeit with substantial variation in the relationship. Mean path length in Malamud and Turcotte's study increased from F0 to F5 tornadoes, respectively, as follows: 1.5, 5.4, 12.1, 25.3, 44.3, and 64.4 km. Brooks (2004) also showed that path width values increased from a mean of 28 m for F0 tornadoes, to 555 m for F5 tornadoes. Note that these and most other published analyses of tornado path length and width are based on the lengths and widths reported by the NWS storm damage survey teams. Beginning in 1994, widths were reported as the maximum width, whereas prior to 1994, the mean width was recorded. Moreover, it has become clear to the first author that the storm damage survey teams may consider the tornado beginning and ending points to be the very first or last tree or structure that exhibits any damage, even if the more representative levels of tornado damage are hundreds of m away. This likely leads to a substantial overestimation of path lengths for our purposes here (namely to develop some idea of the area of forest damaged, how serious the damage is, and when and where it happens). The upshot is that for forestry or ecology purposes, calculations of the area affected would likely be severely overestimated if based on the length and (maximum) width values drawn from the NWS tornado database.

As a preliminary step toward remedying the data limitations just described, we have begun to quantify tornado damage path widths by directly measuring the width of the damage at regular intervals (currently either every 200 m or every 500 m) along the entirety of a tornado path. We have results from 21 tornadoes to date. In each case, we measured damage width perpendicular to the overall trajectory of the tornado. Of necessity, we considered 'visually substantial' canopy damage, which

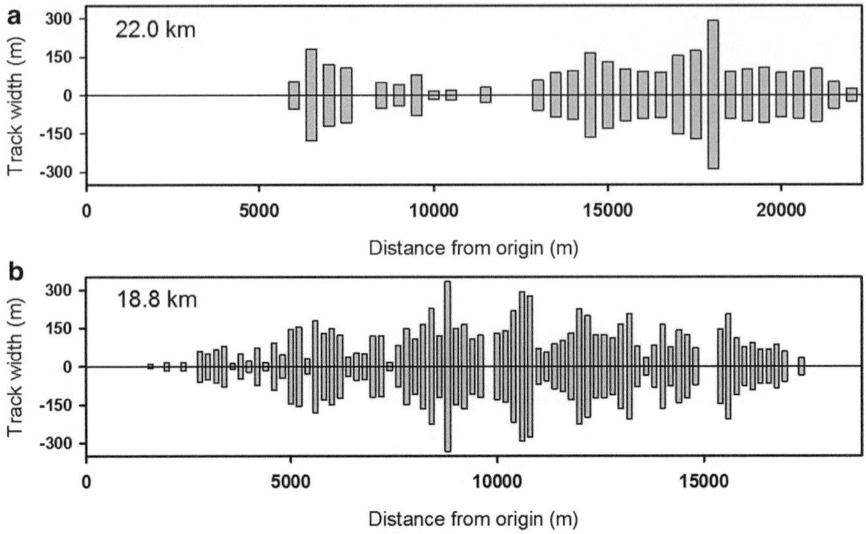

Fig. 5.5 Variation in damage track width in two 2012 tornadoes that passed predominantly through forested landscapes. Width measurements taken every 200 m (**b**), or every 500 m (**a**). Absence of a *vertical* bar at a given distance from origin indicates zero damage width at that point

roughly translates to >25 % of canopy removal by the tornado. Nonforested segments of the tornado path were excluded. Locations where there was no visible damage were recorded as zero width. Figure 5.5 shows two representative tornado tracks with reported track lengths of 22.0 and 18.8 km, respectively; note that width varies greatly along the damage path, with several episodes of widening and narrowing. Note also that at several of our width-sampling points, there was no discernable path, leading us to record a width of zero. Figure 5.6 then presents a summary of our width measurements on 21 tornado tracks, in comparison to the reported path width from the NWS storm damage surveys. Clearly the reported width greatly exceeded the measured width in all cases, sometimes being two – threefold greater than the mean of our width measurements. Mean measured damage path width varied from roughly 100 m to roughly 500 m in our sample.

5.2.2.3 Landscape-Scale Patterns

In contrast to hurricanes, there has been very little research to date on landscape-scale pattern of tornado disturbances in the CHR, or any other regions, in fact. As with several other topics in this chapter, we draw as necessary from studies outside the CHR when patterns and trends are likely to be similar. Foster et al. (1998) pointed out that in general, tornado disturbances are much more linear than other types of large infrequent disturbance, and cover much smaller areas per event than hurricanes or volcanic eruptions. They further noted that although tornadoes are

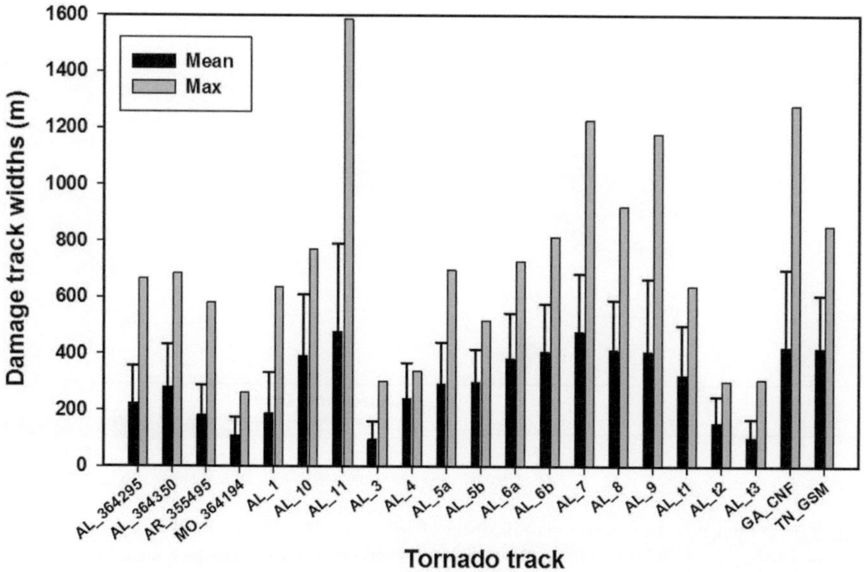

Fig. 5.6 Mean (+SD) (Measured from post-tornado inspection of Google Earth imagery, or our own GIS analyses of aerial photos) and maximum (reported in National Weather Service storm surveys) tornado track widths, for 21 tornado tracks in Georgia, Alabama and Tennessee, formed on 27 April, 2011. X-axis labels are simply tornado track names and have no other significance. Track widths measured from Google Earth or aerial photos consider 'visibly substantial' damage but do not include scattered individual treefalls as part of the damage track

known to leave undamaged segments along the path, this phenomenon does not seem to have a consistent relationship with landforms or topography.

The most detailed work to date has taken place in our lab, led by the second author (Cannon et al. unpubl. data). Two long-track tornadoes that occurred within the late April 2011 outbreak were studied using aerial photos and remote sensing analyses: one in Chattahoochee National Forest (CNF) in northern Georgia, and one in the far western end of Great Smoky Mountains National Park (GSMNP) in Tennessee. These two storms were rated EF3 and EF4 by the NWS. Because the EF-scale is based mainly on damage to man-made structures rather than vegetation, these ratings need to be interpreted cautiously; for example the GSMNP tornado earned its rating on the basis of severe damage to a single metal-truss utility tower (i.e., one point; Godfrey and Peterson 2014). The CNF tornado had a total track length of 65.9 km, but the damage path accounted for roughly 89 % (58.8 km) of that length; a total of 4,493 ha were damaged by this tornado. The GSMNP tornado had a total track length of 31.7 km, but a damage path that was 85 % (26.9 km) of that length.

To quantify the extent, severity, and pattern of tornado damage in these two damage tracks, we used a supervised classification within ArcGIS, based on 20-cm resolution aerial photos (visible spectrum) taken within 2 months of the tornadoes (Cannon et al. unpubl. data). The output of this analysis consisted of all 4-m pixels

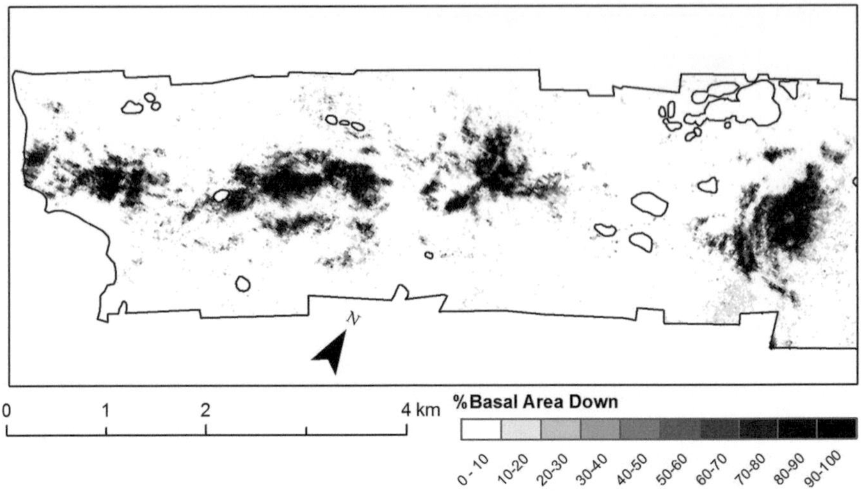

Fig. 5.7 Output of a supervised classification of aerial photos of the 27 April 2011 tornado at Chattahoochee National Forest in northern Georgia showing the westernmost 8 km of the damage track. *White* indicates intact or very slightly damaged forest; *darker shading* indicates more severe damage

classified into one of five levels of disturbance severity; severity was defined as amount of canopy disruption, and therefore included both uprooted and trunk broken trees (Fig. 5.7). Note that in Fig. 5.7, 'basal area down' refers to total basal area (BA) of trees that were either uprooted or trunk broken, which are the two primary mechanisms by which trees are removed from the canopy. After the classification pixels were aggregated and averaged in 20 m × 20 m grid cells to derive the final severity maps. Several important qualitative features of the damage severity maps should be mentioned prior to examining the quantitative details. First, these damage tracks, and many (most?) other tornado tracks do not simply create a binary 'damaged vs. undamaged' swath as might a lawnmower through grass. The tornado vortex commonly changes in size and rotational velocity, leaving areas with light or no damage that might incorrectly imply that the vortex has lifted off the ground, and has complex interactions with topography. Second, despite visual impressions, we reinforce earlier researcher's cautionary notes (Foster et al. 1997; Cooper-Ellis et al. 1999) that the most severe damage is typically a rather small fraction of total affected area. Third, following findings by Turner et al. (1994) after the 1988 Yellowstone fires, the great majority of the disturbed landscape has abundant biological legacies (cf. Franklin et al. 2000) and is seldom far from remnants of intact or lightly damaged forest, thus providing access for propagules and mobile organisms.

As with many other types of disturbance, the largest fraction of affected area in both tornadoes experienced low severity of disturbance (compare to Table 5.1 for hurricanes). For example, 83.9 % of the CNF tornado track lost <20 % of BA, whereas only 3.5 % experienced >80 % BA loss. Trends were similar in the GSMNP track: 70 % of area had <20 % BA loss, whereas 5 % had >80 % BA loss. Another pattern consistent among both tornado tracks was that the total area damaged was surprisingly evenly distributed among damage severity classes, partly due to the patch characteristics described below.

In both tornadoes, complex damage patterns resulted from interactions between tornado tracks and topography, defying some of the expected generalizations. Middle elevations and south- or southeast-facing slopes suffered the greatest damage levels, probably because they bore the brunt of direct tornado impact. In contrast to the easily identified 'sheltered vs exposed' dichotomy for landscapes subject to hurricanes, it is impossible to predict any location as 'sheltered' from a tornado, even if its path were known prior to the event. This is because although it is common for tornadoes to leave steep valley bottoms undamaged, particularly if the path is perpendicular to the long axis of the valley, there are also many cases of tornadoes tracking up and down valley slopes to cause damage along the bottoms as well as adjacent ridges. In addition, the paths of tornadoes can be quite erratic, with many known cases of the damage path apparently getting 'entrained' and tracking lengthwise up valleys. Thus the small size and erratic behavior of tornadoes precludes identifying sheltered and exposed locations that can be identified with some confidence in the context of hurricane (or derecho; see below) damage.

Both of the 2011 tornado tracks we studied created patches of widely varying size, shape and severity (Figs. 5.7 and 5.8; Table 5.2). In this context, we define severity as proportion of pre-disturbance BA that was downed (uprooted or trunk broken), while recognizing that alternative definitions are also possible. For example, in the CNF tornado track analysis, a total of 4,866 disturbed patches (having

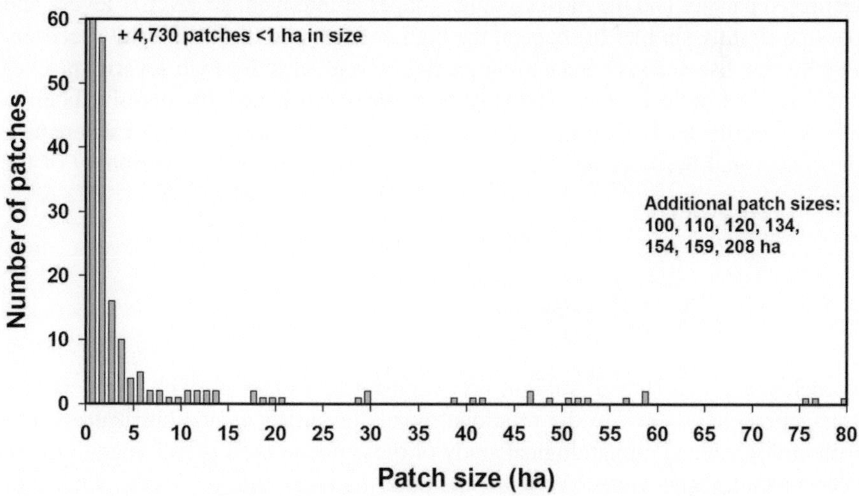

Fig. 5.8 Damaged patch size-frequency histogram for tornado damage swath in Chattahoochee National Forest, northern Georgia. The '4,730' in the *upper left* indicates frequency of patches <1.0 ha in size. The seven largest patch sizes are reported in the note at *upper right* in order to avoid extending the x-axis excessively. See text for details

Table 5.2 Disturbed patch characteristics in five wind-damaged sites in CHR forests

Study	Number of patches	Mean patch size (ha)	Minimum patch size detected or reported (ha)	Maximum patch size (ha)	Predominant aspect
Rebertus and Meier (2001)	38	0.79	0.05	20.1	North, East
McNab et al. (2004)	30	0.69	0.10	3.9	Southeast
Xi et al. (2008a)	109	0.12	0.006	0.80	–
Cannon et al. unpubl. – CNF	4,866	0.57	0.04	207.4	–
Cannon et al. unpubl. – GSMNP	2,487	1.01	0.04	497.8	–

CNF Chattahoochee National Forest, Georgia, *GSMNP* Great Smoky Mountains National Park, Tennessee

>10 % B.A. loss) were identified, with ~97 % of those being <1 ha (Fig. 5.8); an additional 1.8 % were 1–5 ha, and the largest single patch was 207.4 ha (Table 5.2, Cannon et al. unpubl. data). We analyzed patch characteristics with the Fragstats software, and found that patch size increased but patch number decreased with damage severity (i.e., with increasing severity patches were larger but fewer). In this analysis, plots with 0–20 % of B.A. downed (as inferred from the air photos) were considered 'low severity'; those with 21–40 % B.A. downed were considered 'intermediate,' and those with >40 % BA downed were considered 'high severity.' Damaged patches had the most complex shapes at intermediate severity levels, and became slightly simpler in shape at the high and low end of the severity spectrum. Finally, the distance between similar patches was greatest for high severity patches and least for low severity patches (Fig. 5.7). We propose the term 'dissolved bull's-eye' to describe the typical patch damage pattern where damage is most severe near the center, and becomes less severe and more dissected near the periphery of the patch.

5.2.3 *Derecho Disturbance*

Derechos are widespread, straight-line windstorms associated with thunderstorm convection (in contrast to the rotational winds in tornadoes and hurricanes). The meteorological and climatological study of these phenomena is still young (15–18 years) (Ashley and Mote 2005). On the other hand, several ecological studies of forest damage and recovery from derechos have been completed. These provide a solid evidential foundation for generalizations about the impact of derechos, with the qualifier that much of the damage and recovery work has taken place outside of the strict definition of CHR forest types.

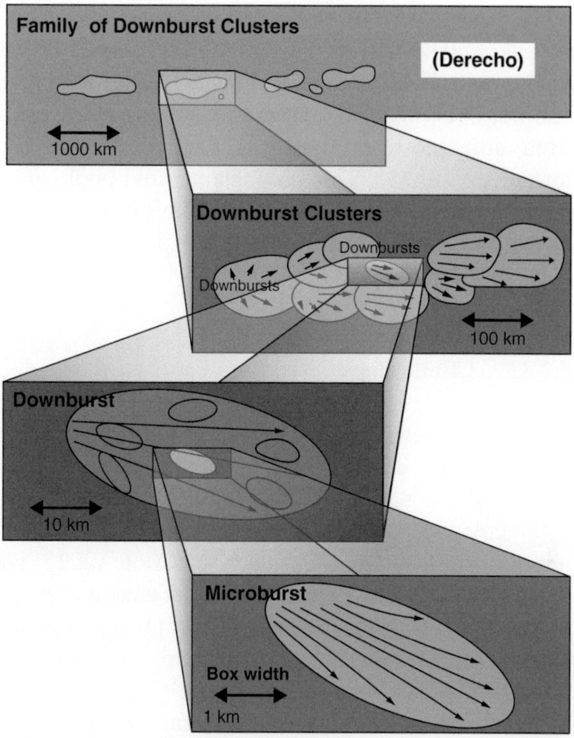

Fig. 5.9 Schematic diagram illustrating hierarchical nature of derecho systems (From Corfidi et al. undated)

Derechos are best understood as a hierarchical phenomenon (Fig. 5.9) in which the entire derecho is composed of several downburst clusters, which in turn are composed of multiple downbursts. The downbursts themselves have dimensions of about 8–10 km and may last for several minutes; these may in turn be further decomposed into individual microbursts. Technically, a derecho is defined as "a concentrated area of convectively induced wind gusts greater than 26 m per second that has a major axis length of 400 km or more" (Ashley and Mote 2005). Note in particular the criterion of a minimum damage swath length of 400 km; as such, derechos are arbitrarily distinguished from somewhat smaller (i.e., 40–400 km length) downburst clusters that are functionally similar. Obviously these storms occur over a large spatial scale; the recent July 2012 derecho over the central eastern USA encompassed much of the Ohio Valley.

5.2.3.1 Formation and Movement

Derechos often form in association with mesoscale features called quasi-linear convective systems, which are essentially long, continuous lines of strong thunderstorms. Evaporative cooling within downdrafts develops a strong cold pool at the surface, which propagates outward from the initial convection, lifting warm and moist air along the resulting gust front and producing more convection. The cycle repeats itself serially as the system moves forward. Thus, a derecho can be considered to be a series of discrete, downwind-developing individual thunderstorm cells (Corfidi et al. undated). The strengthening cold pool induces strong inflow of air from aloft (the rear inflow jet), which carries higher momentum air downward toward the surface. This propagating cold pool can push outward, resulting in a bowed representation (i.e., a 'bow echo') in a radar display (Corfidi et al. undated). Derechos typically propagate at 80 km per hour or greater (perhaps over 110 km per hour), but the unidirectional surface winds can reach well over 160 km per hour.

5.2.3.2 Climatology

Although derechos can and do occur in any month, researchers recognize a primary and secondary seasonal peak in derecho frequency. The primary peak, called warm-season derechos, is from May through August (69 % of events), and the smaller secondary peak, called cool-season derechos, is in October-December (Corfidi et al. undated). In their review, Ashley and Mote (2005) identified 377 derechos in the 18-year period from 1986–2003, for an annual average of nearly 21 over the continental USA. As with all severe weather phenomena, there is substantial year-to-year variation, with their 18-year climatology documenting a maximum of 42 events (1998) and a minimum of 2 events (1988).

Most derecho events produce surface winds that are less than catastrophic, but the more powerful storms can have winds equivalent to a category 4 hurricane. For example, the 16 July 1980 derecho event over Illinois and Michigan had 243 km per hour winds (Ashley and Mote 2005).

The geographical distribution of derechos is broadly similar to that of tornadoes (Fig. 5.10). Research has identified two primary 'corridors' of derecho frequency, each associated with a different season. Warm-season events tend to be concentrated over central Oklahoma, Arkansas and Missouri, with a lobe of the high-frequency isopleth extending to the states surrounding southern Lake Michigan (southern Wisconsin, northern Illinois, and northern Indiana). Cool-season events are concentrated in the lower Mississippi Valley, and northeastward in the vicinity of the Ohio River. Despite these concentrations over the south-central USA, many of the ecological studies have been associated with derechos in Wisconsin (e.g., Dunn et al. 1983; Canham and Loucks 1984), Minnesota (Peterson 2004; Rich et al. 2007; Moser and Nelson 2009), Pennsylvania (Evans et al. 2007; Peterson et al. 2013), and New York (Jenkins 1995; Robinson and Zappieri 1999; Canham et al. 2001). Post-derecho studies within the CHR include Leach (2003), Peterson and Leach (2008a, b), and Holzmueller et al. (2012).

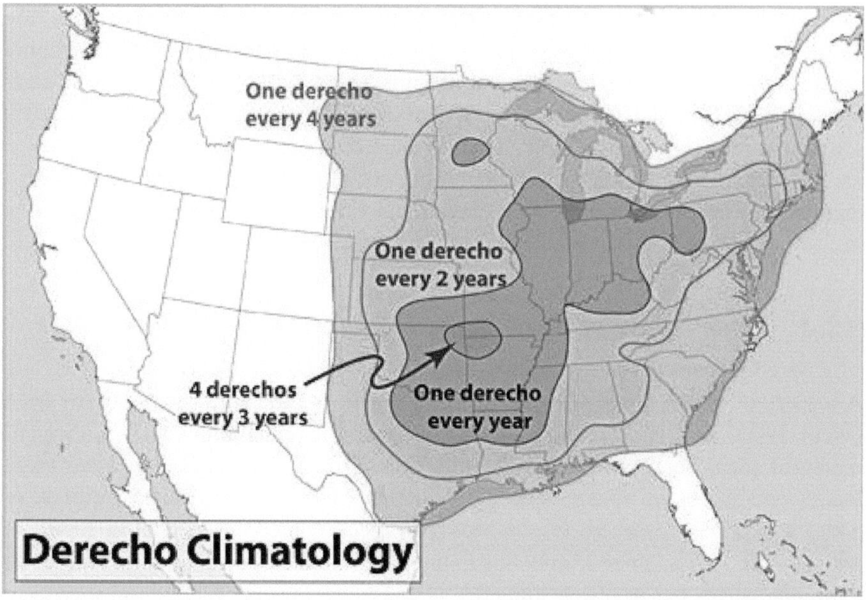

Fig. 5.10 Geographic distribution of frequency of derechos in the eastern USA (From NOAA Storm Prediction Center, 'About Derechos' FAQ website, accessed 10 December, 2014)

5.2.3.3 Landscape-Scale Patterns

Derecho damage can approach small hurricanes in spatial scale. A derecho in 1977 over Minnesota and northern Wisconsin had a path 266 km long and 27 km wide, virtually leveling 240,000 ha of forest (Canham and Loucks 1984). Just beyond the northern limit of CHR, a derecho over western Pennsylvania in July of 2003 damaged 5,000 ha of the 258,000 ha examined by Evans et al. (2007). At a more modest scale, a 5 May 1999 derecho in west-central Tennessee damaged ~3,000 ha within Natchez Trace State Forest alone (Peterson and Leach 2008a).

The size and distribution of damage patches that result from derechos has been examined in several cases, although none within the CHR. Canham and Loucks (1984) used historical reconstruction based on land survey records for northern Wisconsin to estimate that the highest-frequency size class of disturbance patches was <6 ha. In the July 2003 Pennsylvania event, damaged patches ranged from single treefall gaps to 114 ha, although mean and median patch sizes were 4.78 and 2.18 ha, respectively (Evans et al. 2007).

In several respects, the interaction of derecho winds with local topography parallels that of hurricane winds, with damage concentrated on exposed locations facing the direction of origin of the winds. Similar to what might be expected after a hurricane, Jenkins (1995) studied impacts of a 1995 derecho even in the Adirondack

Mountains of New York, noting that nearly all of the larger disturbed patches "are on the tops, west, or northwest slopes of hills or ridges. The air photos and satellite images look like someone spray-painted the ridge tops, pointing the can east-southeast and tipping it down at about a forty-five degree angle." Evans et al. (2007) found that the 2003 Pennsylvania derecho caused greater damage with increases in elevation, and with increases in topographic position relative to surroundings, and that damage decreased with greater elevational variability (rougher topography).

5.2.4 *Other Types of Wind Disturbance*

A variety of storm types affect many forests across the central and eastern USA (Webb 1999). In particular, we wish to focus attention on mountain waves, a type of meteorological phenomenon that is well known in other mountainous areas (e.g., Rocky Mountains or Sierra Nevada Mountains where they are called Chinook or Santa Ana winds, respectively), but was only recently recognized (Gaffin 2009) for the southern Appalachian Mountains in the CHR. To date, forest damage caused by mountain waves has not been studied.

Mountain waves are anomalously high winds in the foothills of mountain ranges. They occur where strong winds flow nearly perpendicularly toward a mountain range with a relatively gentle windward slope and relatively steep lee slope. The incoming winds are lifted by the windward slope, and if they encounter a stable air mass over the crest of the mountains, are redirected back down the lee slope in the form of waves. Experts often call attention to the analog of a standing wave in fast-flowing rivers on the downstream side of large boulders (Kemp 2010).

In the vicinity of the southern Appalachians, the Smoky Mountains in GSMNP provide an unusually high, unbroken ridgeline that exceeds 1,500 m for more than 56 km; notably (and contrary to the generally north-south trending Appalachian range as a whole), the Smokies ridgeline is almost exactly east-west, meaning mountain waves could form on the north and western sides when there are strong southerly winds, or on the south and east sides when there are northerly winds. The very limited data available to date (e.g., Gaffin 2009; J Renfro, GSMNP, pers. comm.) suggest that in GSMNP, mountain waves are usually on the northwest side of the mountains.

In the southern Appalachians, mountain wave winds are concentrated along the foothills. Their actual size and geographic distribution are currently unknown, but they can be exceedingly intense events. Three recent events in or near the northwest part of GSMNP produced documented wind speeds of 162, 171, and 177 km per hour (Kemp 2010). Such winds are thought to produce forest damage at a scale of tens to perhaps a few 100 ha per event (J Renfro, pers. comm.), although no quantitative data on impacts to forests exists. Mountain wave frequency (not all as severe as the extreme cases just mentioned) is indicated from long-term monitoring at the Cove Mountain meteorological station within GSMNP, where 52 mountain waves were recorded between 1999 and 2007 (Gaffin 2009).

5.3 Impacts at the Stand and Tree Scale

5.3.1 Stand- and Tree-scale Patterns

At the tree- and stand-scale there are well-established trends in damage patterns that generalize across types of wind event. Perhaps most general is the increasing probability of tree damage with size, documented in most wind-damage studies within the CHR (Peterson 2007; Xi et al. 2008a), but this relationship is not universal (e.g., Chapman et al. 2008; Oswalt and Oswalt 2008; Harcombe et al. 2009). Even when not examined directly, the increasing vulnerability among larger trees is widely confirmed by most studies reporting greater BA loss compared to density loss (Greenberg and McNab 1998; Peterson and Leach 2008a; Busing et al. 2009; Holzmueller et al. 2012).

Forest type or species composition is not necessarily a consistently good predictor of damage patterns, even when species differ in their vulnerability to wind damage, suggesting that abiotic variables (either site or storm features) sometimes play more important roles in determining damage. In northwestern Pennsylvania, Evans et al. (2007) found that forest type was not generally a useful predictor of wind damage level, although red maple (*Acer rubrum*) stands on wet sites were consistently more heavily damaged than other forest types. Instead, they reported that structural features such as mean tree size or age were especially important predictors (Evans et al. 2007). Nevertheless, many studies have reported that levels of damage varied among species (e.g., Greenberg and McNab 1998; Leach 2003; Busing et al. 2009). Notably, not all of these studies separated the species effect itself from obvious interacting factors such as tree size or age, leaving open the question of how widespread the species differences really are. Peterson (2007), in a review of several studies, was able to statistically isolate species effects from size effects, showing that there were modest but consistent differences among species. For example, sugar maple (*A. saccharum*) consistently emerged as more resistant than many other species. However, other studies have shown that interspecific differences can be inconsistent among studies. Xi et al. (2008a) found that loblolly pine (*Pinus taeda*) had high risk of damage in a North Carolina tornado, but low risk relative to other species in two hurricanes (Hugo and Fran). The authors attributed these differences to stand structures causing the focal species to be either more exposed or more sheltered.

The type of tree damage from severe winds is variously categorized, but most researchers recognize an important distinction between standing undamaged trees, lightly damaged trees (defoliated, small branches broken), uprooted trees, and those with broken trunks. In particular, uprooted trees create treefall root pits and root plates (also called root balls or mounds), and these have several implications for subsequent community and ecosystem dynamics that are not induced by trees with broken trunks (Beatty 1984; Schaetzl et al. 1989). The pit/mound complex diversifies microenvironmental conditions, providing a variety of establishment microsites for seeds and thus facilitating increased diversity among tree seedlings and herb-layer species (Peterson et al. 1990; Harrington and Bluhm 2001; Collins and Battaglia 2002). These microsites also have soil respiration rates distinct from the surrounding intact soil (Millikin and Bowden 1996).

Treefall pit/mound sizes are consistently correlated with tree diameter at breast height (DBH). A synthesis of studies across 10 sites and >1,000 pit/mound complexes in six states (Sobhani et al. 2014) found that a single relationship explained roughly 54 % of the variation in size of root pits: ln pit size = –4.859 + (1.561 * ln DBH) (pit size in m^2 and DBH in cm). Similarly, a single relationship explained 56 % of variation in root mound size: ln mound size = –4.909 + (1.460 * ln DBH). Coverage by recently-formed treefall pit/mound complexes has been reported in a number of locations, and is typically less than 10 or 11 %. For example the area of recent pit/mound complexes was estimated at 5.9 % after a tornado in the Blue Ridge Mountain ecoregion of northern Georgia (Sobhani et al. 2014); 1.6–4.3 % in hurricane gaps of western North Carolina (Greenberg and McNab 1998); 2.0 % after a derecho in west-central Tennessee (Leach 2003); and 2.4–6.0 % after a tornado in Arkansas (Phillips et al. 2008). The coverage of these microsites can in some cases be directly linked to greater vulnerability of particular species. For example, Battaglia et al. (1999) found 4.9 % coverage of pit/mound complexes in a pine-bottomland hardwood stand where the dominant pines were heavily damaged, but only 1.7 % coverage in a nearby bottomland hardwood stand without canopy pines.

The type of damage to trees is widely considered to be strongly influenced by soil saturation and therefore precipitation preceding and during storm events. For example, Xi et al. (2008a) documented that a large proportion of fallen trees were uprooted at Duke Forest following Hurricane Fran, which caused >22 cm of rain in the 48 h preceding high winds. In contrast, the relatively drier Hurricane Hugo caused more trunk breakage than uprooting (see similar trends in Busing et al. 2009). They further contrasted these two hurricanes with tree damage following a 1988 tornado with little precipitation at nearby Umstead State Park, where 70 % of the fallen trees had trunk breakage (Xi et al. 2008a). Greenberg and McNab (1998) reported more than twice as much uprooting as trunk breakage after Hurricane Opal (15 cm of rain at their study site), which appears consistent with this trend (see also Rebertus and Meier 2001). Similarly, Peterson and Rebertus (1997) contrasted tree damage across swamp, bottomland, and upland sites in southeastern Missouri after a tornado, and found trunk breakage to be threefold more common overall, despite greater uprooting than trunk breakage in the swamp site (see also Cowden et al. 2014 for another example with predominantly trunk breakage). Platt et al. (2000) found that slash pine (*P. elliottii*) stands in south Florida damaged by Hurricane Andrew experienced more (often twice as much) trunk breakage than uprooting; while Andrew was quite dry for a hurricane, they attributed the tree damage pattern in part to tree roots being anchored in shallow limestone bedrock. Additionally, in four southern Appalachian (Blue Ridge Mountain ecoregion) post-tornado sites studied by Peterson and Snyder (unpubl. data), uprooting was the most prevalent form of treefall in three of the four sites.

The prevailing mode of treefall can differ substantially for the same species among sites, as well as among species within a site. For example, Greenberg and McNab (1998) note that almost all scarlet oak (*Quercus coccinea*) in their western North Carolina study uprooted, but that 90 % of the fallen scarlet oak from an eastern Kentucky site were snapped (Romme and Martin 1982). Peterson (2007), in describing patterns from central Tennessee (Interior Plateau ecoregion), found that northern red oak (*Q. rubra*) had a six-fold greater probability of uprooting than trunk breaking, compared to either chestnut oak (*Q. montana*) or white oak (*Q. alba*). Therefore we conclude precipitation and soil saturation alone cannot fully explain why trees uproot versus break; other factors such as soil depth, bedrock fracturing, presence or absence of impermeable soil layers, and tree wood properties must also be considered.

The differences in resistance or vulnerability among species are illustrated at the stand scale by the extent of change in species composition and diversity after wind disturbance. How disturbance effects are interpreted depends partly on the interval between wind disturbance and sampling; a wind disturbance that decreases or removes several dominant species from the canopy would be seen as having a negative effect on diversity if sampling was conducted soon after the disturbance. However, if advance regeneration saplings or newly established seedlings were sampled some years later, then the disturbance would be seen as resulting in increased diversity. Greenberg and McNab (1998) found quite small changes in species composition of their five gaps in western North Carolina. Xi et al. (2008a) emphasized that Hurricane Fran caused substantial structural change at Duke Forest, but little immediate compositional change.

Species diversity appears to be either unaffected or negatively affected by wind disturbance, and we propose that the magnitude of the negative effect increases with severity of wind disturbance. In Greenberg and McNab's (1998) study, species diversity (Shannon's index) changed very slightly and inconsistently. Cowden et al. (2014) examined changes in a north-central Alabama site (Sipsey Wilderness) after an EF1 tornado, and found no significant difference in species diversity between undamaged controls and moderately-damaged plots. The study by Leach (2003) after a 1999 derecho over west-central Tennessee and by Holzmueller et al. (2012) after a derecho over southern Illinois; both showed little change in species diversity, and were both of moderate severity of damage. Peterson and Rebertus (1997) reported that of nine large plots damaged in a Missouri tornado, seven saw declines in Shannon diversity, and seven also lost at least one pre-disturbance tree species. Figure 5.11 shows the immediate change in Shannon's diversity index due entirely to removal of the fallen trees, on a plot-by-plot basis across four sites (Peterson unpubl. data). Although there is a hint that very modest wind damage severities sometimes result in slightly increased diversity (probably due to reduction of dominance by the more abundant species), diversity steadily declines as wind damage severity increases. Thus in terms of immediate changes, wind disturbance will most often decrease diversity among the canopy tree species.

Finally, also at the stand scale, wind disturbances may advance, set back, or have little effect on the successional status of the species assemblage. Given that larger

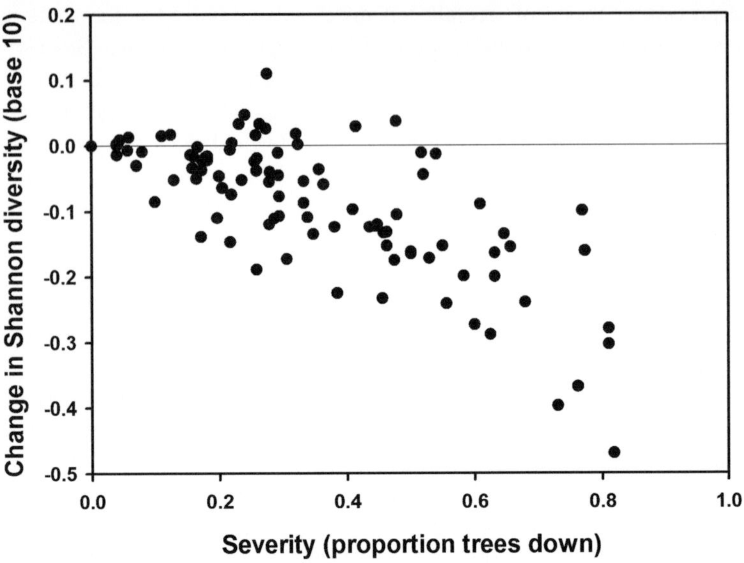

Fig. 5.11 Immediate change (pre-storm to post-storm) in canopy tree diversity from four CHR locations. Diversity calculated as Shannon's H' index, base 10, using density as the abundance measure. Post-storm calculation uses standing trees after uprooted and trunk-broken trees removed. Alternative formatting of figure with severity defined as proportion of basal area down gives very similar results. One location previously reported in Peterson and Rebertus (1997); data from remaining locations unpublished. Points are 20 × 20 m or 30 × 30 m plots

trees are generally more damaged than smaller trees, canopy dominant species tend to be differentially removed or decreased by wind disturbance. When the canopy dominants are pioneer (e.g., black locust (*Robinia pseudoacacia*) or yellow-poplar (*Liriodendron tulipifera*)) or mid-successional (e.g., most oaks and hickories (*Carya* spp.)) species, as is common in CHR secondary forests, the net result is to leave an assemblage with greater dominance by late-successional species (Nelson et al. 2008; Xi et al. 2008b; Holzmueller et al. 2012). In some cases, a pre-disturbance dominant species is effectively completely removed from the canopy and lacks regeneration, causing a more pronounced shift, as was documented in the case of loblolly pine after Hurricane Hugo by Battaglia et al. (1999). In contrast, when primary forests dominated by late-successional species are disturbed, the immediate net effect is neutral (e.g., Held et al. 1998) but succession may eventually be set to an earlier stage if a new cohort of pioneer or mid-successional seedlings becomes established (e.g., Dunn et al. 1983). Xi et al. (2008a) described a composite of the above patterns, where heavy damage to dominant pines shifted some stands to more advanced stages of succession, whereas less heavily damaged hardwood stands were set back to an earlier stage. We propose a broad conceptual model for secondary forests, in which low- to moderate-intensity wind damage advances succession by removing some of the pre-storm canopy dominants and releasing later-successional subcanopy and sapling stems, whereas high severity damage sets succession back to an earlier stage by sufficiently opening the canopy and removing subcanopy vegetation so that early-successional species can establish. The high-severity component of this model has been demonstrated in several cases (some outside of the CHR), wherein entirely new cohorts of early-successional species establish, rather than simply release of advanced regeneration or regrowth of surviving canopy individuals (Dunn et al. 1983; Peterson and Pickett 1995; Nelson et al. 2008).

5.3.2 *Ecosystem Effects of Wind Disturbance: Carbon Cycling and Budgets*

Compared to community ecology and forestry-related research, ecosystem-oriented research in major wind-disturbed areas has been very limited. Recent syntheses of carbon cycling effects of various types of disturbance (Amiro et al. 2010; Goetz et al. 2012; Kasischke et al. 2013) point to only a single integrated (flux tower) study after wind disturbance in temperate forests of North America (Li et al. 2007), although such work has been conducted in Europe (Thurig et al. 2005), in mangroves off the Florida keys (Barr et al. 2012), and in a few tropical sites (Vargas and Allen 2008). Indeed, among the topics considered in this chapter, the ecosystem effects of wind disturbance is least understood, yet important because regional and continental-scale trends in carbon budgets are known to be substantially influenced by major disturbance (Kurz et al. 2008). Unfortunately, space limitations constrain what can be discussed in this chapter and we therefore cannot do justice to this topic. Until a review and synthesis focused on these types of disturbances is written we direct interested readers to the existing reviews mentioned above.

5.4 Summary and Conclusions

Wind disturbance has been the subject of a great deal of research in temperate forests of eastern North America, although somewhat less within the CHR than other nearby regions. Although far from complete, the available information is sufficient to roughly define a wind disturbance regime, and thus provide benchmarks to guide management decisions. The climatology and geographic distribution of hurricanes and tornadoes is well established, with roughly inverse east-west trends (hurricanes decreasing and tornadoes increasing) across the CHR. Derechos have been studied for a much shorter period, so the climatologies and geographic patterns are perhaps a bit less firmly grounded, but the existing approximately 15–18 year time frame is sufficient to define first approximations. Mountain waves are newly-recognized phenomena that are locally important in the southern Appalachians, and their climatology remains poorly known. Combining all of the above, the Arkansas Valley, Ouachita Mountains, Interior Plateau, and Southwest Appalachians ecoregions have the greatest risk of wind damage from tornadoes and derechos, whereas weakening hurricanes are an important disturbance, especially in the Piedmont ecoregion. The Central Appalachians have probably the lowest rate of wind disturbance among the CHR ecoregions.

The distribution of patch sizes created by wind disturbances is one topic where informal visual impressions may greatly mislead the unwary. Multiple studies attest that the great majority of patches are quite small (e.g., <1–2 ha), even though a few may be much larger (to several tens of hectares); the empirical distribution of sizes is approximated by a negative exponential. This is counter to most observers' visual impressions of wind-disturbed areas (cf. Foster et al. 1997).

Severe wind events have strong interactions with topography; these are most easily predicted for hurricanes, somewhat less easily for derechos, and rather poorly for tornadoes. For the larger-scale storms (hurricanes and derechos), knowledge of the peak wind direction is sufficient to identify locations on the landscape that will be sheltered and thus experience less or no wind damage. For example, on the right (east) side of a northward-moving hurricane, winds will be predominantly from the south, causing most damage on south-facing slopes and flat areas, and leaving north-facing slopes undisturbed. Such predictions are precluded by the erratic behavior of tornadoes. Similarly, hurricanes and derechos often cause greater damage at higher elevations (but see counter-example in McNab et al. 2004), whereas tornadoes may or may not track up and down valley slopes, and sometimes even follow the long axis of valleys for several km.

At the stand and tree scale, wind damage is rather unpredictable for a particular tree or stand, but much more predictable in the aggregate (e.g., Peterson 2004). Larger trees are consistently more damaged than smaller trees, and there are often differences among species in damage at particular sites. The interspecific differences, however, appear to be contingent on absolute size, relative size, stand density, and edaphic factors, so broad trends among species are harder to identify. Despite such caveats, numerous studies identify sugar maple as consistently less damaged than surrounding trees, whereas species in the red oak group often suffer greater damage than their neighbors. In bottomland contexts, water tupelo (*Nyssa aquatica*) and bald cypress (*Taxodium distichum*) frequently suffer less damage than other species in the stand. The concentration of damage in dominant size classes means that wind disturbance will typically advance succession in secondary stands; in contrast, a severe disturbance may set back succession to an earlier stage in primary forest stands. Immediate changes in species composition are often subtle in comparison to structural change, although very severe wind events do in fact lead to major changes in species composition. Species diversity is also often changed very little, but when disturbance is sufficiently severe, diversity is most often decreased.

In sum, the trends identified above should enable forest managers to identify which forests are most likely to experience wind disturbance, and what type of storm is most likely at any particular locale. Known sizes of damaged patches, along with knowledge of size and species effects on tree vulnerability and resistance to wind, should facilitate informed decision-making that allows management actions to follow trends seen in natural disturbances.

Acknowledgements Much of our recent work was sponsored by grants AGS-1141926 and DEB-1143511 from the Physical and Dynamic Meteorology, and Population and Community Ecology programs of the National Science Foundation; earlier funding came from a cooperative agreement between University of Georgia and the USDA Forest Service, Southern Research Station. For some of this work, JBC was supported by a Climate Change Youth Initiative fellowship from the National Park Service. We thank the many field assistants over >20 years whose labor in difficult conditions provided the raw data for some of the trends reported here. Tornado track widths were laboriously measured on the aerial photos by Michael Bailey, Somer Rowe, and Tiffany Erickson.

References

Agee, E. M., & Hendricks, A. (2011). An assessment of the climatology of Florida hurricane-induced tornadoes (HITs): Technology versus meteorology. *Journal of Climate, 24*, 5218–5222.

Allen, M. S., Thapa, V., Arévalo, J. R., & Palmer, M. W. (2012). Windstorm damage and forest recovery: Accelerated succession, stand structure, and spatial pattern over 25 years in two Minnesota forests. *Plant Ecology, 213*, 1833–1842.

Amiro, B. D., Barr, A. G., Barr, J. G., Black, T. A., Bracho, R., Brown, M., Chen, J., Clark, K. L., Davis, K. J., Desai, A. R., Dore, S., Engel, V., Fuentes, J. D., Goldstein, A. H., Goulden, M. L., Kolb, T. E., Lavigne, M. B., Law, B. E., Margolis, H. A., Martin, T., McCaughey, J. H., Misson, L., Montes-Helu, M., Noormets, A., Randerson, J. T., Starr, G., & Xiao, J. (2010). Ecosystem carbon dioxide fluxes after disturbance in forests of North America. *Journal of Geophysical Research, 115*, G00K02. doi:10.1029/2010JG001390.

Arévalo, J. R., DeCoster, J. K., McAlister, S. D., & Palmer, M. W. (2000). Changes in two Minnesota forests during 14 years following catastrophic windthrow. *Journal of Vegetation Science, 11*, 833–840.

Ashley, W. S., & Mote, T. L. (2005). Derecho hazards in the United States. *Bulletin of the American Meteorological Society, 86*, 1577–1592.

Barr, J. G., Engel, V., Smith, T. J., & Fuentes, J. D. (2012). Hurricane disturbance and recovery of energy balance, CO_2 fluxes and canopy structure in a mangrove forest of the Florida everglades. *Agricultural and Forest Meteorology, 153*, 54–66.

Battaglia, L. L., Sharitz, R. R., & Minchin, P. R. (1999). Patterns of seedling and overstory composition along a gradient of hurricane disturbance in an old-growth bottomland hardwood community. *Canadian Journal of Forest Research, 29*, 144–156.

Beatty, S. W. (1984). Influence of microtopography and canopy species on spatial patterns of forest understory plants. *Ecology, 65*, 1406–1419.

Boose, E. R., Foster, D. R., & Fluet, M. (1994). Hurricane impacts to tropical and temperate forest landscapes. *Ecological Monographs, 64*, 369–400.

Brewer, R., & Merritt, P. G. (1978). Wind throw and tree replacement in a climax beech-maple forest. *Oikos, 30*, 149–152.

Brooks, H. E. (2004). On the relationship of tornado path length and width to intensity. *Weather and Forecasting, 19*, 310–319.

Busing, R. T., White, R. D., Harmon, M. E., & White, P. S. (2009). Hurricane disturbance in a temperate deciduous forest: Patch dynamics, tree mortality, and coarse woody detritus. *Plant Ecology, 201*, 351–363.

Canham, C. D., & Loucks, O. L. (1984). Catastrophic windthrow in the presettlement forests of Wisconsin. *Ecology, 65*, 803–809.

Canham, C. D., Papaik, M. J., & Latty, E. F. (2001). Interspecific variation in susceptibility to windthrow as a function of tree size and storm severity for northern temperate tree species. *Canadian Journal of Forest Research, 31*, 1–10.

Carpino, E. (1998). *Ecological determinants of hurricane damage in a southeastern Piedmont forest*. Dissertation, Duke University, Durham

Chapman, E. L., Chambers, J. Q., Ribbeck, K. F., Baker, D. B., Tobler, M. A., Zeng, H., & White, D. A. (2008). Hurricane Katrina impacts on forest trees of Louisiana's Pearl River basin. *Forest Ecology and Management, 256*, 883–889.

Collins, B. S., & Battaglia, L. L. (2002). Microenvironmental heterogeneity and *Quercus michauxii* regeneration in experimental gaps. *Forest Ecology and Management, 155*, 279–290.

Cooper-Ellis, S., Foster, D. R., Carlton, G., & Lezberg, A. (1999). Forest response to catastrophic wind: Results from an experimental hurricane. *Ecology, 80*, 2683–2696.

Corfidi, S., Evans, J. S., & Johns, R. H. (Undated). *About derechos*. NOAA Storm Prediction Center webpage. http://www.spc.noaa.gov/misc/AbtDerechos/derechofacts.htm. Accessed 10 Dec 2014.

Cowden, M. M., Hart, J. L., Schweitzer, C. J., & Dey, D. C. (2014). Effects of intermediate-scale wind disturbance on composition, structure, and succession in *Quercus* stands: Implications for natural disturbance-based silviculture. *Forest Ecology and Management, 330*, 240–251.

Dahal, D., Liu, S., & Oeding, J. (2014). The carbon cycle and hurricanes in the United States between 1900 and 2011. *Natural Scientific Reports, 4*, 5197.

Davies-Jones, R., Trapp, R. J., & Bluestein, H. B. (2001). Tornadoes and tornadic storms. *Meteorological Monographs, 28*, 167–222.

Duever, M. J., & McCollom, J. M. (1993). *Hurricane Hugo effects on old-growth floodplain forest communities at Four Holes Swamp, South Carolina*. Proceedings of the Seventh Biennial Southern Silvicultural Research Conference, Mobile, pp. 197–202.

Dunn, C. P., Guntenspergen, G. R., & Dorney, J. R. (1983). Catastrophic wind disturbance in an old-growth hemlock-hardwood forest, Wisconsin. *Canadian Journal of Botany, 61*, 211–217.

Edwards, R., LaDue, J. G., Ferree, J. T., Sharfenberg, K., Maier, C., & Coulbourne, W. L. (2013). Tornado intensity estimation: Past, present and future. *Bulletin of the American Meteorological Society, 94*, 641–653.

Elliott, K. J., Hitchcock, S. L., & Krueger, L. (2002). Vegetation response to large scale disturbance in a southern Appalachian forest: Hurricane Opal and salvage logging. *Bulletin of the Torrey Botanical Club, 129*, 48–59.

Evans, A. M., Camp, A. E., Tyrrell, M. L., & Riely, C. C. (2007). Biotic and abiotic influences on wind disturbance in forests of NW Pennsylvania, USA. *Forest Ecology and Management, 245*, 44–53.

Everham, E. M., & Brokaw, N. V. L. (1996). Forest damage and recovery from catastrophic wind. *The Botanical Review, 62*, 113–185.

Foster, D. R. (1988). Species and stand response to catastrophic wind in central New England, USA. *Journal of Ecology, 76*, 135–151.

Foster, D. R., & Boose, E. R. (1992). Patterns of forest damage resulting from catastrophic wind in central New England, USA. *Journal of Ecology, 80*, 79–98.

Foster, D. R., & Boose, E. R. (1995). Hurricane disturbance regimes in temperate and tropical forest ecosystems. In M. P. Coutts & J. Grace (Eds.), *Wind and trees*. Cambridge: Cambridge University Press.

Foster, D. R., Aber, J. D., Melillo, J. M., Bowden, R. D., & Bazzaz, F. A. (1997). Forest response to disturbance and anthropogenic stress. *BioScience, 47*, 437–445.

Foster, D. R., Knight, D. H., & Franklin, J. F. (1998). Landscape patterns and legacies resulting from large, infrequent forest disturbances. *Ecosystems, 1*, 497–510.

Franklin, J. F., Lindenmayer, D. B., MacMahon, J. A., McKee, A., Magnusson, J., Perry, D. A., Waide, R., & Foster, D. R. (2000). Threads of continuity: Ecosystem disturbance, biological legacies and ecosystem recovery. *Conservation Biology in Practice, 1*, 8–16.

Frelich, L. E., & Ostuno, E. J. (2012). Estimating wind speeds of convective storms from tree damage. *Electronic Journal of Severe Storms Meteorology, 7*, 1–19.

Fujita, T. T. (1971). *Proposed characterization of tornadoes and hurricanes by area and intensity* (Satellite and mesometeorology research project report 91). Chicago: University of Chicago.

Gaffin, D. M. (2009). On high winds and foehn warming associated with mountain-wave events in the western foothills of the southern Appalachian Mountains. *Weather and Forecasting, 24*, 53–75.

Glitzenstein, J. S., & Harcombe, P. A. (1988). Effects of the December 1983 tornado on forest vegetation of the Big Thicket, southeast Texas, USA. *Forest Ecology and Management, 25*, 269–290.

Godfrey, C. M., Peterson, C. J. (2014). Estimating enhanced Fujita scale levels based on forest damage severity. Preprints, Special symposium on Severe local storms: The current state of the science and understanding impacts. American Meteorological Society Annual meeting, Atlanta

Goetz, S. J., Bond-Lamberty, B., Law, B. E., Hicke, J. A., Huang, C., Houghton, R. A., McNulty, S., O'Halloran, T., Harmon, M., Meddens, A. J. H., Pfeifer, E. M., Mildrexler, D., & Kasischke, E. S. (2012). Observations and assessment of forest carbon dynamics following disturbance in North America. *Journal of Geophysical Research, 117*, G02022. doi:10.1029/2011JG001733.

Greenberg, C. H., & McNab, W. H. (1998). Forest disturbance in hurricane-related downbursts in the Appalachian Mountains of North Carolina. *Forest Ecology and Management, 104*, 179–191.

Gresham, C. A., Williams, T. M., & Lipscomb, D. J. (1991). Hurricane Hugo wind damage to southeastern U.S. coastal forest tree species. *Biotropica, 23*, 420–426.

Grinsted, A., Moore, J. C., & Jevrejeva, S. (2013). Projected Atlantic hurricane surge threat from rising temperatures. *Proceedings of the National Academy of Sciences, 110*, 5369–5773.

Harcombe, P. A., Mann Leipzig, L. E., & Elsik, I. S. (2009). Effects of Hurricane Rita on three long-term forest study plots in East Texas, USA. *Wetlands, 29*, 88–100.

Harrington, T. B., & Bluhm, A. A. (2001). Tree regeneration responses to microsite characteristics following a severe tornado in the Georgia Piedmont, USA. *Forest Ecology and Management, 140*, 265–275.

Held, M. E., & Bryant, W. S. (1989). Recovery in a tornado-damaged hardwood forest in northern Kentucky. *Transactions of the Kentucky Academy of Science, 50*, 16–21.

Held, M. E., & Winstead, J. E. (1976). Structure and composition of a climax forest system in Boone County, Kentucky. *Transactions of the Kentucky Academy of Science, 37*, 57–67.

Held, M. E., Jones-Held, S., & Winstead, J. E. (1998). Forest community structure and tornado damage in an old-growth system in northern Kentucky. *Castanea, 63*, 474–481.

Holzmueller, E. J., Gibson, D. J., & Suchecki, P. F. (2012). Accelerated succession following an intense wind storm in an oak-dominated forest. *Forest Ecology and Management, 279*, 141–146.

Hook, D. J., Buford, M. A., & Williams, T. M. (1991). Impact of Hurricane Hugo on the South Carolina coastal plain forest. *Journal of Coastal Research, 8*, 291–300.

Jenkins, J. (1995). *Notes on the Adirondack blowdown of July 15th, 1995*. Wildlands Conservation Society Working Paper No. 5.

Karstens, C. D., Gallus, W. A., Lee, B. D., & Finley, C. A. (2013). Analysis of tornado-induced tree fall using aerial photography from the Joplin, Missouri, and Tuscaloosa-Birmingham, Alabama, tornadoes of 2011. *Journal of Applied Meteorology and Climatology, 52*, 1049–1068.

Kasischke, E. S., Amiro, B. D., Barger, N. N., French, N. H., Goetz, S. J., Grosse, G., Harmon, M. E., Hicke, J. A., Liu, S., & Masek, J. G. (2013). Impacts of disturbance on the terrestrial carbon budget of North America. *Journal of Geophysical Research – Biogeosciences, 118*, 303–316.

Kemp, S. (2010). Mighty mountain waves. *Smokies Life, 4*, 22–27.

Kribel, J. R. G., & Ware, S. (2014). Hurricane-caused tree loss on permanent plots in a temperate hardwoods forest. *Castanea, 79*, 1–7.

Kupfer, J. A., Myers, A. T., McLane, S. E., & Melton, G. N. (2008). Patterns of forest damage in a southern Mississippi landscape caused by Hurricane Katrina. *Ecosystems, 11*, 45–60.

Kurz, W. A., Stinson, G., Rampley, G. J., Dymond, C. C., & Neilson, E. T. (2008). Risk of natural disturbances makes future contribution of Canada's forests to the global carbon cycle highly uncertain. *Proceedings of the National Academy of Sciences, 105*, 1551–1555.

Laing, J. M., Shear, T. H., & Blazich, F. A. (2011). How management strategies have affected Atlantic white-cedar forest recovery after massive wind damage in the Great Dismal Swamp. *Forest Ecology and Management, 262*, 1337–1344.

Lang, K. D., Schulte, L. A., & Guntenspergen, G. R. (2009). Windthrow and salvage logging in an old-growth hemlock-northern hardwoods forest. *Forest Ecology and Management, 259*, 56–64.

Landsea, C. (2014). *Frequently asked questions*. NOAA Hurricane Research Division website. http://www.aoml.noaa.gov/hrd/tcfaq/tcfaqHED.html. Accessed 30 Jun 2014.

Leach, A. D. (2003). *Influence of microsite and salvage logging on post-windthrow recovery in three southeastern US forests*. Thesis, University of Georgia, Athens

Li, J., Powell, T. L., Seiler, T. J., Johnson, D. P., Anderson, H. P., Bracho, R., Hungate, B. A., Hinkle, C. R., & Drake, B. G. (2007). Impacts of Hurricane Frances on Florida scrub-oak ecosystem processes: Defoliation, net CO_2 exchange and interactions with elevated CO_2. *Global Change Biology, 13*, 1101–1113.

Malamud, B. D., & Turcotte, D. L. (2012). Statistics of severe tornadoes and severe tornado outbreaks. *Atmospheric Chemistry and Physics, 12*, 8459–8473.

Matlack, G. R., Gleeson, S. K., & Good, R. E. (1993). Treefall in a mixed oak-pine coastal plain forest: Immediate and historical causation. *Ecology, 74*, 1559–1566.

McNab, W. H., Greenberg, C. H., & Berg, E. C. (2004). Landscape distribution and characteristics of large hurricane-related canopy gaps in a southern Appalachian watershed. *Forest Ecology and Management, 196*, 435–447.

Millikin, C. S., & Bowden, R. D. (1996). Soil respiration in pits and mounds following an experimental forest blowdown. *Soil Science Society of America Journal, 60*, 1951–1953.

Moore, T. W., & Dixon, R. W. (2011). Climatology of tornadoes associated with Gulf Coast-landfalling hurricanes. *Geographical Review, 101*, 371–395.

Moser, W. K., & Nelson, M. D. (2009). Windstorm damage in Boundary Waters Canoe Area Wilderness (Minnesota, USA): Evaluating landscape-level risk factors. *Baltic Forestry, 15*, 248–254.

Myster, R. W., & Malahy, M. P. (2010). Tornado effects on damage, resprouting and spatial heterogeneity in the cross timbers ecotone of Oklahoma, USA. *Journal of Plant Ecology, 3*, 157–163.

Negron-Juarez, R., Baker, D. B., Chambers, J. Q., Hurtt, G. C., & Goosem, S. (2014). Multi-scale sensitivity of Landsat and MODIS to forest disturbance associated with tropical cyclones. *Remote Sensing of Environment, 140*, 679–689.

Nelson, J. L., Groninger, J. W., Battaglia, L. L., & Ruffner, C. M. (2008). Bottomland hardwood forest vegetation and soils recovery following a tornado and salvage logging. *Forest Ecology and Management, 256*, 388–395.

NOAA/NCDC (National Oceanic and Atmospheric Administration/National Climatic Data Center). (Undated a). *Hurricanes*. http://www.ncdc.noaa.gov/img/climate/severeweather/hur5004sm.jpg. Accessed 10 Dec 2014.

NOAA/NCDC (National Oceanic and Atmospheric Administration/National Climatic Data Center). (Undated b). *US Tornado climatology*. http://www.ncdc.noaa.gov/climate-information/extreme-events/us-tornado-climatology. Accessed 10 Dec 2014.

Oswalt, S. N., & Oswalt, C. M. (2008). Relationships between common forest metrics and realized impacts of Hurricane Katrina on forest resources in Mississippi. *Forest Ecology and Management, 255*, 1692–1700.

Palmer, M. W., McAlister, S. D., Arévalo, J. R., & DeCoster, J. K. (2000). Changes in the understory during 14 years following catastrophic windthrow in two Minnesota forests. *Journal of Vegetation Science, 11*, 841–854.

Perera, A. H., Buse, L. J., & Weber, M. G. (Eds.). (2004). *Emulating natural forest landscape disturbances: Concepts and applications*. New York: Columbia University Press.

Peterson, C. J. (2000). Damage and recovery of tree species after two different tornadoes in the same old-growth forest: a comparison of infrequent wind disturbances. *Forest Ecology and Management, 35*, 237–252.

Peterson, C. J. (2004). Within-stand variation in windthrow in southern boreal forests of Minnesota: Is it predictable? *Canadian Journal of Forest Research, 34*, 365–375.

Peterson, C. J. (2007). Consistent influence of tree diameter and species on damage in nine eastern North America tornado blowdowns. *Forest Ecology and Management, 250*, 96–108.

Peterson, C. J., & Godfrey, C. M. (2014). *Side-by-side tree and house damage in the May 2013 Moore, OK EF5 tornado: Lessons for the enhanced Fujita scale. Preprints, special symposium on severe local storms: The current state of the science and understanding impacts*. Atlanta: American Meteorological Society Annual Conference.

Peterson, C. J., & Leach, A. D. (2008a). Limited salvage logging effects on forest regeneration after moderate-severity windthrow. *Ecological Applications, 18*, 407–420.

Peterson, C. J., & Leach, A. D. (2008b). Salvage logging after windthrow alters microsite diversity, abundance and environment, but not vegetation. *Forestry, 81*, 361–376.

Peterson, C. J., & Pickett, S. T. A. (1991). Treefall and resprouting following catastrophic windthrow in an old growth hemlock-hardwoods forest. *Forest Ecology and Management, 42*, 205–217.

Peterson, C. J., & Pickett, S. T. A. (1995). Forest reorganization: A case study in an old-growth forest catastrophic blowdown. *Ecology, 76*, 763–774.

Peterson, C. J., & Rebertus, A. J. (1997). Tornado damage and initial recovery in three adjacent, lowland temperate forests in Missouri. *Journal of Vegetation Science, 8*, 559–564.

Peterson, C. J., Carson, W. P., McCarthy, B. C., & Pickett, S. T. A. (1990). Microsite variation and soil dynamics within newly created treefall pits and mounds. *Oikos, 58*, 39–46.

Peterson, C. J., Krueger, L. M., Royo, A. A., Stark, S., & Carson, W. P. (2013). Disturbance size and severity covary in small and mid-size wind disturbances in Pennsylvania northern hardwoods forests. *Forest Ecology and Management, 302*, 273–279.

Phillips, J. D., Marion, D. A., & Turkington, A. V. (2008). Pedologic and geomorphic impacts of a tornado blowdown event in a mixed pine-hardwood forest. *Catena, 75*, 278–287.

Platt, W. J., Doren, R. F., & Armentano, T. V. (2000). Effects of Hurricane Andrew on stands of slash pine (*Pinus elliottii* var. *densa*) in the everglades region of South Florida (USA). *Plant Ecology, 146*, 43–60.

Prengaman, K. A., Kribel, J. R. G., & Ware, S. (2008). Effects of Hurricane Isabel on a maturing hardwood forest in the virginia coastal plain. *Journal of the Torrey Botanical Society, 135*, 360–366.

Putz, F. E., & Sharitz, R. R. (1991). Hurricane damage to old-growth forest in Congaree Swamp National Monument, South Carolina, USA. *Canadian Journal of Forest Research, 21*, 1765–1770.

Rauber, R. M., Walsh, J. E., & Charlevoix, D. J. (2002). *Severe and hazardous weather*. Dubuque: Kendall/Hunt Publishing.

Rebertus, A. J., & Meier, A. J. (2001). Blowdown dynamics in oak-hickory forests of the Missouri Ozarks. *Journal of the Torrey Botanical Society, 128*, 362–369.

Rentch, J. S. (2010). Relationship between treefall direction, slope-aspect, and wind in eight old-growth oak stands in the central hardwood forest, USA. *Journal of the Torrey Botanical Society, 137*, 391–400.

Rich, R. L., Frelich, L. E., & Reich, P. B. (2007). Wind-throw mortality in the southern boreal forest: Effects of species, diameter and stand age. *Journal of Ecology, 95*, 1261–1273.

Robinson, G., & Zappieri, J. (1999). Conservation policy in time and space: Lessons from divergent approaches to salvage logging on public lands. *Conservation Ecology, 3*, 3 [online].

Romme, W. H., & Martin, W. H. (1982). Natural disturbance by tree falls in old-growth mixed mesophytic forest: Lilley Cornett Woods, Kentucky. In R. N. Muller (Ed.), *Proceedings 16th Central Hardwood Forest Conference*. Lexington: Univ of Kentucky Press.

Schaetzl, R. J., Burns, S. F., Johnson, D. L., & Small, T. W. (1989). Tree uprooting: Review of impacts on forest ecology. *Vegetatio, 79*, 165–176.

Sobhani, V. M., Barrett, M., & Peterson, C. J. (2014). Robust prediction of treefall pit and mound sizes from tree size across 10 forest blowdowns in eastern North America. *Ecosystems, 17*, 837–850.

Stanturf, J. A., Goodrick, S. L., & Outcalt, K. W. (2007). Disturbance and coastal forests: A strategic approach to forest management in hurricane impact zones. *Forest Ecology and Management, 250*, 119–135.

Suckling, P. W., & Ashley, W. S. (2006). Spatial and temporal characteristics of tornado path direction. *The Professional Geographer, 58*, 20–38.

Thurig, E., Palusuo, T., Bucher, J., & Kaufmann, E. (2005). The impact of windthrow on carbon sequestration in Switzerland: A model-based assessment. *Forest Ecology and Management, 210*, 337–350.

Trapp, R. J., Diffenbaugh, N. S., Brooks, H. E., Baldwin, M. E., Robinson, E. D., & Pal, J. S. (2007). Changes in severe thunderstorm environment frequency during the 21st century caused by anthropogenically enhanced global radiative forcing. *Proceedings of the National Academy of Sciences, 104*, 19719–19723.

Turner, M. G., Hargrove, W. W., Gardner, R. G., & Romme, W. H. (1994). Effects of fire on landscape heterogeneity in Yellowstone National Park, Wyoming. *Journal of Vegetation Science, 5*, 731–742.

Vargas, R., & Allen, M. F. (2008). Diel patterns of soil respiration in a tropical forest after hurricane Wilma. *Journal of Geophysical Research, 113*, G03021. doi:10.1029/2007JG00620.

Webb, S. L. (1999). Disturbance by wind in temperate-zone forests. In L. R. Walker (Ed.), *Ecosystems of disturbed ground*. Amsterdam: Elsevier.

White, P. S. (1979). Pattern, process, and natural disturbance in vegetation. *The Botanical Review, 45*, 229–299.

WSEC. (2006). *A recommendation for an enhanced Fujita scale (EF-scale)*. Lubbock: Wind Science and Engineering Center, Texas Tech University.

Wurman, J., Kosiba, L., & Robinson, P. (2013). In situ, Doppler radar, and video observations of the interior structure of a tornado and the wind – damage relationship. *Bulletin of the American Meteorological Society, 94*, 835–846.

Xi, W., & Peet, R. K. (2011). The complexity of catastrophic wind impacts on temperate forests. In A. Lupo (Ed.), *Recent hurricane research: Climate, dynamics, and societal impacts*. Vienna: InTech.

Xi, W., Peet, R. K., Decoster, J. K., & Urban, D. L. (2008a). Tree damage risk factors associated with large, infrequent wind disturbances of Carolina forests. *Forestry, 81*, 317–334.

Xi, W., Peet, R. K., & Urban, D. L. (2008b). Changes in forest structure, species diversity and spatial pattern following hurricane disturbance in a Piedmont North Carolina forest, USA. *Journal of Plant Ecology, 1*, 43–57.

Zeng, H., Chambers, J. Q., Negron-Juarez, R. I., Hurtt, G. C., Baker, D. B., & Powell, M. D. (2009). Impacts of tropical cyclones on US forest tree mortality and carbon flux from 1851 to 2000. *Proceedings of the National Academy of Sciences, 106*, 7888–7892.

Chapter 6
Fire as a Once-Dominant Disturbance Process in the Yellow Pine and Mixed Pine-Hardwood Forests of the Appalachian Mountains

Henri D. Grissino-Mayer

Fire happens almost everyday throughout the year, in some part or other, by Indians for the purpose of rousing game, as also by the lightning.

(William Bartram 1793)

Abstract In the southern and central Appalachian Mountains, dendroecological techniques conducted on fire-scarred trees in the yellow pine and mixed pine-hardwood forests of the Appalachian Mountains have shown that widespread fires burned about once every 7 years (range from 5 to 13 years) from the mid-1700s until 1925–1945 when a policy of widespread fire suppression was introduced and human-ignited fires were greatly reduced. This recent absence of fire has contributed to major changes in tree establishment rates, structural changes in forest stands, and changes in species composition. Major pulses of establishment in the first half of the twentieth century feature tree species that are shade-tolerant and fire-intolerant, replacing species adapted to repeated fires. Southern pine beetles have also dramatically reduced the abundance of yellow pines in xeric upland forests in recent decades. Yellow pines soon may be lost as a major component of Appalachian pine-oak forests as the forest floor develops a thick litter layer covered by ericaceous shrubs that were historically controlled by fire, especially on dry, low quality sites such as ridgetops and south- and southwest-facing slopes. Efforts to restore these pine/mixed hardwood ecosystems by reintroducing fires may be ineffective as land management agencies face forests best characterized as hybrid landscapes where effects of fire are largely unknown.

Keywords Yellow pine/mixed pine-hardwood forest • Fire history • Dendroecology • Tree rings • Fire scars

H.D. Grissino-Mayer (✉)
Department of Geography, University of Tennessee, Knoxville, TN 37996, USA
e-mail: grissino@utk.edu

C.H. Greenberg, B.S. Collins (eds.), *Natural Disturbances and Historic Range of Variation*, Managing Forest Ecosystems 32,
DOI 10.1007/978-3-319-21527-3_6

6.1 Fires in the Eastern USA

Public acceptance of fire (i.e., those caused by lightning and by humans, whether prescribed or not) as an important influence on forest ecology and a forest management tool in the eastern USA can be an uphill battle for foresters and fire scientists (Bright et al. 2007; Melvin 2012). In general, fire is a less important ecosystem process in moister plant communities of eastern forests relative to western forests. This reinforces a misconception that prescribed burning should not be a tool used by land management agencies in any forest type (Matlack 2013). Indeed, some groups vociferously oppose any reintroduction of prescribed fire to southern Appalachian forests (Blue Ridge Environmental Defense League 2013; Georgia Forest Watch 2014). This misunderstanding of the importance of fire in more xeric eastern forests is coupled with management policies dictating that fires, in most cases, should be suppressed (i.e., 'command and control') to prevent damage to forests and, especially, to humans and their dwellings that more and more encroach into wooded areas at the wildland-urban interface (Holling and Meffe 1996; Gaither et al. 2011; Moritz et al. 2014). The public is further indoctrinated on the need for fire suppression by successful but misleading advertisements from land management agencies (Capello 1999) that emphasize how fire is undesirable and adversely impacts wildlife, forest resources, forest health, and local-to-national economies (Jacobson et al. 2001; Gill and Stephens 2009).

Native Americans and then early Euro-American settlers knew the utility of fire. They would regularly set forests and grasslands on fire to ensure successful hunting activities, maintain travel corridors, provide nutrient-rich forage for livestock, keep pests at bay, and improve cultivation of certain forbs, grasses, nuts, and berries for food (Doolittle and Lightsey 1979; Pyne 1984; Putz 2003; Greenberg et al. Chap. 1; Greenberg et al. Chap. 12; Zenner Chap. 14). Early Euro-American settlers in the region likely learned this land management practice from the Native Americans, who learned to burn thousands of years earlier after observing what lightning had done naturally. Some historians believe that Native American burning of forests was so pervasive for so many millennia that entire landscapes ('humanized pyroscapes') arose from this burning with specific species compositions that are reflected to this day (Sauer 1950; Pyne 1984; Denevan 1992; Doolittle 2000; Kay and Simmons 2002; Abrams and Nowacki 2008; see Greenberg et al. Chap. 12). Other scientists contend that Native American burning did not cause a completely humanized pyroscape (Minnich 1988; Swetnam et al. 1999; Vale 2002; Barrett et al. 2005), although certain areas (such as lake margins) did experience increased fire activity caused by humans.

Many agencies acknowledge the need for fire to be introduced back into some central hardwood forest types where fire once operated, to redirect the successional trajectory to again favor fire-tolerant tree species, such as yellow pines (*Pinus* spp.) and oaks (*Quercus* spp.) that once were found in abundance. In the southern Appalachians, yellow pine and mixed pine-hardwood forests historically occurred

Fig. 6.1 North Mountain in the Jefferson National Forest, west of Roanoke, Virginia, looking to the east. The ridgetop trends southwest to northeast. The *arrows* point to the yellow pine/mixed pine-hardwood forest type typically found on more xeric south- and southwest-facing slopes of the southern Appalachians. Mesic hardwood forests separate these xeric sites

on more xeric, lower productivity topographic positions such as ridgetops and south- and southwest-facing slopes (Fig. 6.1). Fields abandoned by Native Americans and (later) by Euro-Americans were often invaded by yellow pines, as well (Greenberg et al. Chap. 1; Tuttle et al. Chap. 10; Wilfahrt et al. Chap. 11). Today, shade-tolerant, fire-intolerant tree species, such as eastern white pine (*P. strobus*), American beech (*Fagus grandifolia*), eastern hemlock (*Tsuga canadensis*), red maple (*Acer rubrum*), and blackgum (*Nyssa sylvatica*) are establishing in these forests (Armbrister 2002; Blankenship and Arthur 2006; DeWeese 2007; Fei and Steiner 2009; Aldrich et al. 2010; van de Gevel et al. 2012; Hart et al. 2012; LaForest 2012; Flatley et al. 2013). These species normally would have been kept in check by frequent fire activity (although individuals of many species can become more tolerant of fire with increasing age). Further, the abundance of yellow pines has been substantially reduced in recent decades due to southern pine beetle (*Dendroctonus frontalis*) outbreaks, further hastening the succession from yellow pine-hardwood forests to hardwood-dominated forests by killing single or groups of overstory pine trees (Nowak et al. Chap. 4). Reintroduction of fire back into yellow pine/mixed pine-hardwood ecosystems requires knowledge of past fire regimes so we can better understand the historical range of variation in fire frequency, seasonality, severity, spatial aspects, and fire-climate relationships (Morgan et al. 1994; Keane et al. 2009). To manage fire in future pine/mixed hardwood forests, we must know how fire affected forests in the past to gain a sense of the bioclimatic 'envelope' in which fire operated while acknowledging the existence of the humanized pyroscape.

6.2 Evidence of Past Fire

Microscopic and macroscopic charcoal (indicative of broad-scale fires and local fires, respectively) is produced during fire events and both make their way into lakes, ponds, and wetlands where they are integrated into the stratigraphy of the sedimentary profile (Whitlock and Larsen 2001; Whitlock and Anderson 2002). Once a core is extracted, the sediment can be sequentially analyzed, the charcoal separated, counted, and analyzed, and inferences made on the history of fire over the entire time span represented by the core once ages are determined for incorporated organic materials from radiocarbon dating. Charcoal in lake sediments from several states in the Central Hardwood Region (CHR) (Tennessee, North Carolina, and Kentucky) led Hazel and Paul Delcourt and their colleagues to conclude that (1) the region experienced considerable fire activity over most of the Holocene, and (2) the abundances and changes in charcoal amounts over time suggested linkages with simultaneous changes in Native American populations and burning practices (Delcourt et al. 1986, 1998; Delcourt and Delcourt 1997, 1998; Ison 2000; Leigh Chap. 8). Kneller and Peteet (1993) found that charcoal content in a core from Brown's Pond in the Ridge and Valley ecoregion of central Virginia showed clearly that the largest amounts occurred during the mid-Holocene when oaks and pines became the dominant tree species in their study area. The dominance of oak in mixed mesophytic forests in western Virginia for more than 6,000 years suggests a fire regime characterized by infrequent, light surface fires (Patterson 2005).

Physical evidence of past fire also can be extracted from soil charcoal (Patterson et al. 1987; Hart et al. 2008; Couillard et al. 2013) although the sequential stratigraphy required for evaluating changes in fire activity over time is rarely preserved due to soil turnover (Carcaillet 2001; Phillips and Lorz 2008). Several such studies have been conducted in the CHR (see also Greenberg et al. Chap. 1; Leigh Chap. 8). Welch (1999) analyzed macroscopic charcoal from soil cores taken from yellow pine-dominated stands on the Cumberland Plateau and found that (1) macroscopic charcoal existed at two sites where fire had not been recorded since federal land acquisition, and (2) fire had occurred at all five study sites in all topographic positions (summit, footslope, etc.) sampled. In a mixed hardwood forest on the Cumberland Plateau in central Tennessee, extracted macroscopic charcoal suggests that fire was common in what is considered today a mesic site (Hart et al. 2008). Charcoal abundances extracted from alluvium from a point bar of Buckeye Creek in West Virginia demonstrated linkages with changes in Native American populations, i.e., fire and charcoal amounts decreased dramatically after Native American abandonment of the watershed (Springer et al. 2010). Fesenmyer and Christensen (2010) obtained 82 radiocarbon ages for soil charcoal collected from the Nantahala National Forest in North Carolina and found that nearly all had dates falling within the past 4,000 years before present. Not only was fire common, but it extended downslope in what would be considered today as mesic hardwood forest. Increased fire activity and fire spread ca. 1,000 years before present suggested burning by Woodland Tradition Native Americans (Fesenmyer and Christensen 2010). Underwood (2013) was further able to identify specific taxa represented by soil charcoal (Scott 2010) by keying in on specific taphonomic and physical attributes. A key finding by Underwood (2013) was the ubiquitous presence of mesic conifer (especially eastern white pine) charcoal in areas currently dominated by yellow pines, suggesting that “the current period of mesophication may not be unique in this ecosystem.”

A third type of physical evidence for past fires can be found near the base of trees in the form of an injury caused by thermal destruction of sensitive tissues found just underneath the bark in the vascular cambium (Zackrisson 1977; Dieterich 1983; Fritts and Swetnam 1989; Smith and Sutherland 1999). The killing of these living tissues by a fire can create a basal injury once the bark sloughs off and exposes the interior xylem. This distinctive feature is known as a ‘catface’ among foresters (Fig. 6.2) and can be found if clues are sought in certain topographic settings. For

Fig. 6.2 A fire-scarred Table Mountain pine snag found on North Mountain in the Jefferson National Forest west of Roanoke, Virginia, showing a large fire-scarred cavity on the upslope side of the trunk. Notice the dense understory of mountain laurel that now surround the snag

Fig. 6.3 Dense thickets of mountain laurel now dominate Appalachian upland forests where fire has been absent since the early decades of the twentieth century. This thicket on Reddish Knob in the George Washington National Forest hides a Table Mountain pine snag with numerous fire scars, barely visible in the center of the photo. The repeated fires that scarred this pine would have kept the mountain laurel in check

example, fire-scarred pines often are found in areas where thickets of mountain laurel (*Kalmia latifolia*) exist (Fig. 6.3). This shrub usually indicates a more xeric site where past fires may have had significant effects on local vegetation. Fire-scarred pines are preferred because, once scarred, the resin exuded by pines around the injury facilitates scarring by future fire events. Careful examination of the trunk usually reveals that the basal wound actually has several noticeable vertically-aligned ridges that indicate the tree was able to record not just one, but several past fire events (Fig. 6.4). My colleagues and I have found individual trees, both living and dead, that recorded over 10 fire events across a 200 year period (Fig. 6.5). Trees are imperfect recorders of fires, however, because not every fire event will result in a fire scar on the tree trunk. Therefore, many fire-scarred trees are collected in a study area to ensure a more complete census of past fire events is obtained.

Fig. 6.4 Dr. Charles Lafon of Texas A&M University next to a felled Table Mountain pine snag found on Reddish Knob in the George Washington National Forest of Virginia, showing multiple *vertical* ridges on the tree trunk that indicate scarring by multiple fires in the past

Fig. 6.5 A cross section of a fire-scarred Table Mountain pine from Kelly Mountain in the Blue Ridge Mountains showing 11 fire scars recorded in a period of less than 200 years. The outermost scar in the *upper/lower left* of the photograph dates to 1921

6.3 Fire and Fire-Adapted Ecosystems

Further evidence that fire was historically important is found in the long list of specific adaptations by trees to repeated fire events (Wright and Bailey 1982; Keeley 2012; He et al. 2013). A well-known adaptation is the release of seeds in response to an environmental process, an adaptation termed 'serotiny.' When that process is fire, the adaptation is termed 'pyriscence' (Lamont et al. 1991). Both gymnosperms and angiosperms can possess pyriscence. In conifers, high temperatures melt away the resin that coats and binds the cone scales together (Wright and Bailey 1982; Pyne 1984; Keeley and Zedler 1998). Once the scales are opened, the seed can be released. Because fires may expose bare mineral soil, the seed has a greater chance of successfully germinating (Bramble and Goddard 1942; Wright and Bailey 1982; Kellman and Tackaberry 1993), although seeds from some tree species in the southeastern USA do best with a thin to moderate layer of duff (Waldrop et al. 2002; Jenkins et al. 2011). Among the yellow pine species in the CHR, Table Mountain (*P. pungens*) pine is distinctly serotinous while Virginia pine (*P. virginiana)* and pitch pine (*P. rigida*) are considered semi-serotinous, meaning that the resin does not bind the cone scales as tightly as it does in Table Mountain pine cones. This allows these cones to open soon after cone maturation under the right light and heat requirements found in certain topographic settings (Burns and Honkala 1990). Other adaptations displayed by trees in the CHR (especially yellow pines) include: thick bark that insulates against lethal temperatures during a wildfire; self-pruning of branches to remove ladder fuels that can carry fire to the crown; sensitive terminal buds protected by succulent, fire-resistant foliage; and the ability of some trees (such as shortleaf pine (*P. echinata*) saplings) to rapidly resprout from basal shoots after a fire (Keeley and Zedler 1998; Schwilk and Ackerly 2001; He et al. 2013).

Such adaptations likely evolved in response to repeated wildfires over millions of years (Scott 2000; Bond et al. 2005; Pausas and Keeley 2009; Keeley 2012). Pines are among the most fire-adapted of all tree species, with adaptations that occurred over the past ca. 150 million years (Keeley 2012). The phylogenetic distribution of pines based on fire adaptations suggests that pines in the two subgenera (*Strobus* = *Haploxylon* = five-needled pines = white pines; *Pinus* = *Diploxylon* = 1–3 needled pines = yellow pines) evolved by two different adaptive strategies (Schwilk and Ackerly 2001; Keeley 2012). The *Strobus* pines evolved to prefer sites considered environmentally stressful (except for eastern white pine in the eastern USA and sugar pine (*P. lambertiana*) in the western USA). Most pine species, however, adapted to evolve in fire-prone ecosystems, and this includes all of the *Diploxylon* pines that grow in the Appalachian Mountains corridor, e.g., pitch pine, shortleaf pine, Virginia pine, and Table Mountain pine. Keeley (2012) suggested that southern pine species (except eastern white pine) can be classified under the 'fire-tolerator pine syndrome' where survival after low and moderate severity fires depends on the pines having adapted specific traits. Some scientists assert that certain plant species (especially pines and grasses) create environmental conditions that actually favor the return of fire to minimize competition from other species (Mutch 1970; Platt et al. 1988; Keeley and Zedler 1998; Keeley 2012).

The association between fire and plants argues against ecosystems being 'humanized pyroscapes.' On the one hand, lightning was "the predominant source of ignition over evolutionary time" (Pyne 1984) long before purposeful human ignitions of fire, which necessitated that plants either adapt to repeated fires caused by lightning or die out. Further, lightning-caused fire likely contributed to the current distribution of biomes we see today, especially the relative proportion of grasslands, woodlands, and forests (Bond and Keeley 2005; Bond et al. 2005). Although lightning-caused fires in the Appalachians today constitute a small percentage of all fire events (Lafon and Grissino-Mayer 2007; see Greenberg et al. Chap. 1, Table 1.6), their effects on some forest types and in some topographic positions can be considerable (Cohen et al. 2007). On the other hand, the overprint of human-caused ignitions and disturbances hinders or perhaps makes impossible our evaluation of lightning-only fire regimes. Once humans arrived, natural fire ignitions became heavily augmented by human ignitions and forests later were influenced by human practices that led to fire exclusion (Stambaugh and Guyette 2006; McEwan et al. 2011; Brose et al. 2013; Leigh Chap. 8; Greenberg et al. Chap. 12). We may never know the specific metrics that characterize a 'natural' fire regime in the southern Appalachians prior to human arrival unless we find portions of landscapes that escaped the effects of human disturbances, an unlikely scenario.

6.4 Dendrochronology, Fire History, and Stand Dynamics in the Southeastern USA

Dendrochronology uses tree rings dated with precision to their exact calendar years of formation to better understand the spatiotemporal properties of physical processes that shaped the earth's surface. The use of tree-ring dating to better understand fire history had its origins when dendrochronologists observed that non-lethal fires would leave distinct and identifiable fire scars embedded within the tree-ring record (Zackrisson 1977; Dieterich and Swetnam 1984). A fire would kill only a portion of the living tissues inside the bark along the basal portion of the tree trunk, allowing the still-living tree to grow over the scar in subsequent years, thus preserving the fire scar. Identifying the year in which the tree ring formed therefore identifies the year in which the fire occurred. If the catface contained multiple such scars, then fire frequency and an estimate of the mean fire interval could be obtained, yielding a more accurate depiction of past fire activity over several centuries. Analyzing numerous such trees in an area would lead to a better understanding of fire history over a wider geographic area (Dieterich 1983; Fritts and Swetnam 1989). Combining the fire histories for many sites within a region can provide insights on the ability of fire to spread, possible ignition sources, and the influence of climate on fire occurrence (Swetnam et al. 1999; Grissino-Mayer and Swetnam 2000; Swetnam and Baisan 2003).

The application of tree rings to date fire scars from tree species in the eastern USA was slow to gain traction, but a landmark study performed by Harmon (1982)

Fig. 6.6 A fire-scarred pitch pine on the Cooper Road Trail in GSMNP showing the original section removed with a hand saw by Harmon (1982). We found this tree after noticing a slope covered with mountain laurel thickets, which usually indicate a fire-prone area where fire-scarred pines may be found

examined fire-scarred pines in Great Smoky Mountains National Park (GSMNP) (Fig. 6.6) and was the first to use tree rings to estimate fire frequency anywhere in the eastern USA (in this study, tree rings were counted/aged and not crossdated with annual precision). Based on 115 fire scars from 43 pines, Harmon found that fires occurred about once every 13 years during the mid-19th and 20th centuries and attributed these fires to human ignitions (also see Tuttle et al. Chap. 10). Such studies remained uncommon until a ground-breaking fire history study was conducted in 1993 based on samples from 16 multiple-scarred Table Mountain pines collected from southwestern Virginia (Sutherland et al. 1995). In this study, dendrochronological crossdating techniques were applied for the first time to obtain precise dates of the fire events with sub-annual (i.e., seasonal) precision. Sutherland et al. (1995) found that fires burned about once every 10 years from ca. 1800 until ca. 1940 and proved that the relatively little-known Table Mountain pine had tree rings and fire scars that could be crossdated.

An important set of complementary analyses to fire-scar investigations focuses on understanding forest stand dynamics, such as age structure (changes in tree establishment over time), stand structure (changes in numbers of seedlings, saplings, and adult trees over time), and stand composition (changes in species composition over time). These properties can reveal details on past fire activity not found in the fire-scar record. For example, Waldrop et al. (2000) found that Table Mountain pines growing in ridgetop, pine/hardwood communities were regenerating well into the middle of the twentieth century, but stopped regenerating with the beginning of successful fire exclusion practices such that no pines were found younger than 50 years old. Instead, the understory became covered with shrubs (particularly mountain laurel) that normally would have been reduced by repeated fires. Brose et al. (2002) found that most oaks and pines at study sites in northern Georgia became established in the early decades of the twentieth century after major disturbances that included both low- and high-intensity fires. Waldrop et al. (2003) observed that chestnut oak (*Q. montana*) was replacing Table Mountain pine in ridgetop pine communities of the southern Appalachian Mountains, caused by the absence of fire that would have maintained the fire-adapted pines and reduced the understory cover of dense mountain laurel thickets. They concluded that pine stands were created and maintained by frequent lower-intensity fires, as higher-intensity fires could prove detrimental to successful regeneration. Jenkins et al. (2011), however, found that yellow pine regeneration in the xeric pine/oak forests of the western portion of GSMNP improved with increasing fire severity, with greatest regeneration occurring after medium to high severity fires, but noted that higher-severity fires likely were uncommon in the late 1800s and early 1900s.

The link between fire severity and tree establishment is crucial because stand age structure can provide clues about fire severity during the 'lifespan' of those trees. Despite their great ecological value, fire scars provide only a partial history of fires. A fire that leaves a visible scar on the tree trunk must, by definition, have been low intensity at the base of that tree because it damaged a portion of the trunk, but did not kill the tree. Hence, fire scars cannot provide information on more intense, higher-severity fires because such fires would have killed the majority or all trees in an area (Smith and Sutherland 1999). Higher severity fires are evident when establishment dates of trees suggest an even-aged stand that initiated after a major disturbance. Overall, however, such higher intensity fires are not common in the pine/mixed hardwood upland forests of the Appalachians as most stands are uneven-aged, suggesting recruitment of individuals over time is continuous.

6.5 Changes in Fire Activity

The majority of investigations to date have established that fires pre-dating the twentieth century were a frequent disturbance in pine/mixed hardwood forests of the Appalachian Mountains until human alterations (especially removal of humans as an ignition source), alongside other human-related (e.g., fire suppression) and natural disturbances, caused a major reduction in fire activity with subsequent changes to forest stand dynamics and succession. The most common question posed by scientists and land managers is: "How often did fires occur prior to the decline in fire occurrence?" Several recent studies show that fires were more frequent than Harmon's initial estimates, and these studies now seem to converge on how often fire occurred. These dendroecological studies focused on the more 'ecologically important' fires by filtering out fire events when a single tree was affected by fire in a study area, usually by analyzing only those fire events where 10 or 25 % of the trees sampled had been scarred. Here, I report metrics based only on these more ecologically important fire events, i.e., those fires that may have been more extensive and/or more severe than others (Swetnam and Baisan 2003; Aldrich et al. 2010; Flatley et al. 2013).

At four ridgetop forests in the Jefferson National Forest of Virginia dominated by Table Mountain pine, fires occurred about once every 5–10 years beginning in the mid- to late 1700s, the length of the fire-scar record at these sites (DeWeese 2007). At three sites in the drier western portion of GSMNP, ecologically important fires occurred about once every 5–7 years (LaForest 2012). Near Cades Cove, an historical settlement in GSMNP, fires occurred once every 3–6 years (Feathers 2010) which likely reflects a higher number of ignitions from human settlement. At three sites in the George Washington National Forest in Virginia, ecologically important fires occurred once every 6–8 years (Aldrich et al. 2010, 2014). Flatley et al. (2013) found that ecologically important wildfires occurred about once every 5 years at Licklog Ridge in GSMNP, once every 7–8 years at House Mountain near Knoxville, Tennessee, and once every 6–7 years at Linville Gorge in North Carolina. An important finding was the ability of fire to spread throughout the pine-dominated stand that was sampled, and apparently through the intervening, more mesic oak-dominated hardwood forests between the pine stands. This latter finding is supported by fire-scarred hardwood trees found occasionally in the pine stands sampled (but not as common as fire-scarred pines) as well as in the intervening mesic hardwood stands (Fig. 6.7). This burning of more mesic (north-facing) slopes and coves was also documented in a lightning-caused fire in GSMNP (the Collins II Fire in April 1999)

Fig. 6.7 Fire-scarred hardwoods are common in the upland pine/mixed hardwood forests of the southern Appalachians but have not yet been the focus of fire history studies. Extracting fire history information from such hardwoods is more challenging than from upland pines

(Cohen et al. 2007). At Peters Mountain, Virginia, Hoss et al. (2008) found that ecologically important fires occurred about once every 12–13 years, an important finding that may help preserve the endangered Peters Mountain mallow (*Iliamna corei*). Based on these studies, fires occurred fairly often in the upland pine/mixed hardwood forests of the Appalachian Mountains, about once every 7 years, with average intervals between major fires ranging from a low of 5 years to a high of 13 years.

Other important properties of past fire regimes were also noted. First, within individual sites, considerable variability existed around these average intervals, from a low of 2 years to a high of 25 years, suggesting that the historical range of variation in fire activity of some eastern forest types is great and may be necessary to ensure successful tree establishment during a lengthy fire-free period. Fires that occur too frequently could kill regeneration before the trees can attain a height above the lethal flame height. Second, several studies clearly established that fires occurred predominantly after tree growth had ceased for the growing season (dormant season fires) or occurred in the earliest weeks or months once the tree broke dormancy (early growing season fires) (DeWeese 2007; LaForest 2012; Flatley et al. 2013; Aldrich et al. 2014), confirming that fires historically occurred during the fall or late winter/early spring months. Third, the influence of Euro-American settlement on fire activity may be time and place specific. For example, LaForest (2012) found that fires increased in frequency once Euro-Americans settled near today's GSMNP, from once every 6–12 years prior to 1834 to about once every 2–3 years after settlement. Other studies showed minimal effects of human settlement (Flatley et al. 2013). Fourth, in nearly all studies, no convincing evidence was found that higher severity fires characterized these fire regimes and might have caused major pulses of tree establishment. Evidence suggests that fires historically were low-intensity/severity disturbances that maintained forests rather than creating new ones. Finally, fires abruptly terminated between 1925 and 1945 in the upland forests of the southeastern USA. This cessation occurred most likely from changes in cultural burning practices and other fire exclusion activities (fire suppression, forest fragmentation, road building, etc.), effectively removing fire as a dominant force in upland yellow pine/mixed hardwood forests of the southern Appalachians.

6.6 Changes in Forests of the Southern Appalachians

The majority of studies suggest that major changes in yellow pine/pine-mixed hardwood forest structure, species composition, and successional pathways have occurred in southern Appalachian forests primarily due to a lack of fire combined with other interacting natural and anthropogenic disturbances (such as logging, tree diseases, livestock grazing, etc.). For example, mid-seral oak and pine species were maintained by periodic fires during the 18th and 19th centuries in an old-growth white pine/mixed oak forest in southern West Virginia, but fire absence during the twentieth century caused a proliferation of fire-intolerant tree species (Abrams et al.

1995). In a follow-up study, Abrams et al. (1998) found that northern red oak (*Q. rubra*) had continuously recruited during the 19th and into the 20th centuries. This species is fairly shade-intolerant and requires openings in the canopy for successful recruitment, which suggests that northern red oak likely persisted as a canopy dominant due to periodic disturbances, such as fire, opening the canopy. In the western, more xeric portions of GSMNP, pines and oaks established regularly until ca. 1945, but regeneration was nearly absent in species from both genera because of the lack of fire since then (Harrod and White 1999). This absence of fire contributed to establishment and eventual co-dominance by fire-intolerant species, especially young red maples (Harrod and White 1999), in a closed-canopy forest with high tree density as opposed to the open woodlands once maintained by frequent (human-ignited) fires. These findings were corroborated by Lafon and Kutac (2003) who observed that lack of fire in stands dominated by Table Mountain pines promoted an increased density of hardwood trees and stimulated hardwood invasion. Lafon and Kutac (2003) concluded with a powerful prediction: "Without fire, long-term maintenance of Table Mountain pine is improbable."

More recent studies in Appalachian forests further documented the impacts caused by the removal of human-ignited fires. Nine pine-dominated stands in Georgia, South Carolina, and Tennessee showed (1) no evidence of high-severity wildfires in the past; and (2) little pine regeneration at any of the nine sites in the last 25–50 years after near-continuous pine recruitment during the prior 300 years (Brose and Waldrop 2006). At several sites in the Jefferson National Forest of Virginia, forest stands currently dominated in the overstory by Table Mountain pine will likely transition to fire-intolerant hardwood species, such as red maple and black gum, in the future without disturbances (DeWeese 2007). At a site in the George Washington National Forest of Virginia, fire activity ceased following a major wildfire in 1930, which permitted fire-intolerant hardwoods to establish over the next several decades (Aldrich et al. 2010). The lack of successful pine and oak regeneration was linked to reduced fire activity in stands now dominated by an understory of fire-intolerant shrubs (Brose and Waldrop 2010). In GSMNP, LaForest (2012) found that wildfires ceased simultaneously with park establishment in 1934, after which fire-intolerant species (such as red maple and eastern white pine) established. She also found that yellow pine regeneration was weak and dominated by Virginia pine in stands formerly dominated by Table Mountain pine. Dye (2013) collected tree demographic data from 30 random plots in an old-growth forest in the Chattahoochee National Forest of northern Georgia and found not a single yellow pine seedling and few yellow pine saplings, even though large-diameter shortleaf and pitch pines were present in the overstory. Red maple, however, was abundant in the understory. Hessl et al. (2011) noted that ridgetop pine communities at a site in West Virginia appeared resilient to both high fire activity (as shown by establishment during a period of high fire activity) and fire exclusion (as shown by establishment after fire cessation ca. 1953), suggesting Table Mountain pine can, in certain topographic settings, regenerate in the absence of fire (see also Barden 2000). McEwan et al. (2011) proposed the 'multiple interacting ecosystem drivers hypothesis' to explain the transition from oak- to maple-dominated forests, which included changes in fire activity, major regime shifts in climate, changes in land use patterns, loss of foundation species, and changes in wildlife populations. In general, the removal of humans as the dominant source of fires has caused changes in yellow pine/pine-mixed hardwood forests of the southern Appalachians, but these changes could signify a return of these forests to conditions that existed prior to large-scale human augmentation of fire activity.

6.7 Climate as a Potential Driver of Wildfire Activity

Fires and climate are intricately linked. Studies performed in the western USA, for example, have firmly established relationships between fires and above-average rainfall in years preceding a fire year followed by severe drought during the fire year (Grissino-Mayer and Swetnam 2000; Swetnam and Baisan 2003). New insights on fire-climate relationships have been established by investigating shorter-term, interannual climate patterns such as El Niño/La Niña events (Swetnam and Betancourt 1990; Margolis and Balmat 2009) as well as longer-term interdecadal climate patterns, such as the Pacific Decadal Oscillation (Norman and Taylor 2003; Gill and Taylor 2010). In some cases, the influence of Native Americans on some landscapes of the western USA was suggested when fires did not appear related to climate patterns (Kaye and Swetnam 1999) or when changes in fire activity coincided with changes in Native American movement, land use, or warfare patterns (Savage and Swetnam 1990; Seklecki et al. 1996; Morino 1996). In the southern Appalachian Mountains, a weak but still significant relationship exists between drought and lightning-caused fires (Mitchener and Parker 2005).

Most studies that attempted to link fire events reconstructed from the fire-scar record with climate activity studies relied on superposed epoch analysis (SEA), a proven statistical technique for identifying fire-climate associations on interannual time scales in the western USA (Kaye and Swetnam 1999; Grissino-Mayer and Swetnam 2000; Rother and Grissino-Mayer 2014). This technique, however, may be limited in its ability to isolate climatic drivers during periods when humans considerably augmented fire activity. SEA is based on simple averaging of annual values for the selected climate variable in years leading up to and during the year of fire, once all fire years are stacked one on top of the other. This averaging process means that a climatic process that heavily contributed to lightning-ignited fires in some years could be masked by other years in which humans ignited fires when climate ordinarily would not have contributed to lightning-ignited fires.

A better statistical technique was needed that did not rely heavily on averaging. Recent studies have confirmed the viability of a statistical technique called bivariate event analysis (BEA) for establishing linkages between two environmental processes, such as fire and climate (Gavin et al. 2006; Senici et al. 2013; Rother and Grissino-Mayer 2014). BEA is based on Ripley's k-function for analyzing point patterns across space to determine if the points are clustered, dispersed, or random. BEA evaluates the pattern of events across time to determine if two processes are synchronous, asynchronous, or independent of each other. More importantly, no averaging is performed and BEA is potentially capable of isolating relationships between two processes when considerable noise is present.

Examination of relationships between climate and fire activity is now possible in the southeastern USA given the development of a growing network of sites where fire history has been analyzed using the fire-scar record. The initial results using SEA, however, were equivocal. Despite fires having occurred frequently for several centuries in the southeastern USA, no clear relationship could be established linking fires with specific climate patterns. For example, DeWeese (2007) found only a weak relationship between reconstructed Palmer Drought Severity Index (PDSI) and fire activity, even after combining the major fire years at her four sites and filtering this data set to isolate regional fire years that likely would have been driven by regional climate. LaForest (2012) suggested that anthropogenic burning during the period of increasing Euro-American settlement in GSMNP may have overridden the climate influence. Flatley et al. (2013) found no statistically significant relationship between fire events and reconstructed PDSI in the southern Appalachians and also suggested that anthropogenic burning may have masked any climatic influence.

To overcome the issues inherent with SEA, I used BEA to test relationships between the more widespread fire events at Brush Mountain, Virginia (DeWeese 2007) and various climate variables, and found two variables that stood out as potentially driving wildfire activity in the southern Appalachians. First, fire events at this site were synchronous with positive phases of the El Niño-Southern Oscillation (Fig. 6.8a). This is reasonable given that El Niño events cause drier conditions in a central core area of the southeastern USA and especially in western Virginia, with monthly rainfall deficits averaging 5–13 cm (Halpert 2014; Climate Prediction Center 2014). Furthermore, winter temperatures during El Niño events have consistently ranked among the warmest. These drier and warmer conditions occur because the subtropical jet (and moisture that accompanies it) is diverted southward during El Niño (warm) events, bringing wetter conditions to Florida and coastal regions of the southeastern USA up to North Carolina. Second, fire events at Brush Mountain were synchronous with positive (warm) phases of the North Atlantic Oscillation (NAO) (Fig. 6.8b). This finding is also reasonable because warm Atlantic waters cause warmer than average winter temperatures in the southeastern USA, and warmer winters can lead to an earlier and/or longer spring fire season and fuels that have had more time to lose moisture. These findings suggest that we may be able to isolate climate drivers of wildfire activity in the southeastern USA even though the fire regimes could be heavily augmented by human ignitions.

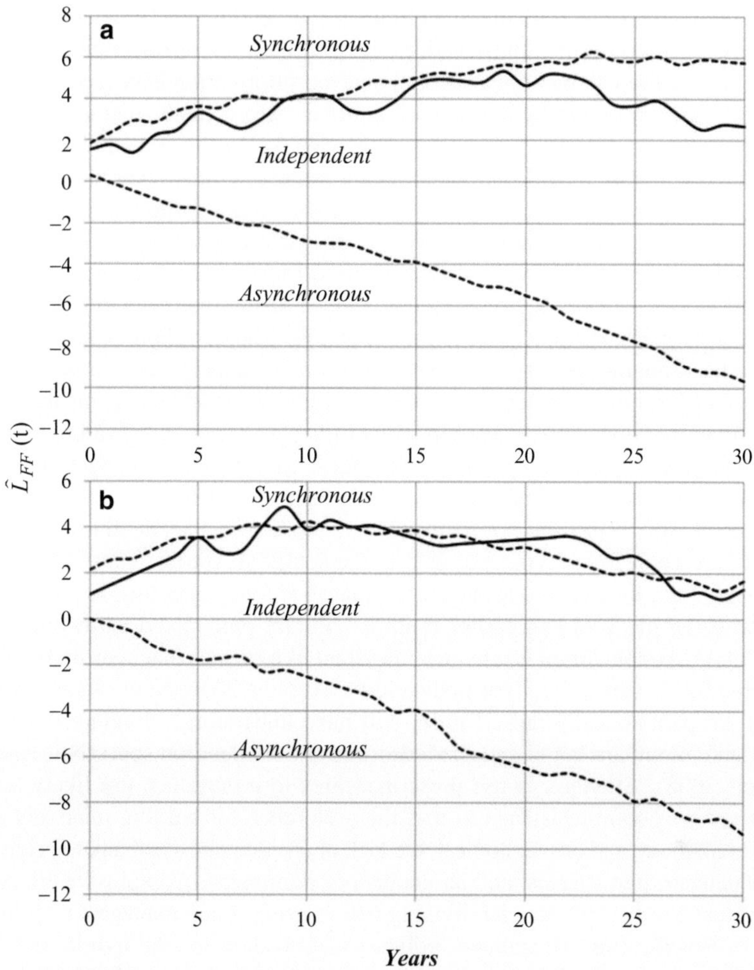

Fig. 6.8 Results from bivariate event analysis showing that fires on Brush Mountain in Virginia were synchronous with (**a**) positive (*warm phase*) El Niño events; and (**b**) positive phases of the North Atlantic Oscillation. *Dashed* lines represent 90 % confidence intervals demarcating synchronous, independent, and asynchronous relationships between climate and fire, while the *solid* line represents L-hat (re-scaled predicted k-function). A significant relationship is indicated when L-hat trends towards and crosses the *upper* or *lower* confidence intervals

6.8 Summary

Whether caused by lightning or humans, fire was once a dominant force in shaping and maintaining the yellow pine/pine-mixed hardwood forest types of the southern Appalachian Mountains for millennia. The removal of humans as the primary source of fire ignitions beginning in the early to mid-twentieth century has interacted with other disturbances to eliminate fire, leading land management agencies to seek methods for reintroducing fire with the goal of restoring these forest types. Before fire can be reintroduced, however, information is needed on fire frequency, intensity, severity, and spatial patterns before major disruptions to fire occurrence began. Dendroecological techniques have documented through the tree-ring based fire-scar record that ecologically important fires occurred about once every 7 years (range from 5 to 13 years) in the upland yellow pine/pine-mixed hardwood forests of the southern Appalachians. These low-severity fires occurred predominantly in the fall and late winter/early spring seasons that coincide with the bimodal fire seasons we experience today. Little evidence has been found to suggest that higher-severity fires occurred in these forests. The cessation of fires is associated with a dramatic decrease in yellow pine establishment and an increase in establishment of fire-intolerant hardwood species (e.g., red maple, blackgum, and eastern white pine) that may reduce the probability of fire occurring in the future (e.g., less flammable fuel types). In some study locations, oak species also appear to be less abundant in landscapes where fire has largely been eliminated.

Is restoration of historic fire frequency and the fire-dependent yellow pine/pine-mixed hardwood forest types possible in the southern Appalachians? The tipping point may have long passed as the forests managed today bear little resemblance to the forests of the 19th and earlier centuries. Future pine/mixed hardwood forests may continue to be dominated by fire-intolerant tree species that contribute to more fire-proof ecosystems (i.e., 'mesophication') (Abrams 2006; Nowacki and Abrams 2008) yet paradoxically these forests will have understories dominated by more flammable mountain laurel and rhododendron (*Rhododendron* spp.) (Nowacki and Abrams 2008). Attempts to use prescribed fires to reintroduce fire likely will be hampered by public resistance to fire, the preference for burning relatively small areas to maintain and ensure control, the lack of an adequate seedbank to bring back fire-dependent tree species, and an understory dominated by explosive ericaceous shrubs that would need manual thinning before firing. Land managers may instead wish to develop fire management policies that account for the hybrid and novel ecosystems that currently dominate our transforming landscapes (Hobbs et al. 2006, 2014).

Acknowledgements This summary on fires in the southern Appalachians would not have been possible without the constant professional companionship of my colleague, Dr. Charles Lafon of Texas A&M University, who provided the initial push to conduct this research. We both benefitted from some outstanding graduate students who participated in this research over the last 15 years: Georgina DeWeese, Serena Aldrich, Lisa LaForest, Will Flatley, Jennifer Flatley, Ian Feathers, Michael Armbrister, Justin Hart, and Alex Dye. I acknowledge Elaine Kennedy Sutherland of the USDA Forest Service for initiating tree-ring based fire history research in the Appalachians in 1993. I thank Mark Harmon of Oregon State University, Beverly Collins of Western Carolina University, Cathryn Greenberg of the USDA Forest Service, and an anonymous reviewer who provided excellent suggestions for improving drafts of this manuscript.

References

Abrams, M. D. (2006). Ecological and ecophysiological attributes and responses to fire in eastern oak forests. In M. B. Dickinson (Ed.), *Fire in eastern oak forests: Delivering science to land managers* (General Technical Report NRS-P-1, pp. 74–89). Newtown Square: USDA Forest Service Northern Research Station.

Abrams, M. D., & Nowacki, G. J. (2008). Native Americans as active and passive promoters of mast and fruit trees in the eastern USA. *Holocene, 18*, 1123–1137.

Abrams, M. D., Orwig, D. A., & Demeo, T. E. (1995). Dendroecological analysis of successional dynamics for a presettlement-origin white-pine-mixed-oak forest in the southern Appalachians, USA. *Journal of Ecology, 83*, 123–133.

Abrams, M. D., Ruffner, C. M., & DeMeo, T. E. (1998). Dendroecology and species co-existence in an old-growth *Quercus-Acer-Tilia* talus slope forest in the central Appalachians, USA. *Forest Ecology and Management, 106*, 9–18.

Aldrich, S. R., Lafon, C. W., Grissino-Mayer, H. D., DeWeese, G. G., & Hoss, J. A. (2010). Three centuries of fire in montane pine-oak stands on a temperate forest landscape. *Applied Vegetation Science, 13*, 36–46.

Aldrich, S. R., Lafon, C. W., Grissino-Mayer, H. D., & DeWeese, G. G. (2014). Fire history and its relations with land use and climate over three centuries in the central Appalachian Mountains, USA. *Journal of Biogeography, 41*, 2093–2104.

Armbrister, M. R. (2002). *Changes in fire regimes and the successional status of Table Mountain pine (Pinus pungens Lamb.) in the southern Appalachians, USA*. Thesis, University of Tennessee, Knoxville.

Barden, L. S. (2000). Population maintenance of *Pinus pungens* Lam. (Table Mountain pine) after a century without fire. *Natural Areas Journal, 20*, 227–233.

Barrett, S. W., Swetnam, T. W., & Baker, W. L. (2005). Indian fire use: Deflating the legend. *Fire Management Today, 65*, 31–34.

Bartram, W. (1793). *Travels through North and South Carolina, Georgia, East and West Florida*. Philadelphia: James and Johnson.

Blankenship, B. A., & Arthur, M. A. (2006). Stand structure over 9 years in burned and fire-excluded oak stands on the Cumberland Plateau, Kentucky. *Forest Ecology and Management, 225*, 134–145.

Blue Ridge Environmental Defense League. (2013). http://www.bredl.org/forests/index.htm. Accessed 9 Dec 2014.

Bond, W. J., & Keeley, J. E. (2005). Fire as a global 'herbivore': The ecology and evolution of flammable ecosystems. *Trends in Ecology and Evolution, 20*, 387–394.

Bond, W. J., Woodward, F. I., & Midgley, G. F. (2005). The global distribution of ecosystems in a world without fire. *The New Phytologist, 165*, 525–538.

Bramble, W. C., & Goddard, M. K. (1942). Effect of animal coaction and seedbed condition on regeneration of pitch pine in the barrens of central Pennsylvania. *Ecology, 29*, 330–335.

Bright, A. D., Newman, P., & Carroll, J. (2007). Context, beliefs, and attitudes toward wildland fire management: An examination of residents of the wildland-urban interface. *Human Ecology Review, 14*, 212–222.

Brose, P. H., & Waldrop, T. A. (2006). Fire and the origin of Table Mountain pine-pitch pine communities in the southern Appalachian Mountains, USA. *Canadian Journal of Forest Research, 36*, 710–718.

Brose, P. H., & Waldrop, T. A. (2010). A dendrochronological analysis of a disturbance-succession model for oak-pine forests of the Appalachian Mountains, USA. *Canadian Journal of Forest Research, 40*, 1373–1385.

Brose, P. H., Tainter, F. H., & Waldrop, T. A. (2002). Regeneration history of three Table Mountain pine/pitch pine stands in northern Georgia. In K. W. Outcalt (Ed.), *Proceedings of 11th biennial southern silvicultural research conference* (General Technical Report SRS-48, pp. 296–301). Asheville: USDA Forest Service Southern Research Station.

Brose, P. H., Dey, D. C., Guyette, R. P., Marschall, J. M., & Stambaugh, M. C. (2013). The influences of drought and humans on the fire regimes of northern Pennsylvania, USA. *Canadian Journal of Forest Research, 43*, 757–767.

Burns, R. M., & Honkala, B. H. (1990). *Silvics of North America, Vol 1: Conifers* (Agricultural Handbook 654). Washington, DC: USDA Forest Service.

Capello, C. (1999). And remember, "Only you can prevent forest fires!" *Fire Management Notes, 59*, 4–5.

Carcaillet, C. (2001). Are Holocene wood-charcoal fragments stratified in alpine and subalpine soils? Evidence from the Alps based on AMS 14C dates. *Holocene, 11*, 231–242.

Climate Prediction Center. (2014). *ENSO impacts on the US*. http://www.cpc.ncep.noaa.gov/products/predictions/threats2/enso/elnino/va_disc.html. Accessed 13 Sept 2014.

Cohen, D., Dellinger, B., Klein, R., & Buchanan, B. (2007). Patterns in lightning-caused fires at Great Smoky Mountains National Park. *Fire Ecology, 3*, 68–82.

Couillard, P. L., Payette, S., & Grondin, P. (2013). Long-term impact of fire on high-altitude balsam fir (*Abies balsamea*) forests in south-central Quebec deduced from soil charcoal. *Canadian Journal of Forest Research, 43*, 188–199.

Delcourt, H. R., & Delcourt, P. A. (1997). Pre-Columbian Native American use of fire on southern Appalachian landscapes. *Conservation Biology, 11*, 1010–1014.

Delcourt, P. A., & Delcourt, H. R. (1998). Influence of prehistoric human-set fires on oak-chestnut forests in the southern Appalachians. *Castanea, 63*, 337–345.

Delcourt, P. A., Delcourt, H. R., Cridlebaugh, P. A., & Chapman, J. (1986). Holocene ethnobotanical and paleoecological record of human impact on vegetation in the Little Tennessee River Valley, Tennessee. *Quaternary Research, 25*, 330–349.

Delcourt, P. A., Delcourt, H. R., Ison, C. R., Sharp, W. E., & Gremillion, K. J. (1998). Prehistoric human use of fire, the eastern agricultural complex, and Appalachian oak-chestnut forests: Paleocology of cliff palace pond, Kentucky. *American Antiquity, 63*, 263–278.

Denevan, W. (1992). The pristine myth: The landscape of the Americas in 1492. *Annals of the Association of American Geographers, 82*, 369–385.

DeWeese, G. G. (2007). *Past fire regimes of Table Mountain pine (Pinus pungens Lamb.) stands in the central Appalachian Mountains, Virginia, USA*. Dissertation, University of Tennessee, Knoxville.

Dieterich, J. H. (1983). Fire history of southwestern mixed conifer forest: A case study. *Forest Ecology and Management, 6*, 13–31.

Dieterich, J. H., & Swetnam, T. W. (1984). Dendrochronology of a fire-scarred ponderosa pine. *Forest Science, 30*, 238–247.

Doolittle, M. L. (2000). *Cultivated landscapes of Native North America*. New York: Oxford University Press.

Doolittle, M. L., & Lightsey, M. L. (1979). *Southern woods-burners: A descriptive analysis* (Research Paper SO-151). New Orleans: USDA Forest Service Southern Forest Experiment Station.

Dye, A. W. (2013). *Stand dynamics and fire history of a southern Appalachian pine-hardwood forest on Rainy Mountain, Chattahoochee National Forest, Georgia, USA*. Thesis, University of Tennessee, Knoxville.

Feathers, I. C. (2010). *Fire history from dendrochronological analysis at two sites near Cades Cove, Great Smoky Mountains National Park, USA*. Thesis, University of Tennessee, Knoxville.

Fei, S. L., & Steiner, K. C. (2009). Rapid capture of growing space by red maple. *Canadian Journal of Forest Research, 39*, 1444–1452.

Fesenmyer, K. A., & Christensen, N. L. (2010). Reconstructing Holocene fire history in a southern Appalachian forest using soil charcoal. *Ecology, 91*, 662–670.

Flatley, W. T., Lafon, C. W., Grissino-Mayer, H. D., & LaForest, L. B. (2013). Fire history and its relation to climate and land use in three southern Appalachian landscapes in the eastern US. *Ecological Applications, 23*, 1250–1266.

Fritts, H. C., & Swetnam, T. W. (1989). Dendroecology: A tool for evaluating variations in past and present forest environments. *Advances in Ecological Research, 19*, 111–188.

Gaither, C. J., Poudyal, N. C., Goodrick, S., Bowker, J. M., Malone, S., & Gan, J. B. (2011). Wildland fire risk and social vulnerability in the southeastern United States: An exploratory spatial data analysis approach. *Forest Policy and Economics, 13*, 24–36.

Gavin, D. G., Hu, F. S., Lertzman, K., & Corbett, P. (2006). Weak climatic control of stand-scale fire history during the late Holocene. *Ecology, 87*, 1722–1732.

Georgia Forest Watch. (2014). http://www.gafw.org/fire.html. Accessed 9 Dec 2014.

Gill, A. M., & Stephens, S. L. (2009). Scientific and social challenges for the management of fire-prone wildland – urban interfaces. *Environmental Research Letters*. doi:10.1088/1748-9326/4/3/034014.

Gill, L., & Taylor, A. H. (2010). Top-down and bottom-up controls on fire regimes along an elevational gradient on the east slope of the Sierra Nevada, California, USA. *Fire Ecology, 5*, 57–75.

Grissino-Mayer, H. D., & Swetnam, T. W. (2000). Century-scale climate forcing of fire regimes in the American southwest. *Holocene, 10*, 213–220.

Halpert, M. (2014). *United States El Niño impacts*.http://www.climate.gov/news-features/blogs/enso/united-states-el-ni%C3%B1o-impacts-0. Accessed 28 Dec 2014.

Harmon, M. (1982). Fire history of the westernmost portion of Great Smoky Mountains National Park. *Bulletin of the Torrey Botanical Club, 109*, 74–79.

Harrod, J. C., & White, R. D. (1999). Age structure and radial growth in xeric pine-oak forests in western Great Smoky Mountains National Park. *Bulletin of the Torrey Botanical Club, 126*, 139–146.

Hart, J. L., Horn, S. P., & Grissino-Mayer, H. D. (2008). Fire history from soil charcoal in a mixed hardwood forest on the Cumberland Plateau. *The Journal of the Torrey Botanical Society, 135*, 401–410.

Hart, J. L., Buchanan, M. L., Clark, S. L., & Torreano, S. J. (2012). Canopy accession strategies and climate-growth relationships in *Acer rubrum*. *Forest Ecology and Management, 282*, 124–132.

He, T., Pausas, J. G., Belcher, C. M., Schwilk, D. W., & Lamont, B. B. (2013). Fire-adapted traits of *Pinus* arose in the fiery Cretaceous. *The New Phytologist, 194*, 751–759.

Hessl, A. E., Saladyga, T., Schuler, T., Clark, P., & Wixom, J. (2011). Fire history from three species on a central Appalachian ridgetop. *Canadian Journal of Forest Research, 41*, 2031–2039.

Hobbs, R. J., Arico, S., Aronson, J., et al. (2006). Novel ecosystems: Theoretical and management aspects of the new ecological world order. *Global Ecology and Biogeography, 15*, 1–7.

Hobbs, R. J., Higgs, E., Hall, C. M., et al. (2014). Managing the whole landscape: Historical, hybrid, and novel ecosystems. *Frontiers in Ecology and the Environment, 12*, 557–564.

Holling, C. S., & Meffe, G. K. (1996). Command and control and the pathology of natural resource management. *Conservation Biology, 10*, 328–337.

Hoss, J. A., Lafon, C. W., Grissino-Mayer, H. D., Aldrich, S. R., & DeWeese, G. G. (2008). Fire history of a temperate forest with an endemic fire-dependent herb. *Physical Geography, 29*, 424–441.

Ison, C. (2000). Fire on the edge: Prehistoric fire along the escarpment zone of the Cumberland Plateau. In D. A. Yaussy (Ed.), *Proceedings of workshop on fire, people, and the central hardwoods landscape* (General Technical Report NE-GTR-274, pp. 36–45). Newtown Square: USDA Forest Service Northeastern Research Station.

Jacobson, S. K., Monroe, M. C., & Marynowski, S. (2001). Fire at the wildland interface: The influence of experience and mass media on public knowledge, attitudes, and behavioral intentions. *Wildlife Society Bulletin, 29*, 929–937.

Jenkins, M. A., Klein, R. N., & McDaniel, V. L. (2011). Yellow pine regeneration as a function of fire severity and post-burn stand structure in the southern Appalachian Mountains. *Forest Ecology and Management, 262*, 681–691.

Kay, C. E., & Simmons, R. F. (Eds.). (2002). *Wilderness and political ecology: Aboriginal influences and the original state of nature*. Salt Lake City: University of Utah Press.

Kaye, M. W., & Swetnam, T. W. (1999). An assessment of fire, climate, and Apache history in the Sacramento Mountains, New Mexico. *Physical Geography, 20*, 305–330.

Keane, R. E., Hessburg, P. F., Landres, P. B., & Swanson, F. J. (2009). The use of historical range and variability (HRV) in landscape management. *Forest Ecology and Management, 258*, 1025–1037.

Keeley, J. E. (2012). Ecology and evolution of pine life histories. *Annals of Forest Science, 69*, 445–453.

Keeley, J. E., & Zedler, P. H. (1998). Evolution of life histories in *Pinus*. In D. M. Richardson (Ed.), *Ecology and biogeography of Pinus* (pp. 219–249). Cambridge: Cambridge University Press.

Kellman, M., & Tackaberry, R. (1993). Disturbance and tree species coexistence in tropical riparian forest fragments. *Global Ecology and Biogeography Letters, 3*, 1–9.

Kneller, M., & Peteet, D. (1993). Late-Quaternary climate in the Ridge and Valley of Virginia, USA: Changes in vegetation and depositional environment. *Quaternary Science Reviews, 12*, 613–628.

Lafon, C. W., & Grissino-Mayer, H. D. (2007). Spatial patterns of fire occurrence in the central Appalachian Mountains and implications for wildland fire management. *Physical Geography, 28*, 1–20.

Lafon, C. W., & Kutac, M. J. (2003). Effects of ice storms, southern pine beetle infestation, and fire on Table Mountain pine forests of southwestern Virginia. *Physical Geography, 24*, 502–519.

LaForest, L. B. (2012). *Fire regimes of lower-elevation forests in Great Smoky Mountains National Park, Tennessee, USA*. Dissertation, University of Tennessee, Knoxville.

Lamont, B. B., Lemaitre, D. C., Cowling, R. M., & Enright, N. J. (1991). Canopy seed storage in woody-plants. *The Botanical Review, 57*, 277–317.

Margolis, E. Q., & Balmat, J. (2009). Fire history and fire-climate relationships along a fire regime gradient in the Santa Fe Municipal Watershed, New Mexico, USA. *Forest Ecology and Management, 258*, 2416–2430.

Matlack, G. R. (2013). Reassessment of the use of fire as a management tool in deciduous forests of eastern North America. *Conservation Biology, 27*, 916–926.

McEwan, R. W., Dyer, J. M., & Pederson, N. (2011). Multiple interacting ecosystem drivers: Toward an encompassing hypothesis of oak forest dynamics across eastern North America. *Ecography, 34*, 244–256.

Melvin, M. A. (2012). *National prescribed fire use survey report*. Coalition of Prescribed Fire Councils, Inc. http://www.prescribedfire.net. Accessed 12 Jan 2015.

Minnich, R. A. (1988). *The biogeography of fire in the San Bernardino Mountains of California*. Berkeley: University of California Press.

Mitchener, L. J., & Parker, A. J. (2005). Climate, lightning, and wildfire in the national forests of the southeastern United States: 1989–1998. *Physical Geography, 26*, 147–162.

Morgan, P., Aplet, G. H., Haufler, J. B., Humphries, H. C., Moore, M. M., & Wilson, W. D. (1994). Historical range of variability: A useful tool for evaluating ecosystem change. *Journal of Sustainable Forestry, 2*, 87–111.

Morino, K. A. (1996). *Reconstruction and interpretation of historical patterns of fire occurrence in the Organ Mountains, New Mexico*. Thesis, University of Arizona, Tucson.

Moritz, M. A., Batllori, E., Bradstock, R. A., et al. (2014). Learning to coexist with wildfire. *Nature, 515*, 58–66.

Mutch, R. W. (1970). Wildland fires and ecosystems – a hypothesis. *Ecology, 51*, 1046–1051.

Norman, S. P., & Taylor, A. H. (2003). Tropical and north Pacific teleconnections influence fire regimes in pine-dominated forests of north-eastern California, USA. *Journal of Biogeography, 30*, 1081–1092.

Nowacki, G. J., & Abrams, M. D. (2008). The demise of fire and "mesophication" of forests in the eastern United States. *BioScience, 58*, 123–138.

Patterson, W. A. (2005). The paleoecology of fire and oaks in eastern forests. In M. B. Dickinson (Ed.), *Fire in eastern oak forests: Delivering science to land managers* (General Technical Report NRS-P-1, pp. 2–19). Newtown Square: USDA Forest Service Northern Research Station.

Patterson, W. A., Edwards, K. J., & Maguire, D. J. (1987). Microscopic charcoal as an indicator of fire. *Quaternary Science Reviews, 6*, 3–23.
Pausas, J. G., & Keeley, J. E. (2009). A burning story, the role of fire in the history of life. *BioScience, 59*, 593–601.
Phillips, J. D., & Lorz, C. (2008). Origins and implications of soil layering. *Earth Science Reviews, 89*, 144–155.
Platt, W. J., Evans, G. W., & Rathbun, S. J. (1988). The population dynamics of a long-lived conifer (*Pinus palustris*). *The American Naturalist, 131*, 491–525.
Putz, F. E. (2003). Are rednecks the unsung heroes of ecosystem management? *Wild Earth, 13*, 10–14.
Pyne, S. J. (1984). *Introduction to wildland fire*. New York: John Wiley and Sons.
Rother, M. T., & Grissino-Mayer, H. D. (2014). Climate influences on fire regimes in ponderosa pine forests of the Zuni Mountains, New Mexico, USA. *Forest Ecology and Management, 322*, 69–77.
Sauer, C. (1950). Grassland climax, fire, and man. *Journal of Range Management, 3*, 16–21.
Savage, M., & Swetnam, T. W. (1990). Early 19th-century fire decline following sheep pasturing in a Navajo ponderosa pine forest. *Ecology, 71*, 2374–2378.
Schwilk, D. W., & Ackerly, D. D. (2001). Flammability and serotiny as strategies: Correlated evolution in pines. *Oikos, 94*, 328–336.
Scott, A. C. (2000). The pre-Quaternary history of fire. *Palaeogeography, Palaeoclimatology, Palaeoecology, 164*, 297–345.
Scott, A. C. (2010). Charcoal recognition, taphonomy and uses in palaeoenvironmental analysis. *Palaeogeography, Palaeoclimatology, Palaeoecology, 291*, 11–39.
Seklecki, M. T., Grissino-Mayer, H. D., & Swetnam, T. W. (1996). Fire history and the possible role of Apache-set fires in the Chiricahua mountains of southeastern Arizona. In P. F. Folliott, L. F. DeBano, M. B. Baker, G. J. Gottfried, G. Solis-Garza, C. B. Edminster, D. G. Neary, L. S. Allen, & R. Hamre (tech coords) *Effects of fire on Madrean Province ecosystems: A symposium proceedings* (pp. 238–246). General Technical Report RM-289, Fort Collins: USDA Forest Service Rocky Mountain Research Station.
Senici, D., Lucas, A., Chen, H. Y. H., Bergeron, Y., Larouche, A., Brossier, B., Blarquez, O., & Ali, A. A. (2013). Multi-millennial fire frequency and tree abundance differ between xeric and mesic boreal forests in central Canada. *Journal of Ecology, 101*, 356–367.
Smith, K. T., & Sutherland, E. K. (1999). Fire-scar formation and compartmentalization in oak. *Canadian Journal of Forest Research, 29*, 166–171.
Springer, G. S., White, D. M., Rowe, H. D., Hardt, B., Mihimdukulasooriya, L. N., Cheng, H., & Edwards, R. L. (2010). Multiproxy evidence from caves of Native Americans altering the overlying landscape during the late Holocene of East-central North America. *Holocene, 20*, 275–283.
Stambaugh, M. C., & Guyette, R. P. (2006). Fire regime of an Ozark wilderness area, Arkansas. *The American Midland Naturalist, 156*, 237–251.
Sutherland, E. K., Grissino-Mayer, H. D., Woodhouse, C. A., et al. (1995). Two centuries of fire in a southwestern Virginia *Pinus pungens* community. In W. L. Myers (Ed.), *Proceedings: Inventory and management in the context of catastrophic events*. University Park: Pennsylvania State University.
Swetnam, T. W., & Baisan, C. H. (2003). Tree-ring reconstructions of fire and climate history in the Sierra Nevada and southwestern United States. In T. T. Veblen, W. L. Baker, G. Montenegro, & T. W. Swetnam (Eds.), *Fire and climatic change in temperate ecosystems of the western Americas* (pp. 154–191). New York: Springer.
Swetnam, T. W., & Betancourt, J. L. (1990). Fire-Southern Oscillation relations in the southwestern United States. *Science, 249*, 1017–1020.
Swetnam, T. W., Allen, C. D., & Betancourt, J. L. (1999). Applied historical ecology: Using the past to manage for the future. *Ecological Applications, 9*, 1189–1206.
Underwood, C. A. (2013). *Fire and forest history in native yellow pine and mixed hardwood-pine communities of the southern Appalachian Mountains, USA*. Dissertation, University of Tennessee, Knoxville.

Vale, T. R. (2002). The pre-European landscape of the United States: Pristine or humanized? In T. R. Vale (Ed.), *Fire, native peoples, and the natural landscape* (pp. 1–39). Washington, DC: Island Press.

van de Gevel, S. L., Spond, M. D., Hart, J. L., White, P. B., Grissino-Mayer, H. D., & Sutton, M. (2012). American chestnut to red oak: Forest dynamics in the Blue Ridge Mountains, North Carolina, USA. *Botany, 90*, 1263–1276.

Waldrop, T. A., Welch, N. T., Brose, P. H., Elliott, K. J., Mohr, H. H., Gray, E. A., Tainter, F. H., & Ellis, L. E. (2000). Current research on restoring ridgetop pine communities with stand replacement fire. In D. A. Yaussey (comp.) *Proceedings of the workshop on fire, people, and the central hardwoods landscape* (pp. 103–109). General Technical Report NE-274, Newtown Square: USDA Forest Service Northeastern Research Station.

Waldrop, T. A., Brose, P. H., Welch, N. T., Mohr, H. H., Gray, E. A., Tainter, F. H., & Ellis, L. E. (2002). High-intensity fires may be unnecessary for stand replacement of table Mountain pine: An overveiw of current research. In K. W. Outcalt (Ed.), *Proceedings of 11th biennial southern silvicultural research conference* (General Technical Report SRS-48, pp. 137–142). Asheville: USDA Forest Service Southern Research Station.

Waldrop, T. A., Brose, P. H., Welch, N. T., Mohr, H. H., Gray, E. A., Tainter, F. H., & Ellis, L. E. (2003). Are crown fires necessary for Table Mountain pine? In K. E. M. Galley, R. C. Klinger, & N. G. Sugihara (Eds.), *Fire conference 2000: The first national congress on fire ecology, prevention, and management* (pp. 157–163). Tallahassee: Tall Timbers Research Station.

Welch, N. T. (1999). Occurrence of fire in southern Appalachian yellow pine forests as indicated by macroscopic charcoal in soil. *Castanea, 64*, 310–317.

Whitlock, C., & Anderson, R. S. (2002). Fire history reconstructions based on sediment records from lakes and wetlands. In T. T. Veblen, W. L. Baker, G. Montenegro, & T. W. Swetnam (Eds.), *Fire and climatic change in temperate ecosystems of the western America* (pp. 3–31). New York: Springer.

Whitlock, C., & Larsen, C. P. S. (2001). Charcoal as a fire proxy. In J. P. Smol, H. J. B. Birks, & W. Last (Eds.), *Tracking environmental change using lake sediments* (Terrestrial, algal, and siliceous indicators, Vol. 3, pp. 75–97). Dordrecht: Kluwer Academic Publishers.

Wright, H. A., & Bailey, A. W. (1982). *Fire ecology: United States and Canada*. New York: Wiley-Interscience.

Zackrisson, O. (1977). Influence of forest fires on the north Swedish boreal forest. *Oikos, 29*, 22–32.

Chapter 7
Ice Storms in Central Hardwood Forests: The Disturbance Regime, Spatial Patterns, and Vegetation Influences

Charles W. Lafon

Abstract Freezing rain occurs more commonly in eastern North America than anywhere else on earth. During a major freezing rain event, or ice storm, heavy ice loads accrete on vegetation and strip branches, snap boles, and uproot trees. Freezing rain commonly develops throughout a large area within a winter storm, and therefore a single storm can damage broad swaths of forest. These disturbances greatly exceed 'background' gap formation in severity and extent, but they do not generate the catastrophic mortality associated with severe fires or windstorms. Ice storms create disturbances of intermediate severity. A paucity of appropriate weather data obscures freezing rain climatology, but some generalizations are possible. Ice storms occur most frequently in the northeastern and midwestern USA, and a zone of high frequency also extends southward along the Appalachian Mountains and Piedmont. In mountainous landscapes, disturbance severity varies strongly over elevational and topographic gradients. Several studies have investigated ice storm damage at the scale of forest stands. Tree species vary in susceptibility, with 'pioneer' species typically sustaining heaviest damage and mortality. Less is known about longer-term consequences for age structure and species composition. The large gaps appear to provide recruitment opportunities for light-demanding colonizers. At the same time, the damage-resistance of slow-growing, stress-tolerant trees such as hemlocks and oaks likely enables persistence in ice-prone forests. The ecological role of ice storms depends, in part, on interactions with fires and other disturbances.

Keywords Forest disturbance • Forest dynamics • Freezing rain • Glaze • Ice storm

---

C.W. Lafon (✉)
Department of Geography, Texas A&M University,
3147 TAMU, College Station, TX 77843, USA
e-mail: clafon@geog.tamu.edu

C.H. Greenberg, B.S. Collins (eds.), *Natural Disturbances and Historic Range of Variation*, Managing Forest Ecosystems 32,
DOI 10.1007/978-3-319-21527-3_7

7.1 Introduction

Once viewed as abnormal events that play a limited role in vegetation development (e.g., Clements 1936), disturbances are seen today as fundamental processes shaping vegetation structure and composition (e.g., Grime 2001). The influence of disturbances is plainly evident in some ecosystems, such as boreal forests, where extensive patches of even-aged trees emerge in the aftermath of infrequent but catastrophic fires or insect outbreaks (Kneeshaw et al. 2011). Such events are uncommon in most temperate and tropical broadleaf forests, however, occurring at intervals of many centuries or millennia (e.g., Frelich 2002). Far more frequent is individual-tree mortality (Hart Chap. 2), which kills up to one or two percent of trees annually and generates a fine-scale mosaic in which a canopy matrix alternates with scattered small gaps (Runkle 1990; Yamamoto 1992). The gaps free enough light and other resources to enable certain trees to regenerate.

Other species, however, including the oaks (*Quercus*) and pines (*Pinus*) that dominate many temperate forests across the northern hemisphere, benefit from larger openings—not necessarily the complete absence of canopy that follows catastrophic events, but at least a thinner stand with broad gaps formed by multiple-tree mortality (Waldrop and Brose 1999; Dey 2002). These moderate forest-thinning events occur frequently enough in some places to affect a stand one or more times within a tree lifespan (Frelich 2002; Wonkka et al. 2013), molding forest species composition, diversity, and structure, and probably exerting selective pressure through their influences on tree survival.

Temperate broadleaf forests would appear especially susceptible to moderate-severity natural disturbances, as they experience a suite of midlatitude and tropical cyclones, thunderstorms, and droughts. Ice storms are one such agent of disturbance. These events result when heavy freezing rain accretes onto trees (Fig. 7.1) and strips their branches, snaps their boles, or uproots the trees entirely (Fig. 7.2). The damage often extends over an entire landscape or region, as freezing rain commonly develops throughout a broad section of a winter storm (Rauber et al. 2001). For example, during February 8–11, 1994, a winter storm produced heavy freezing rain across a large swath of the southeastern USA, including much of the Central Hardwood Region (CHR) (Jones et al. 1997). Ice accumulations of 7.5 cm or more were reported in Virginia, Kentucky, and Mississippi. The storm knocked down power lines and trees in a belt extending from Arkansas and Louisiana northeastward across much of Mississippi, Tennessee, Kentucky, West Virginia, Virginia, North Carolina, and South Carolina.

An ice-storm affected area, however, does not experience homogeneous forest damage. Ice storm disturbances are heterogeneous at multiple spatial scales. This chapter explores the spatial and temporal dimensions of ice storm disturbance across the central hardwood forest of eastern North America. I begin by considering patterns of freezing rain occurrence at the broadest spatial scales—continental to subcontinental—where climate exerts the predominant influence. I then narrow to the finer scales of landscapes and individual stands, where terrain, species

Fig. 7.1 Freezing rain accreted onto tree branches, Mt. Rogers, Virginia

Fig. 7.2 Forest disturbed by an ice storm, Great Smoky Mountains National Park, Tennessee. The trees growing on the slope at the right side of the photograph sustained much damage, as evident from the large number of downed trees visible along the ridge edge of the photograph. In contrast, the slope visible in the background of the photograph (middle and left side) has few downed trees. The differences in forest damage between these different slopes apparently reflect differences in ice accretion on differing aspects

composition, and other non-climatic factors come into focus. Then, having characterized patterns of ice storm damage, I explore the longer-term implications of these periodic disturbances for forest development.

7.2 Latitudinal and Continental Patterns of Freezing Rain

Freezing rain generally develops where warm, moist air is advected over a shallow surface layer of cold, subfreezing air (Stewart and King 1987; Rauber et al. 2001). Rain falling from the overlying warm layer becomes supercooled as it passes through the cold layer below, and freezes onto surface objects with below-freezing temperatures. Most freezing rain events produce minor glazing, which may cause traffic hazards for humans but do not weigh heavily enough on trees to damage them. Major ice storms that produce thick ice accretions are the events that disturb forests, and such events occur relatively infrequently.

Freezing rain is, because of the combination of air masses necessary for its formation, a midlatitude phenomenon. Only in the midlatitudes do subfreezing air masses interact regularly with warm, humid air masses. But not all sections of earth's midlatitude zones are equally suited to freezing rain. These events rarely occur in the southern hemisphere (Bennett 1959), which has no source area for cold, continental air masses that would invade the midlatitudes (Barry and Chorley 2003). In the northern hemisphere, by contrast, the North American and Eurasian continents cover vast expanses at high latitudes, from which continental polar air masses are exported regularly into the midlatitudes during winter (Barry and Chorley 2003). The winter sea ice extending across the entire Arctic Ocean basin also provides a surface that encourages the development of extremely cold arctic air masses, which periodically advance southward into the middle latitudes.

Within the northern hemisphere temperate zone, certain regions are particularly susceptible to the incursion of continental polar and arctic air masses, as a consequence of global atmospheric circulation around the polar vortex in combination with terrain and continental configurations. Central and eastern North America are particularly exposed to these air masses because of undulations in the jet stream (Robinson and Henderson-Sellers 1999), which often develops a northerly or northwesterly flow that directs cold air from central/northern Canada toward the midlatitudes east of the Rocky Mountains. The cold, dense air generally remains east of the Rocky Mountains, pushing southward and eastward across the central plains and into the eastern USA/southeastern Canada, even at times penetrating southward into eastern Mexico and the Gulf of Mexico (Zishka and Smith 1980; Rogers and Rohli 1991).

Eastern Asia sees a similar pattern, with the Siberian high pressure zone generating intensely cold air masses that advance into eastern Asia (Barry and Chorley 2003), although the east-west orientation of the mountains impedes the penetration of cold air into the Asian subtropics. Europe, particularly the east and center, also experiences cold air masses, but the continent's maritime climate generally mitigates the intensity of the cold.

As for the other ingredient in freezing rain—warm, moist air—again it is central and eastern North America that are ideally positioned. The maritime tropical air masses that develop over the Gulf of Mexico and along the Gulf Stream are drawn regularly over the continent within midlatitude cyclones (Robinson and Henderson-Sellers 1999), and this warm air is advected over colder air from the north. The incursion of maritime air ensures that virtually all of the USA east of the Rockies, as well as southern Canada, receives regular precipitation throughout winter. Some of this precipitation comes in the form of freezing rain. Such events occur far less commonly on the Eurasian landmass (Bennett 1959). In eastern Asia, the winters are dry, with the midlatitudes cut off from humid air originating over the tropical Indian Ocean. In Europe, cold incursions are rare, as noted above, and no source of humid tropical air exists. Ice storms, therefore, are largely a disturbance of the temperate forests and temperate-boreal forest transition in central and eastern North America. Thus, they are an important topic for this book on disturbance regimes of the central hardwood forest of the USA.

7.3 Spatial Patterns of Freezing Rain Within North America

Even in North America, the specific meteorological circumstances required for freezing rain—advection of warm moist air over a colder surface layer—develop only occasionally, and most often only in certain favorable places. Where do these events occur most commonly?

Unfortunately, few good meteorological observations of freezing rain exist. In the USA, data on freezing rain occurrence are constrained largely to the precipitation-type records that are reported by the National Weather Service's (NWS) first-order weather stations (Changnon and Creech 2003), which comprise a rather sparse network of stations located primarily at airports. Each weather station records the precipitation type on an hourly basis; and by assembling such reports one can develop a climatology of freezing rain frequency, e.g., the mean annual number of freezing rain events at each station, or the mean annual number of hours during which freezing rain occurs. NWS also operates a larger number of cooperative weather stations, which form a denser network than the first-order stations but provide weather data only on a daily basis. These daily records focus on daily precipitation totals and on diurnal high and low temperatures; the operators are asked to report days when freezing rain occurs, but this request has generally been ignored (Changnon and Creech 2003), with the consequence that a spatially detailed portrayal of freezing rain frequency is not possible.

Climatic analyses based on the first-order stations and the few cooperative stations with sufficient records (Changnon 2003) reveal that during the period 1948–2000, freezing rain occurred at least once per year, on average, everywhere east of about 104° West longitude (roughly the eastern border of Wyoming or New Mexico) and north of the coastal margins along the Gulf of Mexico (e.g., central Alabama; Fig. 7.3). The highest incidence of freezing rain occurs in the midwestern and

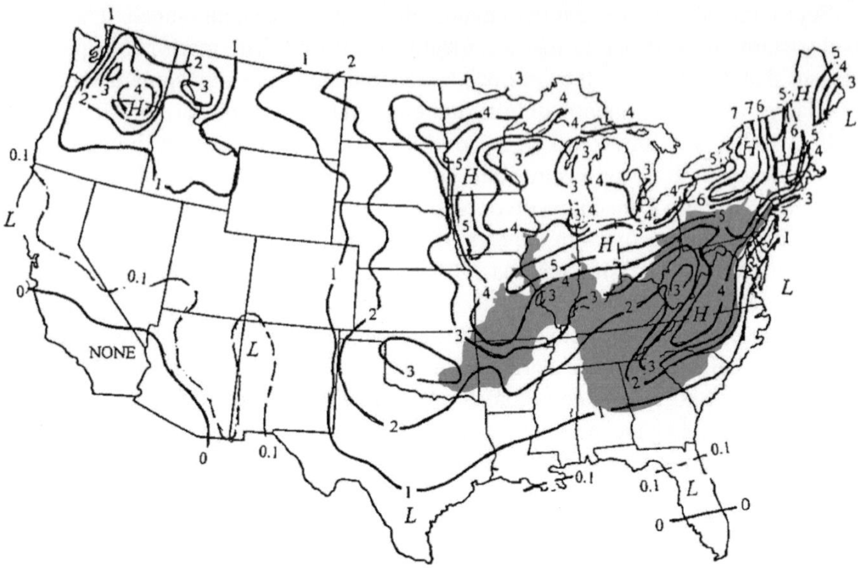

Fig. 7.3 Average annual number of days with freezing rain in the contiguous USA, 1948–2000 (From Changnon (2003)). *Gray* shading indicates the central hardwoods region that is the focus of this book

northeastern USA, along a belt between Missouri and Maine, with over seven events per year over much of New York. A secondary zone of high frequency extends southward along the Appalachian Mountains and Piedmont from Pennsylvania to northern Georgia, with up to four or five events annually (Gay and Davis 1993; Changnon 2003). A disjunct zone also emerges in the Pacific Northwest.

These geographic patterns reflect synoptic-level climatic processes, which control where the appropriate meteorological conditions for freezing rain emerge on a regular basis. Climatologists have identified several synoptic configurations that typically generate freezing rain east of the Rocky Mountains (Rauber et al. 2001). These configurations represent various arrangements between polar/arctic air and maritime tropical air within midlatitude cyclones and anticyclones. In general, the positioning of cold air along northern areas of the USA, combined with the frequent passage of fronts that advect warm, moist air from the south over the colder air, produces the zone of high freezing rain frequency across the Midwest and Northeast.

7.4 Influences of Terrain on Freezing Rain and Ice Storm Damage to Forests

Some freezing rain patterns cannot be understood solely as a response to broad, continental-scale atmospheric circulations, however. Where these circulations encounter elevated terrain, their configurations are altered in such a manner as to impose secondary patterns on the geography of freezing rain. The most prominent of these secondary zones, and the most important for forest disturbances, is the tongue of high freezing rain incidence extending along the Appalachian Mountains and the Piedmont (Fig. 7.3). This zone reflects interactions of high terrain with cold air masses. First, 'cold air damming' results when an anticyclone becomes centered over the northeastern USA, southeastern Canada, or the North Atlantic Ocean, and advects cold air toward the southwest: against the eastern slopes of the Appalachian Mountains (Rauber et al. 2001). The cold, dense air, unable to rise over the mountains, thrusts farther to the southwest and forms a shallow layer of subfreezing air along the eastern edge of the mountains and over the adjacent Piedmont. Often the northeasterly flow (i.e., to the southwest) of cold air from the anticyclone is enhanced by circulation around an Atlantic cyclone offshore of the southeastern USA. A northeasterly low-level jet can develop aloft within the cold dome (Bell and Bosart 1988), helping the cold layer to persist. Warm advection over the dammed cold layer produces ideal conditions for freezing rain in eastern Appalachia and the western/central Piedmont, which correspond to parts of the Piedmont, Blue Ridge Mountains, and Ridge and Valley ecoregions of the CHR (see Greenberg et al. Chap. 1, Fig. 1.1).

Second, 'cold air trapping' arises when a cold air mass is situated over the eastern USA ahead of a cyclone that is approaching from the west (Rauber et al. 2001). Southerly flow (i.e., to the north) ahead of the oncoming cyclone propels warm, moist air into the eastern USA on both the western and eastern sides of the Appalachian Mountains. In the mountains, however, cold air becomes trapped by the terrain and can remain entrenched for many hours within the valleys of the northern and central Appalachian region. Rain falling into this cold, trapped layer can produce heavy icing events, while liquid rain falls on the western side of the mountains. The Tennessee Valley and areas to the northeast, into central West Virginia, see relatively little freezing rain, therefore, even though the mountains and Piedmont just to the east experience some of the greatest frequencies of freezing rain on the continent (Konrad 1998). These striking east-west differences across the Appalachian region likely contribute to more dynamic vegetation conditions in eastern versus western Appalachia, especially given that other disturbance agents, including hurricanes and fires, appear to show similar spatial patterns across the region (Brinkman et al. 1975; Lafon and Grissino-Mayer 2007).

The other secondary zone in terrain-associated freezing rain lies within the Pacific Northwest (Rauber et al. 2001). The Cascade Mountains protect cold air lying over the Columbia Plateau from the warmer maritime air to the west, and when this moist air is advected eastward across the Plateau ahead of a cyclone, some of the cold air can become trapped within the Plateau or surrounding mountains, resulting in freezing rain. This freezing precipitation rarely continues long, as the maritime air rapidly alters the cooler interior air mass until its temperature rises above freezing. Ice storms do not appear, therefore, to constitute an important agent of natural disturbance in the Pacific Northwest.

Terrain affects climate at multiple spatial scales, of course, and the patterning of ice storm disturbance varies strongly at fine scales within hilly or mountainous landscapes. In the Appalachian Mountains, the damage associated with an ice storm is often concentrated in a certain elevation zone. The elevational range of damage depends on the meteorological conditions of the particular storm (Konrad 1998). After the raindrops fall from the overlying warm layer, which has above-freezing temperatures, they pass through the subfreezing air below and become supercooled. They do not achieve supercooling immediately upon entering the cold layer, however, and therefore freezing rain may not form in the highest elevations near the top of the cold layer. Freezing rain may develop, instead, on the mid- or low-elevation slopes or valleys (Boerner et al. 1988), which raindrops reach after passing through a substantial thickness of subfreezing air, cooling as they fall and eventually reaching 0° C. In other cases, temperatures in the valleys and lowest slopes are slightly above freezing, with the result that freezing rain would be concentrated at middle or higher elevations.

Regardless of the disturbance pattern associated with a specific event, it is the long-term pattern of repeated exposure that governs the disturbance regime. Which elevation zone(s) in the Appalachian Mountains sustains the greatest frequency of freezing rain? It is possible that the highest elevations commonly reach above the cold air layer and into the overrunning warm air above, in which case the middle or lower elevations would see more ice storm disturbances. Existing climatic data cannot easily resolve this pattern, but an attempt to estimate fine-scale spatial patterns of freezing rain from cooperative station data (Konrad 1998) suggests that higher elevations generally experience the greatest frequency of freezing rain. This estimate is not strongly conclusive, however, because of data limitations, and the climatic pattern probably cannot be resolved without extensive instrumentation to measure ice thickness. One approach to such instrumentation is described below in the section on ice storm frequency. Other possibilities would be to analyze remotely sensed imagery (e.g., Millward and Kraft 2004) covering forest damage patterns from many events, or using tree rings to identify past ice storm disturbances (e.g., Lafon and Speer 2002) over elevational gradients.

Slope orientation appears to exert a more consistent influence on ice storm disturbance than most other topographic factors. Several field studies note the presence of severe forest damage on east-, northeast-, or southeast-facing slopes (e.g., Whitney and Johnson 1984; Bruederle and Stearns 1985; Lafon et al. 1999), and remote sensing research confirms this pattern for entire landscapes (Millward and Kraft 2004; Stueve et al. 2007; Isaacs et al. 2014). Within a midlatitude cyclone, freezing rain usually develops in a sector of the storm exhibiting easterly, northeasterly, or southeasterly winds (Bernstein 2000). The wind appears to cause greater ice loads to accrete on windward slopes than on other aspects (Lafon et al. 1999). Why these aspects receive more ice is not entirely clear, but several mechanisms may contribute. Precipitation could be orographically enhanced on windward slopes, for example, or supercooling could be more effective. Whatever mechanism(s) generate the pattern, it is commonly a rather sharp one, where heavily damaged forest on windward slopes switches abruptly to relatively undamaged vegetation a few meters away on leeward slopes.

Slope angle has also been found to influence the severity of ice storm disturbance. Steep slopes generally see the heaviest forest damage (e.g., Seischab et al. 1993; Rhoads et al. 2002; Lafon 2006), possibly because of asymmetrical tree growth or shallow rooting on steep slopes. Slope apparently does not exert as strong an influence as aspect or elevation, however (Lafon 2006; Stueve et al. 2007).

To summarize, similar atmosphere-precipitation-terrain interactions seem to emerge repeatedly, exposing particular sites to periodic ice storm disturbance and sparing other sites. This recurrent patterning likely generates spatial differences in forest structure and composition, and sharpens the well known topographic zonation of forests. For example, hardwood stands cover most Appalachian landscapes, but pine-dominated stands commonly occupy west-facing slopes, where moisture stress and (historically) frequent fires thwarted the survival of competing hardwoods (Aldrich et al. 2010). The pines, in turn, are especially vulnerable to ice damage, and ice storms may help confine them to west-facing slopes that generally are sheltered from heavy ice accretion (Lafon et al. 1999). In particular, the sharp boundaries between pine- and hardwood-dominated stands could reflect differences in ice storm climatology between leeward and windward slopes.

7.5 Ice Storm Frequency

The extent to which ice storms play an important role in forest development depends on the frequency with which major, stand-damaging ice storms affect a particular location. Unlike the case for some other disturbance agents, such as fire, floods, or wind, few data exist for estimating ice storm disturbance intervals. A quantitative analysis of disturbance frequency and severity would require ice thickness data, which are not collected routinely in the USA. In Canada, however, such data have been obtained, at least for Québec, which lies within the section of Canada that is most prone to freezing rain: the southeastern part of the country (cf. McKay and Thompson 1969). The utility company Hydro-Québec established a network of 150 Passive Ice Meters (PIMs) on a 50 km grid across Québec during the late 1970s (Laflamme and Periard 1998). A PIM consists of tubes, rods, and plates on which an observer measures ice thickness. A freezing rain climatology based on the PIMs permitted return-interval estimates, for example, the maximum ice thickness probable for a 100-year return period. The return time for ice-induced tree mortality has also been estimated using PIM data combined with a survey of tree damage that resulted from a major ice storm (Proulx and Greene 2001). At the individual-tree scale, tree death from ice storms was estimated to recur, on average, at intervals between about 221 and 295 years in the vicinity of Montreal. These tree deaths do not occur continuously and randomly to individual trees, of course, but instead occur in episodes of multi-tree mortality during periodic major ice storms.

The Québec PIM data suggest that the spatial pattern of ice thickness generally resembles the pattern of freezing rain frequency (Laflamme and Periard 1998), but that the two patterns do not match closely, at least when the pattern is mapped at the

50 km grid resolution within the St. Lawrence River Valley (Fig. 7.4). The mean annual number of freezing rain events peaks at 12 events per year along a 150 km band within the Monts Notre Dame to the southeast of the St. Lawrence River (Fig. 7.4a). Smaller outlying spots with high event frequency (up to 9 events per year) can be found outside these highlands. In comparison, mapping the maximum ice thickness for a 100 year return period (Fig. 7.4b) suggests that the highest values (60 mm) are centered not in the high terrain but along the St. Lawrence River near Montreal. The Monts Notre Dame shows only moderate values (generally about 30 mm), although a zone of high values (up to 50 mm) is mapped for the northwestern slope of the hills and the adjacent St. Lawrence River.

Whether the spatial patterns of freezing rain frequency described above for the USA provide a good representation of geographic patterns in forest disturbance is not clear, therefore, but given the coarse scale at which the pattern has been mapped, the broad picture that emerges is probably reliable. That is, the forests most regularly exposed to ice storm disturbance lie within the midwestern and northeastern USA, and along the Appalachian-Piedmont belt. Support for this generalization can be found in data collected decades ago by utility companies, such as the American Telephone and Telegraph Company (AT&T). AT&T generated field reports on storms that had damaged telephone lines, and which therefore would have been relatively intense (Bennett 1959). According to AT&T data collected between 1911 and 1933, 'heavy storms' (which coated wires with an ice diameter $\geq$1.9 cm) occurred most frequently along a zone that generally corresponds with the Missouri – Maine belt revealed in Fig. 7.3. Specifically, a high-frequency zone extended from Oklahoma/Kansas/Nebraska, on the west, to southern New England, on the east; this region experienced a heavy storm every 3 years or less. This high-frequency zone also included the Mid-Atlantic region from New Jersey southward to southern Virginia, apparently reflecting the influence of Appalachian cold air damming. Hence, the observations of freezing rain frequency (Fig. 7.3) and of 'heavy storm' frequency suggest that within the CHR, ice storms probably have their greatest influence on forest dynamics of the Ozark Highlands, Interior River Valleys and Hills, northern parts of the Appalachian Plateau (including the Central Appalachian and Western Allegheny Plateau ecoregions), northern and central portions of the Ridge and Valley, Blue Ridge Mountains, and the Northern Piedmont, and central Piedmont ecoregions. The least affected areas appear to include the southern and central parts of the CHR (e.g., sections of the Piedmont, Southwestern Appalachians, and Interior Plateau ecoregions) (see Greenberg et al. Chap. 1, Fig. 1.1).

Some attempts have been made to supplement the scarce observations of freezing rain by conducting freezing rain return-interval analyses for the USA using modeled ice thickness. Several models have been created to simulate ice loading on utility lines from heat-balance calculations and observed temperature and rainfall data (e.g., Jones 1998; Makkonen 1998). The Electric Power Research Institute (EPRI) has used one of these (Makkonen 1998) to estimate 50-year return intervals for first-order weather stations across EPRI's Region 2 (Jones et al. 1997), which largely corresponds to the CHR that is the focus for this book. The modeled ice

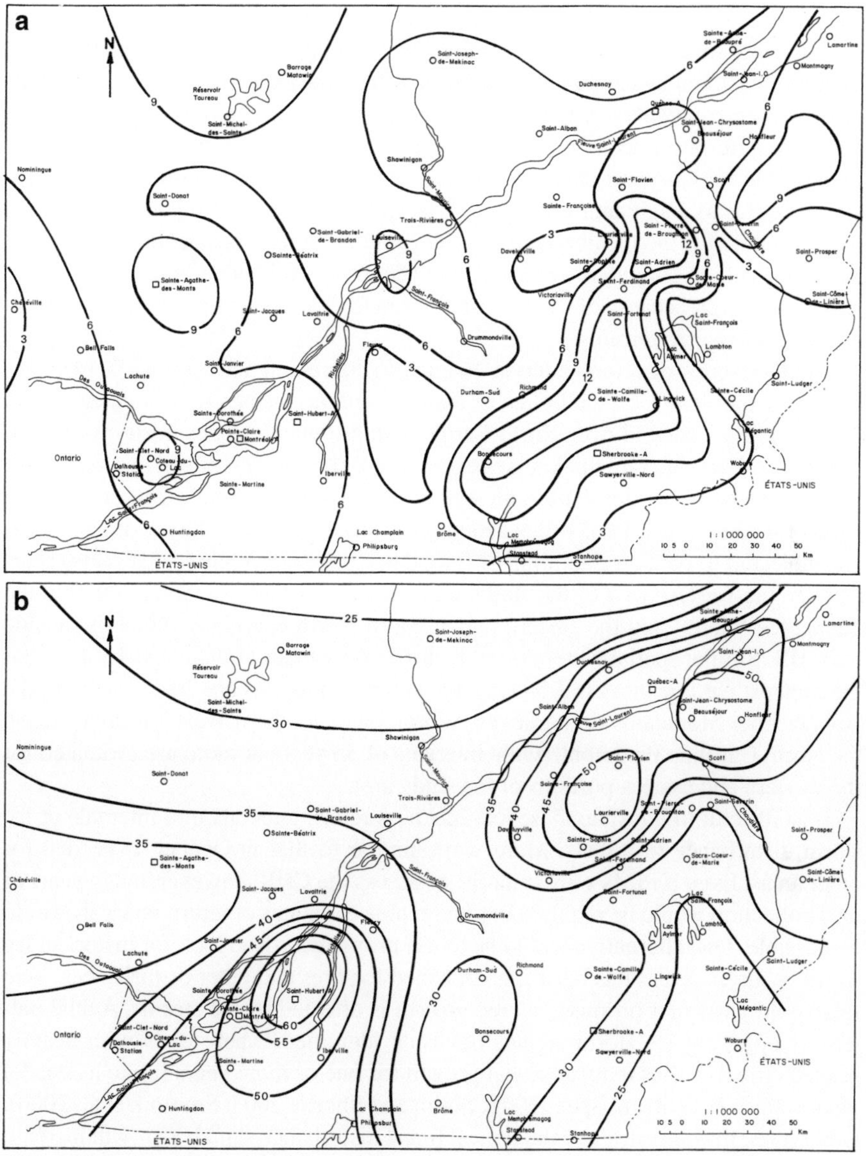

Fig. 7.4 (**a**) Average annual number of freezing rain events in the St. Lawrence Valley, Québec; (**b**) Maximum ice thickness (mm) for a 100-year event in the St. Lawrence Valley, Québec (From Laflamme and Périard (1998))

loads in this case differ from the Hydro-Québec ice thickness data noted above. EPRI's modeled ice loads represent radial ice thickness, that is, the thickness of ice if it were spread evenly around the entire circumference of the utility wire. Radial ice thickness is not directly comparable with simple measurements of ice thickness on one side of a surface. The predicted 50-year radial thicknesses are less than 1.3 cm in the southernmost areas (southern Mississippi and Alabama) and the Great Valley of east Tennessee (southern part of Ridge and Valley ecoregion), which, as discussed above, experiences less freezing rain than surrounding areas because of atmosphere-terrain interactions. Loads of 1.9–2.5 cm characterize most other sections of the region, with over 2.5 cm in southwestern Missouri and northwestern Arkansas (Ozark Highlands ecoregion). The EPRI report warns that their map likely is not reliable for the Cumberland Plateau (Southwestern Appalachians ecoregion) and Appalachian Mountains (Blue Ridge Mountain and Ridge and Valley ecoregions), however, because of strong variations in freezing rain climatology over short distances in rugged terrain.

Another series of return-interval estimates (Fuhrmann and Konrad 2014), conducted by the State Climate Office of North Carolina, focuses on a smaller region. This region is centered over North Carolina, and it encompasses portions of adjacent states. Freezing rain thicknesses exceeding 0.6 cm are predicted to recur at 1–2 year intervals from central Virginia southward to northern South Carolina, i.e., in parts of the Piedmont, Blue Ridge Mountains, and Ridge and Valley ecoregions that are subjected to cold air damming or trapping. Longer intervals (6–10 years) characterize the western part of the study area (Ridge and Valley ecoregion within the Tennessee Valley) and the eastern part (eastern North Carolina), which is outside the CHR. Similar spatial patterns emerge for heavier loads (1.3 cm and 1.9 cm). For the greatest ice thickness (>2.54 cm), the shortest intervals are 20–25 years in the Appalachian Mountains of southwest Virginia and southern West Virginia, and in the North Carolina Piedmont, while intervals of 55 years or more are estimated for the western and eastern portions of the study area.

It is difficult to translate these ice-thickness return intervals into intervals of ice storm disturbance to forests. More work similar to that in Québec is needed for other areas. Even if PIMs were established across the CHR, however, many years of data collection would be required before reliable estimates of return intervals would be possible. One alternative would be to use proxy records to look for historical ice storm occurrences, as has been accomplished for fires and other disturbances. Tree rings hold particular promise, as tree growth is affected by ice storms. A dual pattern of opposing ice storm signals has been identified, where trees with heavily pruned crowns suffer reduced radial growth for one or more years (up to a decade) after a storm (Lafon and Speer 2002; Smith and Shortle 2003; Smolnik et al. 2006). Other trees, in contrast, show the typical forest disturbance signal, a growth increase (Lafon and Speer 2002). This rise in radial growth apparently reflects the release of the tree from competition with surrounding trees that were killed by the storm. By verifying these dendroecological records through independent reports, such as newspaper accounts, a researcher could estimate the frequency of major, forest-disturbing ice storms over approximately the past century. Lafon and Speer (2002), for example, found tree-ring evidence suggesting that ice storms had disturbed an Appalachian forest in southwest Virginia (Ridge and Valley ecoregion) three times between 1914 and 1998: during the winters of 1920–1921, 1978–1979, and 1993–1994. The 1920–1921 signal corresponded with a November 1920 ice storm that was described by *The Roanoke [Virginia] Times and World News* to have destroyed orchards, damaged power lines beyond repair, and wrought havoc on timberlands across western Virginia and southern West Virginia. Local residents claimed that the storm was one of the worst ice storms ever known to have occurred in the region. Similar accounts were found for the major ice storms in January 1979 and in February and March 1994. Therefore, the dendroecological record, supported by independent accounts, indicates that major ice storms disturbed this particular forest at a mean interval of about 25 years during the study period. A longer study period would enable a more reliable estimate, but it would also require sampling older trees and sifting through increasingly sparse newspaper accounts to verify the events. Even if such reconstructions cannot be extended much beyond the early twentieth century, they would appear to offer a useful means to explore elevational gradients of ice storm disturbance in the Appalachian Mountains.

7.6 Tree Damage from Ice Storms

The aftermath of a major ice storm disturbance is a striking scene where uprooted trees lie entangled with snapped trunks amid the arched boles of slender young trees bent toward the ground. Broken tree branches are strewn over the forest floor. In one of the earliest scientific reports of ice storm disturbances, Rhoades (1918) described what he saw on the morning of March 6, 1915, following 24 h of continuous freezing rain and sleet in the Blue Ridge Mountains of North and South Carolina:

> ...the chestnut and oak slopes showed a vast array of whitened, splintered tops and trunks where the limbs had been torn away.... In numerous instances white oak [*Q. alba*], chestnut oak [*Q. montana*], and red oak [*Q. rubra?*] with their strong fibers were broken off sheer at a height of 20 feet [6 m] from the ground.... Second-growth yellow-poplar [*Liriodendron tulipifera*] coming up in dense stands in certain localities had 90 per cent of their number injured beyond recovery.

Yet, standing among all the damaged trees are some largely unscathed trees. The severity of ice storm disturbances rarely approaches the complete leveling of forests that is seen after a catastrophic windstorm (e.g., Peterson et al. Chap. 5). Field studies of ice storm disturbance report various levels of damage. In relatively mild events, less than 10 % of the trees sustained serious injuries such as uprooting, snapped boles, or heavy pruning of the crown (Rebertus et al. 1997). The most severe ice storm disturbances caused serious injuries in more than 30 % of the trees, and even up to 70–80 % of the trees in a stand in some cases (Whitney and Johnson 1984; Duguay et al. 2001; Proulx and Greene 2001; Lafon 2006).

The weight of ice can injure a tree in various manners. Branches commonly are broken, even by relatively small ice loads, and under heavy ice loads some trees lose virtually their entire crown (Whitney and Johnson 1984; Proulx and Greene 2001). Among other trees, the bole is uprooted, snapped, or bent. Some of these trees will recover by growing new branches and foliage, and therefore the disturbance severity cannot be calculated simply on the basis of what proportion of trees was injured.

Delayed mortality may continue for several years after the damage, but ice-induced mortality levels are known for few ice storms because most damage assessments have been conducted within a year of the storm's occurrence. Two studies, which were conducted 2–5 years after ice storm disturbance in the Appalachian Mountains, report mortality levels of 29–38 % (Whitney and Johnson 1984; Lafon 2006).

Whether any consistent patterns of tree damage emerge across ice storms is an important question that bears on the ecological consequences of these events. If certain species repeatedly resist damage better than others, for example, those resistant species may prevail in competition with the less resistant trees. Field studies of ice storm disturbance have been conducted across the temperate forest region of North America, and many of these studies characterize interspecific differences in ice storm damage (e.g., Boerner et al. 1988; Seischab et al. 1993), usually through an ordinal scale (e.g., most resistant, moderately resistant, least resistant). These interspecific rankings generally agree across studies. Eastern hemlock (*Tsuga canadensis*), for example, consistently emerges as a resistant tree, while yellow-poplar usually sustains heavy damage.

The appearance of a general interspecific damage pattern implies an underlying ecological principle. In particular, certain species are well adapted to withstand damage, while others are not. Why should trees differ in their damage resistance? Would it not benefit all species to resist damage? The answer to these questions seems to lie in evolutionary tradeoffs (Loehle 1988). To resist ice storms and other mechanical stresses, it would advantage a tree to construct the most damage-resistant stem and branches possible. The tree could accomplish this by several means, such as growing dense, decay-resistant wood or a short, thick trunk to support heavy ice loading. It could form a crown with only a few stout branches to minimize ice accretion. But these growth strategies would disadvantage the tree in competition with other species. By investing in strong, dense wood, a tree would divert photosynthate from productive tissues (leaves) or reproductive organs to unproductive wood, slowing its height growth and hobbling its competitive ability. Likewise, a sparse crown might accrete small ice loads, but it would also capture little sunlight and carbon dioxide, and would therefore constrain the tree's photosynthetic capacity. A tree cannot simultaneously maximize damage resistance, growth rate, and reproductive output.

Different species have adapted to the competing environmental demands in varying ways, some trees emphasizing resistance, for example, and others specializing in rapid growth (Loehle 1988). These variations are responsible for the emergence of repeated patterns of plant traits that underlie the classification of many species into a few categories of life history strategy (e.g., Grime 2001). Several researchers have observed that early-successional tree species commonly sustain greater ice damage than trees that are common in late succession (e.g., Lemon 1961; Whitney and Johnson 1984; Bruederle and Stearns 1985; Lafon 2006). This observation matches life history theory. For a tree to survive long enough to dominate a late-successional forest, it must invest in traits that confer damage resistance (Loehle 1988). These traits, however, would impede the tree's ability to colonize disturbed sites, as they would limit its growth rate and reproductive output. Therefore, species common in early succession should not be expected to have damage-resistant features that would enable them to withstand ice loading.

To provide a more formal characterization than early- versus late-successional trees, a recent study (Wonkka et al. 2013) classified trees according to Grime's CSR theory (Grime 2001) and examined the response of the different tree strategies to ice storms. Wonkka et al. (2013) analyzed tree damage data from multiple published reports and found that trees characterized as stress-tolerators (Grime's S strategy) consistently showed less damage than the other tree categories. Stress-tolerators include slow-growing, long-lived species, such as eastern hemlock and white oak, which frequently dominate late-successional stands. Their investments in damage-resistant features appear to pay off in terms of low mortality levels. Competitors (Grime's C strategy), in contrast, include yellow-poplar, white pine (*P. strobus*), and other species whose rapid growth and large size enable them to dominate disturbed sites in productive ecosystems. These trees do not withstand heavy ice loading. They commonly lose much of their crowns as limbs are stripped off by the weight of ice. Nonetheless, mortality is low in competitors, at least among yellow-poplar, the most commonly encountered competitor in ice storm studies, because these trees can repair their damaged crowns quickly and continue to grow (Fig. 7.5). It is the trees of more ruderal character (Grime's R strategy) that sustain the greatest

Fig. 7.5 Yellow-poplar tree whose crown and upper trunk were broken by an ice storm in 1978, Giles County, Virginia. The crotch formed where the tree recovered and grew new branches. Many other trees in this stand have a similar form

level of mortality. While no strongly ruderal trees existed in the ice storm datasets analyzed by Wonkka et al., trees showing related strategies such as SR (stress-tolerant ruderals) or CR (competitive ruderals) had high mortality. The yellow pines growing on infertile, fire-prone slopes, for example, sustain heavy damage and do not usually recover. These trees apparently lack the resources necessary to resist or recover from ice storm damage.

7.7 Implications of Ice Storm Disturbances for Forest Development

Ice storms generally are thought to influence forest development in one of two opposing ways: by accelerating successional replacement of pioneer species, or by favoring their continued maintenance in the community (Lemon 1961; Whitney and Johnson 1984; Boerner et al. 1988). The first scenario, accelerated succession, is consistent with the observation that early-successional species (i.e., the ruderals or competitors in Grime's theory) sustain heavier damage than species that typically dominate old-growth forests (i.e., Grime's stress-tolerators). The second scenario, maintenance of shade-intolerant colonizers, is based on the ability of ice storms to create large gaps that may not be filled by lateral extension of canopy trees. If the availability of light and soil resources is enhanced by the loss of competing individuals, ice storms could promote the establishment and canopy accession of shade-intolerant pioneers.

Some evidence exists for ice storms playing both these roles. Studies conducted in the immediate aftermath of ice storms typically reveal heavy damage to pioneer species, as noted above, portending an accelerated replacement of these trees. Some of the studies have also revealed an increase in seedling density or growth rate for pioneer species (Lemon 1961; Whitney and Johnson 1984). If some of these seedlings survive to maturity, the ice storms would have contributed to population maintenance of disturbance-dependent pioneer species.

It is difficult, however, to develop a dataset that examines longer-term forest dynamics associated with ice storm disturbances. The longevity of forest trees and the slow response of forests inhibit such work. One study conducted in Wisconsin (DeSteven et al. 1991) took advantage of permanent forest plots that were established in 1971, 5 years before a major ice storm disturbed the area in 1976. The plots encompassed both windward and leeward slopes within a landscape covered with northern hardwood forest. The forest vegetation within the plots sustained varying levels of damage, with greater disturbance severity on the windward than the leeward slopes. When the plots were resampled in 1987, 11 years after the ice storm, influences of the disturbance were apparent. The main pattern was an overall increase in the dominant, shade-tolerant sugar maple (*Acer saccharum*), indicating that the disturbance had favored the dominant species. Some of the large canopy openings that formed on the windward slope, however, enabled the establishment of shade-intolerant species, as evidenced by an increase of slippery elm (*Ulmus rubra*), white ash (*Fraxinus americana*), American hophornbeam (*Ostrya virginiana*), and black cherry (*Prunus serotina*) in the sapling and/or seedling strata.

Simulation modeling offers a means for investigating the potential influences of recurrent ice storm disturbances on forest dynamics. A gap-modeling study for northern hardwoods forest in the northern Adirondack Mountains of New York (Lafon 2004) suggested that, on slopes exposed repeatedly to heavy ice accretion, ice storms would reduce the abundance of the slowly growing, shade-tolerant eastern hemlock, as compared to its abundance without disturbance. Ice storms would favor its hardwood competitors, especially sugar maple. Although sugar maple itself is a shade-tolerant species typical of old-growth forests, it is a more opportunistic species than hemlock in terms of its more rapid response to disturbance. Additionally, the simulations project a higher abundance of less shade-tolerant species, such as yellow birch (*Betula alleghaniensis*), in the disturbed forest compared to the undisturbed forest, resulting in the maintenance of higher species richness over the long term. These modeling results, combined with field evidence, suggest that ice storms promote forests that are dominated by stress-tolerant species, but with a mix of additional species representing other tree strategies that likely would not be present in the absence of the disturbances (Wonkka et al. 2013).

On actual landscapes, of course, the effects of ice storms are entangled with numerous other factors that influence vegetation structure and composition. For example, on a mesic site, such as the east-facing slope of a cool, moist ravine, forest development would be expected to produce stands dominated by mesophytic, shade-tolerant species such as eastern hemlock, and periodic ice storms likely would enhance species richness by enabling the maintenance of shade-intolerant species, as in the Adirondack simulation. On a droughty slope with slow tree growth, in contrast, periodic ice storms might reduce tree species richness, as few species could recover rapidly enough under such stressful conditions to endure frequent disturbances.

Likewise, the interacting effects of multiple disturbances must be considered if the role of ice storms is to be ascertained. Under the short fire intervals of the past, for example, the large canopy openings created by ice storms likely facilitated the grass/shrub understories that are thought to have characterized many forests and woodlands historically (Harrod et al. 2000; Noss 2013). These openings would have also promoted the establishment of disturbance-dependent trees, such as Table Mountain pine (*P. pungens*) in the Blue Ridge Mountains and Ridge and Valley ecoregions of the Appalachian Mountains (Lafon and Kutac 2003; Brose and Waldrop 2006), thus helping to retain the 'early-successional' nature of the community. Fire exclusion appears to have altered the role of ice storms. In the absence of fire, ice storms accelerate the successional replacement of xerophytic communities by killing overstory trees without clearing the understory or consuming the litter as would a fire (Lafon and Kutac 2003).

The various combinations of ice storm severity/frequency, soil conditions, and other disturbances have contributed to the complexity of vegetation patterning across landscapes, especially in rugged terrain such as the Appalachian Mountains.

At the broader scale of the entire CHR, the pronounced gradients in freezing rain climatology have undoubtedly imprinted themselves upon subcontinental-scale vegetation patterns.

Acknowledgements I thank Steve Croy, Daehyun Kim, Cathryn Greenberg, and Beverly Collins for comments that helped me improve this chapter.

References

Aldrich, S. R., Lafon, C. W., Grissino-Mayer, H. D., DeWeese, G. G., & Hoss, J. A. (2010). Three centuries of fire in montane pine-oak stands on a temperate forest landscape. *Applied Vegetation Science, 13*, 36–46.

Barry, R. G., & Chorley, R. J. (2003). *Atmosphere, weather, and climate* (8th ed.). London: Routledge.

Bell, G. D., & Bosart, L. F. (1988). Appalachian cold-air damming. *Monthly Weather Review, 116*, 137–161.

Bennett, I. (1959). *Glaze: Its meteorology and climatology, geographical distribution, and economic effect* (Technical report EP 105). Natick: Quartermaster Research and Engineering Command. US Army Environmental Protection Research Division.

Bernstein, B. C. (2000). Regional and local influences on freezing drizzle, freezing rain, and ice pellet events. *Weather and Forecasting, 15*, 485–508.

Boerner, R. E. J., Runge, S. D., Cho, D. S., & Kooser, J. G. (1988). Localized Ice storm damage in an Appalachian Plateau Watershed. *The American Midland Naturalist, 119*, 199–208.

Brinkman, W. A. R., Huszar, P. C., Baker, E. J., Baumann, D. D., & Sims, J. H. (1975). *Hurricane hazard in the United States: A research assessment*. Boulder: Institute of Behavioral Science, University of Colorado.

Brose, P. H., & Waldrop, T. A. (2006). Fire and the origin of Table Mountain pine – Pitch pine communities in the southern Appalachian Mountains, USA. *Canadian Journal of Forest Research, 36*, 710–718.

Bruederle, L. P., & Stearns, F. W. (1985). Ice storm damage to a southern Wisconsin mesic forest. *Bulletin of the Torrey Botanical Club, 112*, 167–175.

Changnon, S. A. (2003). Characteristics of ice storms in the United States. *Journal of Applied Meteorology, 42*, 630–639.

Changnon, S. A., & Creech, T. G. (2003). Sources of data on freezing rain and resulting damages. *Journal of Applied Meteorology, 42*, 1514–1518.

Clements, F. E. (1936). Nature and structure of the climax. *Journal of Ecology, 24*, 252–284.

DeSteven, D., Kline, J., & Matthiae, P. E. (1991). Long-Term changes in a Wisconsin *Fagus-Acer* forest in relation to glaze storm disturbance. *Journal of Vegetation Science, 2*, 201–208.

Dey, D. (2002). The ecological basis for oak silviculture in eastern North America. In W. J. McShea & W. M. Healy (Eds.), *Oak forest ecosystems: Ecology and management for wildlife* (pp. 60–79). Baltimore: The Johns Hopkins University Press.

Duguay, S. M., Arii, K., Hooper, M., & Lechowicz, M. J. (2001). Ice storm damage and early recovery in an old-growth forest. *Environmental Monitoring and Assessment, 67*, 97–108.

Frelich, L. E. (2002). *Forest dynamics and disturbance regimes: Studies from temperate evergreen-deciduous forests*. Cambridge: Cambridge University Press.

Fuhrmann, C., & Konrad, C. E. (2014). *A winter weather climatology for the southeastern United States*. Raleigh: State Climate Office of North Carolina, North Carolina State University.

Gay, D. A., & Davis, R. E. (1993). Freezing rain and sleet climatology of the southeastern USA. *Climate Research, 3*, 209–220.

Grime, J. P. (2001). *Plant strategies, vegetation processes and ecosystem properties*. Chichester: John Wiley.

Harrod, J. C., Harmon, M. E., & White, P. S. (2000). Post-fire succession and 20th century reduction in fire frequency on xeric southern Appalachian sites. *Journal of Vegetation Science, 11*, 465–472.

Isaacs, R. E., Stueve, K. M., Lafon, C. W., & Taylor, A. H. (2014). Ice storms generate spatially heterogeneous damage patterns at the watershed scale in forested landscapes. *Ecosphere, 5*, 1–14.

Jones, K. F. (1998). A simple model for freezing rain ice loads. *Atmospheric Research, 46*, 87–97.

Jones, K. F., Muherin, N., & Ryerson, C. C. (1997). *EPRI freezing rain ice mapping project: Region 2, CR-00020*. Hanover: U.S. Army Corps of Engineers, Cold Regions research Engineering Laboratory.

Kneeshaw, D., Bergeron, Y., & Kuuluvainen, T. (2011). Forest ecosystem structure and disturbance dynamics across the circumboreal forest. In A. C. Millington, M. A. Blumler, & U. Schickoff (Eds.), *The sage handbook of biogeography* (pp. 263–280). Los Angeles: Sage.

Konrad, C. E. (1998). An empirical approach for delineating fine scaled spatial patterns of freezing rain in the Appalachian region of the USA. *Climate Research, 10*, 217–227.

Laflamme, J. N., & Periard, G. (1998). The climate of freezing rain over the province of Quebec in Canada: A preliminary analysis. *Atmospheric Research, 46*, 99–111.

Lafon, C. W. (2004). Ice-storm disturbance and long-term forest dynamics in the Adirondack Mountains. *Journal of Vegetation Science, 15*, 267–276.

Lafon, C. W. (2006). Forest disturbance by ice storms in Quercus forests of the southern Appalachian Mountains, USA. *Ecoscience, 13*, 30–43.

Lafon, C. W., & Grissino-Mayer, H. D. (2007). Spatial patterns of fire occurrence in the central Appalachian Mountains and implications for wildland fire management. *Physical Geography, 28*, 1–20.

Lafon, C. W., & Kutac, M. (2003). Effects of ice storms, southern pine beetle infestation, and fire on Table Mountain pine forests of southwestern Virginia. *Physical Geography, 24*, 502–519.

Lafon, C. W., & Speer, J. H. (2002). Using dendrochronology to identify major ice storm events in oak forests of southwestern Virginia. *Climate Research, 20*, 41–54.

Lafon, C. W., Graybeal, D. Y., & Orvis, K. H. (1999). Patterns of ice accumulation and forest disturbance during two ice storms in southwestern Virginia. *Physical Geography, 20*, 97–115.

Lemon, P. C. (1961). Forest ecology of ice storms. *Bulletin of the Torrey Botanical Club, 88*, 21–29.

Loehle, C. (1988). Tree life-history strategies – The role of defenses. *Canadian Journal of Forest Research, 18*, 209–222.

Makkonen, L. (1998). Modeling power line icing in freezing precipitation. *Atmospheric Research, 46*, 131–142.

McKay, G. A., & Thompson, H. A. (1969). Estimating the hazard of ice accretion in Canada from climatological data. *Journal of Applied Meteorology, 8*, 927–935.

Millward, A. A., & Kraft, C. E. (2004). Physical influences of landscape on a large-extent ecological disturbance: The northeastern North American ice storm of 1998. *Landscape Ecology, 19*, 99–111.

Noss, R. F. (2013). *Forgotten grasslands of the South: Natural history and conservation*. Washington, DC: Island Press.

Proulx, O. J., & Greene, D. F. (2001). The relationship between ice thickness and northern hardwood tree damage during ice storms. *Canadian Journal of Forest Research, 31*, 1758–1767.

Rauber, R. M., Olthoff, L. S., Ramamurthy, M. K., Miller, D., & Kunkel, K. E. (2001). A synoptic weather pattern and sounding-based climatology of freezing precipitation in the United States East of the Rocky Mountains. *Journal of Applied Meteorology, 40*, 1724–1747.

Rebertus, A. J., Shifley, S. R., Richards, R. H., & Roovers, L. M. (1997). Ice storm damage to an old-growth oak-hickory forest in Missouri. *The American Midland Naturalist, 137*, 48–61.

Rhoades, V. (1918). Ice storms in the southern Appalachians. *Monthly Weather Review, 46*, 373–374.

Rhoads, A. G., Hamburg, S. P., Fahey, T. J., Siccama, T. G., Hane, E. N., Battles, J., Cogbill, C., Randall, J., & Wilson, G. (2002). Effects of an intense ice storm on the structure of a northern hardwood forest. *Canadian Journal of Forest Research, 32*, 1763–1775.

Robinson, P. J., & Henderson-Sellers, A. (1999). *Contemporary climatology*. Essex: Longman.

Rogers, J. C., & Rohli, R. V. (1991). Florida citrus freezes and polar anticyclones in the Great Plains. *Journal of Climate, 4*, 1103–1113.

Runkle, J. R. (1990). Gap dynamics in an Ohio *Acer-Fagus* forest and speculations on the geography of disturbance. *Canadian Journal of Forest Research, 20*, 632–641.

Seischab, F. K., Bernard, J. M., & Eberle, M. D. (1993). Glaze storm damage to western New York forest communities. *Bulletin of the Torrey Botanical Club, 120*, 64–72.

Smith, K. T., & Shortle, W. C. (2003). Radial growth of hardwoods following the 1998 ice storm in New Hampshire and Maine. *Canadian Journal of Forest Research, 33*, 325–329.

Smolnik, M., Hessl, A., & Colbert, J. J. (2006). Species-specific effects of a 1994 ice storm on radial tree growth in Delaware. *The Journal of the Torrey Botanical Society, 133*, 577–584.

Stewart, R. E., & King, P. (1987). Freezing precipitation in winter storms. *Monthly Weather Review, 115*, 1270–1279.

Stueve, K. M., Lafon, C. W., & Isaacs, R. E. (2007). Spatial patterns of ice storm disturbance on a forested landscape in the Appalachian Mountains, Virginia. *Area, 39*, 20–30.

Waldrop, T. A., & Brose, P. H. (1999). A comparison of fire intensity levels for stand replacement of Table Mountain pine (*Pinus pungens* Lamb.). *Forest Ecology and Management, 113*, 155–166.

Whitney, H. E., & Johnson, W. C. (1984). Ice storms and forest succession in southwestern Virginia. *Bulletin of the Torrey Botanical Club, 111*, 429–437.

Wonkka, C. L., Lafon, C. W., Hutton, C. M., & Joslin, A. J. (2013). A CSR classification of tree life history strategies and implications for ice storm damage. *Oikos, 122*, 209–222.

Yamamoto, S. I. (1992). The Gap theory in forest dynamics. *The Botanical Magazine, Tokyo, 105*, 375–383.

Zishka, K. M., & Smith, P. J. (1980). The climatology of cyclones and anticyclones over North America and surrounding ocean environs for January and July, 1950–77. *Monthly Weather Review, 108*, 387–401.

Chapter 8
Multi-millennial Record of Erosion and Fires in the Southern Blue Ridge Mountains, USA

David S. Leigh

Abstract Bottomland sediments from the southern Blue Ridge Mountains provide a coarse-resolution, multi-millennial stratigraphic record of past regional forest disturbance (soil erosion). This record is represented by 12 separate vertical accretion stratigraphic profiles that have been dated by radiocarbon, luminescence, cesium-137, and correlation methods continuously spanning the past 3,000 years of pre-settlement (pre-dating widespread European American settlement) and post-settlement strata. Post-settlement vertical accretion began in the late 1800s, appears to be about an order of magnitude faster than pre-settlement rates, and is attributable to widespread deforestation for timber harvest, farming, housing development, and other erosive activities of people. Natural, climate-driven, or non-anthropic forest disturbance is subtle and difficult to recognize in pre-settlement deposits. There is no indication that pre-settlement Mississippian and Cherokee agricultural activities accelerated erosion and sedimentation in the region. A continuous 11,244 years before present (BP) vertical accretion record from a meander scar in the Upper Little Tennessee River valley indicates abundant charcoal (prevalent fires) at the very beginning of the Holocene (11,244–10,900 years BP). In contrast, moderate to very low levels of charcoal are apparent over the remaining Holocene until about 2,400 years BP when charcoal influx registers a pronounced increase. These data are consistent with the idea that Native Americans used fire extensively to manage forests and to expanded agricultural activities during Woodland and later cultural periods over the past 3000 years. However, there is no indication that prehistoric intentional use of fire and agriculture caused accelerated erosion and sedimentation.

Keywords Alluvium • Chronostratigraphy • Holocene • Overbank • Charcoal

D.S. Leigh (✉)
Department of Geography, The University of Georgia, Athens, GA 30602-2502, USA
e-mail: dleigh@uga.edu

C.H. Greenberg, B.S. Collins (eds.), *Natural Disturbances and Historic Range of Variation*, Managing Forest Ecosystems 32,
DOI 10.1007/978-3-319-21527-3_8

8.1 Introduction

Forests shelter soils on hillslopes from intense rainfall, runoff, and erosion in the southern Blue Ridge Mountains (SBRM). Without forests these mountains would be badlands, because of the highly erodible surface mantle of saprolite (weathered bedrock). Hydrologists posit that under pristine forest conditions surface erosion is minimal because there is complete infiltration of precipitation and virtually no Hortonian overland flow (i.e., rainfall intensity exceeds soil infiltration capacity); and instead, precipitation is delivered to streams via subsurface pathways and as saturation overland flow (Hewlett et al. 1977; Swift et al. 1988; Committee on Hydrologic Impacts of Forest Management 2008). However, disturbance of forest cover has occurred on a variety of spatial and temporal scales exposing bare soil, reducing infiltration capacities, and resulting in erosion and deposition of sediment on nearby bottomlands. Natural pathways and patterns of erosional disturbance typically are driven by destructive tornado swaths and windblown tree-throws (Peterson et al. Chap. 5), landslides (Wooten et al. Chap. 9), ice storms (Lafon Chap. 7), and fires ignited by lightning (Grissino-Mayer Chap. 6; Greenberg et al. Chap. 1, Table 1.6; Greenberg et al. Chap. 12); and the magnitude and frequency of such events is mediated by variations in climate. In contrast, human-induced disturbance primarily includes timber harvest, land clearing for agriculture and settlement, and fires ignited by humans. Floodplain deposits record both natural and human-induced modes of past forest disturbance via changes in sedimentology, which result from changes in the sediment transport and rainfall-runoff regimes, and which, in turn are mediated by changes in climate and vegetation cover. For example, timber harvest and forest clearance for croplands drastically increase sediment concentrations in streamflow, and floodplain sedimentation rates (Happ 1945; Trimble 1974; Walling 1987; Meade et al. 1990; Saxton and Shiau 1990; Knighton 1998; Walling 1999; Knox 2006) because complete removal of the protective vegetation facilitates dislodgment of soil from raindrop impact, sheetflow, and rilling mechanisms of soil erosion. The SBRM are no exception to this rule, because historical human impacts (largely by timber harvest, small farms, and home development) since the late 1800s following widespread European American settlement (post-settlement time) have created land surfaces conducive to higher sediment yields than the pre-settlement landscapes occupied by indigenous people of the Cherokee, Mississippian, and earlier cultures.

The main purpose of this chapter is to describe and explain how bottomland alluvial sediments record past fluctuations in erosional forest disturbance of the SBRM. Paleobotanical records from wetland peat deposits are very important sedimentary sources of information about past forest composition and disturbance (Delcourt and Delcourt 2004), but they have been scarcely studied in this region and are beyond the scope of this chapter. Instead, emphasis here is on past soil erosion and its ultimate deposition on bottomlands and floodplains, and what sorts of information can be gleaned from these widespread stratified deposits. Numerous examples and analyses are drawn from stratigraphic data originally collected by the author and his former students.

The SBRM physiographic province (see Fig. 1.1) is the late Precambrian and early Paleozoic crystalline terrain of high relief mountains in the Blue Ridge province south of the Roanoke River in Virginia, and includes western North Carolina, eastern Tennessee, and northeastern Georgia (Fenneman 1938; Thornbury 1965). This province is broadly representative of forests in the Central Hardwood Region. Bedrock consists largely of metasedimentary rocks (micaceous gneisses and schists) with some outcrops of basement granites and granitic gneisses (Robinson 1992; Thigpen and Hatcher 2009). This highly dissected mountain range has summits up to around 2000 m above sea level and main valley bottoms at around 600 m above sea level. Hillslopes and summits commonly are mantled by 0–10 m of sandy to clayey micaceous saprolite, whereas colluvium and alluvium cover the footslopes and bottomlands (Hatcher 1988; Southworth et al. 2003). Stream and river channels typically exhibit meandering pool and riffle morphology with bed sediment composed of fine to very coarse gravel. Floodplains and recent terraces commonly have a thick vertical accretion mantle (0.5–2.0 m) overlying much thinner lateral accretion and channel lag bedload materials. Much of the data presented in this paper were drawn from the Upper Little Tennessee River valley between Franklin, North Carolina and Mountain City, Georgia (Fig. 8.1), which is centrally located in the SBRM.

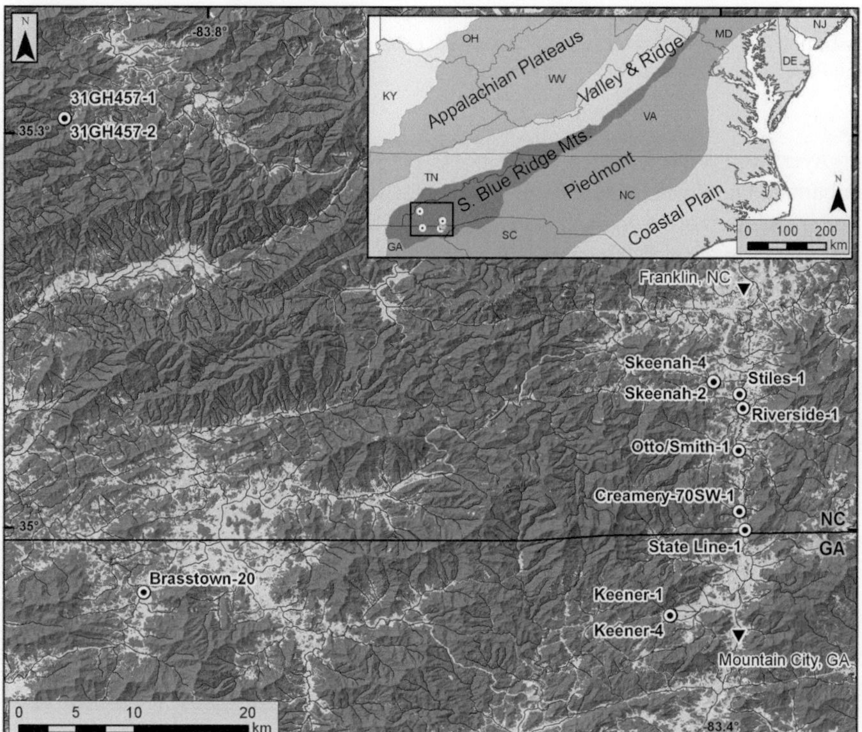

Fig. 8.1 Map showing the southern Blue Ridge Mountains, the Upper Little Tennessee River valley, and other sample locations mentioned in the text

Fig. 8.2 Photographic plate XVB from Glenn (1911), entitled "A well-dissected part of the Asheville Plateau" illustrating the deforested and cultivated appearance of hillslopes in the southern Blue Ridge Mountains in the early 1900s

Ayers and Ashe (1905) and Glenn (1911) were perhaps the first to extensively document human-induced erosive damages to forests and related floodplain deposition in the southern Appalachians. Ayers and Ashe (1905) note that "The three agencies that have wrought changes in the forests of the Southern Appalachians are fires, lumbering, and clearing of lands for farming… During heavy rains the earth of freshly burned or freshly plowed land is rapidly washed away. The streams from such lands are often more than half earth, and the amount of soil thus eroded every year is enormous." Glenn (1911) provides excellent examples, including photographs of eroded land, flooding, and sedimentation (Figs. 8.2 and 8.3) within specific watersheds that had experienced deforestation. For example, he states "Floods on the Tuckasegee are said by old residents, some of whom have known the river for 60 years, to be generally higher, to come more suddenly, to disappear quicker, and to be much more destructive than 50 years ago. Both the river and its tributaries are said to be much muddier during floods and for a considerable time afterwards than they formerly were…" and "…the stream may be overloaded with waste and forced to drop a part of it until it fills its channel. It then flows here and there in irregular shifting meanders across its flood plain, depositing great quantities of waste and building up or aggrading the flood plain. Such flood plains rapidly become useless for agriculture." (Glenn 1911).

The early 1900s were arguably the most erosive decades in history, but it is important to realize that even during the heyday of timber harvest during 1900–1910

Fig. 8.3 A USDA photo from 1901 AD of "A seven acre field on Scott's Creek, which was very badly washed by the freshet of 1901." Scott's Creek flows through Sylva, North Carolina. The photo was taken by E. Block and reproduced by David Leigh (April 2010) from US Archives photo 134Er18, caption label "25316 Inf – Influences, Floods, Effects, Land Damage, North Carolina" (with handwritten number 25540). Notice the thick deposits of fresh sandy alluvium along the fence line to the left and a thin drape of fine sediment along with flood gravels across much of the floodplain. Also, notice the steep deforested hillslopes in the background

the landscape of the SBRM region still retained about 70 % forest cover at any given point in time as clear cut areas were quick to sprout saplings and begin successional reforestation (Ayers and Ashe 1905; Holmes 1911). Thus, only fractional deforestation is necessary to produce very noticeable changes in the geomorphic system, a point which was recently affirmed by Price and Leigh (2006a, b).

Considerable attention has been placed on post-settlement alluvium in the Piedmont province of the southern Appalachian region (see Greenberg et al. Chap. 1, Fig. 1.1) where row-crop agriculture has been more extensive than in the mountains (Fuller 1934; Happ 1945; Trimble 1974; Costa 1975; Jacobson and Coleman 1986; Meade et al. 1990; Ruhlman and Nutter 1999; Lichtenstein 2003; Jackson et al. 2005; Ambers et al. 2006; Hupp et al. 2013; Lecce and Pavlowsky 2014), and these studies generally indicate that about 1–2 m of post-settlement alluvium was delivered to river bottomlands over the span of a couple of centuries. However, very little follow-up to Glenn's (1911) study has been done in the much less agricultural and more forested SBRM. Price and Leigh (2006a, b) provided examples of how

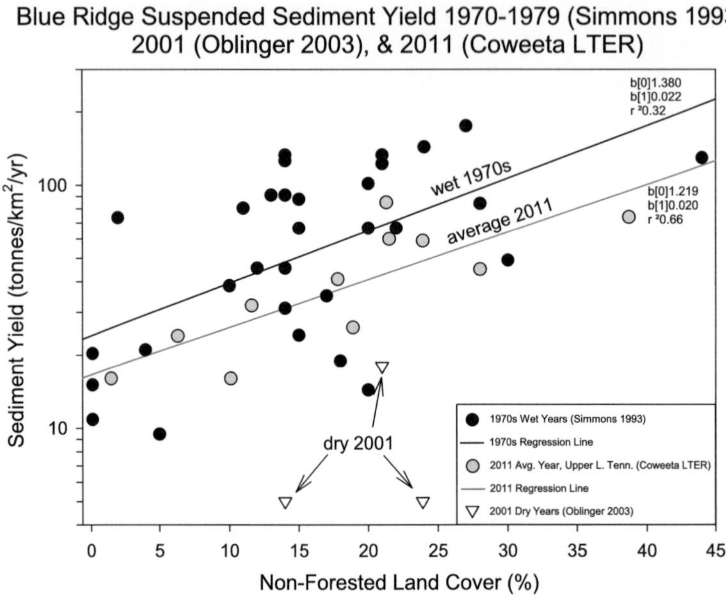

Fig. 8.4 Sediment yield data of Simmons (1993) for samples collected in the southern Blue Ridge Mountains of western North Carolina in 1970–1979, with additional points from the Upper Little Tennessee River valley collected in 2001 (Oblinger 2003) and 2011 (Coweeta LTER unpubl). Separate linear regression lines are for the 1970s (*black*) and water year 2011 (*grey*) with respective parameters and R^2 values stated to the right

relatively modest levels of deforestation relate to significant changes in stream channel morphology and increased levels of turbidity and suspended sediment concentrations in tributaries to the Upper Little Tennessee River. They include radiocarbon-dated stratigraphic sections showing 40–50 cm of post-settlement alluvium. Miller et al. (2005) examined 111 years of reservoir sediment in the SBRM of western North Carolina and found that sedimentation rates progressively increased with fastest rates in the 1990s and 2000s, presumably in association with the widespread development of new house sites.

Simmons (1993) collected hundreds of suspended sediment observations from the SBRM from 1970 to 1979 as part of a statewide study entitled, "*Sediment Characteristics of North Carolina Streams, 1970–1979*." He stratified land use characteristics of Blue Ridge watersheds into four categories, including fully forested (F), mostly forested with minor development (D), non-agricultural rural (N), and agricultural rural (R) types. His results (Fig. 8.4; Table 8.1) clearly illustrate how the *less*-forested watersheds generate considerably more sediment than fully forested ('pristine') watersheds. It is significant to note that Simmons' three 'pristine' watersheds generated an average of 15 tonnes per km^2 per year of suspended sediment (individual observations of 9.5, 15.1, 20.3 tonnes per km^2 per year), and this illustrates the significance of sediment yield from *natural* disturbances in the fully

Table 8.1 Mean and standard deviation values of specific sediment yield of Blue Ridge Mountain streams in North Carolina, 1970–1979 (From Simmons 1993)

Land use class	Sediment yield (tonnes/km^2/year)	n =	±1 SD (tonnes)
F – Entirely forested ('pristine')	15	3	5
D – Mostly forested with minor development	45	4	22
N – Rural with non-agricultural development	79	12	54
R – Rural with agricultural development	82	12	39

forested watersheds (primarily from landslides, tree-throws, and stream bank erosion sources of sediment).

Meteorological events mediated by climate are the main drivers of variations in natural erosive disturbances of forests, but recent attention has been cast on Native Americans as agents of prehistoric erosion and sedimentation. Stinchcomb et al. (2011) argued that indigenous people of the Mississippian culture (ca. 1,000–460 years BP) were responsible for a 50 % reduction in forest cover in tributaries to the Delaware River in eastern Pennsylvania and western New York (largely for maize (*Zea mays*) cultivation), and they claim to recognize the first floodplain sediment attributed to erosive human impact on the landscape (sensu stricto 'legacy sediment' of James 2013) at circa 1100–1600 AD. They referred to it stratigraphically as 'pre-Colonial sediment' which is distinguished by faster sedimentation rates than previous deposits, and they indicated that it may be recognizable throughout eastern North America. Delcourt and Delcourt (2004) argue that Native Americans had been managing forests for thousands of years in the southern Appalachians by use of fire to clear understories for nut harvest and game management, and to clear plots for agriculture (also see Grissino-Mayer Chap. 6; Greenberg et al. Chap. 12). Thus, indigenous people could have caused significant enough erosive impact on the land to be expressed in the pre-settlement stratigraphic record.

Although the focus of this book is on 'natural' disturbances, which would primarily relate to erosion and sedimentation prior to widespread European American settlement (pre-settlement time) (but see Greenberg et al. Chap. 1), this chapter emphasizes sedimentation both before and after settlement, because the later represents the end-member of maximum forest disturbance. In fact, the expression of forest disturbance usually is very subtle in the pre-settlement stratigraphic record, compared to the post-settlement record. If the ultimate objective is to manage forests within the realm of natural disturbance, then it is still useful to understand a frame of reference that includes the pronounced effects of human-induced disturbance that are clearly outside of the natural range of variation. Furthermore, since it has been argued that indigenous people caused significant disturbance to forests via intentional burning and cultivation for agriculture (see Greenberg et al. Chap. 12), understanding sedimentary products of post-settlement forest disturbance is critical for framing the possible impact that indigenous people had on the sedimentary record. Finally, from an academic perspective, "...understanding the timescales and pathways for response and recovery of rivers and floodplains to episodic changes in erosion and sedimentation has been a long standing issue in fluvial geomorphology" (Knox 2006).

8.2 Distinction Between Pre-settlement Versus Post-settlement Strata

Significant human impact on the fluvial system in North America did not begin until after immigration and settlement of large numbers of non-indigenous people, and the time-stratigraphic designations before and after such settlement have been referred to as pre-settlement and post-settlement units (Trimble 1974; Jacobson and Coleman 1986; Knox 1972, 1977, 1987; Wilkinson and McElroy 2007), though *prehistoric* and *historic* time-stratigraphic designations also are commonly used. In many localities the stratigraphic boundary between pre- and post-settlement alluvium is demarcated by a buried paleosol where the upper surface of the buried A horizon represents the soil surface that existed and was buried by the first post-settlement alluvium (Fig. 8.5). Leigh (2010) established a date of 1870 AD (±10 years) for this stratigraphic boundary in the Upper Little Tennessee valley, primarily based on correlation with historical records and census data (Leigh 2010) indicating that population, improved land in farms, and value of products in manufacturing in the region did not begin to expand until about 1870 AD (Fig. 8.6). Also, intensive

Fig. 8.5 Photograph of the river cutbank stratigraphic section at the Riverside-1 site along the Little Tennessee River showing the stratigraphic boundary of pre- and post-settlement alluvium atop a prominent buried A horizon, which dates to circa AD 1870

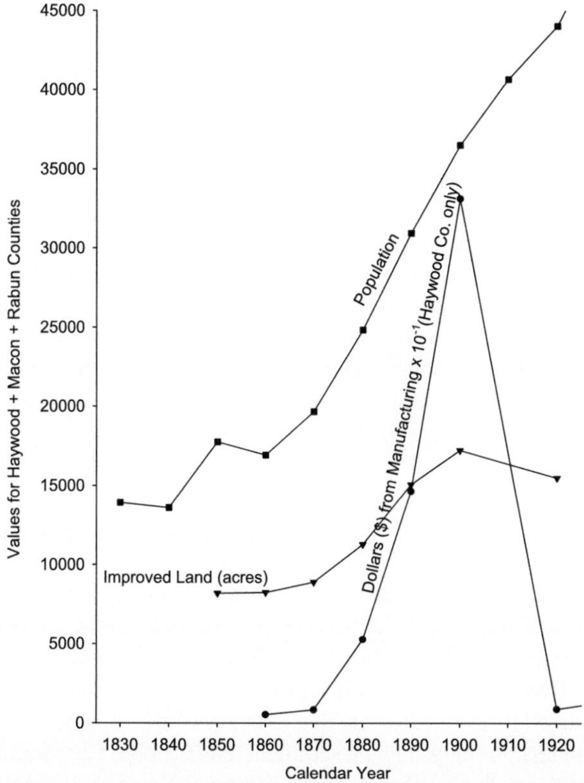

Fig. 8.6 United States Census data for Haywood and Macon Counties, North Carolina and Rabun County, Georgia showing pronounced expansion of human presence after 1870 AD. These counties are the only ones proximal to the Upper Little Tennessee River valley that have complete census records extending back to at least 1860 AD. Dollars from manufacturing line for Haywood County mostly represents timber products

and widespread timber harvest did not begin until after the railroads entered the mountains west of Asheville in the 1880s, which greatly opened accessibility to this rugged terrain. Except for the use of mechanized cultivation equipment and domesticated livestock, sparse European-American settlers prior to 1870 probably used the land in a subsistence fashion similar to their indigenous predecessors. Considering the enormous indigenous population decline due to disease after European contact circa 1540 (Hudson 2002) and the federally mandated removal of Cherokee people in 1835, it is quite possible that the population densities (including both indigenous and non-indigenous people) immediately prior to 1870 were comparable to the pre-1540 indigenous population densities. Indeed, the 1860 USA Census tallied 6004 people in Macon County (1344 km^2) or 4.5 people per km^2, which is much less than estimates of Mississippian indigenous population densities of 16–17 people per km^2 reported by Delcourt and Delcourt (2004) for the lower Little Tennessee River valley in Tennessee.

Many studies have found significant differences in the character of overbank alluvium of post-settlement versus pre-settlement periods, such as in sediment textures (particle size), sedimentology (bedding structures), and sedimentation rates (Lecce 1997; Knox 1977, 1987, 2001, 2006; Benedetti 2003; Leigh 2010). That is, post-settlement overbank alluvium generally is coarser, redder in color (more oxidized), and has sedimentation rates that greatly exceed those of pre-settlement time (Orbock et al. 1993; Knox 2001, 2006; Benedetti 2003).

In addition to the physical characteristics, chemical and paleobotanical characteristics can distinguish post-settlement from pre-settlement alluvium. For example, the lead and zinc content in overbank sediments increased greatly in the upper Mississippi River valley during the 1800s due to mining activities (Knox 1987, 2006). Mercury and gold concentrations are distinctively higher in post-settlement alluvium of the gold belt spanning from northern Georgia through western North Carolina, because gold mining that began in the 1830s involved mercury amalgamation to recover gold from sediment sluices (Leigh 1994, 1997; Lecce et al. 2008). However, such chemical indicators of human agency commonly do not enter into the stratigraphic record until several decades subsequent to the stratigraphic boundary between pre-settlement and post-settlement alluvium, and therefore are not very useful for discriminating the initial onset (or lower boundary) of the post-settlement stratigraphic unit. Pollen preserved within the sediment can help to distinguish the boundary by recognition of the influx of non-native species and weeds indicative of land clearance (Delcourt and Delcourt 2004), but pollen preservation can be poor within the oxidized fluvial sediments due to decay of the organic matter.

Usage of the terminology 'legacy sediment' has become very popular within the last decade, and usually it is intended to denote sediment resulting from distinct human-induced erosion of the landscape. Thus, it can be synonymous with the chronostratigraphic terminology of post-settlement alluvium. However, James (2013) recently clarified the definition of 'legacy sediment' to relate to any recognizable human impact. Indeed, one of the important questions posed in this chapter is whether or not prehistoric Native Americans engaged in sufficient human impact on the landscape to register a sedimentary signature of 'legacy sediment.' Thus, pre- and post-settlement are used throughout this chapter, because they are chronostratigraphic terms that have more precise meaning and distinguish a time of significant population growth of non-indigenous settlers to the region whose land use practices were very erosive.

8.3 Methods

Research presented here relies on 12 stratigraphic sections where both the pre-settlement and post-settlement units occur in the exact same stratigraphic section, and where both units have been well dated by a combination of radiocarbon, luminescence, ^{137}Cs, or correlation dating techniques. The majority of these dated stratigraphic sections are from the Upper Little Tennessee River catchment in northeast Georgia and western North Carolina between Franklin, North Carolina and Mountain City, Georgia (Fig. 8.1, Table 8.2) where they were collected in association with the Coweeta Long-Term Ecological Research (Coweeta LTER) program (http://coweeta.uga.edu/). Exceptions include the Brasstown stratigraphic section (Leigh 1996), which came from the valley of Brasstown Creek in the upper Hiawassee drainage of northern Georgia, and another two sections (at archaeological site 31GH457) that came from the valley of West Buffalo Creek, a tributary to Lake Santeetlah near Robbinsville, North Carolina (Leigh 2011). Ten of these 12 stratigraphic sections are composed entirely of late Holocene alluvium younger than 3,500 years BP, which is most appropriate for comparison of the recent post-settlement alluvium to its immediate predecessor of pre-settlement alluvium. Other stratigraphic sections have been collected in the region (Leigh and Webb 2006; Leigh and Rogers 2007), but do not contain the both the pre-settlement and post-settlement strata in the same section.

Table 8.2 Location and attributes of study sites

Site name	Latitude (decimal degrees)	Longitude (decimal degrees)	Drainage area (km^2)	Profile type	Reference
Keener-1	34.933053°	−83.437982°	7	7.6 cm giddings core	Price and Leigh (2006a), Wang and Leigh (2012)
Keener-4	34.932923°	−83.438349°	7	7.6 cm giddings core	Price and Leigh (2006a)
Skeenah-2	35.110998°	−83.404813°	15	7.6 cm giddings core	Price and Leigh (2006a)
Skeenah-4	35.110858°	−83.404828°	15	7.6 cm giddings core	Price and Leigh (2006a)
31GH457-1	35.310047°	−83.912289°	28	7.6 cm hand auger	Leigh (2011)
31GH457-2	35.310247°	−83.912211°	28	7.6 cm hand auger	Leigh (2011)
Brasstown-20	34.949807°	−83.850598°	65	Backhoe pit #20	Leigh (1996)
State Line-1	34.998003°	−83.380416°	146	Right stream cutbank	Leigh (2007), Wang and Leigh (2012)
Creamery-70SW-1	35.012233°	−83.384814°	171	7.6 cm giddings core	Leigh (2007)
Otto/Smith-1	35.058677°	−83.385510°	213	Left stream cutbank	Leigh (2007)
Riverside-1	35.091003°	−83.382106°	313	Right stream cutbank	Leigh (2007), Wang and Leigh (2012)
Stiles-1	35.101464°	−83.384691°	319	7.6 cm giddings core	Leigh this paper

Radiocarbon dating methods mainly involved samples of small angular charcoal or bulk sediment (micro-charcoal fractions and humus) using the accelerator mass spectrometry (AMS) method at the University of Georgia's Center for applied isotope studies. In a few instances large pieces of uncarbonized wood were dated by the conventional scintillation counting method, and the Beta Analytic, Inc. lab was used for the two samples at the Brasstown-20 site (Table 8.3). Charcoal fragments were cleaned and leached of possible carbonates with an acid-alkali-acid pretreatment (1 N HCl-1 N NaOH-1 N HCl) prior to ^{14}C dating. Bulk sediment material was dated following ultrasonic dispersion and sieving through a 125 μm mesh and cleaning with 1 N HCl to remove possible carbonates. Calendar year calibrations were calculated using the program CLAM (Blaauw 2010) based on the IntCal-13 calibration curve (Reimer et al. 2013) and the delta ^{13}C corrected ^{14}C ages. The calendar years before present (years BP) reference 1950 AD as 'present.' Sedimentation rates were calculated by linear interpolation with the program CLAM ('classical' age modeling of Blaauw 2010), which considers the probability distributions of the separate calibrated ages to produce the most probable match for linear interpolation between the two samples. Thus, CLAM produces long-term-average sedimentation rates. Only vertical-accretion facies were used for estimates of sedimentation rates. In cases where lateral accretion facies were dated beneath the vertical accretion, that date was assigned as the basal boundary of the vertical accretion unit, because lateral accretion deposition occurs very rapidly and immediately prior to being overlain by vertical accretion deposits (Bridge 2003).

Although it is possible that charcoal may be detritus and can thereby produce erroneous ages for the actual time of sedimentation, care was taken to selectively date the largest angular fragments of charcoal that exhibit minimal traits of abrasion and rounding. One paired set of radiocarbon samples from the Riverside site (both from 142 to 144.5 cm depth) included dating results from the <125 μm bulk sediment versus charcoal isolated from the <125 μm fraction by flotation separation in sodium polytungstate at 1.7 g per cm^3. Both samples produced almost exactly the same age, 1,650 ± 25 versus 1,710 ± 25 ^{14}C years BP, respectively (Table 8.3), and they exhibit overlap at 1,555–1,615 years BP upon calendar year calibrations with two standard deviations. Although, comprehensive conclusions cannot be drawn from only one sample, this single example demonstrates that charcoal is not necessarily much older than associated humus in overbank sediments. It is sensible that charcoal is slightly older than humus, because it has great potential for inheriting ages older than the actual time of sedimentation (i.e., charcoal may have been from tree rings that were hundreds of years older than tree death, transit time of charcoal from the fire site to the deposition site may have taken hundreds of years). Therefore, the bulk sediment date was preferentially used for age-depth modeling in this case and it yielded a best estimate of 1,552 years BP (Table 8.3).

Luminescence dating was applied in a few instances when radiocarbon dating was not feasible, based on the samples of Wang and Leigh (2012). Samples were collected by pounding light-tight PVC tubes (about 15 cm long by 4 cm diameter) into the outcrop and sealing the ends with duct tape upon removal. Ages were measured from sediment extracted from the center of the tubes at the University of Georgia Luminescence Dating Laboratory using the single-aliquot regenerative-dose proto-

Table 8.3 Raw data for samples used for age determinations in each individual stratigraphic section or site

Site	Lab #	Sample depth (cm)	Plot point depth (cm)	Unit & facies	Fraction of VA unit thickness (0=bottom, 1=top)	Dating method	Material dated	Material isotopic del ^{13}C	^{14}C years BP	Non-14C age estimate cal. year BP	±1 SD	Best age Est. (cal. years BP)	Inter-sample rate (mm/year)
Keener-1	n.a.	0	0.0	2VA	1.00	GS	Surface 2005	n.a.	n.a.	−55	1	−56	n.a.
Keener-1	n.a.	12–14	13.0	2VA	0.70	Cs137	Flood sediment	n.a.	n.a.	−13	10	−13	3.02
Keener-1	n.a.	43	43.0	1/2VA	0 or 1	CORR	Top of Ab horizon	n.a.	n.a.	80	10	79	3.26
Keener-1	UGA-09OSL-672	105–109	107.0	1VA	0.36	OSL	Sandy loam	n.a.	n.a.	1300	50	1,297	0.53
Keener-1	UGA-14484	125	125.0	1VA	0.18	C14	Charcoal	−26.25	1,620	n.a.	40	1,474	1.02
Keener-1	UGA-14485	167	143.0	1LA	0.00	C14	Acorn (uncarb.)	−25.79	1,630	n.a.	40	1,546	2.53
Keener-4	n.a.	0	0.0	2VA	1.00	GS	Surface 2005	n.a.	n.a.	−55	1	−55	n.a.
Keener-4	n.a.	44	44.0	1/2VA	0 or 1	CORR	Top of Ab horizon	n.a.	n.a.	80	10	80	3.24
Keener-4	UGA-14482	133	106.0	1VA	0.00	C14	Charcoal	−26.44	600	n.a.	40	599	1.19
Skeenah-2	n.a.	0	0.0	2VA	1.00	GS	Surface 2003	n.a.	n.a.	−53	1	−53	n.a.
Skeenah-2	n.a.	10–15	13.0	2VA	0.52	Cs137	Flood sediment	n.a.	n.a.	−13	10	−13	3.23
Skeenah-2	n.a.	27	27.0	1/2VA	0 or 1	CORR	Top of Ab horizon	n.a.	n.a.	80	10	80	1.51
Skeenah-2	UGA-13068	113	81.0	1LA	0.00	C14	Wood	−25.31	1,580	n.a.	40	1,469	0.39
Skeenah-4	n.a.	0	0.0	2VA	1.00	GS	Surface 2003	n.a.	n.a.	−53	1	−53	n.a.
Skeenah-4	n.a.	40	40.0	1,2VA	0 or 1	CORR	Top of Ab horizon	n.a.	n.a.	80	10	80	3.00
Skeenah-4	UGA-13067	125	112.0	1LA	0.00	C14	Wood	−24.9	2440	n.a.	280	2,494	0.30
31GH457-1	n.a.	0	0.0	2VA	1.00	GS	Surface 2009	n.a.	n.a.	−59	1	−59	n.a.
31GH457-1	n.a.	56	56.0	1/2VA	0.00	CORR	Top of Ab horizon	n.a.	n.a.	80	10	80	4.02

(continued)

Table 8.3 (continued)

Site	Lab #	Sample depth (cm)	Plot point depth (cm)	Unit & facies	Fraction of VA unit thickness (0 = bottom, 1 = top)	Dating method	Material dated	Material isotopic del ^{13}C	^{14}C years BP	Non-14C age estimate cal. year BP	±1 SD	Best age Est. (cal. years BP)	Inter-sample rate (mm/year)
31GH457-1	UGAMS-8461	158	158.0	1VA	0.14	C14	Charcoal	–26.7	5,270	n.a.	30	6,057	0.17
31GH457-2	n.a.	0	0.0	2VA	1.00	GS	Surface 2009	n.a.	n.a.	–59	1	–59	n.a.
31GH457-2	n.a.	37	37.0	1/2VA	0 or 1	CORR	Top of Ab horizon	n.a.	n.a.	80	10	80	2.67
31GH457-2	UGAMS-8462	77	77.0	1VA	0.73	C14	Charcoal	–27.1	3,170	n.a.	25	3,401	0.12
Brasstown-20	n.a.	0	0.0	2VA	1.00	GS	Surface 1993	n.a.	n.a.	–43	1	–43	n.a.
Brasstown-20	n.a.	40	40.0	1/2VA	0 or 1	CORR	Top of Ab horizon	n.a.	n.a.	80	10	80	3.26
Brasstown-20	BETA-65972	49–51	50.0	1VA	0.78	C14	Charcoal	n.a.	640	n.a.	60	609	0.19
Brasstown-20	BETA-65973	172–175	96.0	1VA	0.00	C14	Wood/limb (uncarb.)	n.a.	1,880	n.a.	70	1,811	0.38
State Line-1	n.a.	0	0.0	2VA	1.00	CORR	Surface 2005	n.a.	n.a.	–55	1	–55	n.a.
State Line-1	n.a.	73	73.0	2VA	0.54	Cs137	Flood sediment	n.a.	n.a.	–13	10	–13	17.24
State Line-1	n.a.	144	144.0	2VA	0.09	CORR	Flood sediment	n.a.	n.a.	48	10	48	11.77
State Line-1	n.a.	159	159.0	1/2VA	0 or 1	CORR	Top of Ab horizon	n.a.	n.a.	80	10	80	4.65
State Line-1	UGA-09OSL-670	225	225.0	1VA	0.50	OSL	Sandy loam sediment	n.a.	n.a.	1100	50	1,100	0.65
State Line-1	UGA-14480	312	290.0	1LA	0.00	C14	Leaf stem	–26.18	1,380	n.a.	40	1,301	3.24
Creamery-70SW-1	n.a.	0	0.0	2VA	1.00	GS	Surface 2006	n.a.	n.a.	–56	1	–56	n.a.
Creamery-70SW-1	n.a.	37.5–42.5	40.0	2VA	0.69	Cs137	Flood sediment	n.a.	n.a.	–13	10	–13	9.26

Creamery-70SW-1	n.a.	129	129.0	1/2VA	0 or 1	CORR	Top of Ab horizon	n.a.	n.a.	80	10	80	9.62
Creamery-70SW-1	UGAMS-03501	134	134.0	1VA	0.88	C14	Charcoal	−26.7	895	n.a.	25	829	0.07
Creamery-70SW-1	UGAMS-17623	145–147	146.0	1VA	0.60	C14	<125 um sediment	−26.3	1,300	n.a.	20	1,238	0.29
Creamery-70SW-1	UGAMS-17624	204–211	172.0	1LA	0.00	C14	Charcoal	−26.1	1,400	n.a.	20	1,309	3.64
Otto/Smith-1	n.a.	0	0.0	2VA	1.00	GS	Surface 2006	n.a.	n.a.	−56	1	−56	n.a.
Otto/Smith-1	n.a.	62.5–67.5	65.0	2VA	0.68	Cs137	Flood sediment	n.a.	n.a.	−13	10	−13	15.15
Otto/Smith-1	n.a.	204	204.0	1/2VA	0 or 1	CORR	Top of Ab horizon	n.a.	n.a.	80	10	80	14.93
Otto/Smith-1	UGAMS-01602	308	308.0	1VA	0.10	C14	Charcoal	−26.24	3,110	n.a.	60	3,309	0.32
Riverside-1	n.a.	0	0.0	2VA	1.00	GS	Surface 2005	n.a.	n.a.	−55	1	−55	n.a.
Riverside-1	n.a.	40–45	42.5	2VA	0.65	Cs137	Flood sediment	n.a.	n.a.	−13	10	−12	9.90
Riverside-1	n.a.	122	122.0	1/2VA	0 or 1	CORR	Top of Ab horizon	n.a.	n.a.	80	10	80	8.70
Riverside-1	UGAMS-10393	142–144.5	143.3	1VA	0.82	C14	<125 um sediment	−24.9	1,650	n.a.	25	1,552	0.14
Riverside-1	UGAMS-9709	142–144.5	143.3	1VA	0.82	C14	Charcoal in silt fraction	−26.3	1,710	n.a.	25	n.a.	n.a.
Riverside-1	UGA-09OSL-669	198–202	200.0	1VA	0.34	OSL	Sandy loam sediment	n.a.	n.a.	2400	150	2,371	0.70
Riverside-1	UGA-9054	303–305	240.0	1LA	0.00	C14	Leaf	−26.43	2,530	n.a.	23	2,634	1.52
Stiles-1	n.a.	0	0.0	2VA	1.00	CORR	Soil surface in 2009	n.a.	n.a.	−59	1	−59	n.a.
Stiles-1	n.a.	15	15.0	2VA	0.73	Cs137	Flood sediment	n.a.	n.a.	−13	10	−13	3.26

(continued)

Table 8.3 (continued)

Site	Lab #	Sample depth (cm)	Plot point depth (cm)	Unit & facies	Fraction of VA unit thickness (0 = bottom, 1 = top)	Dating method	Material dated	Material isotopic del ^{13}C	^{14}C years BP	Non-14C age estimate cal. year BP	±1 SD	Best age Est. (cal. years BP)	Inter-sample rate (mm/year)
Stiles-1	n.a.	55	55.0	1/2VA	0 or 1	CORR	Top of Ab horizon	n.a.	n.a.	80	10	80	4.27
Stiles-1	UGAMS-9644	67–68	67.5	1VA	0.74	C14	<125 um sediment	−26.1	1270	n.a.	20	1,179	0.11
Stiles-1	UGAMS-9635	99–101	100.0	1VA	0.08	C14	Charcoal	−26.1	2,230	n.a.	25	2,230	0.32
Stiles-1	UGAMS-17408	103–105	104.0	1VA	0.00	C14	<125 um sediment	−26.9	2,870	n.a.	25	2,991	0.05
Stiles-1	UGAMS-12165	111–114	113.0	1VA	0.73	C14	<125 um sediment	−26.69	4,420	n.a.	25	5,011	0.05
Stiles-1	UGAMS-17625	129–132	131.0	1VA	0.64	C14	<125 um sediment	−28.3	5,960	n.a.	25	6,791	0.10
Stiles-1	UGAMS-12164	219–220	219.5	1VA	0.23	C14	Wood	−29.14	9,510	n.a.	30	10,842	0.22
Stiles-1	UGAMS-3978	294	267.5	1LA	0.00	C14	Nut hull (uncarb.)	−27.4	9,850	n.a.	30	11,244	1.20

Abbreviations: *1LA* pre-settlement lateral accretion, *1VA* pre-settlement vertical accretion, *1LA* post-settlement lateral accretion, *2VA* post-settlement vertical accretion, *C14* radiocarbon, *Cs137* cesium-137, *CORR* correlation, *GS* ground surface, *OSL* optically stimulated luminescence, *n.a.* not applicable, *uncarb.* uncarbonized col (Murray and Wintle 2000). The 125–250 μm quartz grains were dated following isolation by flotation in sodium polytungstate and etching in hydrofluoric acid. Light stimulation and emission readings were made with a RISØ® system according to the protocol specified in Wang (2010).

The radioactive ^{137}Cs content of sediments was measured by gamma spectrometry with an Ortec® digital spectrometer in the University of Georgia Geomorphology Laboratory, which is coupled to a 5 cm diameter high purity germanium crystal. Count times of 5,000–10,000 s were used to achieve analytical errors <10 %. The maximum content of ^{137}Cs in the sediment column was assumed to represent the year 1963 AD (Walling and He 1997). In several cases, the 1963 timeline was corroborated by the obvious presence of immediately overlying sandy flood deposits of the 1964 flood, which is the largest flood on record at the USGS gauge (#03500000) on the Upper Little Tennessee River near Prentiss, Tennessee, with records extending from 1945 to present.

Additional correlation dating (age established for a particular feature or stratum in one place is equated to a similar feature or stratum in another place) was used at the State Line site to establish the age of the 1902 flood deposit based on knowledge from old newspaper accounts that the Tallulah Falls Railroad grade crossed the river from Georgia into North Carolina in 1906 AD. That is, the original railroad grade stands only 30–40 m from the radiocarbon dated stratigraphic section, and upon drilling through the artificial fill of the railroad grade there was only approximately 20 cm of post-settlement alluvium that was noted beneath the artificial fill dirt and atop the undisturbed buried A horizon of circa 1870 AD. In addition, a very sandy flood layer occurred at about 5 cm below the artificial fill, which therefore probably corresponds to the 1902 flood, which is the largest flood on record at another USGS gauge (#03507000) on the Upper Little Tennessee River at Judson that collected records from 1897 to 1944.

Charcoal was separated from the >125 μm fraction from all close-interval samples at the Stiles-1 site to produce a time-series of charcoal accumulation rates (pieces per cm^2 per year) or CHAR (Higuera et al. 2009), incorporating the counts, bulk density, and sedimentation rates in a manner similar to that specified by Higuera et al. (2009). Once the >125 μm fraction was sieved and dried, the charcoal particles were counted under a binocular microscope at 40× magnification. Bulk density was measured by weighing selected oven-dried clods, sealing them with plastic spray paint, and immersing the samples in a graduated cylinder of water to determine the volume. Sedimentation rates were calculated by CLAM as described above and then incorporated into the calculation of CHAR.

8.4 Comparison of Pre- and Post-settlement Rates of Sedimentation

Results of age versus depth modeling and calculations of weighted mean values of long-term-average sedimentation rates from the 12 stratigraphic sections show that the post-settlement sedimentation rates were about an order of magnitude faster than latest Holocene pre-settlement rates (Tables 8.3 and 8.4; Fig. 8.7). Pre-settlement rates generally are in the range from 0.1 to 1.0 mm per year, whereas post-settlement rates are in the range from 1 to 10 mm per year, and the median values of the weighted averages of each unit are 0.4 and 3.6 mm per year, respectively. A paired *t*-test indicates that the sedimentation rates at these 12 sites are significantly different (t-statistic of 4.29 and a one-tailed significance level of 0.0006). The ratio of post-settlement to pre-settlement rates derived from individual sites ranged from 2.7 to 46.9 with a median value of 7.8 and a mean value of 12.9 (Table 8.4). Such disparity of pre- versus post-settlement rates is on par with areas of intensive past agriculture, such as the Upper Mississippi River valley region (Knox 2006) and the southern Piedmont (Happ 1945; Trimble 1974, 1975), which is somewhat surprising given the mostly forested condition of the SBRM. However, the Blue Ridge topography is considerably more sloping than those regions, which probably facilitates comparable amounts of erosion and sedimentation.

There is a pronounced difference in both the sedimentation rates and thicknesses of post-settlement alluvium from tributary watersheds draining less than 65 km^2 versus those along the main channel of the Upper Little Tennessee River draining 146–319 km^2. The mean of the long-term-average sedimentation rates of the seven smaller tributaries is 2.9 mm per year versus 9.4 mm per year from five sites along the main channel; and the average post-settlement stratigraphic unit thickness of the tributaries is 41 cm versus 134 cm along the main channel. This disparity may be attributed to the fact that the tributaries have been exhibiting high rates of lateral migration and stream bank erosion since the early 1900s, whereas lateral migration of the main channel has been minimal (Leigh 2010). Such migration enhances sediment concentrations downstream along the main channel because erosion of tributary stream bank sediment is added to the hillslope-derived load. Also, the higher gradient of the tributary streams is more effective in transferring the sediment across the floodplain and further downstream than the lower-gradient main channel. Furthermore, the laterally migrating tributaries have established a new floodplain that inherently reduces vertical accretion rates on the historical terrace that contains much of the post-settlement alluvium. This uneven pattern of sediment erosion, transfer, and deposition adheres to the pattern that Knox (1987, 2006) recognized in tributaries to the upper Mississippi River in southwestern Wisconsin and to Trimble's (1993) 'distributed sediment budget.' Recent real-time observations of sediment concentrations in Upper Little Tennessee tributaries versus the main channel show the same sort of disparity (Rhett Jackson, pers. comm.). Although there are small, but similar, differences in tributary versus main channel sedimentation rates and thicknesses for the pre-settlement alluvium, they are nowhere near the same magnitude of those observed in the post-settlement alluvium (Table 8.4).

Another important difference in pre-settlement versus post-settlement sedimentation rates (in cases where there are three or more age control points within each stratigraphic unit) is that the pre-settlement unit exhibits an exponential decline in sedimentation rates with time, which is the norm for most vertical accretion floodplain sediments (Bridge 2003), whereas the post-settlement unit exhibits a constant linear rate or even an exponential increase with time (Fig. 8.8). This is well

Table 8.4 Basal ages, stratigraphic unit thickness, weighted average sedimentation rates for the entire unit, and ratios of post-settlement/pre-settlement and post-1963/1870-1963 sedimentation rates

Site name	Basal age of pre-settlement VA (cal years BP)[a, b, c]	Post-settlement thickness (cm)	Pre-settlement thickness (cm)	Post-settlement weighted average (mm/year)	Pre-settlement weighted avg. (mm/year)	Post-/pre- ratio of sed. rates	Post-1963/1870-1963 ratio of sed. rates
Keener-1	1,546	43	100	3.2	0.9	3.5	0.9
Keener-4	599	44	62	3.2	1.2	2.7	n.a.
Skeenah-2	1,469	27	54	2.3	0.4	6.0	2.1
Skeenah-4	2,494	40	72	1.6	0.4	4.0	n.a.
31GH457-1	>6,057 (at 14 %)	56	119	4.0	0.2	23.6	n.a.
31GH457-2	>>3,401 (at 73 %)	37	148	2.7	0.1	22.3	n.a.
Brasstown-20	1,811	40	59	3.3	0.3	9.5	n.a.
State Line-1	1,301	159	131	9.5	1.9	4.9	1.5
Creamery-70SW-1	1,309	129	43	9.5	2.3	4.2	1.0
Otto/Smith-1	>3,309 (at 10 %)	204	116	15.0	0.3	46.9	1.0
Riverside-1	2,634	122	118	9.1	0.9	10.4	1.1
Stiles-1	11,244	55	213	4.0	0.2	16.3	0.8
	Median. =	50	108	3.6	0.4	7.8	1.0
	Average.=	80	103	5.6	0.8	12.9	1.2
	±1 St. Dev. =	58	49	4.1	0.7	13.0	0.5
	Tributary avg. =	41	88	2.9	0.5	10.2	1.5
	Main channel avg. =	134	124	9.4	1.1	16.5	1.1

[a]% refers to the percentage of stratigraphic thickness measured upward from the base of unit
[b]VA is the abbreviation for vertical accretion facies
[c]>indicates that a basal age was not determined

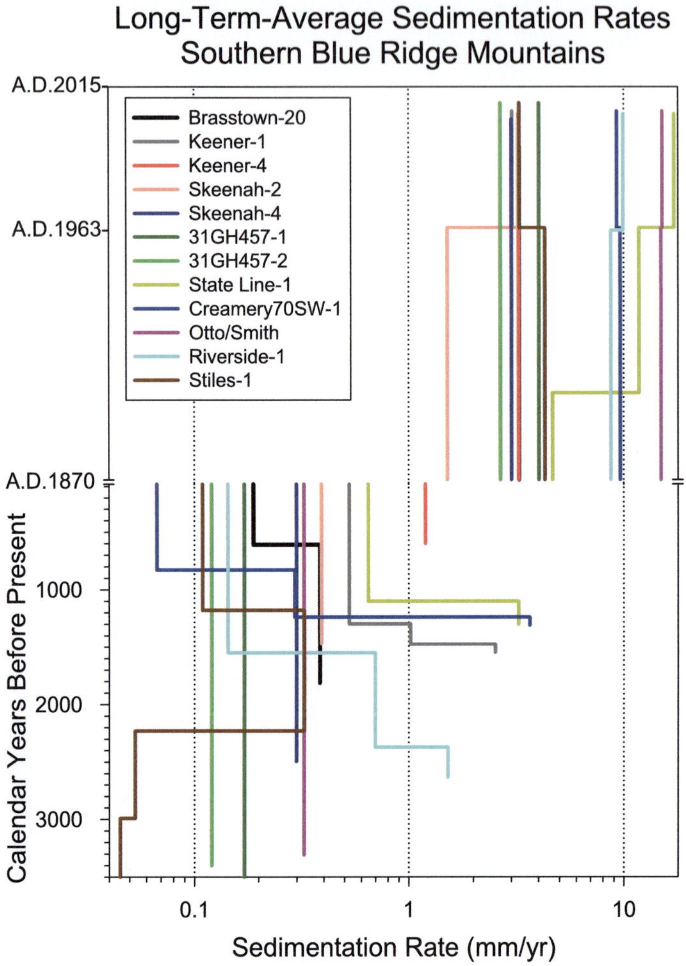

Fig. 8.7 Sedimentation rate plots for the 12 sites. Each site represents a profile of vertical accretion floodplain sediments from stream valleys in the southern Blue Ridge Mountains (see Fig. 8.1) that accumulated over the past 3,500 years. Line ends and right-angle changes in line segments represent age control points

illustrated by the ratio of post-1963 to pre-1963 post-settlement sedimentation rates given that the median ratio is 1.0 and the average ratio is 1.2 (Table 8.4). This indicates that erosive processes of the deforested landscape have remained vigorous since the 1960s, even though much of the deforested land is covered in grasses and shrubs; this is consistent with the findings of Miller et al. (2005) and suggestive that second home development and road/driveway construction is generating considerable sediment. Lateral erosion of the tributaries also contributes to sustained sediment supply to the main stem of the river. Alternatively, the sustained high rates of sedimentation post-1963 may indicate intrinsic mechanisms of the fluvial system,

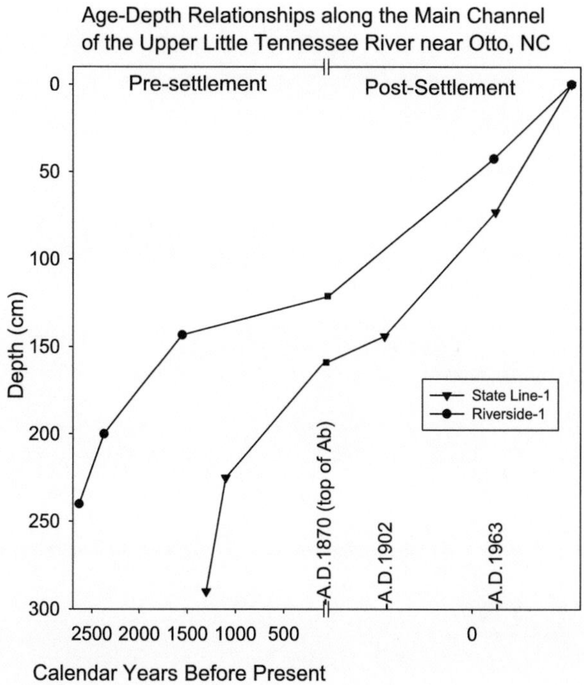

Fig. 8.8 Age-depth plots of the pre-settlement versus post-settlement units at the Riverside-1 and State Line-1 sites illustrating the progressively slower sedimentation rates (decreasing slopes of lines) with more recent age for the pre-settlement unit versus faster sedimentation with time (increasing slopes of lines) for the post-settlement unit. The square point on each plot represents the 1870 AD top of the buried A horizon

namely stream bank erosion and natural levee progradation, so that when sample sites progressively become nearer to the channel due to lateral erosion, this closer proximity to the channel inherently favors faster sedimentation rates (Bridge 2003). However, the Keener-1, Skeenah-2, and Creamery-70mSW-1 sites are not situated on prograding levees, but they still exhibit post-1963 to pre-1963 ratios of about 0.9–2.1 (Table 8.4).

8.5 Natural Disturbance Expressed by Fluvial and Colluvial Sediment

Natural disturbances in the forested mountains typically are not obvious in terms of macro-stratigraphy and long-term-average sedimentation rates, unless a distinctive landslide or debris flow event is preserved in the stratigraphic record. Natural disturbances in the Blue Ridge Mountains that produce recognizable sedimentary records primarily are driven by climatic variations that modulate mass wasting; that is,

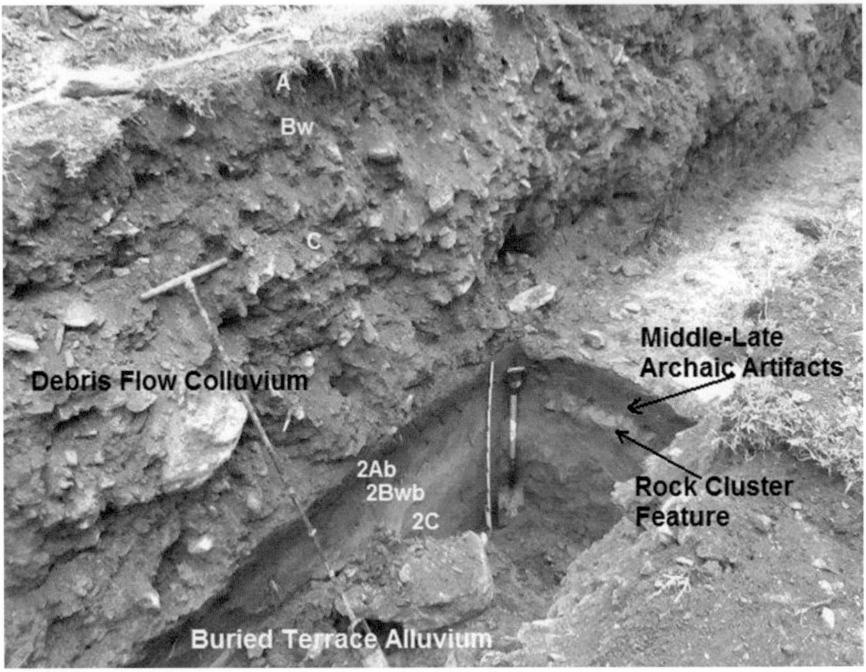

Fig. 8.9 Debris flow deposits (coarse upper stratum) overlying river alluvium at the Ferebee picnic area along the Nantahala River (From Leigh 2011)

wet periods generate more mass wasting events (landslides and debris flows) than dry periods (see Wooten et al. Chap. 9) and mass wasting releases huge volumes of sediment to the fluvial system. In addition, wet periods may stimulate more lateral bank erosion (causing riparian forest disturbance) and thus increase sediment yield from erosion of terraced stream banks (Rogers and Leigh 2013).

An example illustrating that mass wasting events do indeed become deposited within floodplains was documented by Leigh (2009) at the Ferebee picnic area along the Nantahala River in western Carolina (Fig. 8.9). While this sort of evidence clearly indicates that mass wasting definitely has occurred on nearby hillslopes, it is not sufficient to construct any sort of time series for the entire watershed in order to understand variation in the magnitude and frequency of mass wasting events through time. Wooten et al. (Chap. 9) indicate that regional debris flow events in the Blue Ridge province have an historic recurrence frequency of approximately 25–29 years, driven by the passage of tropical storms and their heavy rains. Also, Eaton (2003a, b) was able to construct a millennial-scale time series of mass wasting for the Rapidan River basin in Virginia. He argued that debris flows and landslides are the dominant modes of Holocene sediment production in forests of the SBRM, whereas overland flow and surface erosion are relatively insignificant under native forest cover. Eaton indicated at least a 2500 year recurrence frequency for debris flow activity within any individual tributary of the upper Rapidan catchment, although he was unable to relate debris flow frequency to Holocene paleoclimate drivers of variation in magnitude and frequency through time.

Broad inferences about mass wasting frequency in relation to late Pleistocene and Holocene fluctuations in paleoclimate have been made in a few cases. Delcourt and Delcourt (2004) indicate that a marked reduction in fluvial sediment supply to the lower Little Tennessee River valley occurred at some time between 7,800 and 4,000 years BP in response to a change in sediment flux from forested watersheds, largely related to mass wasting events. In support of their idea Leigh and Webb (2006) suggested that relatively fast footslope sedimentation rates during the early Holocene versus late Holocene possibly were driven by wetter paleoclimate conditions in the early Holocene. Such inferences are broad and somewhat speculative, but they have firm grounding in theory. That is, wetter climatic periods, especially those related to prolonged heavy rains associated with tropical storm events tend to trigger a higher frequency of landslides and debris flows (Wooten et al. Chap. 9) that should be recognizable in the stratigraphic record. However, the temporal resolution needed to derive significant variations in sedimentation rates through the Holocene is difficult and expensive to obtain because of the numerous samples required.

A modern example of sediment yield from forested watersheds in relation to climatic fluctuations is provided by Simmons' data set collected during the relatively wet 1970s (Simmons 1993) when compared to data collected from the same region (Fig. 8.4) during average or drier climatic periods (Coweeta LTER unpubl; Oblinger 2003). Oblinger (2003) observed that annual sediment concentrations and yields were an order of magnitude less during their relatively dry study period in 2001 compared to Simmons' estimates from the relatively wet 1970s. In contrast, Simmons' 1970s sediment yields are only slightly higher than those observed in water-year 2011 by researchers of the Coweeta LTER (Fig. 8.4), which was a slightly above average year in terms of wetness. Also, the sediment yields for the least disturbed watersheds ('pristine' forested watersheds of Simmons 1993) are very comparable to those fully forested sites observed in water year 2011 by the Coweeta LTER researchers.

A forward stepwise linear regression was applied to evaluate the combined sediment yields of the three time periods mentioned above by using three independent variables or predictors: (1) percent nonforest cover values published in Simmons' (1993) for the 1970s and tallies from the National Land Cover dataset for the 2001 and 2011 datasets for the Oblinger and Coweeta LTER data, respectively; (2) catchment drainage area; and (3) measured annual-mean runoff at the USGS Prentiss gauge (USGS #03500000) as an indicator of wetness. The wetness variable of annual-mean runoff essentially is a 'dummy variable' (values of 12.7 m^3 per second for the 1970s; 5.44 m^3 per second for 2001; and 8.97 m^3 per second for water year 2011). The results found all three independent variables to be highly significant at probability levels less than 0.01 and the three-variable model explains 52.5 % of the variance in sediment yield values of the three combined data sets at a probability level of <0.001 and an F-statistic of 15.114 (Table 8.5). Drainage area explained the largest portion of variance (24 %), followed by the percent of non-forested land (15 %),

Table 8.5 Multiple linear regression model explaining 52.5 % of the variation in annual sediment yield of Blue Ridge Mountains streams

Variable	Coefficient	*t*-test value of coefficient validity or significance	Probability (ratio) that coefficient is invalid or not significant	Proportion of variation explained (partial R^2)
Y-intercept	−77.903	−2.911	0.006	Not applicable
x1: drainage basin size or area, DA (km^2)	0.0187	2.926	0.006	0.241
x2: percent of basin that is not forested	1.891	3.6	<0.001	0.15
x3: annual average daily runoff from entire basin (m^3/s)	8.425	3.971	<0.001	0.134
			Total explained (total R^2) =	0.525

Model equation: Sediment yield (tonnes/km^2/year) = −77.903 + (0.0187 * DA (km^2)) + (1.891 * % Non-forest) + (8.425 * Runoff (m^3/s))
Number of samples in the regression model = 43
F-statistic of overall model strength = 15.114
Standard error of estimate = 31.131 (t/km^2/year)
Shapiro-Wilk normality test score (which passed) = 0.624

and then average runoff or wetness (13 %). The fact that drainage area was such a strong predictor is consistent with the observation above concerning the distributed sediment budget, because the dependent variable of sediment yield (tonnes per km^2 per year) is already normalized for contributing drainage area. Overall, the results indicate that the climatic 'dummy variable' is essentially as good a predictor as the percent of nonforest cover and reinforce the idea that subtle variations in climate can have a significant influence on sediment yield. This is consistent with Knox's (1993) observation that modest changes in climate can have a significant influence on flood magnitude and frequency.

Little to no evidence of pre-settlement erosional forest disturbance can be discerned from the 12 stratigraphic sites identified in this paper. There are some minor variations in sedimentation rates at individual sites, but it is impossible to determine whether those were driven by intrinsic mechanisms of the fluvial system or by extrinsic mechanisms, such as climate change. This result stems from the fact that it is very difficult to obtain the spatial and temporal resolution from long-term average sedimentation rates that allows discrimination of the ultimate drivers of natural forest disturbance. Also, it is apparent from pollen records (Delcourt et al. 1986; Delcourt and Delcourt 1988) that forest cover has persisted throughout the Holocene (mostly deciduous with evergreens at high elevation), so that vegetation (and probably climate) did not vary enough to favor big changes in sediment yield and runoff. However, temporally continuous pollen records for the entire Holocene are lacking for the SBRM. Indeed, other places where large geomorphic variations have been noted are within ecotonal zones where pronounced shifts from grassland to forest have occurred (e.g., Knox 1983).

In summary, paleoclimatic changes in rainfall delivery that lead to increased soil wetness make hillslopes more susceptible to erosion (especially by mass wasting) and are agents that theoretically should be reflected in the stratigraphic record. However, in practice the scale of observation in the stratigraphic record, along with the medium to large watershed scale that derives sediment from many different tributaries, is generally too coarse to resolve climatically-driven disturbances. Only the most pronounced variations in sediment yield (driven largely by mass-wasting) are crudely discernable in the pre-settlement stratigraphic record.

8.6 Were Prehistoric Native Americans Agents of Hillslope Erosion?

There is recent interest about the extent to which prehistoric Native Americans altered natural vegetation and environmental conditions. Abrams and Nowacki (2008) contend that indigenous land management affected vegetation 'ubiquitously.' In contrast it has been argued that indigenous impact varied widely with some regions exhibiting no discernable human impacts (Vale 2002). Most recently, Munoz et al. (2014) document the spatial pattern of the late prehistoric human impacts to be 'localized and heterogeneous' for eastern North America, and they challenge the idea that the prehistoric Native American's impact on vegetation was widespread and ubiquitous. Instead, they argue that indigenous land use was patchy, including spatially varied niches of 'undomesticated woodland management' and agricultural land. Furthermore, they stress the idea that human impacts were temporally dynamic.

Delcourt and Delcourt (1988, 1997, 1998, 2004) and Delcourt et al. (1986, 1998) advanced the idea that prehistoric Native American impacts on southern Appalachia were significant, primarily in terms of managed silviculture that involved burning the understory of nut-bearing trees (American chestnut (*Castanea dentata*), oak (*Quercus* spp.), hickory (*Carya* spp.)) and by cultivation of maize, squash and gourds (*Cucurbita* spp.), beans (*Phaseolus* spp.), and other crops. Forests are thought to have been disturbed and managed by Native Americans for edible mast as early as 4,000–5,000 years BP (Delcourt and Delcourt 2004). Cultivation of squash and gourds is known to have occurred as early as 5,200 years BP (Delcourt and Delcourt 2004), and widespread domestication and cultivation of goosefoot (*Chenopodium* spp.) is known to have occurred by 3,500 years BP in eastern North America (Smith and Cowan 1987). Later, cultivation of maize is documented in the Little Tennessee River valley by 1,700 years BP, but it did not become a mainstay of the diet until 1,000 years BP (Delcourt and Delcourt 2004). Beans were introduced into the lower Little Tennessee valley at about 600 years BP (Delcourt and Delcourt 2004). In summary, indigenous silvicultural and agricultural activities in the Little Tennessee valley were established by at least 4,000–5,000 years BP and these activities progressively increased though the late Holocene, culminating with widespread dependence on agriculture during the Mississippian cultural period circa 1,000–500 years BP.

The question is not whether indigenous people were managing and altering the ecosystem and vegetation patterns of the SBRM, but whether or not their use of the land was erosive enough to register 'legacy sediment' in stratigraphic sections. James (2011) provides a nice summary of indigenous land use patterns in the eastern USA, and he concludes that indigenous practices were minimally destructive to soil and slope stability, as no mechanized methods were practiced (no plows, wheels, or metal tools) and agriculture was focused on flat alluvial bottomlands. In stark contrast, Stinchcomb et al. (2011) promote the idea that indigenous people caused significant erosion and bottomland sedimentation circa 1100–1600 AD and registered the first 'legacy sediment' (sensu James 2013). They extrapolated their example of 'pre-Colonial sediment' from a tributary of the Delaware River valley to all of eastern North America, while acknowledging that "future research efforts should focus on mapping the chronological and geographic range."

Six of the 12 geochronologies presented above contain at least three age control points spanning 2,500 years BP to 80 years BP (1870 AD) that facilitate evaluation of the question of acceleration of sedimentation rates by indigenous people in the SBRM (Fig. 8.10; Table 8.3). The past 2,500 years encompasses the Middle Woodland, Late Woodland, Mississippian, and Cherokee prehistoric cultural phases,

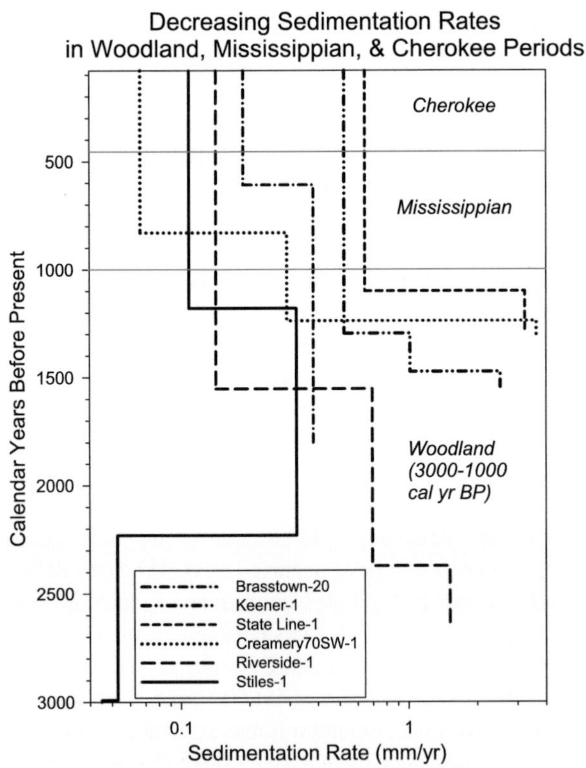

Fig. 8.10 Sedimentation rate plots that have two or more age control points between 2,500 years BP and 80 years BP (1870 AD) that enable evaluation of whether indigenous people caused increased sedimentation rates during the Woodland, Mississippian, and Cherokee cultural periods

and during this time agricultural expansion is known to have occurred (Delcourt and Delcourt 2004), with maximum pre-settlement extent of cropland occurring during the Mississippian and Cherokee phases. Furthermore, the Upper Little Tennessee River valley is known to have been a hub of the prehistoric Cherokee culture (Hudson 2002). Results show that all six stratigraphic sections fail to indicate any increase in sedimentation rates during the Mississippian or Cherokee phases; in contrast, all six sections show a pronounced *decrease* in sedimentation rates from the Middle Woodland through Cherokee phases. Although the resolution of the dating on these six sections is somewhat coarse, all six fail to support the idea that Mississippian or Cherokee peoples accelerated bottomland sedimentation rates. While lack of support does not mean that indigenous human-induced erosion and accelerated sedimentation did not happen anywhere, it does indicate that it was not prevalent enough to be commonly observed in the stratigraphic record like the post-settlement stratum.

The Stiles site shows an interesting increase in sedimentation rates at 2,230–1,179 years BP during the Middle Woodland period (Fig. 8.10; Table 8.3), but thereafter a decrease is apparent. This Middle Woodland increase may be attributed to human agency, but on the other hand it is quite possible that it simply results from intrinsic shifting of the Upper Little Tennessee River. That is, the Stiles meander scar sedimentation site may have been far away from the active channel until 2,230 years BP when the active channel avulsed and shifted its position closer to the site (near it present position), thus automatically increasing the delivery of overbank sediments. The current position of the modern channel, and the fact that sedimentation rates at the Stiles site *decrease* after 2,230 years BP, are consistent with this intrinsic mechanism. Also, it seems that if it were attributed to human agency, then the higher Middle Woodland sedimentation rates would have persisted into Mississippian and Cherokee periods.

The above example of intrinsic mechanisms causing change in fluvial systems raises the point that, in addition to human agencies, there are other important drivers that can cause changes in fluvial sedimentation rates, namely intrinsic channel shifting and climate change. Climate change is particularly germane to the discourse about indigenous land use influences on erosion and sedimentation. The Medieval Climate Anomaly (MCA, circa 800–1300 AD) and the Little Ice Age (LIA circa 1400–1800 AD) exhibited significant shifts from relatively warm to cold climate conditions, respectively (Bradley and Jones 1995; Cronin et al. 2003; Grove and Switsur 1994), at exactly the time when indigenous agricultural land use was rapidly expanding. Although such MCA temperature shifts have not been documented in the southeastern USA, Wang and Leigh (2012) noted significant evidence (coarser sand sedimentology) of increased flood magnitudes in the Blue Ridge Mountains at the beginning and end of the MCA, but no changes in sedimentation rates were observed. They concluded that the Upper Little Tennessee valley may have been influenced by a relatively wet climate and large floods at the beginning and end of the MCA, circa 1,150–1,350 years BP and 650–900 years BP. Stinchcomb et al. (2011) acknowledged that global climate change associated with the transition from the MCA to the LIA could have played a role in depositing their 'pre-Colonial sediment,' but they discounted that explanation and favored the interpretation of indigenous human-induced erosion.

In summary, no compelling evidence has been recovered in the SBRM to indicate that indigenous people caused accelerated erosion and increased alluvial sedimentation rates. Of course, there might be some localized areas of accelerated prehistoric bottomland sedimentation rates that were proximal to agricultural fields (which have yet to be found), but preliminary evidence suggests that relatively slow, or 'natural,' rates of sedimentation were normally occurring on river and stream bottomlands, despite significant disturbance of the native forest in support of mast harvest, hunting, and agriculture during pre-settlement time. It is likely that most of the pre-settlement agricultural disturbance of the forest occurred in relatively low-gradient alluvial bottomlands where sediment erosion and transport was physically improbable. Furthermore, indigenous population densities probably were small enough to negate the need to expand agricultural fields on to steep hillslopes. Delcourt and Delcourt (2004) estimated (based on Baden's 1987 model) that Mississippian population densities in the lower Little Tennessee River valley were 16–17 persons per km^2 for entire watersheds and 52–56 persons per km^2 if concentrated over cultivated alluvial bottomlands. However, values of 16–17 persons per km^2 equates to the 1983 AD population density in Macon County, North Carolina, which suggests the Delcourt's estimates for the Mississippian period may be high. Baden's (1987) model indicated that minimum caloric needs for food required 0.1–0.6 ha per person. Using those values, even if 56 persons per km^2 of alluvial bottomland is assumed, then that equates to 1.8 ha of bottomland available per person (100 ha divided by 56 persons = 1.8 ha per person) or at least three times Baden's (1987) presumed bottomland requirement for subsistence. Thus, one could contend that available bottomland was not a limiting factor for indigenous Mississippian and Cherokee agriculture, so they did not need to deforest and cultivate steep erosive hillslopes. Baden (1987) assumed that nutrient depletion was a key factor in the progressively decreasing productivity of agricultural lands, which ultimately limited and depleted the soil resource. However, it is not clear whether he was considering nutrient replenishment from annual floods, weathering, and additions by humans. Indeed, Bartram speaks of the bottomlands in the Upper Little Tennessee River valley as appearing very fertile in 1775 AD.

8.7 Past Fires and Charcoal Records in Alluvium

Like many parts of the USA, suppression of human-ignited fires has been the norm for the SBRM during the past century (see Grissino-Mayer Chap. 6). However, fire routinely was used to clear land by early settlers during the 1700s and 1800s (Jurgelski 2008) and by Native Americans during pre-settlement times (Delcourt and Delcourt 2004) (also see Grissino-Mayer Chap. 6; Greenberg et al. Chap. 12). Bartram's 1775 travels (Van Doren 1928) through the SBRM describe many grassy meadows and open grassy understories in woodlands indicative of frequent low-intensity fires. Paleoecological studies have shown that fires were an important component of the pre-settlement landscape (Delcourt and Delcourt 2004). Delcourt and Delcourt (2004), as well as Delcourt and Delcourt (2004), assert that human use of fire caused distinct changes in forest composition and structure after 3,000 years BP, which favored fire-adapted mast-bearing trees such as oaks, chestnuts, hickories, and walnuts (*Juglans* spp.), as well as a patchwork of meadows and clearings initiated by fire. Fesenmyer and Christensen (2010) radiocarbon dated 83 samples from 18 separate soil pits (0–30 cm depth) on hillslopes bounding the upper Nantahala River valley in Macon County, North Carolina in an effort to construct a pre-settlement stand-level fire history. They found that "fires were frequent over the past 4,000 years, and their frequency appears to have increased significantly about 1,200 years before present (YBP), coinciding with the advent of the Mississippian Native American culture."

With respect to sediment yield, fires can greatly increase surface erosion by removal of the protective vegetation cover and in some cases by producing a hydrophobic surface that reduces infiltration and enhances overland flow (Stine 2013). However, low-severity and prescribed fires generally produce small effects on erosion and sedimentation (Committee on Hydrologic Impacts of Forest Management 2008). Indeed, Elliot and Vose (2005) found no significant increase in total suspended solids following prescribed restoration fires in the Conasauga watershed of northwest Georgia and southeast Tennessee, and noted that several other studies in the southeastern USA report little to no soil erosion following light- to moderate-intensity fires. Neary and Currier (1982) measured the total suspended solids derived from watersheds that were intensively burned in April 1978 during an 'abnormally hot' wildfire in the SBRM of South Carolina, and observed no significant difference between the forested 'control' watershed and burned watersheds. Thus, there is no compelling evidence to indicate that intentional pre-settlement use of fire for removal of understory and for garden clearings would have increased sediment yield and be reflected in the stratigraphic record. Recall that the six stratigraphic sections noted above that have sufficient chronologies to evaluate sedimentation rates after 2,500 years BP generally show decreasing sedimentation rates. Thus, it is not possible to infer that increased use of fire by Native Americans led to increased rates of pre-settlement erosion and sedimentation.

One of the 12 stratigraphic sections noted above, the Stiles site, has a continuous sedimentary history for the entire Holocene (11,244 years BP basal age of vertical accretion). Charcoal fragments larger than 125 μm were sieved and counted from a continuous column of closely spaced samples representing a time series of charcoal frequency or the CHAR metric of Higuera et al. (2009). The CHAR metric relies on the long-term sedimentation chronology derived from radiocarbon dates and the age-depth model (CLAM) described above. The results (Fig. 8.11) show a very pronounced increase in charcoal concentrations at about 2,400 years BP, whereas the majority of the record prior to then exhibits little to moderate evidence of fire except for the earliest Holocene (circa 11,244–10,900 years BP). This increase in charcoal concentrations after 2,400 years BP is consistent with the assertion that

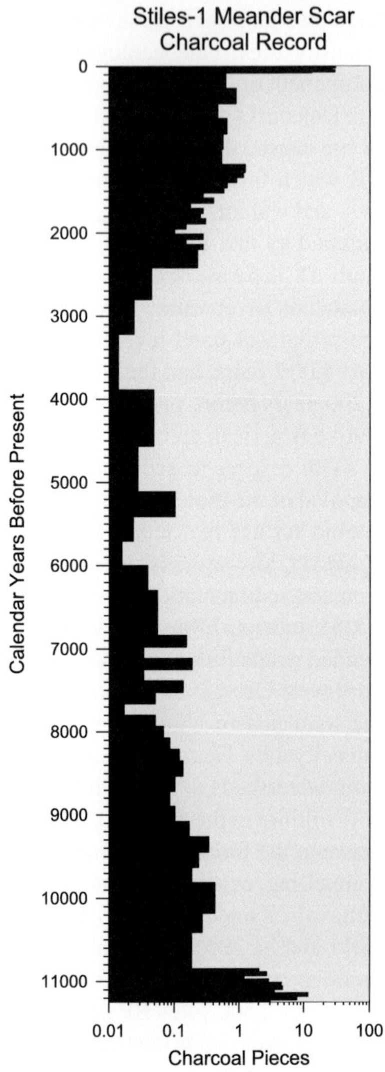

Fig. 8.11 Charcoal accumulation rate (pieces >125 μm per cm^2 per year) versus age (years BP) for the entire Holocene at the Stiles site, which is a cutoff meander scar between Otto and Franklin, North Carolina in the Upper Little Tennessee River valley

human-ignited fires became increasingly more prevalent after 3,000 years BP (Delcourt and Delcourt 2004, p. 85), as there is no known paleoclimate change at 2,000–3,000 years BP in the southeastern USA to entertain as an alternative explanation involving lightning ignitions. The high frequency of charcoal at the very beginning of the Holocene is more difficult to reconcile, although it is a known period of rapid climate change that may have involved frequent lightning ignitions. Unfortunately, there currently are insufficient data to support either lightning- or human-ignitions in the earliest Holocene.

In summary, pre-settlement and post-settlement fires have not been demonstrated to register distinct changes in sedimentation rates in the SBRM, and modern prescribed low-intensity burns do not appear to be erosive. The Stiles stratigraphic section confirms that fire was a definite component of the pre-settlement landscape, though unevenly through time, and it strongly suggests that fires became much more frequent in association with expansion of agriculture during the Woodland and later prehistoric cultural periods as suggested by others. The charcoal record from the Stiles site reinforces the idea that stratigraphic chronologies of past fire provide valuable insight about the importance of fire in pre-settlement forests of the SBRM. However, when coupled with sedimentation rate analysis there is no indication that forest fires caused any significant changes in hillslope erosion and bottomland sedimentation rates.

8.8 Conclusions

In the SBRM, vertical accretion alluvium provides a coarse-resolution proxy for past erosional forest disturbance that clearly discriminates the end members of fully forested to approximately 30 % regional deforestation, as illustrated by pre-settlement versus post-settlement alluvial stratigraphic records. Post-settlement sedimentation rates are about one order of magnitude greater than pre-settlement rates (typically ranging from 1.0–10 mm per year versus 0.1–1.0 mm per year, respectively), and this is consistent with post- versus pre-settlement stratigraphy recognized nationally (Knox 2006) and even globally (Wilkinson and McElroy 2007). Clearly, the stratigraphy and sedimentation rates of the post-settlement alluvium provide a geologic and stratigraphic record in support of the newly proposed Anthropocene epoch (Crutzen 2002; Zalasiewicz et al. 2010). Under 'natural' conditions that are not mediated by humans, only the most pronounced variations in sediment yield (driven largely by mass-wasting) are crudely discerned in the pre-settlement stratigraphic record. Indigenous human impacts of soil erosion and changes in bottomland sedimentation rates are *not* recognizable in the pre-settlement stratigraphic record of the SBRM. However, the use of fire for forest management and agricultural activities by Woodland, Mississippian, and Cherokee prehistoric cultures is apparent in the fluvial stratigraphic record via charcoal preserved in the stratigraphic sections.

Acknowledgements Jake McDonald, Katie Price, James Rogers and Lixin Wang assisted with data collection, and Paul Webb and Bruce Idol of TRC-Environmental, Inc. facilitated data collection at site 31GH457. Rhett Jackson provided water year 2011 sediment yield data from the Coweeta Long Term Ecological Research program (LTER). Much of this research was conducted in association with National Science Foundation Grants to the Coweeta LTER (awards DEB-0823923, DEB-0218001) and grants from the University of Georgia Research Foundation, Inc. Jake McDonald and Katie Price provided useful editorial remarks on earlier drafts of this manuscript. Beverly Collins, Cathryn Greenberg and two anonymous reviewers provided useful remarks and changes that improved this manuscript. Much thanks is extended to landowners who granted permission to collect data.

References

Abrams, M. D., & Nowacki, G. J. (2008). Native Americans as active and passive promoters of mast and fruit trees in the eastern USA. *The Holocene, 18*, 1123–1137.

Ambers, R. K. R., Druckenbrod, D. L., & Ambers, C. P. (2006). Geomorphic response to historical agriculture at Monument Hill in the Blue Ridge foothills of central Virginia. *Catena, 65*, 49–60.

Ayers, H. B., & Ashe, W. W. (1905). *The southern Appalachian forests* (US Geological Survey professional paper, 37). Reston: US Geological Survey.

Baden, W. W. (1987) *A dynamic model of stability and change in Mississippian agricultural systems*. Dissertation, University of Tennessee.

Benedetti, M. M. (2003). Controls on overbank deposition in the upper Mississippi River. *Geomorphology, 56*, 271–290.

Blaauw, M. (2010). Methods and code for 'classical' age-modelling of radiocarbon sequences. *Quaternay Geochronology, 5*, 512.

Bradley, R. S., & Jones, P. D. (1995). Recent developments in studies of climate since AD1500. In R. S. Bradley & P. D. Jones (Eds.), *Climate since AD 1500 (revised edn)*. London: Routledge.

Bridge, J. S. (2003). *Rivers and floodplains: Forms, processes, and sedimentary record*. Malden: Blackwell.

Committee on Hydrologic Impacts of Forest Management. (2008). *Hydrologic effects of a changing forest landscape*. Washington, DC: National Academies Press.

Costa, J. E. (1975). Effects of agriculture on erosion and sedimentation in the Piedmont province, Maryland. *Geological Society of America Bulletin, 86*, 1281–1286.

Cronin, T. M., Dwyer, G. S., Kamiya, T., Schwede, S., & Willard, D. A. (2003). Medieval warm period, Little Ice Age and 20th century temperature variability from Chesapeake Bay. *Global and Planetary Change, 36*, 17–29.

Crutzen, P. J. (2002). The geology of mankind. *Nature, 415*, 23–24.

Delcourt, H. R., & Delcourt, P. A. (1988). Quaternary landscape ecology: Relevant scales in space and time. *Landscape Ecology, 2*, 23–44.

Delcourt, H. R., & Delcourt, P. A. (1997). Pre-Columbian Native American use of fire on southern Appalachian landscapes. *Conservation Biology, 11*, 1010–1014.

Delcourt, P. A., & Delcourt, H. R. (1998). The influence of prehistoric human-set fires on oak-chestnut forests in the southern Appalachians. *Castanea, 63*, 337–345.

Delcourt, P. A., & Delcourt, H. R. (2004). *Prehistoric Native Americans and ecological change: Human ecosystems in eastern North America since the Pleistocene*. Cambridge: Cambridge University Press.

Delcourt, P. A., Delcourt, H. R., Cridlebaugh, P. A., & Chapman, J. (1986). Holocene ethnobotanical and paleoecological record of human impact on vegetation in the Little Tennessee River Valley, Tennessee. *Quaternary Research, 25*, 330–349.

Delcourt, P. A., Delcourt, H. R., Ison, C. R., Sharp, W. E., & Gremillion, K. J. (1998). Prehistoric human use of fire, the Eastern Agricultural Complex and Appalachian oak-chestnut forests: Paleoecology of Cliff Palace Pond, Kentucky. *American Antiquity, 63*, 263–278.

Eaton, L. S., Morgan, B. A., Kochel, C. R., & Howard, A. D. (2003a). Quaternary deposits and landscape evolution of the central Blue Ridge of Virginia. *Geomorphology, 56*, 139–154.

Eaton, L. S., Morgan, B. A., Kochel, R. C., & Howard, A. D. (2003b). Role of debris flows in long-term landscape denudation in the Central Appalachians of Virginia. *Geology, 31*, 339–342.

Elliot, K. J., & Vose, J. M. (2005). Initial effects of prescribed fire on quality of soil solution and streamwater in the southern Appalachian Mountains. *Sothern Journal Applied Forest, 29*, 5–15.

Fenneman, N. M. (1938). *Physiography of eastern United States*. New York: McGraw-Hill.

Fesenmyer, K. A., & Christensen, N. L., Jr. (2010). Reconstructing Holocene fire history in a southern Appalachian forest using soil charcoal. *Ecology, 91*, 662–670.

Fuller, G. L. (1934). Charting the effects of erosion in the old plantation belt of the southern Piedmont. *Transactions American Geophysical Union, 15*(II), 495–500.

Glenn LC (1911) Denudation and erosion in the southern Appalachian region and the Monongahela Basin. USGS Prof Pap 72, US Geol Surv, Reston

Grove, J. M., & Switsur, R. (1994). Glacial geological evidence for the Medieval Warm Period. *Climate Change, 26*, 143–169.

Happ, S. C. (1945). Sedimentation in South Carolina Piedmont valleys. *American Journal of Science, 243*, 113–126.

Hatcher, R. D., Jr. (1988). Bedrock geology and regional geologic setting of Coweeta Hydrologic Laboratory in the eastern Blue Ridge. In W. T. Swank & D. A. Crossley Jr. (Eds.), *Forest hydrology and ecology at Coweeta* (Ecological studies). New York: Springer.

Hewlett, J. D., Fortson, J. C., & Cunningham, G. B. (1977). The effect of rainfall intensity on storm flow and peak discharge from forest land. *Water Resources Research, 13*, 259–265.

Higuera, P. E., Brubaker, L. B., Anderson, P. M., Hu, F. S., & Brown, T. A. (2009). Vegetation mediated the impacts of postglacial climate change on fire regimes in the south central Brooks Range, Alaska. *Ecological Monographs, 56*, 201–219.

Holmes, J. S. (1911). *Forest conditions in western North Carolina* (The North Carolina Geological and Economic Survey Bulletin, 23). Raleigh: Edwards and Broughton Printing.

Hudson, C. M. (2002). *The transformation of the Southeastern Indians, 1540–1760*. Jackson: University Press of Mississippi.

Hupp, C. R., Noe, G. B., Schenk, E. R., & Benthem, A. J. (2013). Recent and historic sediment dynamics along Difficult Run, a suburban Virginia Piedmont stream. *Geomorphology, 180*(181), 156–169.

Jackson, C. R., Martin, J. K., Leigh, D. S., & West, L. T. (2005). A southeastern Piedmont watershed sediment budget: Evidence for a multi-millennial agricultural legacy. *Journal of Soil and Water Conservation, 60*, 3–3.

Jacobson, R. B., & Coleman, D. J. (1986). Stratigraphy and recent evolution of Maryland Piedmont flood plains. *American Journal of Science, 286*, 617–637.

James, L. A. (2011). Contrasting geomorphic impacts of pre- and post-Columbian land-use changes in Anglo America. *Physical Geography, 32*, 399–422.

James, L. A. (2013). Legacy sediment: Definitions and processes of episodically produced anthropogenic sediment. *Anthropocene, 2*, 16.

Jurgelski, W. M. (2008). Burning seasons, burning bans: Fire in the southern Appalachian Mountains, 1750–2000. *Appalachian Journal, 35*, 170.

Knighton, D. (1998). *Fluvial forms and processes: A new perspective*. New York: Arnold.

Knox, J. C. (1972). Valley alluviation in southwestern Wisconsin. *Annals of the Association of American Geographers, 62*, 401–410.

Knox, J. C. (1977). Human impacts on Wisconsin stream channels. *Annals of the Association of American Geographers, 67*, 323–342.

Knox, J. C. (1983). Responses of river systems to Holocene climates. In H. E. Wright (Ed.), *Late quaternary environments of the United States, vol 2, the Holocene*. Harlow Essex: Longman Group Ltd.

Knox, J. C. (1987). Historical valley floor sedimentation in the Upper Mississippi Valley. *Annals of the Association of American Geographers, 77*, 224–244.

Knox, J. C. (1993). Large increases in flood magnitude in response to modest changes in climate. *Nature, 361*, 430–432.

Knox, J. C. (2001). Agricultural influence on landscape sensitivity in the Upper Mississippi River Valley. *Catena, 42*, 193–224.

Knox, J. C. (2006). Floodplain sedimentation in the Upper Mississippi Valley; Natural versus human accelerated. *Geomorphology, 79*, 286–310.

Lecce, S. A. (1997). Spatial patterns of historical overbank sedimentation and floodplain evolution, Blue River, Wisconsin. *Geomorphology, 18*, 265–277.

Lecce, S. A., & Pavlowsky, R. T. (2014). Floodplain storage of sediment contaminated by mercury and copper from historic gold mining at Gold Hill, North Carolina, USA. *Geomorphology, 206*, 122–132.

Lecce, S., Pavlowsky, R., & Schlomer, G. (2008). Mercury contamination of active channel sediment and floodplain deposits from historic gold mining at Gold Hill, North Carolina, USA. *Environmental Geology, 55*, 113–121.

Leigh, D. S. (1994). Mercury contamination and floodplain sedimentation from former gold mines in north Georgia. *Journal of the American Water Resources Association, 30*, 739.

Leigh, D. S. (1996). Soil chronosequence of Brasstown Creek, Blue Ridge Mountains, USA. *Catena, 26*, 99–114.

Leigh, D. S. (1997). Mercury-tainted overbank sediment from past gold mining in north Georgia, USA. *Environmental Geology, 30*, 244–251.

Leigh, D. S. (2007). *Human influence on floodplain sedimentation in the Upper Little Tennessee River Valley, Southern Blue Ridge Mountains, USA*. Paper presented at the Annual Meeting of the Association of American Geographers, April 2007, San Francisco, CA.

Leigh, D. S. (2009). Geomorphic assessment of the Ferebee Memorial Park at "Bridge-3" along the Nantahala River, US Highway 74, Swain County, North Carolina. In T. Benyshek (Ed.), *Archaeological deep testing and geomorphic study for the proposed replacement of Bridge 3 on US 19/74 across the Nantahala River*. TIP B-4286 Addendum, Swain County, North Carolina. Submitted to NC Dept Transp, Office of Hum Environ, 1583 Mail Service Center, Raleigh, North Carolina 27699, Submitted by: TRC Environmental Corp, Raleigh.

Leigh, D. S. (2010). Morphology and channel evolution of small streams in the southern Blue Ridge Mountains of western North Carolina. *Southeastern Geographer, 50*, 397–421.

Leigh, D. S. (2011). Geomorphic assessment of Site 31GH457 near Bridge 117 along West Buffalo Creek, Graham County, North Carolina. Rep to TRC-Environmental, Raleigh, NC, In B. Idol (Ed.), *Archaeological excavations at 31GH457 for the replacement of Bridge 117 on SR 1123 over West Buffalo Creek, Graham County, North Carolina*. TIP B-4123, WBS Number 32572.1.1, NC Dept Transp Div 14, TRC Project 171390, TRC-Environmental Corp, Raleigh.

Leigh, D. S., & Rogers, J. C. (2007). *Geomorphic analysis of stream channels in the southern Appalachian Highlands* (University of Georgia Geomorphology Laboratory Technology Research Report, 3). Athens: University of Georgia.

Leigh, D. S., & Webb, P. A. (2006). Holocene erosion, sedimentation, and stratigraphy at Raven Fork, southern Blue Ridge Mountains, USA. *Geomorphology, 78*, 161–177.

Lichtenstein, K. P. (2003). *Historic sedimentation and allostratigraphy of the South Fork Broad River, northeast Georgia, United States*. Thesis, University of Georgia.

Meade, R. H., Yuzyk, T. R., & Day, T. J. (1990). Movement and storage of sediment in rivers of the United States and Canada. In M. G. Wolman (Ed.), *Surface water hydrology. The geology of North America* (Vol. O-1). Boulder: Geological Society of America.

Miller, J. R., Lord, M., Yurkovich, S., Mackin, G., & Kolenbrander, L. (2005). Historical trends in sedimentation rates and sediment provenance, Fairfield Lake, western North Carolina. *Journal of the American Water Resources Association, 41*, 1053–1075.

Munoz, S. E., Mladenoff, D. J., Schroeder, S., & Williams, J. W. (2014). Defining the spatial patterns of historical land use associated with the indigenous societies of eastern North America. *Journal of Biogeography, 41*, 2195–2210.

Murray, A. S., & Wintle, A. G. (2000). Luminescence dating of quartz using an improved single-aliquot regenerative-dose protocol. *Radiation Measurements, 32*, 57–73.

Neary, D. G., & Currier, J. B. (1982). Impact of wildfire and watershed restoration on water quality in South Carolina's Blue Ridge Mountains. *Sothern Journal of Forestry, 6*, 81–90.

Oblinger, C. J. (2003). *Suspended sediment and bed load in three tributaries to Lake Emory in the upper Little Tennessee River basin, North Carolina, 2000-02. US Geological* (Water Resources Investigations Report, 03-4194). Raleigh: Survey.

Orbock, M. S., Ritter, D. F., Kochel, R. C., & Miller, J. R. (1993). Fluvial responses to land-use changes and climatic variations within the Drury Creek watershed, southern Illinois. *Geomorphology, 6*, 309.

Price, K., & Leigh, D. S. (2006a). Morphological and sedimentological responses of streams to human impact in the southern Blue Ridge Mountains, USA. *Geomorphology, 78*, 142–160.

Price, K., & Leigh, D. S. (2006b). Comparative water quality of lightly- and moderately-impacted streams in the southern Blue Ridge Mountains, USA. *Environmental Monitoring and Assessment, 120*, 269–300.

Reimer, P. J., Bard, E., Bayliss, A., Beck, J. W., Blackwell, P. G., Ramsey, C. B., Buck, C. E., Hai, C., Edwards, R. L., Friedrich, M., Grootes, P. M., Guilderson, T. P., Haflidason, H., Hajdas, I., Hatté, C., Heaton, T. J., Hoffmann, D. L., Hogg, A. G., Hughen, K. A., & Kaiser, K. F. (2013). INTCAL13 and MARINE13 radiocarbon age calibration curves 0–50,000 years cal BP. *Radiocarbon, 55*, 1869–1887.

Robinson, G. R., Jr. (1992). *Bedrock geology and mineral resources of the Knoxville 1 ° × 2 ° quadrangle Tennessee, North Carolina, and South Carolina* (US Geological Survey bulletin, 1979). Denver: US Geological Survey.

Rogers, J. C., & Leigh, D. S. (2013). Modeling stream-bank erosion in the southern Blue Ridge Mountains. *Physical Geography, 34*, 354–372.

Ruhlman, M. B., & Nutter, W. L. (1999). Channel morphology evolution and overbank flow in the Georgia Piedmont. *Journal of American Water Resources Association, 35*, 277–290.

Saxton, K. E., & Shiau, S. Y. (1990). Surface waters of North America; Influence of land and vegetation on streamflow. In M. G. Wolman (Ed.), *The geology of North America* (Vol. O-1, pp. 55–80). Boulder: Geological Society of America.

Simmons, C. E. (1993). *Sediment characteristics of North Carolina streams, 1970–79* (US Geological Survey water-supply paper, 2364). Denver: US Geological Survey.

Smith, B. D., & Cowan, C. W. (1987). Domesticated *Chenopodium* in prehistoric eastern North America: New accelerator dates from eastern Kentucky. *American Antiquity, 52*, 355–357.

Southworth, S., Schultz, A., Denenny, D., & Triplett, J. (2003). *Surficial geologic map of the great smoky mountains National Park Region, Tennessee and North Carolina* (Open file report, 03–081). Reston: US Geological Survey.

Stinchcomb, G. E., Messner, T. C., Driese, S. G., Nordt, L. C., & Stewart, R. M. (2011). Pre-colonial (AD 1100–1600) sedimentation related to prehistoric maize agriculture and climate change in eastern North America. *Geology, 39*, 363–366.

Stine, M. B. (2013). Fire as a geomorphic agent. In J. Shroder, D. R. Butler, & C. R. Hupp (Eds.), *Treatise on geomorphology, vol. 12, ecogeomorphology*. San Diego: Academic.

Swift, L. W., Jr., Cunningham, G. B., & Douglass, J. E. (1988). Climatology and hydrology. In W. T. Swank & D. A. Crossley Jr. (Eds.), *Forest hydrology and ecology at Coweeta* (Ecological studies, Vol. 66). New York: Springer.

Thigpen, J. R., & Hatcher, R. D. (2009). *Geologic map of the western Blue Ridge and portions of the eastern Blue Ridge and Valley and Ridge Provinces in southeast Tennessee, southwest North Carolina, and northern Georgia* (Map and chart series (Geological Society of America) product code, MCH097F). Boulder: Geological Society of America. http://ngmdb.usgs.gov/Prodesc/proddesc_87098.htm

Thornbury, W. D. (1965). *Regional geomorphology of the United States*. New York: Wiley.

Trimble, S. W. (1974). *Man-induced soil erosion on the southern Piedmont, 1700–1970*. Ankeny: Soil Conservation Society of America.

Trimble, S. W. (1975). A volumetric estimate of man-induced erosion on the southern Piedmont. *USDA, Agricultural Research Service Publication, S40*, 142–145.

Trimble, S. W. (1993). The distributed sediment budget model and watershed management in the Paleozoic Plateau of the upper Midwestern United-States. *Physical Geography, 14*, 285–303.

Vale, T. R. (2002). The pre-European landscape of the United States: Pristine or humanized? In T. R. Vale (Ed.), *Fire, native peoples, and the natural landscape*. Washington, DC: Island Press.

Van Doren, M. (1928). *The travels of William Bartram*. New York: Dover Publications.

Walling, D. E. (1987). Rainfall, runoff and erosion of the land: A global view. In K. J. Gregory (Ed.), *Energetics of physical environment*. New York: Wiley.

Walling, D. E. (1999). Linking land use, erosion and sediment yields in river basins. *Hydrobiologia, 410*, 223–240.

Walling, D. E., & He, Q. (1997). Use of fallout (super 137) Cs in investigations of overbank sediment deposition on river floodplains. *Catena, 29*, 263–282.
Wang, L. (2010). *Anthropic signatures and paleofloods in alluvium of the upper Little Tennessee River valley, southern Blue Ridge Mountains*. Thesis, University of Georgia.
Wang, L., & Leigh, D. S. (2012). Late-Holocene paleofloods in the Upper Little Tennessee River valley, southern Blue Ridge Mountains, USA. *Holocene, 22*, 1061.
Wilkinson, B. H., & McElroy, B. J. (2007). The impact of humans on continental erosion and sedimentation. *Geological Society of America Bulletin, 119*, 140–156.
Zalasiewicz, J., Williams, M., Steffen, W., & Crutzen, P. (2010). The new world of the Anthropocene. *Environmental Science and Technology, 44*, 2228–2231.

Chapter 9
Frequency and Magnitude of Selected Historical Landslide Events in the Southern Appalachian Highlands of North Carolina and Virginia: Relationships to Rainfall, Geological and Ecohydrological Controls, and Effects

Richard M. Wooten, Anne C. Witt, Chelcy F. Miniat, Tristram C. Hales, and Jennifer L. Aldred

Abstract Landsliding is a recurring process in the southern Appalachian Highlands (SAH) region of the Central Hardwood Region. Debris flows, dominant among landslide processes in the SAH, are triggered when rainfall increases pore-water pressures in steep, soil-mantled slopes. Storms that trigger hundreds of debris flows occur about every 9 years and those that generate thousands occur about every 25 years. Rainfall from cyclonic storms triggered hundreds to thousands of debris

R.M. Wooten (✉)
Geohazards and Engineering Geology, North Carolina Geological Survey,
2090 US Highway 70, Swannanoa, NC 28778, USA
e-mail: Rick.Wooten@ncdenr.gov

A.C. Witt
Virginia Department of Mines, Minerals and Energy, Division of Geology and Mineral Resources, 900 Natural Resources Drive, Suite 500, Charlottesville, VA 22903, USA
e-mail: anne.witt@dmme.virginia.gov

C.F. Miniat
Coweeta Hydrologic Lab, Center for Forest Watershed Research, USDA Forest Service, Southern Research Station, 3160 Coweeta Lab Road, Otto, NC 28763, USA
e-mail: cfminiat@fs.fed.us

T.C. Hales
Hillslope Geomorphology, School of Earth and Ocean Sciences, Cardiff University, Main Building, Park Place, Cardiff CF10 3AT, UK
e-mail: halest@cf.ac.uk

J.L. Aldred
Department of Geography and Earth Sciences, University of North Carolina at Charlotte, 9201 University City Blvd., Charlotte, NC 28223, USA
e-mail: jaldred@uncc.edu

C.H. Greenberg, B.S. Collins (eds.), *Natural Disturbances and Historic Range of Variation*, Managing Forest Ecosystems 32,
DOI 10.1007/978-3-319-21527-3_9

flows in 1916, 1940, 1969, 1977, 1985, and 2004. Debris flows have caused loss of life and property, and severely affected forest lands by altering forest structure and disrupting aquatic ecosystems. Forests on mountain slopes are critical in mitigating the impacts of recurring landslide events. Forest cover is an important stabilizing factor on hillslopes by intercepting precipitation, increasing evapotranspiration, and reinforcing roots. Precipitation and hillslope-scale landforms have a controlling effect on soil moisture, root strength, and debris flow hazards. Anthropogenic influences have increased the frequency of mass wasting for a given storm event above historical natural levels through changes in vegetation and disturbances on mountain slopes. Climate change that results in increased occurrences of high intensity rainfall through more frequent storms, or higher intensity storms, would also be expected to increase the frequency of debris flows and other forms of mass-wasting in the SAH. The interdisciplinary technical and scientific capacity exists to investigate, analyze, identify and delineate landslide prone areas of the landscape with increasing reliability.

Keywords Debris flow • Ecohydrological • Landslide • Blue Ridge Mountains • Southern Appalachian Highlands

9.1 Introduction

Landsliding is a recurring process of mass wasting and sediment transport in the landscape evolution of the southern Appalachian Highlands (SAH) of the USA. The SAH encompasses the Blue Ridge Mountains, and adjoining mountainous and high relief areas of the Central Appalachians, Ridge and Valley, and Piedmont Ecoregions (Greenberg et al. Chap. 1, Fig. 1.1). In comparing the relative frequency of disturbances among ecoregions of the Central Hardwood Region (CHR), White et al. (2011) found that the remnants of hurricanes (tropical cyclones) and, consequently, landslides are more common disturbances in the Blue Ridge Mountains, Ridge and Valley, and Central Appalachians ecoregions. Here we concentrate on the Blue Ridge Mountains and adjacent Piedmont of North Carolina and Virginia where previous and recent landslide mapping and studies have helped quantify the extent and magnitude of major historical landslide events. These events have caused loss of life, damage and destruction of homes, property and transportation networks, and have had major impacts on forest structure and hydrologic systems (Fig. 9.1). Future debris flow events in the SAH are certain and will have similar impacts.

The term landslide refers to a variety of gravity-driven ground movements of soil and/or rock materials. Landslides may be swift and catastrophic (i.e., rockfalls and debris flows) or may travel slowly and incrementally downslope (i.e., some soil slides). Landslide incidence and susceptibility occurs in nearly all of the high-relief areas of the USA including the CHR (Fig. 9.2). Although many types of landslides occur throughout the SAH, debris flow is the dominant landslide process in the Blue Ridge Mountains of North Carolina and Virginia, and the SAH, and will be the

Fig. 9.1 Damaged and destroyed homes, and debris flow erosion and deposition along the run out zone (lower track) of the September 16, 2004 Peeks Creek debris flow in Macon County, North Carolina. The debris flow triggered by rainfall from Hurricane Ivan claimed five lives and destroyed 16 homes (September 19, 2004 NCGS photo). Refer to Fig. 9.17 for location

focus of this chapter. A debris flow is a water-laden (i.e., liquefied) moving mass of rock fragments and soil (debris) in which the majority of soil particles are sand-sized or larger (Cruden and Varnes 1996). Debris flows move rapidly downslope, attaining speeds in excess of 50 km per hour, and are capable of destroying or damaging everything in their paths. A typical debris flow pathway consists of an upper initiation site or source area, a main track or path along a drainage way or stream channel, and a lower depositional area or run out zone on mountain footslopes (Clark 1987; Cruden and Varnes 1996). The present SAH landscape includes many remnants of prehistoric (Pleistocene and older) debris flow deposits (Leigh Chap. 8). These features are typically composite, recording multiple episodes of prehistoric and historic mass wasting in mountain coves and foot slopes. Importantly, these deposits indicate areas that may be affected by future debris flow activity, as modern debris flows generally deposit sediment in areas occupied by past debris flow deposits.

The primary trigger for debris flows is heavy rainfall (generally greater than 125–250 mm in 24 h) that results in excess pore-water pressures in relatively thin soil on steep slopes. From 1916 to 2006 seven major cyclonic storms tracked over the SAH, setting off hundreds to thousands of debris flows in multi-county areas in the North Carolina, Virginia and West Virginia. In addition, rainfall associated with

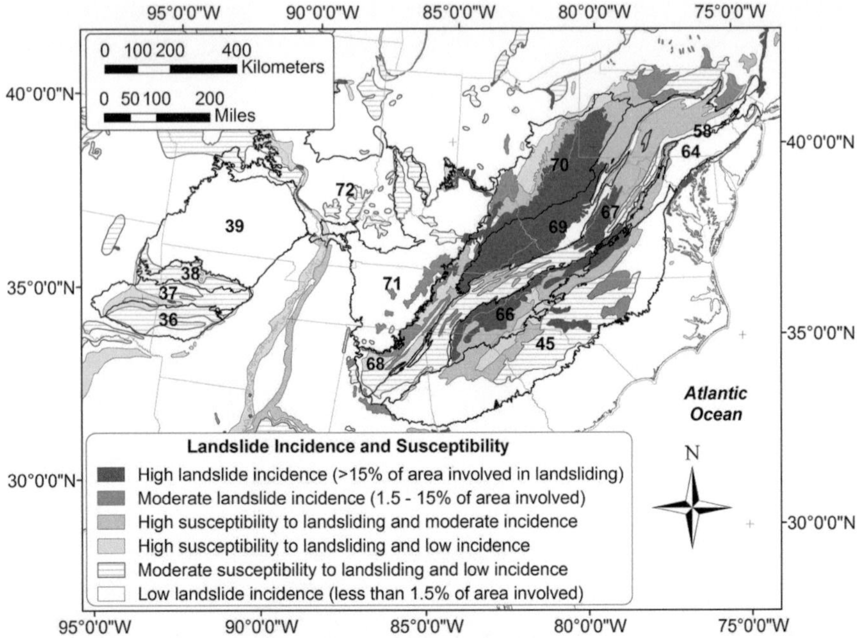

Fig. 9.2 Generalized map of landslide incidence and susceptibility (From Godt (1997) shown for the CHR and ecoregions within it. Within the CHR the Appalachians have the overall highest landslide incidence and susceptibility. Landslide incidence is the percentage of the area involved in landsliding. Susceptibility is defined as the probable degree of the areal response of rocks and soil to natural or artificial cutting or loading of slopes, or to anomalously high precipitation. Ecoregions shown with *bold outlines*: 36 = Ouachita Mountains, 37 = Arkansas Valley, 38 = Boston Mountains, 39 = Ozark Highlands, 45 = Piedmont, 58 = Northeastern Highlands, 64 = Northern Piedmont, 66 = Blue Ridge Mountains, 67 = Ridge and Valley, 68 = Southwestern Appalachians, 69 = Central Appalachians, 70 = Western Allegheny Plateau, 71 = Interior Plateau, 72 = Interior Valleys and Hills

low pressure systems, and localized storms, especially when coincident with periods of above average rainfall can trigger tens to hundreds of debris flows. From 1876 to 2013, at least 16 of these storm events generated tens to hundreds of debris flows.

The majority of landslides in the SAH occur in the Blue Ridge Mountains of North Carolina and Tennessee, the northern Blue Ridge Mountains of Virginia, and the Ridge and Valley of Virginia and West Virginia (Figs. 9.3 and 9.4). The concentration of landslide activity in the North Carolina Blue Ridge Mountains and adjacent portions of the Great Smoky Mountains National Park (GSMNP) in Tennessee is partly the result of the high relief and ruggedness of the terrain, and partly owing to the more frequent impacts of cyclonic storms in this region (9 of 13 storms). Although there are fewer documented landslide events for the Blue Ridge Mountains, and Ridge and Valley of Virginia and West Virginia, rainfall events there have

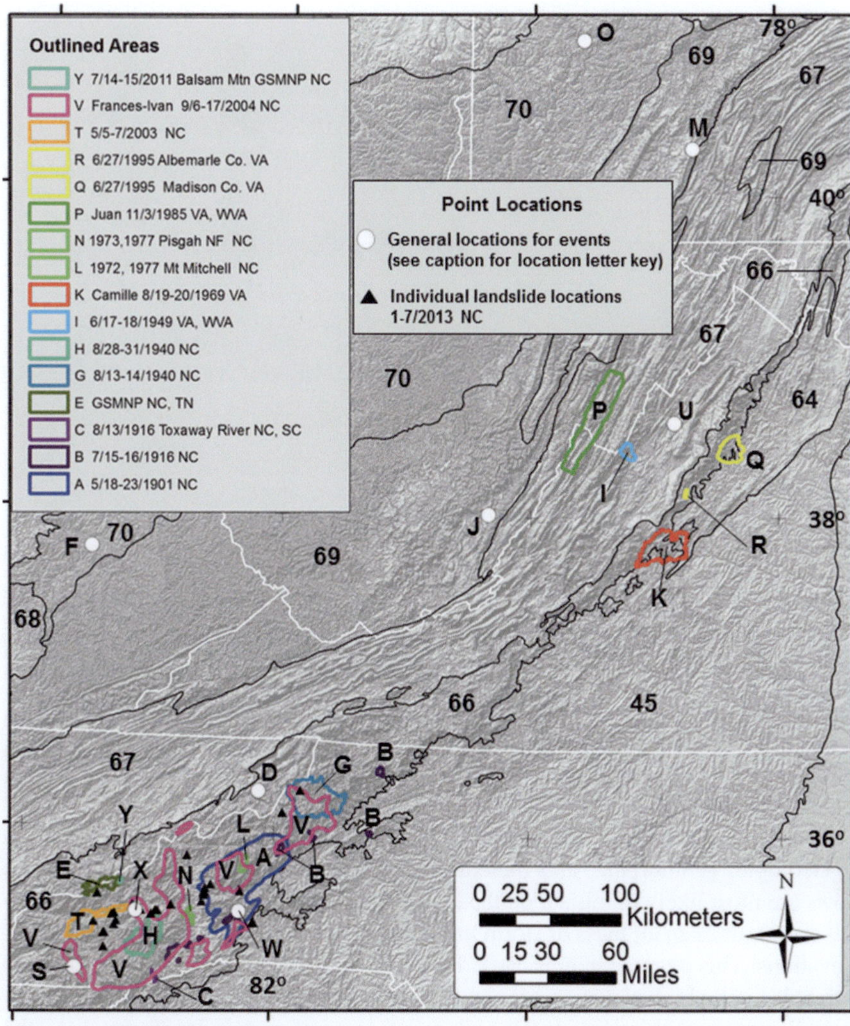

Fig. 9.3 Shaded relief map showing ecoregions and areas of selected past debris flow events in the SAH and other selected locations in the CHR. Lettered locations correspond to events in Table 9.1. General point locations: D=6/3/1924 Carter Co. TN, F=7/4-5/1939 KY, J=Camille 8/19-20/1969 Greenbrier C. WV, M=7/19/1977 PA, O=8/14/1980 PA, S=Opal 10/3-5/1995 NC, U=Isabel 9/18-19/2003 NC, W=Cindy 7/7/2005 NC, X=Ernesto 8/3/2006 NC. Ecoregion number designation: 45=Piedmont, 64=Northern Piedmont, 66=Blue Ridge Mountains, 67=Ridge and Valley, 68=Southwestern Appalachians, 69=Central Appalachians, 70=Western Allegheny Plateau, 71=Interior Plateau

triggered the greatest numbers of documented landslides. The remnants of Hurricane Camille in 1969 generated a total of 5,377 documented landslides (mainly debris flows) in Virginia and West Virginia making it the largest magnitude, well-documented landslide event in the SAH.

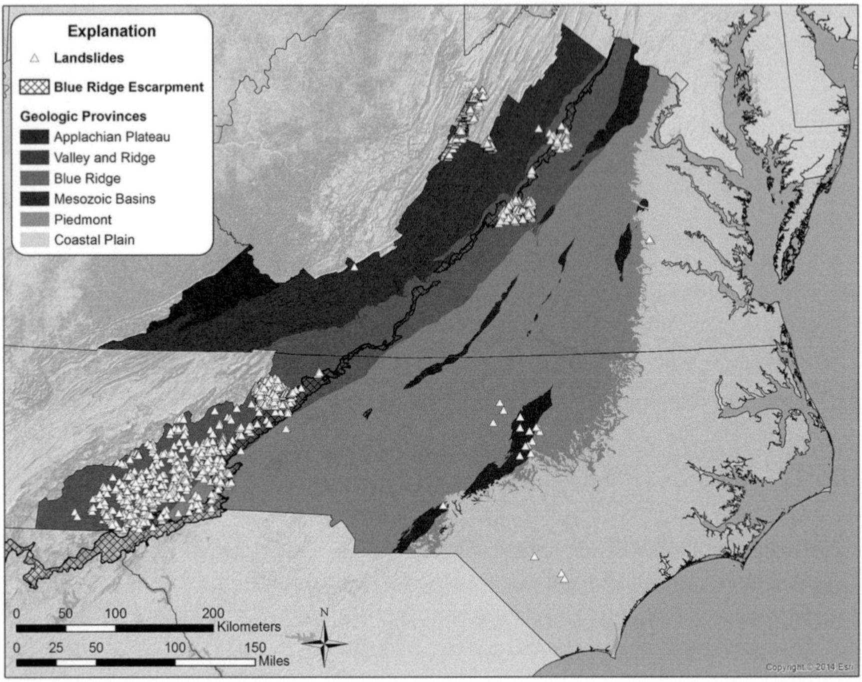

Fig. 9.4 Geologic provinces of North Carolina and Virginia, the Blue Ridge Escarpment, and landslide locations in landslide geodatabases of the North Carolina Geological Survey and Virginia Department of Mines Minerals and Energy. Clustered distribution of landslides results from detailed mapping in some areas of major landslide events, and incomplete mapping in other areas

Geologic, geomorphic, and meteorological conditions influence where debris flows are most likely to initiate on the landscape. Orographic enhancement of rainfall can occur as an air mass or storm moves over a high mountain range from lower elevations. This phenomenon is caused by the lifting and cooling of the air mass as it travels over a high elevation area and often produces excess precipitation. Heavy rainfall, when combined with the high-relief areas of certain landforms (i.e., the multi-basin scale Blue Ridge Escarpment and the watershed-scale Nantahala Mountains Escarpment) and erosional reentrants into them, are more prone to debris flow activity. Geologically, intersecting bedrock structural discontinuities (e.g., fracture, foliation, and bedding planes) and differential weathering control the locations and subsurface morphologies of convergent landforms (i.e., colluvial hollows) where debris flows typically initiate. Less frequently, soil on planar or divergent (i.e., convex) slopes such as ridge noses, also controlled by bedrock discontinuities, serve as debris flow initiation zones. Bedrock geology is dynamically coupled with hillslope geomorphology, hydrology, soil, and vegetation, all of which influence hillslope stability. In addition to these factors, ground disturbance from human activity, such as poorly constructed or maintained cut and fill slopes, and drainage systems, can further destabilize hillsides making them more susceptible to damaging debris flows.

Forest cover is an important stabilizing factor on steep upland hillslopes through precipitation interception, evapotranspiration, and root biomass (e.g., root reinforcement). Debris flows often initiate where the reinforcing ability of plant roots is at a minimum, either through reduced root biomass and/or tensile strength, and/or less connectivity between roots and the bedrock substrate. The reinforcement provided by the roots of forest plants reflects hillslope-scale differences in belowground biomass and tensile strength, and is directly affected by precipitation and soil moisture. Systematic differences in forest structure driven by differences in soil moisture and nutrient distributions, combined with the expansion of weakly rooted species such as the shrub Rhododendron (*Rhododendron maximum*), appear to influence the size of individual landslides and possibly regional landsliding events. Studies of SAH woody species show responses of root tensile strength to changes in precipitation and soil moisture. Roots are weaker in convergent compared to divergent geomorphic features, and roots in wetter soils, i.e., after rain events, become weaker compared to when soils are drier. Precipitation, therefore, has a controlling effect on soil moisture, root tensile strength, and debris flow hazards. Although forest cover is beneficial, forested slopes are a common location for debris flows triggered by storm events in the SAH.

9.1.1 Methods

In North Carolina and Virginia, earlier landslide mapping has been integrated into a geographic information system (GIS) environment for ease of data entry and for statistical analyses. Field studies and the development of similar statewide, GIS-based, landslide geodatabases by the North Carolina Geological Survey (NCGS) and the Virginia Department of Mines, Minerals and Energy – Division of Geology and Mineral Resources (DGMR), capture and help to quantify the frequency and severity of debris flow events of various magnitudes in the SAH (Fuemmeler et al. 2008; Bauer et al. 2012; Witt and Heller 2014). The NCGS geodatabase currently has over 3,400 landslide points and over 3,200 landslide deposit (mainly debris) polygons, located primarily in the four counties with completed landslide hazard maps. The DGMR geodatabase currently has over 5,200 landslide points and associated data. Digital elevation models, including those derived from LiDAR, coupled with archival aerial photography and recent orthophotography have advanced the capability to identify, map, and analyze prehistoric and historic landslide features within the context of the current landscapes and land covers. Currently, the NCGS and DGMR do not actively map landslides and landslide deposits, but landslide features are added to the geodatabases on an as-needed basis

A compilation of existing landslide information and new mapping of landslide features in a GIS environment by the NCGS began in 2003 in the western North Carolina Blue Ridge Mountains (Wooten et al. 2005) and was funded in part by the Federal Emergency Management Agency (FEMA). New mapping and data collection in North Carolina included a geologic hazards inventory along the North

Carolina portion of Blue Ridge Parkway (Latham et al. 2009), and completion of landslide hazard maps for Macon, Watauga, Buncombe and Henderson Counties (Wooten et al. 2006, 2008b, 2009b, 2011). The NCGS has also responded to requests for technical assistance in over 85 landslide events in which field data were collected during investigations. In a 1-year FEMA funded pilot project, the Virginia DGMR mapped landslides and prehistoric landslide deposits in Page County, Virginia (Witt and Heller 2012, 2013, 2014; Witt et al. in press).

9.2 General Geologic and Geomorphic Setting

The bedrock geology of the Blue Ridge Mountains in North Carolina and Virginia and the adjacent Piedmont includes metasedimentary slate, phyllite, marble, schist, and gneiss, and metaigneous amphibolite and greenstone (metabasalt), granitic gneiss, and relatively unmetamorphosed granitic rock (Fig. 9.4). Protoliths of these rocks were deposited or crystallized during distinct periods spanning the last 1.8 billion years (Hatcher 2010; Ownby et al. 2004). The oldest rocks, Mesoproterozoic gneisses, are highly metamorphosed igneous and sedimentary rocks formed between 1.8 and 1 billion years ago. These gneisses comprised the edge of the ancient North American craton upon which early Paleozoic sediments were deposited in rift and ocean basins. During the Paleozoic, at least three continental collisional events subjected all of these rocks to high temperatures and differential pressures, creating complex folding, faulting, and widespread metamorphism. Igneous activity associated with the Paleozoic orogenies emplaced numerous granitic plutons into the surrounding country rock. Thrusting along low angle faults folded and transported these rock packages tens to hundreds of km to the northwest, placing them on top of and deforming younger, low-grade to unmetamorphosed, folded and faulted Paleozoic sedimentary rocks of the Valley and Ridge province (Hatcher 1989). The multiple episodes of metamorphism, folding, thrust faulting, and fracturing during the southern Appalachian orogen have resulted in complex ductile (e.g., foliation and shear zones) and brittle (e.g., fractures) bedrock structures that are reflected in the topography at scales from a single outcrop to the region.

Later, Cenozoic uplift and subsequent post-orogenic erosion and denudation, enhanced by climatic variations from repeated glacial and interglacial intervals, have resulted in many of the Quaternary geomorphic features seen in the Appalachian Mountains today (Kochel and Johnson 1984; Kochel 1987,1990; Soller and Mills 1991). The most extensive regional landform in the SAH is the Blue Ridge Escarpment (BRE) (Hack 1982; Clark 1993), which is also referred to the as the eastern Blue Ridge front in Virginia. This high relief, erosional feature extends from northeast Georgia to northwest Virginia, generally corresponds with the Eastern Continental Divide, and marks the boundary between the mountainous Blue Ridge Mountains and the rolling foothills of the Piedmont physiographic province to the east (Fig. 9.4).

Ancient and modern deposits from debris flows and other types of landslides on mountain footslopes and in coves record a long history of mass wasting from early Miocene to the present, reflecting the ongoing landscape evolution of the SAH. These accumulations of unconsolidated, matix- or clast-supported, clay- to boulder-sized sediment (i.e., composite diamictons) are referred to by various names, including debris fans, alluvial fans, and piedmont cove deposits (Kochel and Johnson 1984; Kochel 1987, 1990; Mills 1982, 1998; Mills et al. 1987; Whittecar and Ryter 1992; Mills and Allison 1995a, b). The morphology and composition of these deposits varies greatly depending on their age, topographic setting, and past and present depositional processes. These deposits are typically composites of several generations of debris flows reworked and incised by alluvial action. Individual fan-shaped deposits occur at outlets of first and second order drainages. Coalescing fans can form continuous apron-like deposits, along footslopes, and fill valley floors where topographically constrained (Fig. 9.5). These deposits collectively will be referred to here as debris deposits or debris fans. In North Carolina and Virginia, large debris deposits are often found in the Blue Ridge Mountains, but extend into

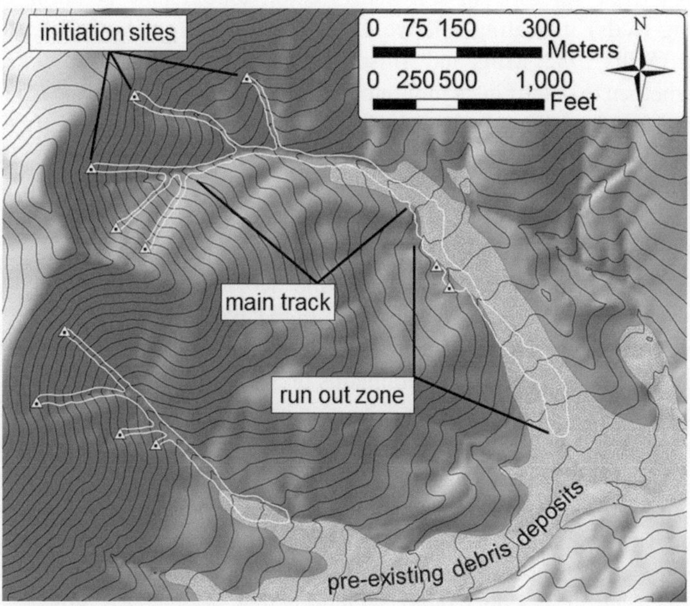

Fig. 9.5 Tracks of 1977 debris flows (*white outlines*) and pre-existing debris deposits (stippled polygons) in the Rocky Branch area, Bent Creek Experimental Forest, Pisgah National Forest, in Buncombe County North Carolina. Initiation sites (*triangles*), main track, and run out zone labels show typical components of a debris flow. The 1977 debris flows deposited material in areas underlain by pre-existing debris flow deposits. Topographic contours (*black lines*) and the shaded relief map are derived from a 6 m-pixel resolution LiDAR digital elevation model (DEM). Contour interval = 6.1 m (20 ft). Elevation ranges from 1,152 to 762 m, with lower elevations in the southeast portion of the map (Derived from Wooten et al. 2009b) (Location N in Fig. 9.3 and Table 9.1)

the adjoining Piedmont along footslopes of the BRE. These deposits also occur along the eastern border of the Ridge and Valley in Virginia.

Recent dating of a large, deeply weathered debris fan in the Big Levels 7.5-min quadrangle in the western Virginia Blue Ridge Mountains was completed using the cosmogenic$^{26}Al/^{10}Be$ burial decay method (Heller et al. 2014). Two samples determined the age of the fan to be early Miocene: sampling at 20 m below the land surface yielded an age of 7.94 ± 2.4 Mega-annum (one million years), while a higher fan layer 15 m below the surface yielded an age of 6.90 ± 1.7 Mega-annum. This age range is much older that those derived using radiocarbon dating on slope deposits exposed by a 1995 storm in Madison County, Virginia (Eaton et al. 2003a) and on debris fans in Nelson County, Virginia (Kochel and Johnson 1984). Eaton et al. (2003a) found that stratified slope deposits in Madison County were formed in the late Pleistocene (15.8–27.4 Kilo-annum, 1,000 years), while maximum ages for debris flow deposits were found to be >50 Kilo-annum. Kochel and Johnson (1984) dated basal units of debris fans in the Davis Creek area of Nelson County to the start of the Holocene (10.7 Kilo-annum).

In the North Carolina Blue Ridge Mountains, large, composite debris deposits are characterized by variation in relative fan-surface ages as reflected by physical differences (e.g., variation in soil matrix color, topographic position, and clast weathering rinds) indicating relative ages that range from Early Pleistocene to Holocene (Mills 1982; Mills and Allison 1995a). Mills and Allison (1995b) used paleomagnetism to determine a minimum relative age of early Pleistocene (78 Kilo-annum) for weathered debris deposits in Watauga County. Subsequently Mills and Grainger (2002) used cosmogenic$^{26}Al/^{10}Be$ to date a debris fan deposit on the slopes of Rich Mountain, also in Watauga County, as early Pleistocene (1.45 ± 0.17 Mega-annum). Late Holocene debris deposits are also present in the region. Radiocarbon dating of charcoal beneath a debris flow deposit at a site near the Nantahala River in Swain County, North Carolina, indicates the debris flow postdates 4,441–4,797 years before present (Leigh 2009; see also Leigh Chap. 8).

Older debris deposits can indicate areas that can be affected by modern debris flows. In the eastern Blue Ridge Mountains of Virginia heavy rainfall in 1969 and 1995 triggered debris flows on older debris fan material in Nelson County (Williams and Guy 1973; Kochel and Johnson 1984) and Madison County (Morgan et al. 1997; Eaton et al. 2003a), respectively. Field studies and mapping in Macon County in the North Carolina Blue Ridge Mountains determined that debris deposits of various ages cover about 4,500 ha or 3.3 % of the land area there (Wooten et al. 2006). These pre-existing debris deposits were identified at all of the 62 relatively recent debris-flow sites in Macon County, evidence that prior debris flow events had occurred at the same locations, in many cases multiple times (Wooten et al. 2008a). The common occurrence of modern debris flow deposition in areas with past debris flow deposits was also identified by landslide mapping in Watauga, Buncombe and Henderson Counties in the North Carolina Blue Ridge Mountains (Wooten et al. 2008b, 2009b, 2011). These studies found that modern debris flows typically affect areas where streams have incised into, or flow around the margins of older deposits. In relatively rare cases modern debris deposition has occurred on fan surfaces outside of stream channels (Fig. 9.5).

On the west slope of the Virginia Blue Ridge Mountains, debris fans tend to be larger, better developed, and more weathered than those along the eastern Blue Ridge front. These west-facing fans appear to be dominated by alluvial and fluvial processes, having more of the characteristics of a braided stream deposit (Kochel 1990; Heller et al. 2014). Debris flow activity tends to be confined to the upper-to-middle reaches of these fans and to active channels (Whittecar and Ryter 1992; Eaton et al. 2003a; Heller et al. 2014). In the western portion of Page County, no modern landslides (post-1950) were found within or along ancient debris fans originating from the western flank of the Blue Ridge Mountains (Witt and Heller 2013, 2014).

9.3 Temporal Frequency and Magnitude of Debris Flow Events Related to Regional and Localized Rainfall Events

9.3.1 Rainfall Scenarios

Building on the work of Scott (1972), Clark (1987), Witt (2005), and Wieczorek et al. (2009), we have compiled existing data and reported new data for a total of 31 landslide events listed in Table 9.1, and shown graphically in Fig. 9.6. Figure 9.3 shows the general locations for the major events. In summary, tropical cyclones and an extratropical cyclone tracked over the SAH, setting off tens to thousands of debris flows in multi-county areas in North Carolina in 1916, 1940, 1977 and 2004, and in Virginia and West Virginia in 1969 and 1985. These cyclonic storms have resulted in the most widespread and numerous flooding and landslide events in the SAH. From 1916 to 2004, 13 cyclonic storms have impacted the SAH, on average, approximately every 7 years. Five of these storms (Agnes, Opal, Isabel, Cindy and Ernesto) generated relatively few landslides in the region. Although no landslides were reported for the July 7, 1916 tropical cyclone, it was significant because it created high antecedent moisture prior to a July 15–16, 1916 tropical cyclone, which was the storm of record for the French Broad watershed of North Carolina.

The short duration between Hurricanes Frances and Ivan in September 2004 was significant in that it established a pattern of back-to-back major storms within 6–20 days of each other causing flooding and triggering debris flows in Blue Ridge Mountains of North Carolina. Three such scenarios have occurred over an 88-year period from 1916 to 2004 (Witt 2005). Work by the US Geological Survey (1949), Tennessee Valley Authority (1964), Scott (1972), and Witt (2005) established that similar weather patterns had triggered regional flooding and debris flows in July of 1916 and August 1940. Following Frances and Ivan, the average frequency of such weather scenarios in western North Carolina is 29 years.

In addition to tropical cyclones, rainfall associated with low pressure systems and localized storms, especially when coincident with periods of above normal rainfall, have triggered from a few to hundreds of landslides in each of the 18 events

Table 9.1 Selected examples of landslide (mainly debris flow) events induced by rainfall in the SAH, and other selected locations in the Central Hardwood Region. Letters in the first column correspond with locations shown in Fig. 9.3. *TC* tropical cyclone, *ETC* extratropical cyclone, *EANR* period of extended of above normal rainfall. Number in parentheses in 'Total Area of Tracks' column represents number of landslides used to determine track area where less than the total number of reported landslides. Other abbreviations used in the table: *NCGS* NC Geological Survey, *TVA* Tennessee Valley Authority, *USDA* US Department of Agriculture, *USGS* US Geological Survey

Lettered location Fig. 9.3	Type of rainfall event	Rainfall total (mm)	Time of most intense rainfall	Landslide dates mm/day/year	Landslides reported	Total track area (ha)	General location ecoregion	Reported landslide locations	Notes – References
–	Storm	165	?	6/15/1876				Macon, Jackson Co., NC	Clingman 1877; Witt 2005
A	Storm	228	?	5/21-22/1901	>17		Blue Ridge Mountains, Piedmont	Buncombe, Henderson, Mitchell, McDowell Co., NC	Holmes 1917; Myers 1902; Scott 1972; USDA 1902; Witt 2005
–	TC (unnamed)		1 day	7/9/1916 (rainfall)	0		Blue Ridge Mountains	Western NC	Henry 1916; Scott 1972
B	TC (unnamed)	564	24 h	7/15-16/1916	Many, >45		Blue Ridge Mountains	Western NC	Bell 1916; Scott 1972; Witt 2005
C	Storm	584	24 h	8/13/1916	1	122.1	Blue Ridge Mountains	Toxaway River, Transylvania Co., NC	Dam failure – debris flow; Wooten et al. 2003a, b
D	Thunderstorm	380	2 days	6/13/1924	~100		Blue Ridge Mountains	Carter Co., TN	Slides from 8/13-14/1940 storm also here (Hack and Goodlett 1960); Scott, 1972
E	Cloudburst	305	4 h	8/4-5/1938	>100		Blue Ridge Mountains	Webb Mountain, TN (GSMNP)	Within concentrated debris flow area in Southworth et al. 2012; Scott 1972
F	Thunderstorm	508	2 days	7/4-5/1939	4		Western Allegheny Plateau	Wolfe, Breathitt Co., KY	Total number of landslides not reported; Schrader, 1945

G	TC (unnamed)	254	6 h	8/13-14/1940	2,120	368	Blue Ridge Mountains	Watauga, Ashe, Allegheny, Avery, Wilkes, Caldwell Co., NC; Unicoi County, TN	USGS 1949; Wieczorek et al. 2004; Witt et al. 2007a; Wooten et al. 2008b
H	Storm	330	2 days	8/28-31/1940	>200			Macon, Jackson (location 6) and Haywood Counties, NC	TVA 1940; USGS 1949; NCGS landslide geodatabase
–	Storm	381	24 h	Winter 1942	Numerous		Blue Ridge Mountains	Watauga Co., NC	Gryta and Bartholomew 1983
I	Cloudburst	400	1 day	6/17-18/1949	466		Ridge and Valley	Augusta and Rockingham Co., VA; Highland County, WV	Hack and Goodlett 1960
E	Cloudburst	100	1 h	9/1/1951			Blue Ridge Mountains	Mt. LeConte, TN (GSMNP)	Within concentrated debris flow area in Southworth et al. 2012; Bogucki 1976; Clark et al. 1987
J	TC Camille	635	8 h	8/19-20/1969 9/5-6/1969	1,584		Central Appalachians	Spring Creek, Greenbrier County, WV	Schneider 1973
K	TC Camille	710–800	8 h	8/19-20/1969	3,793	1,200	Blue Ridge Mountains, Piedmont	Albemarle, Amherst, Nelson Co., VA	Williams and Guy 1971, 1973; Morgan et al 1999a
L	TC Agnes	203	2 days	6/20-21/1972	?		Blue Ridge Mountains	Mt. Mitchell area (location L), Waterrock Knob, Haywood Co., NC.	Bailey et al. 1975; Clark 1987
H, N	Thunderstorms	196	1 h	5/26-28/1973	Many, 8 documented		Blue Ridge Mountains	Tuckasegee watershed, Jackson Co., NC (location H), Pisgah NF, NC (location M)	Young 1973; Zeedyk 1973; NCGS landslide geodatabase

(continued)

Table 9.1 (continued)

Lettered location Fig. 9.3	Type of rainfall event	Rainfall total (mm)	Time of most intense rainfall	Landslide dates mm/day/year	Landslides reported	Total track area (ha)	General location ecoregion	Reported landslide locations	Notes – References
M	Thunderstorm or Cloudburst	300	9 h	7/19-20/1977	Several hundred		Central Appalachians	Johnstown area, western PA	Pomeroy, 1980
L	ETC (unnamed)	300	2 days	11/5-7/1977	13	25.2	Blue Ridge Mountains	Mt. Mitchell, Black Mtns., Yancey Co., NC	Eschner and Patrick 1982; Wooten et al. this chapter
N	ETC (unnamed)	150	2 days	11/5-7/1977	83	32.8	Blue Ridge Mountains	Pisgah NF, Buncombe, Henderson Co., NC.	Neary and Swift 1987; Pomeroy 1991; Otteman 2001; Wooten et al. 2009b, 2011
O	Thunderstorm	115	14 days	8/14/1980	62		Western Allegheny Plateau	East Brady along Allegeny River, PA	Pomeroy 1984
P	TC Juan	350	3 days	11/3-5/1985	3,000		Ridge and Valley	Pendleton Co., WV; Highland Co., VA	Jacobson, 1993
–	Storm	60	1 day	12/23/1990	4	2.57	Blue Rige Mountains	Nantahala NF in Swain, Cherokee, Clay Co., NC	Wooten et al. 2007
Q	Storm	770	14 h	6/27/1995	629	600	Blue Rige Mountains	Madison Co., VA	Eaton et al. 2004; Morgan et al. 1997, 1999b; Wieczorek et al. 1995, 2000
R	Storm	635	1 day	6/27/1995	72	49.8	Blue Ridge Mountains	Albemarle Co., VA	Morgan and Wieczorek 1996
S	TC Opal	274	1 h	10/3-5/1995	2	1.4	Blue Ridge Mountains	Poplar Cove, Macon Co., NC	Wooten et al. 2008a
T	Storm	187	1 day	5/5-7/2003	10	4.5	Blue Ridge Mountains	Swain Co., NC	Wooten and Latham 2004

U	TC Isabel	513	1 day	9/18-19/2003	6		Ridge and Valley	Shennandoah Valley, VA	Wieczorek et al. 2004
V	TC Frances	599	2 days	9/6-8/2004	400+	81.3 (184)	Blue Ridge Mountains	Western NC	Reported landslide total for Frances and Ivan; Collins 2008; Wooten et al. 2008a; Wieczorek et al. 2009; Collins 2014; Roth 2015
	TC Ivan	432	2 days	9/16-17/2004					
W	TC Cindy	110	2 days	7/7/2005	1	0.05	Blue Ridge Mountains	Buncombe Co., NC	Rockslide; Wooten et al. 2007
X	TC Ernesto	168	12 h	8/31/2006	1	1.1	Blue Ridge Mountains	Haywood Co., NC	Wooten and Latham 2006
–	Storm	152	24 h	1/7/2009	2	1.5	Blue Ridge Mountains	Haywood Co., NC	Embankment failures-debris flows; Wooten et al. 2009a
–	Storm	35	4 h	2/5/2010	1	2.6	Blue Ridge Mountains	Maggie Valley, Haywood Co., NC	Witt et al. 2012; Wooten et al. this chapter
Y	Thunderstorm	125	4 h	7/14-15/2011	21	12.9	Blue Ridge Mountains	Balsam Mountain, NC (GSMNP)	Lee et al. 2011; Miller et al. 2012; Wooten et al. this chapter
Point locations	Storm (within EANR period)	388	2 days	1/14-17/2013	Numerous 268 confirmed	5.2 (18)	Blue Ridge Mountains	Western NC	Gibbs 2013; Jennifer Bauer personal comm.; Wooten et al. 2014, and this chapter
Point locations	Storm (within EANR period)	168	2 days	5/4-6/2013	Numerous 50 confirmed	1.3 (8)	Blue Ridge Mountains	Western NC	Gibbs 2013; Wooten et al. this chapter
Point locations	Storm (within EANR period)	250	2 days	7/2-9/2013	Numerous 9 confirmed	5.0 (7)	Blue Ridge Mountains	Western NC	Wooten et al. 2014; and this chapter

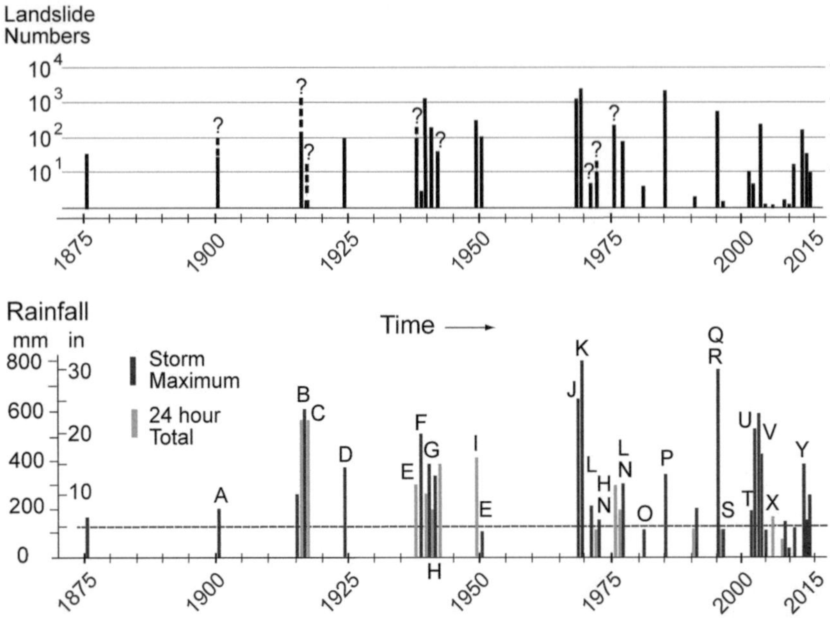

Fig. 9.6 Chart showing the landslide numbers and rainfall associated with tropical cyclones and other storms that triggered landslides in the SAH and other selected locations in the CHR. The 24-h rainfall threshold of 125 mm for triggering debris flows from Eschner and Patrick (1982) shown by a *dashed line*. Single letter in parentheses corresponds with locations in Fig. 9.3, and Table 9.1 (Adapted and expanded from Wooten et al. 2007)

documented for the SAH in Table 9.1. Six events in this category each generated 100 or more reported landslides. The June 27, 1995 storm event in Madison County produced 629 landslides (mainly debris flows) making it the largest event in this category. During a period of above normal rainfall throughout western North Carolina from January to August 2013, four storms collectively triggered more than 300 landslides. Landslides from three of these storms are documented here. Undoubtedly many other landslide events of this nature have occurred throughout the SAH that are not reported here (e.g., Crawford 2014) and not documented in the literature.

9.3.2 *Characterization of Triggering Rainfall*

Total storm rainfall is an important factor in debris flow initiation. As can be seen from Table 9.1 and Fig. 9.6, the majority of events fall within 125–250 mm per day precipitation thresholds presented by Eschner and Patrick (1982) needed to generate debris flows on forested slopes in the SAH. Fuhrmann et al. (2008) found that landslide activity in western North Carolina is strongly related to antecedent precipitation over a 90-day period. Other studies in the SAH of North Carolina and

Virginia have demonstrated that rainfall rate (intensity) and duration is a critical factor in debris flow initiation. Areas of high-intensity rainfall promote the development of debris flows and slides as evident in the 1969 Nelson County storm (Williams and Guy 1973) and the 1995 Madison County storm (Wieczorek et al. 2000). Neary and Swift (1987) concluded that rainfall rates on the order of 90–100 mm per hour (188 mm storm total) initiated debris flows in the Bent Creek area near Asheville, North Carolina, during a November 3–5, 1977 storm, but they do not report specific durations associated with these rates. Wieczorek et al. (2000, 2009) present a rainfall intensity-duration threshold curve for the Central Blue Ridge Mountains of Virginia that ranges from approximately 90 mm per hour for 1 h, to 10 mm per hour for 24 h. Wieczorek et al.(2004) reported that 254 mm of rain within 6 h (42 mm per hour average) triggered over 700 debris flows during the August 13–14, 1940 storm in the Deep Gap area of Watauga County, North Carolina, a value that plots above the Virginia threshold curve. An average rate of 25 mm per hour for the 4.65 h period of peak cumulative rainfall (5.5 mm per hour for the 33.2 h storm total) during Ivan preceded the Peeks Creek and Wayah debris flows in Macon County, North Carolina (Wooten et al. 2008a). This value falls below the Virginia curve threshold; however, the North Carolina debris flows occurred with high antecedent moisture conditions from the passage of the remnants of Hurricane Frances the previous week. An average rainfall rate of 57 mm per hour for 2 h generated the 2011 Balsam Mountain debris flows in the GSMNP (Miller et al. 2012; Tao and Barros 2014), a value that plots below the Virginia threshold curve.

In two North Carolina cases, less rainfall was required to generate debris flows on slopes with evidence of prior instability related to human activity, when compared to debris flows generated on unmodified, forested slopes as described above (Wooten et al. 2009a, 2010b). In 2009, peak rainfall of ~6.4 mm per hour for 1.2 h (3.3 mm per hour for 23.5 h storm total) triggered a debris flow in fill material that destroyed a home. In 2010, peak rainfall of ~3.4 mm per hour for 2 h (4.4 mm per hour for 13.3 h rain total on snow) contributed to a retaining wall failure that mobilized into a debris flow which damaged three homes (Witt et al. 2012). This limited number of cases indicates that the destabilizing effects of human activity likely decreases the requisite rainfall needed to initiate debris flows on some modified slopes in contrast with rainfall amounts needed to generate debris flows on forested slopes not modified by human activity.

9.4 Summary of Selected Major Historical Events

9.4.1 July 1916: North Carolina

The storm of record for the French Broad watershed at Asheville occurred on July 15–16, 1916 when a hurricane made landfall near Charleston, South Carolina and moved northwest over western North Carolina causing extensive flooding and triggering numerous landslides (Bell 1916; Holmes 1917; Scott 1972; Witt 2005).

The storm set the 24-h rainfall record for North Carolina of 564 mm at Altapass on the crest of the BRE in Mitchell County. Preceding the mid-July storm, a tropical cyclone produced 100–250 mm of rain over western North Carolina on July 8 and 9, 1916 (Henry 1916; Scott 1972). Although no landslides were reported for this early July storm, it created high antecedent moisture conditions in advance of the July 15–16, 1916 storm. While only 45 landslides were reported for this storm, they occurred over a widespread extent. Landslides were reported in a 200 km-long corridor of the Blue Ridge Mountains and Piedmont from Brevard in Transylvania County, northeast to Basin Creek in Alleghany and Wilkes Counties (in what is now Doughton Park) and were the direct cause of in 22 fatalities. Devastated by fatalities and destruction from flooding and landslides, the Basin Creek community never recovered.

9.4.2 August 13–17, 1940: North Carolina

During August 10–17, 1940, the remnants of a landfalling hurricane caused flooding throughout much of the southeastern USA (US Geological Survey 1949). Rainfall, totaling 340 mm in Watauga and adjacent counties in the North Carolina Blue Ridge Mountains, triggered numerous debris flows during August 13–14, 1940 (Fig. 9.7), during which time as much as 254 mm of rain may have fallen within a 6 h period (Wieczorek et al. 2004). Landslides caused 14 deaths, damaged or destroyed 32 structures in Watauga County, and destroyed transportation networks there and in neighboring counties (Witt et al. 2007a). Over 700 debris flows triggered by this storm were identified in the Deep Gap area of southeastern Watauga County by Wieczorek et al. (2004). Upon completion of the landslide hazard maps for Watauga County, 2,120 landslides, mainly debris flows and debris slides, attributed to this storm were identified and mapped (Wooten et al. 2008b). Of the 2,120 landslides, 2,099 occurred throughout Watauga County, but were generally concentrated in the Deep Gap area (Fig. 9.7) and in a highly dissected, mountainous area in the northwest part of the county. The remaining 21 debris flows occurred in adjacent portions of Wilkes and Ashe Counties mainly along the BRE. Given the magnitude of the debris flow event in Watauga County, and the widespread nature of the heavy rainfall along other nearby areas of the BRE, we speculate that this storm likely triggered many more debris flows in northwestern North Carolina.

Debris flows ranged widely in size from 2 m wide and 12–15 m long, upwards to 60–90 m wide and 400–800 m long. The largest debris flows were in an area of 600 m of relief in the Deep Gap area of the BRE, where the longest track measured from 1940 aerial photography was nearly 2,100 m long (Witt et al. 2008). Within Watauga County, the total area of mapped debris flow tracks was 368 ha in mainly first order and some second order drainages (Witt et al. 2008; Wooten et al 2008b). Although 368 ha is only about 0.5 % of the 819.5 km^2 (81,950 ha) area of Watauga County, it is a significant component of the riparian area. Many of the 1940 debris flows deposited sediment in footslope areas where pre-existing debris deposits were mapped (Wooten et al. 2008b). Examination of the locations of the 1940 debris flow

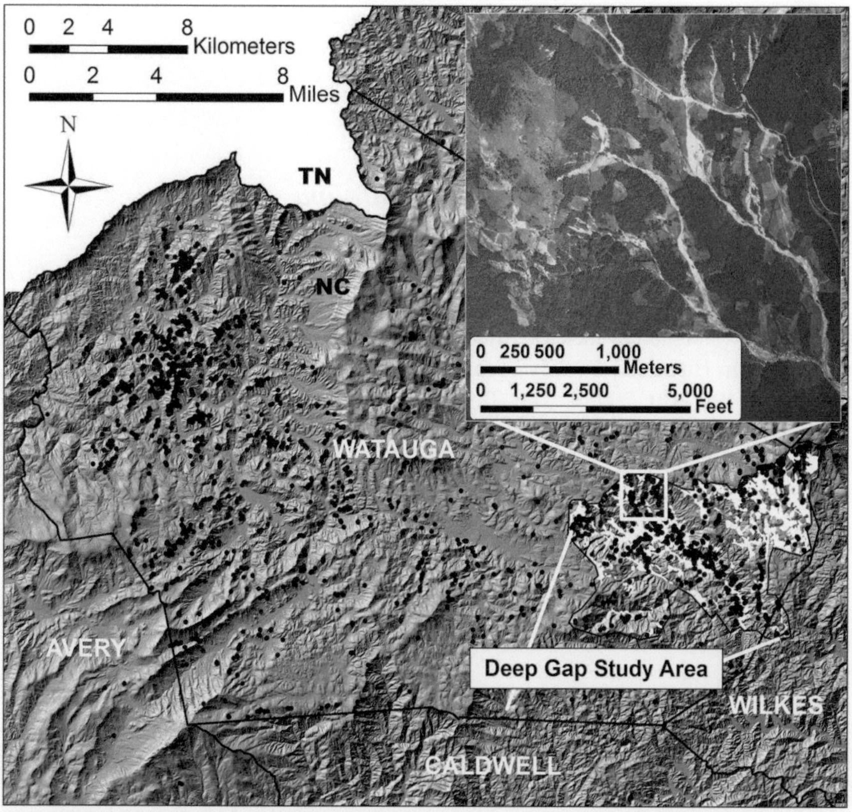

Fig. 9.7 Shaded relief map of Watauga County and the Deep Gap study area (see also Sect. 9.8), *black dots* are point locations for the over 2,100 landslide (mainly debris flow) initiation sites in Watauga and adjacent Wilkes Counties triggered by the August 13–14, 1940 tropical cyclone (from Wooten et al. 2008b). Inset image *upper right* shows debris flow tracks visible as linear high reflectance areas on September 29, 1940 aerial photography. Unforested slopes in the Deep Gap study area are shown in white; remaining slopes within the Deep Gap study area are forested. Shaded relief map derived from a 6 m-pixel resolution LiDAR DEM (Location G in Fig. 9.3 and Table 9.1)

tracks with 2005 orthophotography revealed two findings with respect to current land use patterns and the built environment: (1) since 1940, 136 structures, mainly residences, have been built in the tracks of 1940 debris flows; and (2) 521 tracks of 1940 debris flows cross existing roads (Witt et al. 2007a).

Additional mapping identified another 154 landslides of various types that had occurred in Watauga County since 1940 (Wooten et al. 2008b). Nearly 60 of these landslides were triggered by the remnants of Hurricanes Frances and Ivan in September 2004. Fifteen of the September 2004 debris flow sites were located at, or very near to 1940 initiation sites. Although some of the reactivated sites had been modified by human activity (e.g., fills) since 1940, this finding indicates that the recurrence interval for some landslide prone sites could be on the order of decades, rather than on millennial scales as found in the Virginia Blue Ridge Mountains (Kochel 1987; Eaton et al. 2003a).

9.4.3 August 28–30, 1940: North Carolina

A second major storm struck western North Carolina in August 1940. This storm, a low-pressure system, occurred during August 28–31, and affected mainly the Little Tennessee watershed of Jackson, Macon and Swain Counties, and the French Broad watershed in Haywood County (Tennessee Valley Authority 1940; US Geological Survey 1949; Witt 2005). The Tennessee Valley Authority (Tennessee Valley Authority 1940) reported the heaviest rainfall in the headwaters of the Tuckasegee River in Jackson County where 241–305 mm fell over a 24 h period. The high intensity rainfall over a relatively small area of 388 km^2 triggered more than 200 debris flows which claimed six lives in Jackson and Haywood Counties (Tennessee Valley Authority 1940). High antecedent moisture conditions from the earlier mid-August 1940 likely contributed to the severity of the flooding and debris flows near the center of the late August storm.

9.4.4 August 19–20, 1969 Camille: Virginia, West Virginia

The landslides and flooding associated with the remnants of Hurricane Camille on the night of August 19–20, 1969, was one of the worst natural disasters experienced in Virginia. Approximately 710 mm of rain fell in a span of roughly 8 h, mostly in rural, forested Nelson County in the Blue Ridge Mountains of central Virginia (Williams and Guy 1973). Debris flows and slides permanently altered the landscape and created deep scars in mountainsides that are still visible on aerial photography today. The storm caused the deaths of over 150 people, the majority of whom were killed by blunt force impact related directly to landslides (Simpson and Simpson 1970).

Based on mapping by Morgan et al. (1999a) and Bartholomew (1977), over 3,700 landslides (mostly debris flows and slides) have been identified as occurring during the August 1969 storm. The greatest concentration of debris flows occurred in Nelson County (Figs. 9.8 and 9.9), covering approximately 40 % of the county. Slides were also identified in northern Amherst County and southern Albemarle County. In total, approximately 1,200 ha were damaged by landslide scarring and deposition. Most of this area is comprised of agricultural land and temperate broadleaf and mixed forests with varieties of oak, poplar, and ash (Williams and Guy 1971). The total area stated here is probably a low estimate, as the most recent mapping of these debris flows occurred in 1999 at a scale of 1:24,000 (Morgan et al.

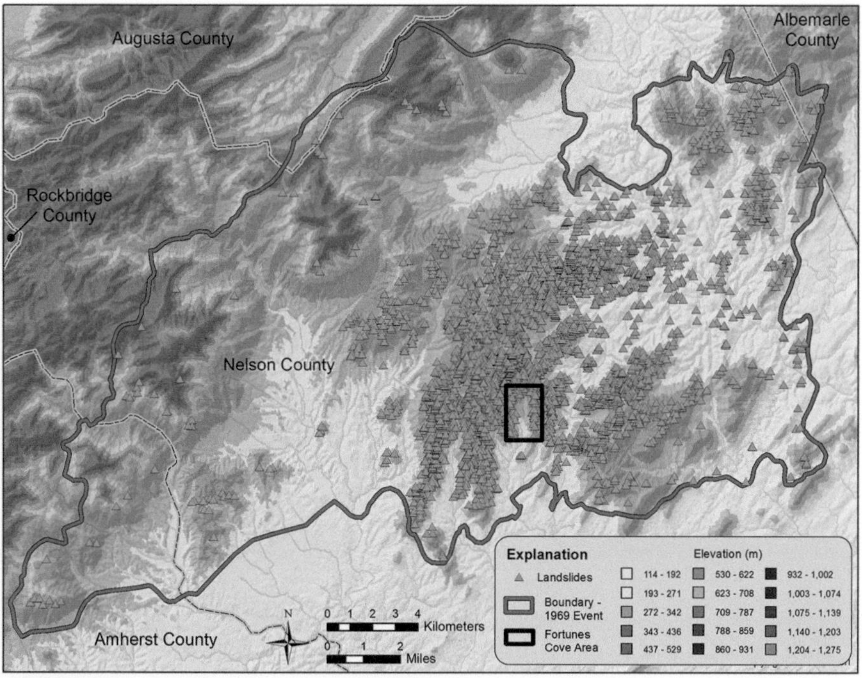

Fig. 9.8 Landslide (mainly debris flow) initiation sites triggered by Hurricane Camille from August 19–20, 1969 in Nelson County Virginia. Camille triggered over 3,700 landslides in this area, impacting 1,200 ha. Map base is a shaded relief map, color-coded by elevation derived from a 30 m digital elevation model. *Black outline* shows the location of the Fortunes Cove area in Fig. 9.9 (Location K in Fig. 9.3 and Table 9.1)

1999a). Re-evaluation of aerial photography, coupled with LiDAR mapping when it becomes available, will likely increase the total amount of land area disturbed.

The greatest number of debris flows occurred along Davis Creek where nearly every first- and second-order stream, and many mountain coves, failed. Along this drainage, over 400 coalescing debris flows and slides destroyed 290 ha. Of the 25 houses along this drainage, 23 were destroyed (Williams and Guy 1971). Along Davis Creek and other heavily damaged drainages, headscarps and scour in the upper portion of the debris flow track frequently left the bedrock exposed. At the headscarps, vegetative cover, including large trees, was completely removed and boulders up to 3 m in length were transported (Williams and Guy 1973). Further downstream, excessive stream discharge scoured even small drainages deeply. The amount of sediment transport and denudation from hillsides and drainages in this area was enormous for a single event. Williams and Guy (1973) studied three watersheds to the south and west of Davis Creek and extrapolated the average denudation for these areas to be approximately 360–500 mm. In comparison, average denudation rates in the Blue Ridge Mountains are estimated to be 150–360 mm per 1,000 years (Judson and Ritter 1964).

Fig. 9.9 Outlines of numerous debris flow tracks that occurred during Camille from August 19 to 20, 1969 and affected nearly every drainage in the Fortunes Cove area of Nelson County Virginia. The outlines are superimposed onto 2009 orthophotography illustrating the progress of vegetative recovery along the debris flow tracks since 1969 (Location K in Fig. 9.3 and Table 9.1)

Forest recovery at the debris flow sites in Nelson County varies depending on the morphology of the headscarp scar (Fig. 9.9). Where bedrock was exposed along debris flow initiation sites, forest recovery has been exceedingly slow and scars are still visible on the landscape today. Until colluvium fills in these areas, vegetation will not return. Where debris flows occurred within colluvium, forest regrowth occurred quickly and may be fully reestablished today (Kochel 1987).

Schneider (1973) reported 1,584 landslides during Camille in the central Appalachians of Greenbrier County, West Virginia. When combined with the 3,793 landslides documented in Virginia, a total of 5,377 landslides resulted from Camille, making it the storm that triggered the greatest number of documented landslides in the SAH.

9.4.5 *November 5–7, 1977: North Carolina*

During November 2–7, 1977 an extratropical cyclone that originated in the Gulf of Mexico passed over western North Carolina causing extensive flooding and triggering debris flows over a multi-county region (Neary and Swift 1987). Although the total storm rainfall in the area was 150 mm, intense convective downpours on the night of November 5–6, 1977 set off debris flows in the Bent Creek Experimental Forest within the Pisgah National Forest near Asheville in southwestern Buncombe County (Neary and Swift 1987) (Figs. 9.5, 9.10, and 9.11). Pomeroy (1991) mapped 72 debris flows in the Bent Creek area, and Otteman (2001) incorporated his mapping into GIS as part of a study of the area's debris flow susceptibility. Landslide mapping in Buncombe County (Wooten et al. 2009b) and adjacent Henderson County (Wooten et al. 2011) identified 11 additional debris flows attributed to this event bringing the total number of debris flows from the 1977 storm in this area to 83.

Rainfall from the 1977 storm also triggered debris flows on the slopes of Mount Mitchell and the Black Mountains in the Pisgah National Forest (Eschner and Patrick 1982). Total rainfall from the storm was 300 mm in the vicinity of Mount Mitchell (Neary and Swift 1987) which at elevation 2,037 m is the highest peak in eastern North America. The upper tracks from these debris flows are still visible on the southeast facing slopes of Mt. Mitchell and the Black Mountains (Fig. 9.12). Silver (2003) relates a personal account of a resident who witnessed one of these debris flows along Shuford Creek that originated on Celo Knob. The upper portion of the track of the Shuford Creek debris flow is visible in 2013 aerial photography. An evaluation of several vintages of aerial photography dating from 1993 to 2013 reveal at least three episodes of debris flows occurring on the southeast-facing

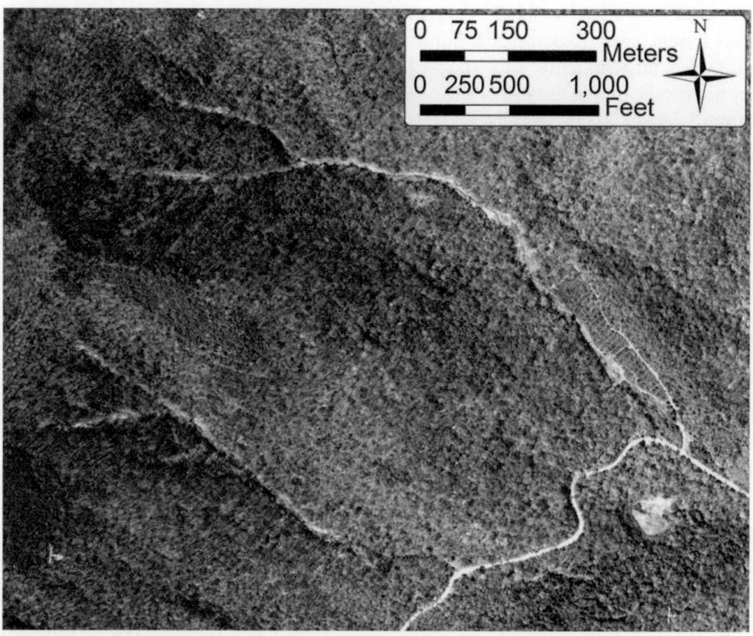

Fig. 9.10 Sparsely vegetated tracks of 1977 debris flows visible in 1983 aerial photography of the Rocky Branch area, Bent Creek Experimental Forest, Pisgah National Forest, in Buncombe County North Carolina. Elevation ranges from 1,152 to 762 m, with lower elevations in the southeast portion of the map. Same view as Fig. 9.5 (Location N in Fig. 9.3 and Table 9.1)

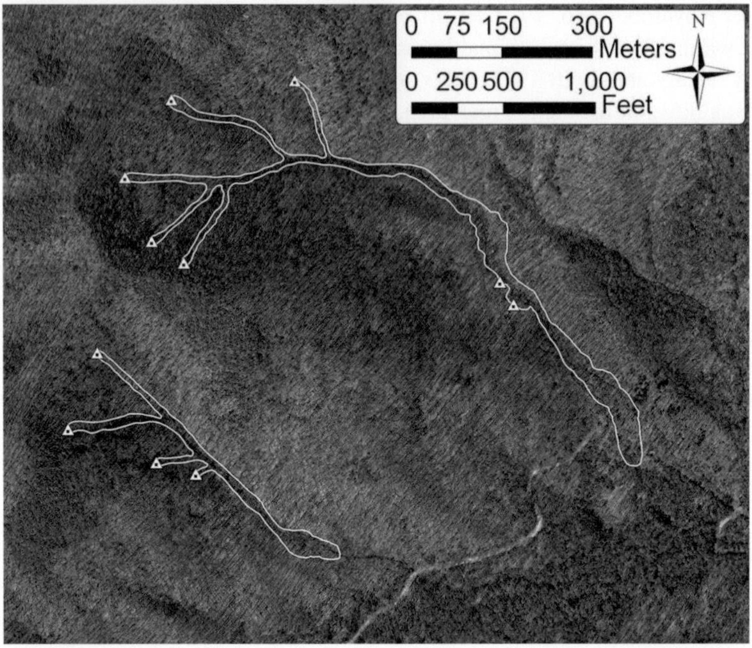

Fig. 9.11 Outlines of tracks of 1977 debris flows visible in 2010 orthophotography of the Rocky Branch area, Bent Creek Experimental Forest, Pisgah National Forest, in Buncombe County North Carolina Debris flow initiation sites shown by triangles (From Wooten et al. 2009b). Vegetative recovery has progressed in the tracks since 1977, and since 1983 as visible in Fig. 9.10. Elevation ranges from 1,152 to 762 m, with lower elevations in the southeast portion of the map. Same view as Figs. 9.5 and 9.10 (Location N in Fig. 9.3 and Table 9.1)

slopes of Mount Mitchell and the Black Mountains. These flows appear to correspond with the occurrence of four separate tropical cyclones in 1972, 1977, 1994 or 1995, and 2004. Clark (1987) reported debris flows during a hurricane in June of 1972 (Agnes) in the area of Mount Mitchell, although the exact location was not specified. As many as 13 of the debris flow tracks are attributable to the 1977 event. Three tracks may correspond with the passage of tropical cyclone Beryl in 1994 or Opal in 1995, and one track probably corresponds with the remnants of Hurricanes Frances and Ivan in September 2004.

The total area of the 83 debris flow tracks for the November 1977 event for the Bent Creek area is 32.8 ha. The 13 tracks on Mount Mitchell and the Black Mountains affected 25.2 ha resulting in a total of 58 ha for the November 1977 event. As in other study areas, the 1977 and 2004 debris flows in the Bent Creek area deposited sediment in areas of pre-existing debris deposits. Neary et al. (1986) point out that although debris avalanching (flows) are destructive events and are major contributors to long term erosion rates, they lead to formation of some of the more productive forest soils. Although the tracks of the 1977 debris flows are still discernable in the Bent Creek area where maximum elevations are on the order of

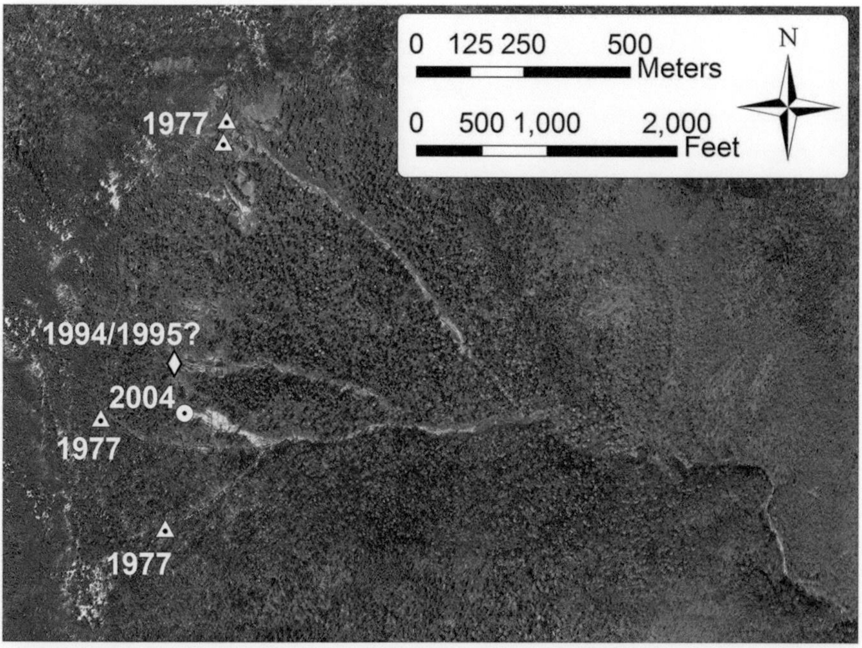

Fig. 9.12 Sparsely vegetated upper portions of tracks from three generations of debris flows in the Black Mountains near Mount Mitchell, North Carolina shown on 2010 orthophotography (initiation sites: *1977 = triangles, 1994/1995? = diamond, 2004 = circle*). Debris flow recurrence at steep, high elevation sites can be on decadal time scales; whereas vegetative recovery can be on decadal to centennial time scales. Downslope direction is from west (elev. 2,005 m) to east (elev. 1,260 m) (Location L on Fig. 9.3)

1,150 m; the tracks of 1977 debris flows are more readily visible on the upper slopes of Mount Mitchell and the Black Mountains were elevations range from 2,037 to 1,500 m.

9.4.6 *June 27, 1995: Madison County and Albemarle County Virginia*

On June 27, 1995, a series of severe rainstorms struck the high relief areas of the Blue Ridge Mountains in central Virginia. Approximately 630–770 mm of rain fell over a period of 16 h causing severe flooding and debris flows and slides in rural Madison County in northwestern Virginia (Fig. 9.13) (Morgan et al. 1997). Flooding and landslides destroyed or damaged 1,700–2,000 residential buildings; property damage was estimated to be $112 million. One fatality was also attributed to a debris flow (Wieczorek et al. 1995).

Landslides related to the June 1995 storm were originally mapped by Morgan et al. (1999b). To improve the inventory of debris flows and tracks for the Madison

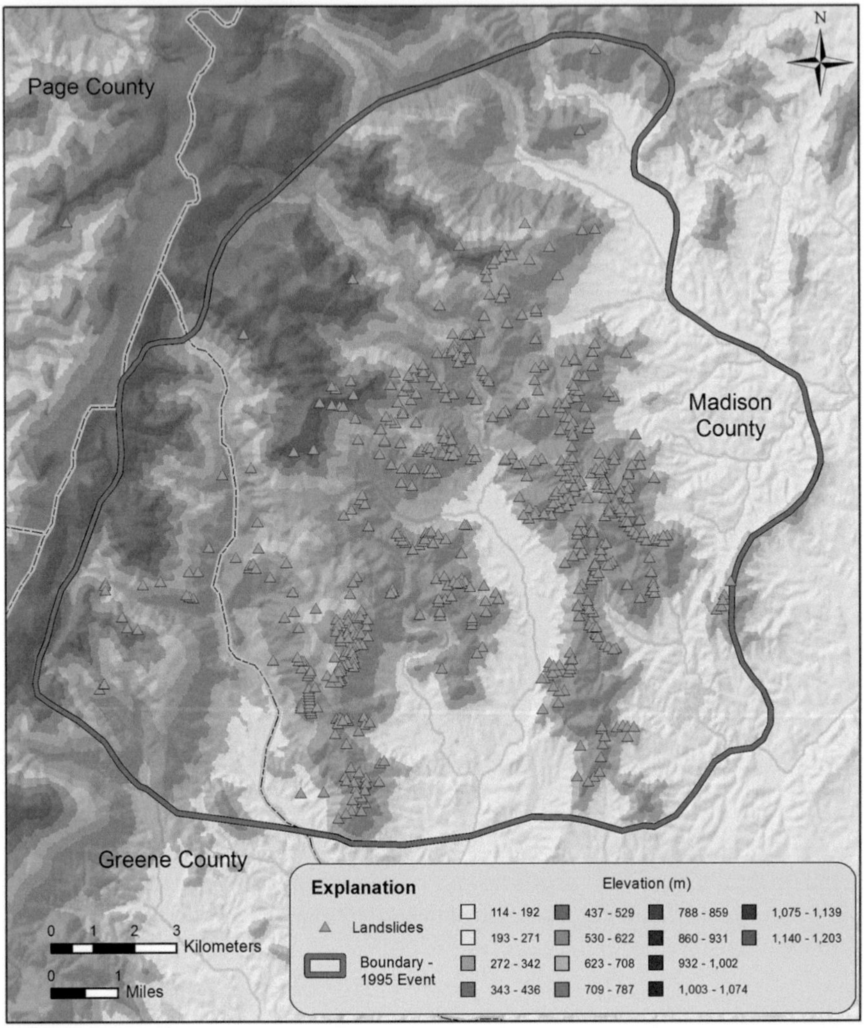

Fig. 9.13 Point locations for 629 landslide (mainly debris flow) sites triggered by the June 27, 1995 storm in Madison County Virginia. Debris flows inundated approximately 600 ha of land here. Map base is a shaded relief map, color-coded by elevation derived from a 30 m digital elevation model (Landslide locations from Morgan and Wieczorek (1996)) (Location Q on Fig. 9.3 and Table 9.1)

County area, we digitized the Morgan et al. (1999b) mapping and identified additional debris flows using 1998 infrared photography and 2002 orthophotography. Based on this work, a total of 629 landslide headscarps or initiation sites were found in both Madison County and northern Greene County occupying an area of about 240 km^2 (24,000 ha) Approximately 600 ha of land were inundated by debris flows and slides.

The largest debris flow occurred along Kinsey Run which damaged a total of 43 ha and contains 29 individual headscarps from multiple coalescing debris flows. The volume of deposited debris attributed to the Kinsey Run debris flow was estimated to be 570,000 m^3 (Mazza and Wieczorek 1997). Much like in the 1969 Nelson County storm, the upper portions of the debris flows exposed bedrock and denuded hillsides of soil, causing significant sediment transport during a single catastrophic event. Eaton et al. (2003a) estimated that the average basin-denudation rates for the upland areas in Madison County were approximately 330 mm during the storm, accounting for 27–65 % of the long-term denudation that would have occurred in 2,500 years.

Numerous debris flows stripped vegetation from the hillsides, sometimes along the entire length of larger debris flows. The rapidly moving flows had sufficient force to snap meter-wide trees at their base (Wieczorek et al. 2000). This vegetative material added significantly to the volume of debris, causing log jams and backups along paths (Wieczorek et al. 2000). The recovery rate of various plant species along the Kinsey Run debris flow was studied and it was found that pioneer species like black locust (*Robinia pseudoacacia*) and (non-native) tree-of-heaven (*Ailanthus altissima*) were the first woody plant types to reestablish and compete with other native species (Eaton and Reynolds 2002).

Landslides were also reported later in the evening on June 27 about 45 km southwest of Madison County along the North Fork of the Moormans River within Shenandoah National Park in western Albemarle County. While no official storm totals exist, eye-witness accounts indicate that rainfall totals varied from 279 to 635 mm (Morgan and Wieczorek 1996; Eaton et al. 2003b). Mapping and field work completed by Morgan and Wieczorek (1996) identified 72 initiation sites of debris flows and slides, many of which were hidden by thick forest cover. Additional interpretation of the area using 1997 infrared photography and 2002 orthophotography allowed for more detailed mapping of individual tracks. Within the 13 km^2 (1,300 ha) watershed, we calculated that a nearly 50 ha area was inundated by debris flows, many of which coalesced into Moormans River. Debris surged downstream into the Sugar Hollow Reservoir, the main water source for Charlottesville, reducing its holding capacity by 15 % (Eaton et al. 2003b).

9.4.7 September 2004 Frances and Ivan: North Carolina

In September 2004, intense rainfall from the remnants of Hurricanes Frances (Sept. 7–8) and Ivan (Sept. 16–17) triggered at least 400 landslides that caused five deaths, destroyed at least 27 residential buildings, and disrupted transportation corridors throughout western North Carolina (Collins 2008, 2014; Witt 2005; Wooten et al. 2005, 2007). Nineteen western North Carolina counties were federally declared disaster areas as a result of flooding and landslide damage from the storms. Known landslide events occurred in a 200 km-long swath in the Blue Ridge Mountains from Macon County northeast to Watauga County North Carolina. Major damage

Fig. 9.14 Large woody debris and boulder deposits along the track the Bear Drive Creek fill failure-debris flow that occurred during Hurricane Frances, September 6–8, 2004. The debris flow initiated as a fill failure on the Blue Ridge Parkway and scoured 8.61 ha along a 2.44 km-long track. Location is in the Pisgah National Forest near Curtis Creek approximately 2 km downslope from the debris flow initiation site. Direction of flow to the right. Geologist at bottom right for scale (November 17, 2004 NCGS photo)

occurred on the Blue Ridge Parkway including three major debris flows that initiated from fill failures that scoured destructive paths downslope into the Pisgah National Forest (Collins 2008; Latham et al. 2009). Figure 9.14 shows imbricated boulder deposits and large woody debris along the track of the Bear Drive Creek fill failure-debris flow near Curtis Creek. Originating at elevations of 1,116, 1,412 (Bear Drive Creek), and 1,349 m along the crest of BRE, these debris flows scoured tracks 2.44, 3.05, and 3.27 km-long, creating canopy gaps of 6.1, 8.0 and 8.61 ha respectively in the headwaters of the Catawba River.

The deadliest of the September 2004 debris flows occurred along Peeks Creek which resulted in 5 deaths (including an unborn child), 2 serious injuries requiring amputation, and 16 destroyed residences (Latham et al. 2006, Witt 2005, Wooten et al. 2006, 2008a). Residents reported that the debris flow occurred at about 2110 EST on September 16, 2014, corresponding to the time of heaviest rainfall from a spiral rain band as it passed over Macon County. The debris flow began on the steep (33–55°) forested slopes of Fishhawk Mountain and traveled 1.5 km through the Nantahala National Forest before entering private land where the fatalities and destroyed homes occurred in the run out zone (Fig. 9.1). The 3.6 km-long track of the debris flow covered an area of 13.8 ha from the upper, northeast-facing slopes of Fishhawk Mountain (Fig. 9.17) downstream to the Cullasaja River. Calculated estimates of a peak velocity of 14.8 m per second, and discharge values ranging from 1,275 m^3 per second to 1,980 m^3 per second for the debris flow attest to its destructive power. Pre-existing debris deposits exposed along the path of the Peeks Creek debris flow indicate that at least two debris flows had occurred before the September 2004 event. Clingman (1877) reported 'water spouts' on the southwest and northeast sides of Fishhawk Mountain in 1876 that, from his description, were likely debris flows (Witt 2005).

Of the 48 landslides attributed to Frances and Ivan in Macon County, 33 were debris flows that initiated on relatively undisturbed forested slopes, mainly on the Nantahala National Forest. The total area of the 27 mapped debris flow tracks throughout Macon County is 26.9 ha, with over half of that being the 13.8 ha of the Peeks Creek debris flow. The tracks of the 184 mapped debris flows in western North Carolina attributed to Frances and Ivan constitute a total disturbed area of 83.1 ha. Of this total, 45.6 ha (54.8 %) resulted from debris flows that originated on slopes modified by human activity, whereas 37.1 ha (44.6 %) resulted from those that originated on forested slopes not modified by human activity.

9.4.8 July 14–15, 2011 Balsam Mountain Debris Flows: North Carolina

On the night of July 14–15, 2011, a stationary thunderstorm storm over the GSMNP caused flash flooding in Straight Fork and triggered debris flows on Balsam Mountain (Miller et al. 2012). The flash flooding resulted in major damage to the Eastern Band of Cherokee Indians fish hatchery on Straight Fork where damage estimates ranged from $30,000 to $50,000 (Lee et al. 2011). Rainfall measurements made from the high elevation Duke University GSMNP rain gage network (Tao et al., 2012) indicate that the storm produced intense rainfall of 125 mm in a 4-h period (Miller et al. 2012) and set off 21 debris flows on the slopes of Balsam Mountain (elevation 1827 m), upstream of Straight Fork (Fig. 9.15). Here the high elevation peaks and watershed divides of GSMNP likely contributed to orographic forcing of rainfall (Tao and Barros 2014) which in combination with steep slopes predisposes these locations to debris flow activity.

In September 2011, NCGS geologists made field investigations of three debris flows in the Gunter Fork watershed on the northeast slopes of the Balsam Mountain (Tao and Barros 2014). An additional 18 debris flows were also identified on the southwest slopes of Balsam Mountain through the analysis of 2013 aerial photography. Scoured tracks, downed trees in initiation sites, and accumulations of large woody debris along the tracks are visible in the 2013 imagery. The 18 debris flows on the southwest side of Balsam Mountain all fed in into the upper reaches of Balsam Corner Creek or its tributaries which then flow into Straight Fork. National Park Service staff confirmed that the damage from debris flows in Balsam Corner Creek occurred in the July 14–15, 2011 event. Southworth et al. (2012) previously

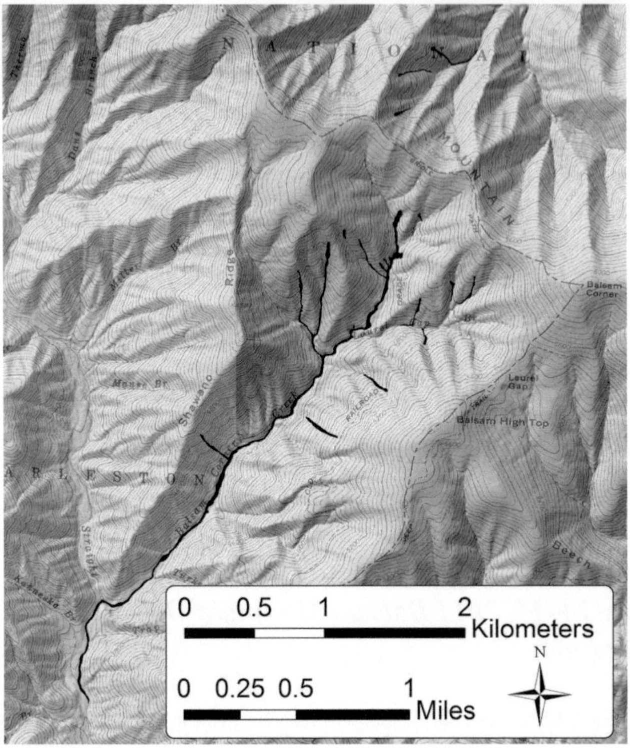

Fig. 9.15 Tracks of 21 debris flows (*black*) triggered by the July 14–15, 2011 storm near Balsam Mountain, GSMNP, North Carolina. Coalescing debris flows scoured a 4.6 km reach of Balsam Corner Creek. Map base is an excerpt of the USGS 7.5-min Luftee Knob quadrangle superimposed onto a 6 m-pixel resolution LiDAR Hillshade map (Location Y on Fig. 9.3 and Table 9.1)

mapped numerous pre-existing debris flows in the headwaters of Straight Fork immediately west of the headwaters of Balsam Corner Creek indicating that the area is prone to debris flow activity.

The track area for the Balsam Mountain debris flows measured from aerial photography is 13 ha, with 12 ha of that total contributed by the Balsam Corner debris flows. The damage to the riparian area from the main debris flow visible in the aerial imagery extends 4.6 km along Balsam Corner Creek. The debris flows in Balsam Corner Creek are significant not only because of the riparian damage along 12.9 ha of tracks, but because they likely contributed to the flooding at the Cherokee Fish hatchery 13.3 km downstream, and 760 m lower in elevation from the initiation sites. This event demonstrates that summer thunderstorms with the potential to trigger debris flows and flash flooding, can develop with little, if any, warning and cause significant damage to resources and communities located several km downstream.

9.4.9 2013 Extended Period of Above Normal Rainfall: North Carolina

Record amounts of rain fell in western North Carolina in between January and August of 2013. By the end of August, the National Weather Service had recorded 1,730 mm of cumulative rainfall for the year at the Asheville airport, 585 mm above a 30-year normal (Fig. 9.16). From July through August 2013, 335 reported landslides of various types occurred throughout western North Carolina (Gibbs 2013; Wooten et al. 2014). Six storms within this period triggered landslides, and the storm events of January 14–18, July 2–7 and July 27, 2013 were federally declared disasters for landslides and flooding. During this timeframe the NCGS, in response to requests for technical assistance, investigated 33 of the reported landslides that resulted in 5 destroyed or condemned homes, and damage to 4 other homes and 24 roads.

Information to date indicates that the vast majority of the 335 reported slope failures involved slopes modified by human activity, mainly embankment slope failures that mobilized into damaging debris flows. Only two landslides that occurred during this period are known to have originated on natural (i.e., unmodified by human activity) slopes. This finding indicates that although record rainfall amounts occurred throughout the region, rainfall was mostly below thresholds necessary to trigger slope failures on forested slopes not modified by human activity.

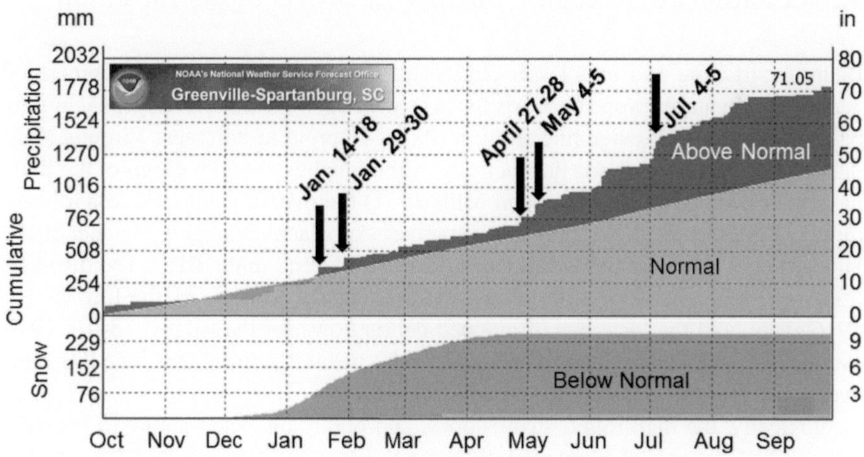

Fig. 9.16 Precipitation chart for the period October 2012 through September 2013 showing rainfall departure from the 30-year normal at the Asheville Regional Airport (KAVL) determined by the National Weather Service. The January 14–18, 2013 event began an extended period of above normal rainfall. *Black arrows* denote rainfall events that triggered landslides throughout western North Carolina (Note: Snow season typically ends in April. The graph shows below normal snow accumulation remained constant after April)

Two debris flows that occurred during this period are noteworthy with respect to impacts of forest structure and riparian areas. On December 16, 2013 a major debris flow occurred on US 441 in the GSMNP, cutting off the main transportation route through the Park from Cherokee, North Carolina to Gatlinburg Tennessee. Accounts by National Park Service staff and field investigations by the NCGS indicate that a debris slide began in road fill and mobilized into a rapidly moving debris flow. About 68,000 m^3 of road fill and the underlying colluvial deposits were involved in the debris flow that removed vegetation and scoured 1.4 ha of slopes that drain into the Beech Flats Prong of the Oconaluftee River. One of the two landslides that initiated on naturally forested slopes not modified by human activity was the July 4, 2013 debris flow on the Nantahala National Forest along Herron Branch, a tributary to the Tuckasegee River in western Jackson County. The scoured riparian area of the 2.1 km-long track was about 3.4 ha as mapped from field investigations. The debris flow did not damage any structures, but large dams of woody debris remained in the track near private homes built along Herron Branch.

A similar, but lesser period of above normal rainfall occurred between September 2009 and February 2010 (Bauer et al. 2010). Western North Carolina received about 1,060 mm of rain, approximately 400 mm above normal. Rain events during this period triggered over 40 landslides; 15 of those investigated by the NCGS were on slopes modified by human activity.

9.5 Landform-Geologic Controls on Debris Flows Initiation

Bedrock structure and to a lesser degree lithology influence the development of geomorphic features prone to debris flows at a variety of scales. At a regional, multi-basin scale the BRE extending from northeast Georgia to northeast Virginia (Fig. 9.4) makes it prone to debris flow activity. Although the geologic origins of the BRE have long been debated (Soller and Mills 1991) the overall southwest to northeast trend of major geologic features in the southern Appalachian orogen (Hatcher 2010; Hibbard et al. 2006) strongly influence the parallel trend in the BRE. The distribution of the generalized locations of areas affected by debris flows for the July 15–16, 1916 event, and the locations of mapped debris flows for the August 13–14, 1940 event in North Carolina, and the August 19–20, 1969 (Camille), the June 27, 1995 events in Virginia generally correspond with the BRE (Figs. 9.3 and 9.4). Geologically, the high relief, steep slopes, and highly dissected nature (a possible reflection of the spatial frequency of bedrock discontinuities) of the BRE make it susceptible to debris flows. Orographic forcing of rainfall along the BRE is shown by the greater rainfall totals along the BRE as compared to the surrounding regions for the storms of July 15–6, 1916 (Scott 1972, Witt 2005), August 10–17, 1940 (US Geological Survey 1949; Wieczorek et al. 2000, 2004; Witt 2005) and June 27, 1995 (Wieczorek et al. 2000, 2004). Steep topography in Madison County, Virginia along the Blue Ridge Mountains likely favored the development of heavy rainfall during the 1995 storm due to orographic lifting (Pontrelli et al. 1999), although the relation between orographic lifting and the heavy rainfall in Nelson County during Hurricane Camille remains unclear (Williams and Guy 1973).

Similar geologic controls on the configuration of landforms prone to debris flows occur in the mountain headwaters of river basins and individual watersheds. For example, the majority (25 of 33) of the debris flows related to the September 2004 rainfall from Frances and Ivan in Macon County, North Carolina occurred on the steep eastern flanks of the Nantahala Mountains Escarpment (NME) (Wooten et al. 2008a). Here, the 100 km^2, 25 km-long NME forms an abrupt topographic rise that contains the headwaters of east-flowing tributaries of the Little Tennessee River (Fig. 9.17). The main northwest and north trends, and secondary northeast trends of its different segments parallel numerous topographic lineaments with orientations similar to measured bedrock discontinuities in the area, reflecting the influence of bedrock structures on the NME. Wooten et al. (2008a) used the term 'structural-geomorphic domain' for such features. Orographic forcing of rainfall by the NME occurred during Hurricanes Frances and Ivan. The Mooney Gap rain gage (elevation

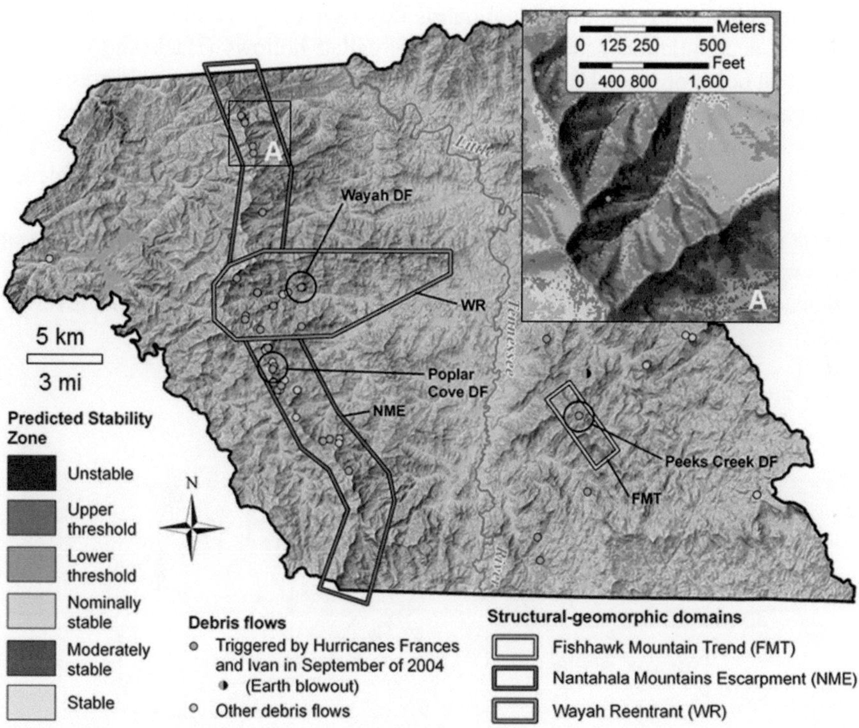

Fig. 9.17 Structural-geomorphic domains in relationship to debris flow locations for Macon County. Map base is a shaded relief 6 m LiDAR DEM overlain by a stability index map (Modified from Sheet 2 of Wooten et al. 2006). Unstable (*purple*) and upper threshold (*red*) stability zones portrayed on the map highlight the topographic features of the structural-geomorphic domains discussed in the text. Inset *A* shows enlarged area for color resolution (Reproduced from Wooten et al. (2008a)). DF = named debris flow location. Poplar Cove debris flow is location S on Fig. 9.3 and Table 9.1

1,364 m) on the crest of NME at the USDA Forest Service Coweeta Hydrologic Laboratory received 100 mm more rainfall in each storm than did several lower elevation gages in the area (Wooten et al. 2008a). At the watershed scale, the September 2004 debris flows were concentrated within the Wayah Creek and Poplar Cove erosional reentrants on the NME, which are likely controlled by bedrock structures that intersect the NME. The Fishhawk Mountain trend, where the 2004 Peeks Creek debris flow originated, has a similar orientation and configuration of geologic structures as the NME, but is a smaller-scale structural geomorphic domain. Perhaps the best example of where bedrock structure can be related to a concentrated debris flow activity is the Deep Gap area of Watauga County (Fig. 9.7, and Sect. 9.8), where 700 debris flows occurred during the August 13–14, 1940 storm (Wieczorek et al. 2004; Wooten et al. 2008b). Here, Elk Creek and its tributaries form a highly dissected erosional reentrant within the BRE that corresponds with WNW-trending ductile faults (Bryant and Reed 1970) and other WNW-trending topographic lineaments that intersect the BRE (Wooten et al. 2008a, b; Gillon et al. 2009).

At the hillslope scale, individual debris flows typically originate in convergent colluvial accumulation zones or catchments called hollows (Hack and Goodlett 1960) that occur on steep hillslopes above the highest extent of channelized streams in mountainous terrain (Fig. 9.18). Colluvial hollows are widely recognized geomorphic features known to be initiation sites for debris flows triggered by heavy rainfall in mountainous terrain (Kochel 1987; Reneau and Dietrich 1987). Residence time of colluvium within individual hollows between successive debris flow events can be more than 20,000 years. The two oldest radiocarbon ages for colluvium in hollows determined by Hales et al. (2011) were 23,989 ± 238 (S.D.) and 23,546 ± 265 radiocarbon years before present in one trench. Samples collected at higher levels in

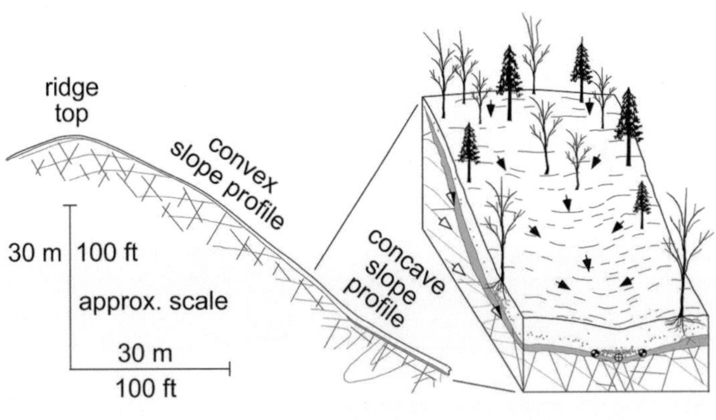

Fig. 9.18 Generalized conceptual model of a hillslope profile (*left*) and hollow (*right*) with colluvial soil layers overlying bedrock (shown with lines depicting curved and planar discontinuities). *Black arrows* depict directions of surface water flow; *black and white arrows* depict shallow groundwater seepage along the colluvium-bedrock contact; *white arrows* depict bedrock fracture flow. *Circles* depict out-of-plane seepage

the same trench had ages that ranged from 4,278 ± 129 to 569 ± 61 radiocarbon years before present. In another pit ages ranged from 8,065 ± 95 to 2,964 ± 112 radiocarbon years before present.

Workers have long recognized the influence of bedrock structure (e.g., planar discontinuities such as fractures, bedding and foliation planes) on the formation hollows in the SAH (Grant 1988; Wooten et al 2008a; Sas and Eaton 2008). Differential weathering of bedrock and enhanced weathering along intersecting discontinuities influences the formation of trough- or wedge-shaped depressions in bedrock surfaces that underlie hollows (Fig. 9.18). In some locations underlain by moderately dipping (less than 45°) layered metasedimentary rocks, the opposing slope (or scarp slope) of the landform is steeper, and contains more colluvial hollows than the slope that coincides with the dip direction (Wooten et al. 2003a). The convergent surface and subsurface geometries of hollows contribute to the accumulation of colluvial soil, which along with the build-up of excess pore-water pressures from infiltrating rainfall and fracture flow along bedrock discontinuities (Sas and Eaton 2008), combine to initiate debris flows. The relationships between topographic convergence in hollows, soil moisture content, and root cohesion are discussed in detail in Sect. 9.7.

Of the 880 debris flows and slides in the NCGS landslide geodatabase where the geomorphic shape of the initiation sites was categorized, 68 % (601) occur on concave slopes or hollows; 18 % (157) are on convex slopes; and 14 % (122) are on planar slopes. Although the majority of debris flows originate in colluvial soil, a lesser number initiate in residual soil derived from the in situ weathering of bedrock. Where detailed studies were done at 21 debris flow initiation sites in western North Carolina Wooten et al. (2012) found that soil at 65 % (15) of the debris flow detachment surfaces to be colluvial soil, and soil at 35 % (6) sites to be residual soil. Steep slopes and relatively thin soil characterize debris flow initiation zones. Slopes at the detailed study sites range from 22° to 40°; inferred detachment depths range from 0.6 to 3 m, which generally correspond with soil depth. Ground slopes (28–44°) and soil depths (0.5–2.4 m) were within similar ranges at 28 debris flow sites investigated in Macon County (Wooten et al. 2008a).

Relationships between bedrock type and debris flow occurrence are less clear. In the Virginia Blue Ridge Mountains, bedrock resistance to weathering contributes to the ruggedness of the topography and thus to the overall steepness of the area. Gryta and Bartholomew (1989) conclude that bedrock lithology is related to preferred debris flow initiation by contributing to topographic relief as a function of resistance to weathering in areas that experience heavy rainfall. In the Nelson County area, a majority of 1969 debris flows initiated in foliated biotite gneisses with steep topography. In contrast, low-relief areas with similar bedrock lithologies did not experience debris flows, even if heavy rainfall was recorded (Gryta and Bartholomew 1989). Morgan et al. (1997) concluded that bedrock type had a negligible effect on debris flow initiation during the 1995 Madison County storm, except in relation to soil and colluvial development. As in Nelson County, the bedrock underlying the areas of debris flows is primarily highly resistant bedrock, commonly granitoids and granitic gneisses, interspersed with high-strain mylonitic zones (Eaton et al. 2004).

There was, however, a minor correlation between the phyllitic metasedimentary units within the Catoctin Formation (metabasalts) and debris flow initiation in western Nelson County, although only a few debris flows failed in this area due to lesser rainfall totals (Gryta and Bartholomew 1989). A similar correlation was noted by Witt and Heller (2014) in Page County in the phyllitic units of the Catoctin Formation; two debris flows and one potential debris flow occurred in phyllites with a foliation dipping to the SE, parallel to slope. Clusters of August 1940 debris flows in western North Carolina correspond with highly dissected areas underlain by Proterozoic granitic gneisses in northwest Watauga County, and in the Deep Gap area of the BRE (Wooten et al. 2008b; Gillon et al. 2009). Further analysis is needed to determine if this apparent correlation results from bedrock lithologies, bedrock structure, meteorological affects, or some combination of factors.

Several investigators have documented slope instability associated with sulfidic rocks in the southern Appalachians. Clark et al. (1987) reported an increased severity of [debris] slides and flows in pyrite-rich rocks of the Anakeesta Formation in the GSMNP. During the May 5–7, 2003 storm in western North Carolina (location T, Fig. 9.3 and Table 9.1) six debris flows originated in embankments constructed with pyrite-bearing rock derived from the underlying sulfidic and graphitic metasedimentary rocks (Wooten and Latham 2004). Latham et al. (2009) reported on rock slides that involved sulfidic and graphitic rocks on the North Carolina portion of the Blue Ridge Parkway.

The weathering of sulfidic rocks can decrease the stability of slopes in several ways (Bryant et al. 2003). Sulfuric acid produced by the breakdown of the iron-sulfide minerals pyrite and pyrrhotite reduces the shear strength of rock and soil. The acid accelerates the rate of rock weathering, and over time the rock fragments in a fill will behave mechanically more like soil, and less like rock. The acid also attacks the clay mineral structure in soil and decreases the cohesion, thereby possibly reducing the shear strength of the soil component of the fill. Sulfidic materials are also susceptible to heaving due to mineral expansion as sulfide minerals oxidize when exposed to moisture. Heaving can increase the porosity and decrease the relative density of the material allowing for more infiltration and the destabilizing build-up of pore-water pressure. Graphite, typically present in sulfidic rocks in western North Carolina, may reduce the shear strength of rock and soil materials.

9.6 Anthropogenic Influences

Anthropogenic influences on hillslopes can have destabilizing effects (causes) which predispose them to slope failures in precipitation events (triggers). Inadequately constructed and maintained fill slopes are a well-documented source of debrisflows in mountainous terrain (Collins 2008; Wooten et al. 2009a, 2014) (Fig. 9.14). Excavations (i.e., cut slopes) into hillsides can also destabilize slopes (Collins 2008; Gillon et al. 2009; Latham et al. 2009); however, debris flows that originate from fill failures typically travel greater distances and impact larger areas. Of 3,267 landslides analyzed in the NCGS landslide geodatabase, 380 (11.6 %) are categorized as cut slope failures, and 421 (12.9 %) are embankment (fill) failures. The remaining 2,466 landslides are categorized as initiating on slopes not modified by ground-disturbing human activity; however 1,752 of those occurred on unforested slopes during the August 1940 storm in Watauga County. Forest cover is an important stabilizing factor, particularly on mountain slopes. This factor and consequences of forest removal are addressed in Sect. 8.8. Given that the vast majority of those unforested slopes likely resulted from human activity, as many as 78 % of the total of 3,267 landslides analyzed may have been influenced in one way or another by humans. Examples of landslides related to ground-disturbing activity follow to help illustrate their spatial impacts on the landscape.

The largest known single debris flow event related to human activity in western North Carolina occurred on August 13, 1916 when the original earthen dam at Lake Toxaway failed when a low pressure system dropped 584 mm of rain over Transylvania County (Wooten et al. 2003a, b). The catastrophic dam failure triggered a debris flow covering a minimum area of 122 ha along an 11.4 km reach of the Toxaway River where it flowed down the BRE and into the adjacent Piedmont of South Carolina (location C Figs. 9.3 and 9.6). The enormous outflow of the breached dam, calculated to be on the order of 8,665 m^3 per second (Wooten et al. 2010a) scoured the steep valley walls and transported boulders as large as 18 m long. Deposits from this event are preserved in Gorges State Park and beneath the upper portion of Lake Jocassee. The debris flow scoured to bedrock a 3.7 km length of the upper reach of the river from Lake Toxaway downstream to Wintergreen Falls, a condition that persists today. Boulder levees and other deposits left by the debris flow below Wintergreen Falls now support vegetation. Initial revegetation of the deposit areas probably began soon after the debris flow. Tree ring studies in Gorges State Park (Wooten et al. 2003a, b; 2004) show 1917 to be the beginning growth year for a pitch pine (*Pinus rigida*) now growing on the 1916 boulder deposits near the confluence of Bearwallow Creek and the Toxaway River.

Three major debris flows that damaged the Blue Ridge Parkway and slopes below on the Pisgah National Forest during Frances and Ivan in September 2004 originated as fill slope failures (Collins 2008, Latham et al. 2009). Collins (2014) assessed 105 of the hundreds of landslides on the Pisgah and Nantahala National Forests triggered by rainfall from the remnants of Frances and Ivan and found that 78 (74 %) were road-related failures, mainly fill slope failures. In 2010 a retaining wall failure that mobilized into a 0.82 km-long debris flow damaged three houses and 2.6 ha of mountain side riparian area in Haywood County, North Carolina (Witt et al. 2012). More recently, during the period of extended above average rainfall throughout western North Carolina in 2013 (Sect. 8.4.9) only two of the reported 335 landslides occurred on slopes not modified by human activity. Although the assessment of the 2013 landslides in western North Carolina is incomplete, this high proportion of landslides related to human activity points to the destabilizing influence that some slope modifications and vegetation removal can have.

9.7 Ecohydrological Controls on Debris Flows Initiation

The historical range of variation in landslides is important to consider in the context that the past may be an adequate predictor of the future. However when novel conditions occur, as are expected with changes in precipitation amount and distribution (Ford et al. 2011) and changes in species' ranges (Hansen et al. 2001; Burrows et al. 2014), it is important to consider the ecological and hydrological controls on landslide initiation. The large majority of landslides that initiate in the SAH are 'shallow' landslides that initiate in the soil column and often form debris flows. The spatial distribution of these shallow landslides and their frequency is strongly controlled by the cohesion of the soils (Crozier et al. 1990; D'Odorico and Fagherazzi 2003). Where soils have no cohesion, landslides are smaller and more frequent than in cohesive soils. Cohesive soils are thicker (i.e., can support a greater volume of soil), require larger precipitation events (usually tropical cyclones) to initiate slides that are of a greater volume (magnitude) (Gabet and Dunne 2003). In the southern Appalachians the steep, colluvial soils that initiate landslides typically have a low soil cohesion, so the cohesive strength provided by the roots (this is technically an apparent cohesion, see Schmidt et al. (2001) for a full derivation) of plants acts as a primary driver of the spatial and temporal distribution of shallow landslides. In addition to the provision of cohesive strength, there are a number of additional ways that vegetation affects landslide initiation including adding weight from the stem and aboveground biomass, altering shallow surface hydrology, and affecting soil structure particularly by adding macropores (Hales et al. 2009).

Because of the important role that vegetation plays in influencing the distribution and thickness of soils, landslides initiate where additional cohesion provided by roots is at a minimum (Roering et al. 2003; Hwang et al. 2015). The role that vegetation plays in controlling the initiation of southern Appalachian debris flows, can be challenging to generalize due to the interactions that occur among forest ecology and physiology, and hydrology (henceforth ecohydrology). Differences among species and forest structures are found not only in transpiration and interception rates (Ford et al. 2012), but also in root system architecture, root density, root strength, and how these change with local soil moisture conditions driven in-part by topography.

The additional soil shear strength provided by roots is a large proportion of total shear strength of the thin (average soil depth is 0.9 m), low (soil) cohesion, colluvial soil in the southern Appalachians (Band et al. 2011; Hales et al. 2009). For example, the apparent cohesion provided by roots can represent up to 100 % of the cohesive strength of hillslope soils (Abernethy and Rutherfurd 2001). Roots add shear strength through a frictional coupling with the surrounding soil particles (Schwarz et al. 2010). The shear strength provided by roots is an 'apparent' cohesion, or an additive force (of a magnitude between 0 and 50 kilopascals) when calculating the overall stability of a slope (Waldron 1977). Before a landslide can initiate, the total tensile strength of all roots crossing the slide plane must be exceeded (Schmidt et al. 2001). This is typically envisioned in terms of a dynamic bundle of roots with different elasticities and tensile strengths that break progressively (Pollen and Simon 2005; Schwarz et al. 2010). Thus higher below-ground biomass imparts higher soil cohesion. However, both the elasticity (Schwarz et al. 2010) and tensile strength (Hales et al. 2009) of roots vary with their diameter. The distribution of biomass by soil depth and root diameter is also is also important (Hales and Miniat 2015). While forest ecologists have long recognized the relationship between below-ground biomass, root diameter distribution and soil resources (Albaugh et al. 1998; Joslin et al. 2000; Keyes and Grier 1981), these relationships have been largely absent in the theoretical considerations of predicting whether slopes will fail.

Debris flows typically initiate on the steep upper slopes (Wooten et al. 2008a), above the highest extent of stream channels (0 order basins), where the topography is dissected into minor ridges called noses, and convergent colluvial accumulation zones (Fig. 9.18) called hollows (Hack and Goodlett 1960). Hollows are wetter and have roots that are weaker, with more even vertical root distributions when compared with noses (Figs. 9.19 and 9.20) (Hales et al. 2009). Although roots are exponentially distributed with depth in hollows and noses (Fig. 9.19), soils in hollows tend to have a higher number of roots at depths greater than 50 cm compared to noses. This greater number is important for slope stability, as the total strength of the soil is dependent upon the total tensile strength of roots crossing the failure plane. Hollows, in this case, would have a greater frequency of roots relative to other parts of the landscape. Tree root tensile strength is controlled by the amount of cellulose within the root structure; hence, larger diameter roots that have a greater number of cellulose microfibrils are stronger (Genet et al. 2005; Hales et al. 2009). This effect is modified by soil moisture content and wood structure (Fig. 9.20) (Hales et al. 2013). In the same way that wet wood is weaker than dry wood, root tensile strength decreases in wetter roots through the breakdown of electrostatic bonds between cellulose fibrils at the microscopic level (Winandy and Rowell 2005). The relationships among geology, topographic convergence, soil moisture content, and hollow stability provides a framework for estimating regional apparent root cohesion and how it might change with changing land cover and land use.

Plants also alter soil moisture primarily through transpiration and interception losses and the development of macropores that increase the transmissivity of the soil and reduce the magnitude of the pore pressures produced by any given storm (Selby 1993). During transpiration, plants remove water from the soil column as it is lost from leaf surfaces in the process of CO_2 uptake. In the soil, there is a concurrent increase in apparent cohesion through the addition of a matric suction force. Matric suction is the capillary stress formed in partially saturated soils and is dependent on the soil moisture content and soil matrix properties (Selby 1993). The total amount of apparent cohesion added to the soil by suction is reduced during large storms, as plants cease to transpire when leaves are wet and atmospheric humidity is near saturation. Interception losses are the amount of precipitation that is intercepted by plant and litter surfaces and subsequently evaporates. These losses reduce the total volume of water added to the soil during precipitation events. During a rain event, these losses introduce a lag in the time taken to reach maximum soil pore water pressure (Keim and Skaugset 2003). Landslide initiation tends to occur during large storms

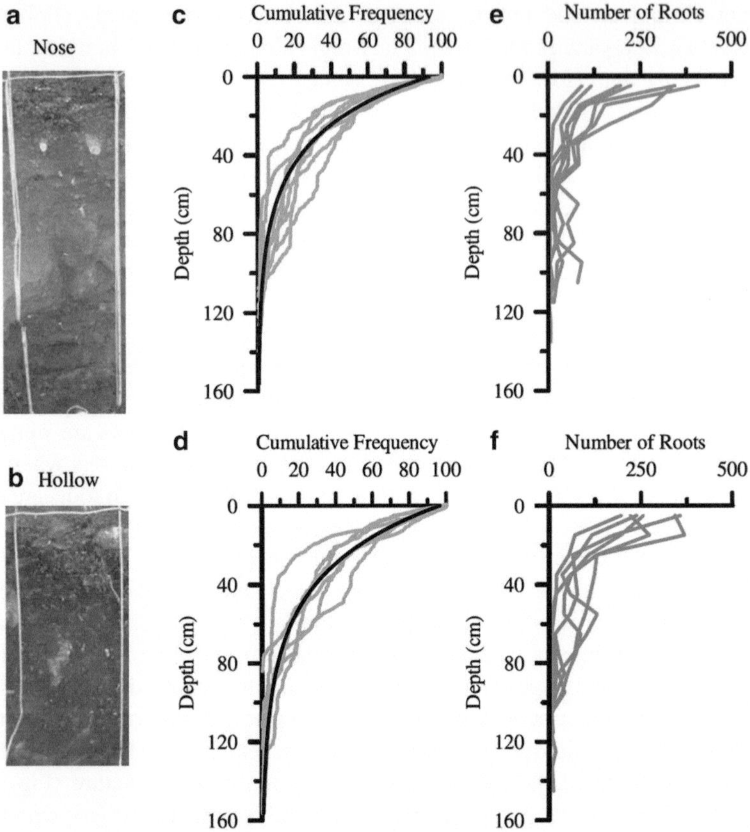

Fig. 9.19 The distribution of roots as a function of soil depth for 16 pits—9 were located on noses and 7 in hollows. (**a** and **b**) Photographs are vertical sections of two northern red oak (*Quercus rubra*) pits dug within 20 m of each other. The diameters of the blue painted roots (high reflectance areas) were measured in an image analysis program to calculate the depth distributions. (**c** and **d**) The cumulative frequency of the number of roots as a function of depth, with gray lines representing individual pits, while black lines are the modeled mean of all pits. (**e** and **f**) The absolute number of roots binned at every 10 cm depth interval, which provides an approximate measure of root area ratio. *Gray lines* are individual pits (From (Hales et al. 2009)

when transpiration is minimal and soils are at or near saturation and suction forces are low (Godt et al. 2009; Montgomery and Dietrich 1994). As a result, plants primarily affect soil shear strength through the added shear strength of roots distributed throughout the soil column rather than through transpiration and interception (Pollen and Simon 2005; Schwarz et al. 2010).

In summary, vegetation is a strong control on size, spatial distribution, and frequency of landsliding in the southern Appalachians. Vegetation serves to both mitigate and promote landslide initiation, with the primary driver of this being the support provided by the roots. Densely rooted forests are more resistant to the effects of large storms, despite having a larger stems that add more weight to the soil, because of the mitigating effects of a large, strong root mass, and more efficient

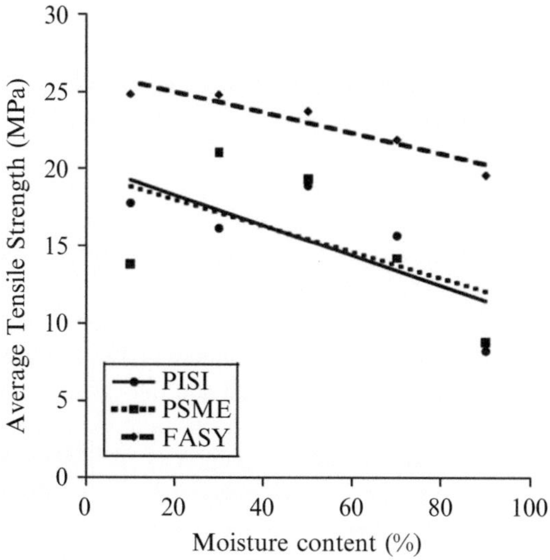

Fig. 9.20 Average root tensile strength as a function of laboratory-controlled root moisture content for species representing two main root xylem architecture types. European beech (FASY; *Fagus sylvatica*), a hardwood with more evolutionary advanced, stronger xylem; and Sitka spruce (PISI; *Picea sitchensis*) and Douglas fir (PSME; *Psuedotsuga menziesii*), coniferous trees with evolutionarily older, weaker xylem. Within each wood type, the wetter the root, the less force required to make the root fail (Modified from Hales et al. 2013)

transpiration. The large roots of trees also increase the porosity and permeability of soils through the development of macropores, reducing the likelihood of high pore pressures required to initiate landslides.

9.7.1 Southern Appalachian Land Use and Slope Stability

Humans have made extensive changes to land cover and land use in the southern Appalachians. Forest types have been affected by land-use change to varying degrees, with high-elevation, northern hardwood forests remaining less affected by development than lower elevation, cove hardwood forests. For example, up to 70 % of the lower elevation cove hardwood forests have been altered by various disturbance processes (Turner et al. 2003). Land-use changes have affected landslide distributions through changes in vegetative land cover. Forests, with deeply rooted trees on cohesive soils have been replaced with agricultural and pasture lands that have shallowly rooted grasses, crops, and shrubs on less cohesive, weak soils. Replacement of trees with grasses, crops and shrubs reduces the apparent root cohesion to values <10 kPa (Schmidt et al. 2001), meaning that shear strength of soils is dependent primarily on the cohesive and frictional properties of the soil particles. Reduction of apparent root cohesion can increase both the number and frequency of shallow landslide events (Gabet and Dunne 2003).

The species and age structure of forests has also changed, with invasive insects and pathogens causing mortality to specific tree species and with forest harvesting (Sakals and Sidle 2004). Recent introductions of two invasive species, the chestnut blight (*Cryphonectria parasitica*), and the hemlock woolly adelgid (*Adelges tsugae*), have functionally removed American chestnut (*Castanea dentata*) and eastern hemlock (*Tsuga canadensis*) trees from southern Appalachian forests. The species that is increasing more than any other in response to eastern hemlock loss is the rhododendron shrub (Ford et al. 2012). As shrubs generally have weaker roots than trees (Hales et al. 2009), soils likely will become weaker over time in these impacted forests. Observational records show that southern Appalachian landslide initiations appear to be strongly concentrated in areas containing this shrub. Additionally, harvesting changes the age structure of the forest and thus affects landslide -susceptibility (Fig. 9.21) (O'Loughlin and Watson 1979; Sidle et al. 2006). After harvest, tree root strength decays exponentially (Schmidt et al. 2001). Young trees with sparse, shallow root systems do not provide as much root reinforcement as mature trees. Therefore, after forest harvest a minima in root reinforcement exists, usually at around 10 years after forest harvest when harvested tree roots have decayed and the young forest is not old enough to provide significant root reinforcement (Fig. 9.21) (Sidle et al. 2006).

Considering all of the interacting factors affecting landslide susceptibility—including vegetation, hydrology, geology—requires an integrated modeling approach. Differences in below-ground biomass and the root strengths of different forest tree and shrub species affect the stability of steeper slopes. A transition from forests to shrub thickets or grasslands means the landscape will support thinner soils and require lower soil pore water pressures to trigger landslides (Gabet and Dunne 2003) (Fig. 9.22). Ecohydrology controls the spatial distribution of root strengths

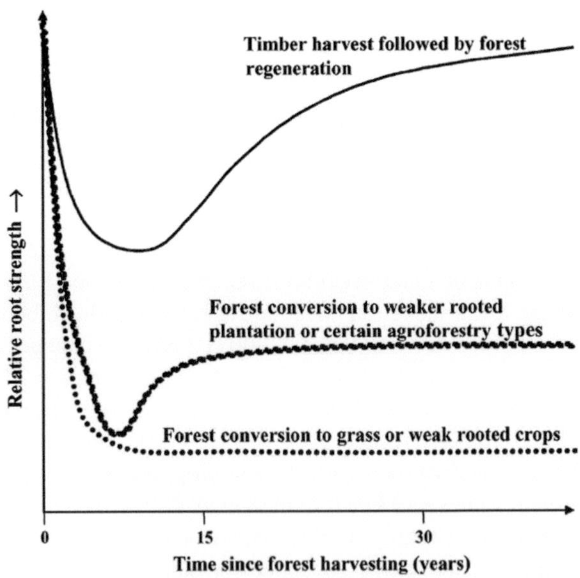

Fig. 9.21 Conceptual diagram showing the temporal response of root cohesion to forest harvest and land use change (From Sidle et al. 2006)

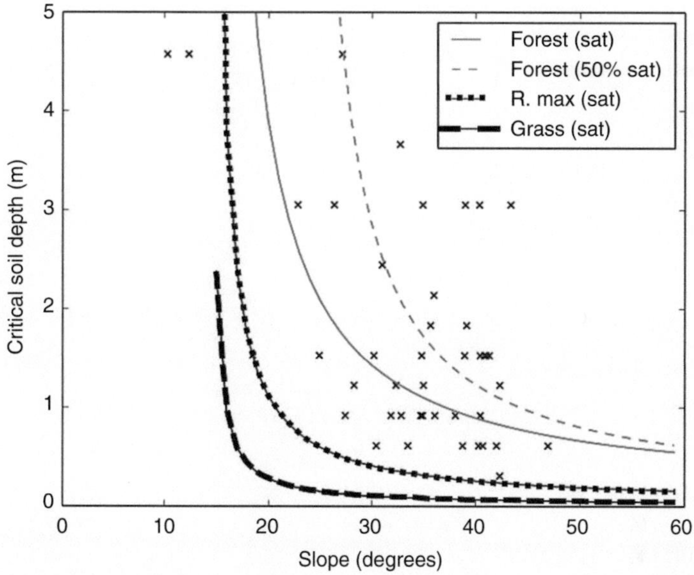

Fig. 9.22 Plot of the relationship between slope and soil depth (measured as the thickness of the landslide scarp) for landslides recorded in the North Carolina landslide database in Macon County, North Carolina. Blue crosses represent each individual landslide event. The four lines represent theoretical failure criteria for different vegetation and hydrologic conditions calculated using the infinite slope model, points that fall above the line should theoretically fail under these conditions. In this model we have maintained constant values of soil cohesion (0 kilopascals) and friction angle (30°) measured using triaxial tests. The *lines* represent failure criteria for average Appalachian hardwood forest (*solid line*, cohesion 6 kilopascals), *Rhododendron maximum*, a common shrub species associated with landslide initiation zones (*dotted line*, 2 kilopascals), and grass (*heavy dashed line*, 0.5 kilopascals) under fully saturated conditions. The *light gray dashed line* is for average Appalachian hardwood forest but with a soil that is 50 % saturated

within forested landscapes, with hollows having the lowest root cohesions. This results in increased landslide potential in these areas (Fig. 9.23). This is an area of significant future research. It is anticipated that more accurate maps of land surfaces and forest structures made using high resolution LiDAR data will lead to more accurate models of root cohesions.

9.8 Vegetation- Debris Flow Relationships: Deep Gap North Carolina

We empirically assessed the spatial relationship between vegetation type and debris flow occurrence in the Deep Gap area of Watauga County and adjoining Wilkes County. The Deep Gap study area was chosen because of the high concentration of debris flows triggered by rainfall from the remnants of a hurricane that passed over

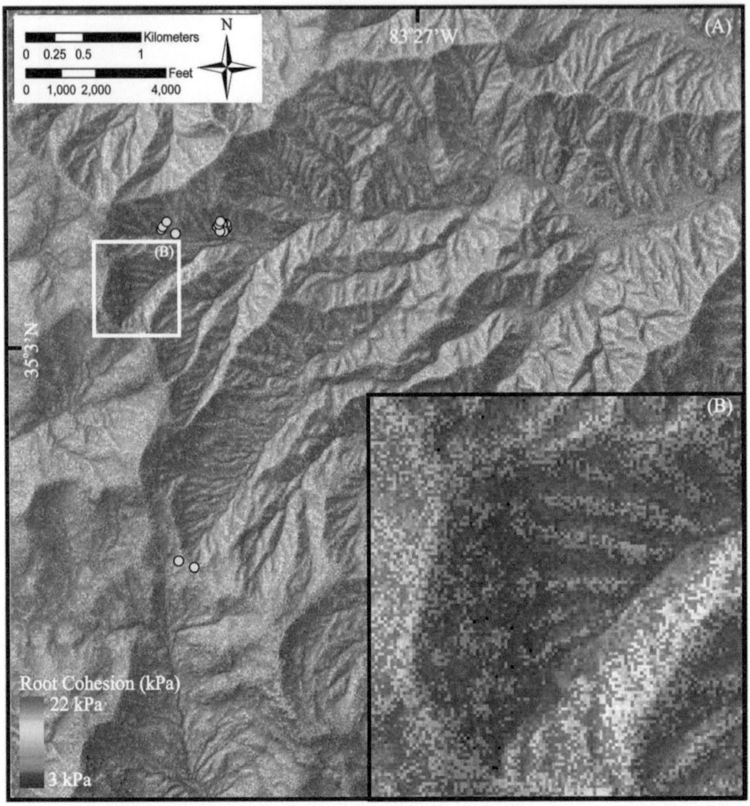

Fig. 9.23 Spatial model of the distribution of root cohesion across Coweeta Hydrological Laboratory in Macon County, North Carolina. *Yellow dots* represent the locations of pits and trenches sampled by Hales et al. (2009). The model is constructed by calculating profile curvature across the catchment. Root apparent cohesion values were assigned using a random uniform distribution of values for each topographic class (nose or hollow)

the area on August 13–14, 1940. The Deep Gap area is a highly dissected erosional reentrant on the BRE that coincides with the surface traces of west-northwest trending ductile thrust faults and topographic lineaments that transect the BRE. Figure 9.7 shows the Deep Gap study on the steep slopes of the BRE, and the landslide initiation sites identified on the landslide hazard maps of Watauga County (Wooten et al. 2008a).

Our approach was to determine the relative spatial frequency of debris flows that initiated on forested slopes versus unforested slopes. As part of landslide hazard mapping in Watauga County (Wooten et al. 2008a), September 29, 1940 vintage black and white aerial photographs (Fig. 9.7) were scanned and georegistered for use in GIS to identify and map the debris flow initiation sites and tracks. Approximately 10 % of the nearly 2100 debris flows, debris slides and blowouts identified as being triggered by the 1940 event were field verified. In this current study, forested slopes were defined as closed canopy, mixed conifer and hardwood

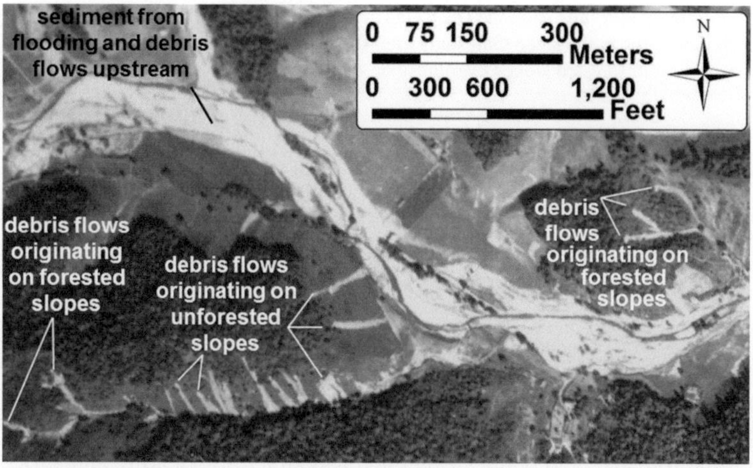

Fig. 9.24 Excerpt of a September 29, 1940 aerial photograph showing examples of August 13–14, 1940 debris flow initiation sites on forested and unforested slopes within the Deep Gap study area (Fig. 9.7). Analysis of 629 debris flow sites in the Deep Gap study area showed a nearly five-fold increase in debris flow initiation sites on unforested slopes over those on forested slopes

forest that were readily identifiable on the aerial photos by the dark color and distinct textural contrasts of foliage. Unforested slopes were defined as open areas that included grasses, shrubs, agricultural fields, areas of recently harvested timber, but also included small areas containing, isolated stands of trees, or bare earth. In general, the unforested slopes occurred in areas more readily suited for clearing and agriculture such as in valleys and toe slopes. Figure 9.7 shows the mapped forested and unforested slopes in the study area. Figure 9.24 shows examples of debris flows originating of forested vs. unforested slopes on 1940 aerial photography of the Deep Gap area.

Potential sources of error include misinterpreting the type of vegetation present at initiation sites and in mapping vegetation patterns. Individual photographs were georegistered, not ortho-rectified; therefore, planimetric areas of the map unit polygons are approximate give the high relief of the area. The numbers of debris flows within forested slopes may be underestimated because they were obscured by the forest canopy. The scoured tracks of debris flows, and sediment deposited in second and third order drainages by debris flows and flood waters obscured the vegetation types present prior to the storm event.

The results of the analysis are given in Table 9.2. Within the 78.2 km^2 study area, 629 debris flow initiation sites were mapped for an average of 8.0 initiation sites/km^2, and a total track area of 18.7 km^2 (1,870 ha). Forested slopes accounted for approximately 63.9 km^2 of the total area, whereas unforested slopes made up a significantly smaller area at approximately 14.3 km^2. On forested slopes there were 234 initiation sites representing 38.4 % of the total sites, resulting in an average of 3.8 sites per km^2. In contrast, 263 sites occurred on unforested slopes yielding an average of 19.5 sites per km^2, nearly a fivefold increase over those on forested

Table 9.2 Tabulated results from the Deep Gap area vegetation and debris flow study. Analysis of 629 debris flow sites in the Deep Gap study area showed a nearly fivefold increase in debris flow initiation sites on unforested slopes over those on forested slopes

	Area km^2	Number of debris flows	Percentage of total debris flows	Average debris flow frequency per km^2
Study area	78.2	629	100	8.0
Forested area	63.9	242	38.4	3.8
Unforested area	14.3	279	44.4	19.5
Boundary of forested – unforested area	–	108	17.2	–

slopes. Interestingly, 108 initiation sites (17.2 % of the total) were mapped as being on the boundary between forested and unforested slopes. This occurrence may reflect a change in slope conditions near the boundaries of forested and cleared land, perhaps the forested land being steeper and less suited for clearing and agriculture. Although there were several potential sources of error in the assessment, the nearly five-fold difference in the frequency of debris flows on unforested slopes over that on forested slopes supports the concept that deep rooted vegetation provides a greater stabilizing influence on slopes than shallow rooted vegetation.

9.9 Debris Flows Effects on Landscape and Forest Structure

Debris flows, in contrast with most other disturbances such as wind, fire, ice, insects and diseases in the SAH remove not only the forest, but the soil and land beneath the forest. Debris flows remove vegetation, scour surface soil, and disrupt aquatic ecosystems, creating linear canopy gaps and early successional habitats along their tracks. As noted in the Introduction chapter (Greenberg et al., Chap. 1) landslides (debris flows) can facilitate species diversity and lead to persistent patches of early successional vegetation in older forests (Seiwa et al. 2013). Debris flows evacuate sediment from their initiation zones and along their scoured paths in upland areas, and deposit it in their run out zones in footslope drainage valleys and channels, commonly in areas of older debris flow deposits (Fig. 9.5). The development of landslide geodatabases and the mapping of debris flows in a GIS (Bauer et al. 2012; Witt and Heller 2014; Crawford 2014) provide the framework for studying the geo-spatial and temporal aspects forest recovery from documented historical debris flow event in the SAH.

An individual debris flow may initiate with soil volumes on the order 10^1–10^2 m^3, but with entrainment and bulking of soil along its path, it may erode and transport in excess of 10^3–10^4 m^3 of soil. Vegetation recovery can begin rapidly in the depositional (i.e., run out) zones of debris flows. Even in the case of catastrophic events like the 11.4 km-long, 122 ha debris flow that resulted from the August 1916 failure of the Lake Toxaway Dam tree growth (pitch pine) began in boulder deposits the following year (Wooten et al. 2003a, b). Sediment from individual debris flows has a relatively short term negative impact on aquatic ecosystems. A positive aspect of debris flow activity over the long term is that in some areas of western North Carolina accumulated deposition in footslope areas leads to the formation of productive forest soils (Neary et al. 1986) at centennial and millennial time frames. Vegetative recovery in the steep, high elevation, upper portions of debris flow tracks, commonly scoured to bedrock (Fig. 9.12), would generally be slower than in debris deposition zones in the lower reaches of drainages and foot slope area (Fig. 9.5).

Major debris flow events induced by tropical cyclones can disturb large areas over multiple watersheds. Mapped debris flows from the August 13–14, 1940 event disturbed a minimum of 368 ha of riparian area in Watauga County not accounting for the large volume of sediment transported downstream below the BRE into Wilkes and Caldwell Counties. Mapped debris flows from Hurricane Camille in the Virginia Blue Ridge Mountains and Piedmont account for a minimum of 1,200 ha of disturbed riparian area, nearly every first and second order drainage in some watersheds. Smaller, though spatially concentrated events such as the Balsam Mountain debris flows triggered by a July 14–15, 2011 thunderstorm resulted in near source disturbances on the order of 13 ha. Impacts from this localized, but intense, storm event extended over 13 km downstream from the debris flow initiation sites.

9.10 Methods and Approaches to Debris Flow Susceptibility Mapping and Modeling

Quantitative methods for landslide hazard mapping geared toward debris flow mapping and susceptibility modeling are well established, readily available, and are continually being improved. Slope stability assessments are important not only for forest and watershed management, but are critical to help protect public health and safety on forest lands and at the interface of forest lands with private property in areas downslope from upland forests (Collins 2014; US Department of Agriculture 2014).

Physically-based computerized models that use a limit-equilibrium approach to slope stability analysis governed by variations of the infinite slope model incorporated into a GIS platform have been developed and in use since the 1990s. In evaluating the computer programs SINMAP (Pack et al. 1998) and SHALSTAB (Montgomery and Dietrich 1994) for use in the North Carolina landslide hazard mapping program Witt et al. (2007a) found the output results to be very similar. SINMAP has been used in North Carolina to model debris flow susceptibility at the county level (Fig. 9.17) (Witt et al. 2007a; Wooten et al. 2007). Modeling using SINMAP coupled with a GIS-based hydrologic routing model and field mapping of debris flows and deposits have been used together to model potential debris flow pathways (Fuemmeler et al. 2008; Witt et al. 2008). These steady-state hydrologic models can be used to forecast the spatial distribution of unstable slopes and landslide occurrence. Models that incorporate transient hydrologic responses (Baum et al. 2002) have the capability to assess spatial and temporal slope stability for varying precipitation scenarios. Morrissey et al. (2004) found such an approach useful in attempting to simulate the locations and timing of debris flow from the 1995 event in Madison County, Virginia.

Next generation models that are being developed and tested include the RHESSys (Regional HydroEcological Simulation System) model in which ecohydrological and geomorphic inputs are coupled to simulate spatial and temporal slope stability. The RHESSys model has undergone development and successfully applied at the watershed scale at the Coweeta Hydrologic Laboratory (Band et al. 2011). Development and application a hydro-mechanical model HILLSOPE FS2 (Lu et al. 2010) that incorporates soil capillary tension to model stability at the hillslope scale is to be tested in western North Carolina (Lewis et al. 2013) on the Nantahala National Forest and Coweeta Hydrologic Laboratory in Macon County, and at the Bent Creek Experimental Forest in Buncombe County. A hydrologic model used with a high elevation rain gage network to predict flood response and debris flow initiation has been applied to GSMNP and adjoining area of western North Carolina (Tao and Barros 2014).

The performance of models used to map slope instability at various spatial scales is highly dependent on the quality of the geologic, soil, geotechnical, hydrologic, and vegetative input parameters, and the quality of the landslide mapping and data used to calibrate them and evaluate their results (Witt et al. 2007b). Landslide and landslide deposit inventories are important not only for identifying areas affected by recent and past landslide activity, but serve as important means to calibrate debris flow susceptibility and run out models. Quality digital elevation data are critical to model performance. Where available 6 m- or higher (e.g., 1 m) pixel resolution LiDAR digital elevation models improve model results over standard 10 or 30 m digital elevation models. These models do not substitute for site-specific stability assessments by qualified earth scientists, but they provide a means for planning purposes to identify potentially at-risk areas where on-site analyses are warranted.

9.11 Summary

The term landslide refers to a variety of gravity-driven movements of soil or rock materials downslope. Landslides may be swift and catastrophic (i.e., rockfalls and debris flows) or may travel slowly and incrementally downslope (i.e., soil slides). Debris flows, the dominant landslide processes in the SAH, are a water-laden (i.e., liquefied) mass of rock fragments and coarse grained soil (debris). Debris flows can attain speeds in excess of 50 km per hour and are capable of destroying or damaging everything in their paths.

Debris flows originate on steep mountain slopes, mainly in areas of convergent topography known as colluvial hollows. Debris flows initiate when the shear stresses on a hillslope exceed the frictional and cohesion shear strength of the hillslope material, which is composed of soil, rock fragments, and roots. Many interrelated, chiefly geologic, factors contribute to debris flow occurrence, including steep slopes, landforms (i.e., their influence on rainfall patterns, runoff and infiltration in convergent topography), bedrock types and structure, soil types (i.e., mechanical and hydrologic properties), and vegetation (i.e., evapotranspiration and root cohesion). Excessive rainfall leading to elevated pore-water pressures (i.e., decreased shear strength) is the primary trigger for debris flows, especially when antecedent moisture conditions are high.

Debris flows and other forms of mass wasting are natural processes of landscape evolution in the SAH. The present landscape includes many remnants of prehistoric (Pleistocene and older) debris flow deposits. Native American and early European settlements were confined mainly to valleys and some mountain footslope areas that include debris flow run-out zones. In the twentieth and twenty-first centuries, however, anthropogenic influences on the landscape have increased the frequency of mass wasting for a given storm event above historical natural levels through changes in vegetation and disturbances on mountain slopes. Where forests with deeply rooted trees have been replaced with agricultural and pasture lands that have shallow-rooted grasses, crops, and shrubs, the stabilizing effects of precipitation interception, evapotranspiration and root cohesion by vegetation have been reduced. In the latter part of the twentieth century increased development on steep mountain slopes has resulted in ground-disturbing human activity where debris flows initiate. Poorly constructed and maintained cut and fill slopes are the primary destabilizing influences that increase the susceptibility of mountain slopes to debris flow initiation. This upslope development pattern combined with increased development in footslope areas within debris flow run-out zones has increased the overall societal risk from damaging debris flows.

Debris flows remove vegetation, remove surface soil and vegetation, and disrupt aquatic ecosystems, creating linear canopy gaps and early successional habitats along their tracks. The area affected by an individual debris flow originating on forested slopes ranges from less than 10 m^2 to 13.8 ha (2004 Peeks Creek debris flow). Coalescing debris flows can impact areas on the order of 16.5 ha or more. Human activity resulted in the largest area affected by a single debris flow in this study where the failure of the Lake Toxaway dam in 1916 disturbed a minimum of 122 ha along the Toxaway River in North and South Carolina. Recovery of areas impacted by debris flows to pre-disturbance conditions can be on decadal and centennial time scales. Vegetative recovery in the steep upper portions of debris flow tracks, commonly scoured to bedrock, would generally be slower than in debris deposition zones in the lower reaches of drainages and foot slope area.

In 1940, 1969 and 1985, each of three tropical cyclones that passed over the SAH generated thousands of documented debris flows. Given the magnitude of the July 15–16, 1916 storm, it is reasonable to conclude that this event triggered thousands of debris flows, bringing the total to four tropical cyclones that have had the greatest impacts on the SAH over the last century. The average frequency of these major events is about 25 years. Where documented by mapping, these catastrophic storms disrupt forest structure and/or hydrologic systems over areas ranging from about 368 to 1,200 ha. Major storms that have each generated hundreds of reported landslides in the SAH have occurred 10 times from 1924 to 2013 for an average frequency of about 9 years. Where documented by mapping, these major storms disrupt forest structure and/or hydrologic systems over areas ranging from about 70 to 600 ha. Collectively these catastrophic and major storms have an average frequency of 7 years over the period 1916–2013.

Sixteen smaller-scale storms that generated less than a hundred of debris flows have a maximum average frequency of 7–8 years over the period from 1876 to 2013. These smaller storms that have induced debris flows are undoubtedly more frequent, and many have not been recorded in the literature. Where documented by mapping, these lesser storms disrupt forest structure and/or hydrologic systems over areas ranging from about 1.4 to 50 ha. Taken together, the 31 landslide events documented here over the period from 1876 to 2013 have an average frequency of about 4 years.

Debris flow occurrence is strongly correlated with antecedent precipitation and rainfall intensity (i.e., rainfall rate and duration). Looking forward, should climate change result in increased occurrences of high intensity rainfall through more frequent storms, or less frequent, but higher intensity storms, then an increased frequency of debris flows and other forms of mass-wasting should be expected in the SAH (see Dale et al. Chap. 13). With regard to the difficulties and uncertainties in predicting the affects that climate change scenarios will have on landslide occurrence, Sidle and Ochiai (2006) conclude that a higher priority should be given to understanding the interactions between land use and landslides, and applying this knowledge in managing mountainous and unstable terrain. Given the importance of the stabilizing influences of forest cover, healthy forests on mountain slopes are critical in mitigating the impacts of recurring landslide events. Reducing losses from landslides are important from the perspectives of ecosystem and infrastructure integrity, but most importantly from the standpoint of public safety.

Quantitative methods for landslide hazard assessment geared toward debris flow mapping and susceptibility modeling are well established, readily available, and are continually being improved. Landslide and landslide deposit inventories are important not only for identifying areas affected by recent and past landslide activity, but serve as important means to calibrate landslide models. These models do not substitute for site-specific stability assessments by qualified earth scientists, but provide a means at the planning level to identify potentially at-risk areas where detailed on-site analyses are warranted. The interdisciplinary technical and scientific capacity exists now to investigate, analyze, identify and delineate landslide prone areas of the landscape with increasing reliability. The March 22, 2014 landslide near Oso, Washington that killed 41 people is yet another reminder of the destructive power of landslides of all types, and the ongoing need to identify and map landslide hazard zones in mountain slopes.

Acknowledgements Many people and agencies have contributed to this effort, but the dedicated work of Jennifer Bauer, Tommy Douglas, Stephen Fuemmeler, Ken Gillon, Rebecca Latham, and Anne Witt of the North Carolina Geological Survey landslide mapping team from 2005 to 2011 made it possible. Reviews by Francis Ashland, Bart Cattanach, Tom Collins, Matt Heller, Kenneth Taylor, and the editors, Beverly Collins and Cathryn Greenberg, resulted in many substantial improvements to the manuscript. Statistical and GIS support by Nick Bozdog and Zach West, and proof-reading and formatting by Rachael Pressley are gratefully acknowledged.

References

Abernethy, B., & Rutherfurd, I. D. (2001). The distribution and strength of riparian tree roots in relation to riverbank reinforcement. *Hydrological Processes, 15*, 63–79.

Albaugh, T. J., Allen, H. L., Dougherty, P. M., Kress, L. W., & King, J. S. (1998). Leaf area and above- and below-ground growth responses of loblolly pine to nutrient and water additions. *Forest Science, 44*, 317–328.

Bailey, J. F., Patterson, J. L., & Paulhus, J. L. H. (1975). Hurricane Agnes rainfall and floods June-July 1972. *United States Geological Survey, 924*, 403.

Band, L. E., Hwang, T., Hales, T. C., Vose, J. M., & Ford, C. R. (2011). Ecosystem processes at the watershed scale: Mapping and modeling ecohydrological controls of landslides. *Geomorphology, 137*, 159–167.

Bartholomew, M. J. (1977). Geology of the Greenfield and Sherando quadrangles, Virginia. *Virginia Division of Mineral Resources, 4*, 43. Charlottesville: Virginia Division of Mineral Resources.

Bauer, J. B., Wooten, R. M., Gillon, K. A., Douglas, T. J., Fuemmeler, S. J., Witt, A. C., & Latham, R. S. (2010). *Your country is falling apart: Response to recent landslides by the North Carolina Geological Survey*. Charleston: Abstract Association of Environmental and Engineering Geologists.

Bauer, J. B., Fuemmeler, S. F., Wooten, R. M., Witt, A. C., Gillon, K. A., & Douglas, T. J. (2012). Landslide hazard mapping in North Carolina – Overview and improvements to the program. In E. Eberhardt et al. (Eds.), *Landslides and engineered slopes: Protecting society through improved understanding. 11th International symposium on landslides and 2nd North American symposium on landslides, June 2–8 2012*. Banff.

Baum, R. L., Savage, W. Z., & Godt, J. W. (2002). *TRIGRS – A Fortran program for transient rainfall infiltration and grid-based regional slope-stability analysis* (US Geological Survey open-file report 02–424). Reston: US Geological Survey.

Bell, W. M. (1916). *The North Carolina flood: July 14,15,16, 1916* (pp. 1–86). Charlotte: Ramsey Library at University of North Carolina Asheville special collections.

Bogucki, D. J. (1976). Debris slides in the Mt. LeConte area, Great Smoky Mountains National Park, USA. *Geografiska Annaler, 58A*, 179–191.

Bryant, B., & Reed, J. C., Jr. (1970). *Geology of the Grandfather Mountain window and vicinity. North Carolina and Tennessee* (US Geological Survey Professional paper 615). Washington, DC: US Government Printing Office.

Bryant, L., Mauldon, M., & Mitchell, J. K. (2003). *Geotechnical problems with pyritic rock and soil* (pp. 1–88). Blacksburg: Center for Geotechnical Practice and Research, Virginia Polytechnic Institute and State University.

Burrows, M. T., Schoeman, D. S., Richardson, A. J., Molinos, J. G., Hoffmann, A., Buckley, L. B., Moore, P. J., Brown, C. J., Bruno, J. F., Duarte, C. M., Halpern, B. S., Hoegh-Guldberg, O., Kappel, C. V., Kiessling, W., O'Connor, M. I., Pandolfi, J. M., Parmesan, C., Sydeman, W. J., Ferrier, S., Williams, K. J., & Poloczanska, E. S. (2014). Geographical limits to species-range shifts are suggested by climate velocity. *Nature, 507*, 492–496.

Clark, G. M. (1987). Debris slide and debris flow historical events in the Appalachians south of the glacial border. In J. E. Costa & G. F. Wieczorek (Eds.), *Debris flows, avalanches: Process, recognition, and mitigation* (Reviews of engineering geology, Vol. 7, pp. 125–138). Boulder: Geological Society of America.

Clark, G. M. (1993). Quaternary geology and geomorphology of part of the Inner Piedmont of the southern Appalachians in the Columbus Promontory upland area, southwestern North Carolina and northwestern South Carolina. In T. L. Davis, R. D. Hatcher Jr (Eds.), *Studies of Inner Piedmont geology with a focus on the Columbus Promontory*. Field trip guidebook. Raleigh: Carolina Geological Society.

Clark, G. M., Ryan, P. T., Jr., & Drumm, E. C. (1987). Debris slides and debris flows on Anakeesta Ridge, Great Smoky Mountains National Park, Tennessee. In A. Schultz & C. Southworth (Eds.), *Landslides of eastern North America* (US Geological Survey Circular 1008, pp. 18–19). Washington, DC: US Government Printing Office.

Clingman, T.L. (1877). Water spouts. In J. Nichols (Ed.), *Selections from the speeches and writings of the Honourable Thomas L. Clingman of North Carolina with Additions and Explanatory Notes*. Raleigh: Afro-Americana Imprint.

Collins, T. K. (2008). Debris flows caused by failure of fill slopes: Early detection, warning, and loss prevention. *Landslides, 5*, 107–119.

Collins, T. K. (2014). Supplemental report geologic hazards assessment, plan revision Nantahala and Pisgah National Forests. http://www.fs.usda.gov/Internet/FSE_DOCUMENTS/stelprd3793002.pdf34p. Accessed 14 July 2014.

Crawford, M. M. (2014). Kentucky Geological Survey landslide inventory: From design to application. *Kentucky Geological Survey Information Circle, 31*, 18.

Crozier, M. J., Vaughan, E. E., & Tippett, J. M. (1990). Relative instability of colluvium-filled bedrock depressions. *Earth Surface Processes and Landforms, 15*, 329–339.

Cruden, D. M., & Varnes, D. J. (1996). Landslide types and processes. In A. K. Turner & R. L. Schuster (Eds.), *Landslides investigation and mitigation* (Special report (National Research Council (U.S.). Transportation Research Board), 247). Washington, DC: National Academy Press.

D'Odorico, P., & Fagherazzi, S. (2003). A probabilistic model of rainfall-triggered shallow landslides in hollows: A long-term analysis. *Water Resources Research, 39*, 1262.

Eaton, L. S., & Reynolds, S. H. (2002). Trends in vegetation recovery of the Kinsey Run Debris Fan, Madison County, Virginia. *Abstracts with Programs- Geological Society of America, 34*, 8.

Eaton, L. S., Morgan, B. A., Kochel, R. C., & Howard, A. D. (2003a). Quaternary deposits and landscape evolution of the central Blue Ridge of Virginia. *Geomorphology, 46*, 139–154.

Eaton, L. S., Morgan, B. A., Kochel, R. C., & Howard, A. D. (2003b). Role of debris flows in long-term landscape denudation in the central Appalachians of Virginia. *Geology, 31*(4), 339–342.

Eaton, L. S., Bailey, C. M., Gilmer, A. K. (2004). The debris flows of Madison County, Virginia. In *Virginia geological field conference guidebook 34th Annual Virginia Geology Field Conference*. Harrisonburg.

Eschner, A. R., & Patrick, J. H. (1982). Debris avalanches in eastern upland forests. *Journal of Forestry, 80*, 343–347.

Ford, C. R., Laseter, S. H., Swank, W. T., & Vose, J. M. (2011). Can forest management be used to sustain water-based ecosystem services in the face of climate change? *Ecological Applications, 21*, 2049–2067.

Ford, C. R., Elliott, K. J., Clinton, B. D., Kloeppel, B. D., & Vose, J. M. (2012). Forest dynamics following eastern hemlock mortality in the southern Appalachians. *Oikos, 121*, 523–536.

Fuemmeler, S. J., Douglas, T. J., Bauer, J. B., Wooten, R. M., Latham, R. S., Witt, A. C., & Gillon, K. A. (2008). Landslide hazard mapping, methodology and examples from North Carolina. In *Abstracts with programs*, Charlotte (Vol. 40, p. 12). Boulder: Geological Society of America.

Fuhrmann, C. M., Konrad, C. E., III, & Band, L. E. (2008). Climatological perspectives on the rainfall characteristics associated with landslides in western North Carolina. *Physical Geography, 29*, 289–305.

Gabet, E. J., & Dunne, T. (2003). A stochastic sediment delivery model for a steep Mediterranean landscape. *Water Resources Research, 39*, 9.

Genet, M., Stokes, A., Salin, F., Mickovski, S. B., Fourcaud, T., Dumail, J.-F., & van Beek, R. (2005). The influence of cellulose content on tensile strength of tree roots. *Plant and Soil, 278*, 1–9.

Gibbs, M. (2013). NC Department of Transportation damage report, NCDOT Div 14. Rockslide conference, Haywood Community College, Clyde, 5 June 2001.

Gillon, K. A., Wooten, R. M., Latham, R. L., Witt, A. W., Douglas, T. J., Bauer, J. B., & Fuemmeler, S. J. (2009). Integrating GIS-based geologic mapping, LiDAR-based lineament analysis and site specific rock slope data to delineate a zone of existing and potential rock slope instability located along the Grandfather Mountain Window-Linville Falls shear zone contact, southern Appalachian Mountains, Watauga County, North Carolina. In 43rd US rock mechanics symposium and 4th US-Can rock mechanics symposium 28 June–1 July 2009 (p. 13), 181, Asheville.

Godt, J. W. (1997). *Digital representation of landslide overview map of the conterminous United States. Open-File Report 97–289, scale 1:4,000,000*. Reston: US Geological Survey.

Godt, J. W., Baum, R. L., & Lu, N. (2009). Landsliding in partially saturated materials. *Geophysical Research Letters, 36*, L02403.

Grant, W. H. (1988). Debris avalanches and the origin of first-order streams. In W. T. Swank & D. A. Crossley Jr. (Eds.), *Ecological studies* (Forest hydrology and ecology at Coweeta, Vol. 6, pp. 103–110). New York: Springer.

Gryta, J. J., & Bartholomew, M. J. (1983). Debris-avalanche type features in Watauga County, North Carolina. In S. E. Lewis (Ed.), *Carolina Geological Society guidebook* (pp. 1–22). Boone: North Carolina Division of Land Resources, 21–23 October, article 5.

Gryta, J. J., & Bartholomew, M. J. (1989). Factors influencing the distribution of debris avalanches associated with the 1969 Hurricane Camille in Nelson County, Virginia. In A. P. Schultz & R. W. Jibson (Eds.), *Landslide processes of the eastern United States and Puerto Rico* (Special paper (Geological Society of America), 236, pp. 15–28). Boulder: Geological Society of America.

Hack, J. T. (1982). *Physiographic divisions and differential uplift in the Piedmont and Blue Ridge* (US Geological Survey professional paper, 1265). Washington, DC: US Government Printing Office.

Hack, J. T., & Goodlett, J. C. (1960). *Geomorphology and forest ecology of a mountain region in the Central Appalachians* (US Geological Survey professional paper, 347). Washington, DC: US Government Printing Office.

Hales, T. C., & Miniat, C. F. (2015). Root reinforcement and slope stability dynamically adjust to soil moisture. *Earth Surface Process and Landforms.*

Hales, T. C., Ford, C. R., Hwang, T., Vose, J. M., & Band, L. E. (2009). Topographic and ecologic controls on root reinforcement. *Journal of Geophysical Research, 114*, F03013.

Hales, T. C., Scharer, K. M., & Wooten, R. M. (2011). Southern Appalachian hillslope erosion rates measured by soil and detrital radiocarbon in hollows. *Geomorphology, 138*, 121–129.

Hales, T. C., Cole-Hawthorne, C., Lovell, L., & Evans, S. (2013). Assessing the accuracy of simple field based root strength measurements. *Plant and Soil, 372*, 553–56.

Hansen, A. J., Neilson, R. P., Dale, V. H., Flather, C. H., Iverson, L. R., Currie, D. J., Shafer, S., Cook, R., & Bartlein, P. J. (2001). Global change in forests: Responses of species, communities, and biomes. *BioScience, 51*, 765–779.

Hatcher, R. D., Jr. (1989). Tectonic synthesis of the US Appalachians. In R. D. Hatcher Jr., W. A. Thomas, & G. W. Viele (Eds.), *The Appalachian-Ouachita orogen in the United States, vol F-2*. Boulder: Geological Society of America.

Hatcher, R. D., Jr. (2010). The Appalachian orogen: A brief summary. In R. P. Tollo, M. J. Bartholomew, J. P. Hibbard, & P. M. Karabinos (Eds.), *From Rodinia to Pangea: The lithotectonic record of the Appalachian region* (Memoir (Geological Society of America), 206, pp. 1–19). Boulder: Geological Society of America.

Heller, M.J., Hancock, G. S., & Carter, M. W. (2014). Structure and geomorphology at the Blue Ridge-Valley and Ridge boundary near Big Levels, Virginia. In *Abstracts with programs*, Geological Society of America Southeast Sect 46:12, Blacksburg. Boulder: Geological Society of America.

Henry, A. J. (1916). Floods in the east Gulf and South Atlantic states, July 1916. *Monthly Weather Review, 44*, 431–498.

Hibbard, J., van Stall, C., Rankin, D., & Williams, H. (2006). *Lithotectonic map of the Appalachian orogen (south), Canada-United States of America* (Map (Geological Survey of Canada), 2069A). Ottawa: Geological Survey of Canada. scale 1:1500000.

Holmes, J. S. (1917). Some notes on the occurrence of landslides. *Journal of the Elisha Mitchell Society, 33*, 100–105.

Hwang, T., Hales, T. C., Miniat, C. F., Vose, J. M., Bolstad, P. V., Miles, B., & Price, K. (2015). Simulating vegetation controls on hurricane-induced shallow landslides with a distributed eco-hydrological model. *J of Geophysical Research-Biogeosciences*. doi:10.1002/2014JG002824.

Jacobson, R. B. (Ed.). (1993). *Geomorphic studies of the storm and flood of November 3–5, 1985, in the Upper Potomac and Cheat River Basins in West Virginia and Virginia* (187p). Bulletin 1981. Reston: US Geological Survey.

Joslin, J. D., Wolfe, M. H., & Hanson, P. J. (2000). Effects of altered water regimes on forest root systems. *New Phytologist, 147*, 117–129.

Judson, S., & Ritter, D. F. (1964). Rates of regional denudation in the United States. *Journal of Geophysical Research, 69*, 3395–3401.

Keim, R. F., & Skaugset, A. R. (2003). Modelling effects of forest canopies on slope stability. *Hydrological Process, 17*, 1457–1467.

Keyes, M. R., & Grier, C. C. (1981). Above- and below-ground net production in 40-year-old Douglas-fir stands on low and high productivity sites. *Canadian Journal of Forest Research, 11*, 599–605.

Kochel, R. C. (1987). Holocene debris flows in central Virginia. In J. E. Costa & G. F. Wieczorek (Eds.), *Debris flows, avalanches: Process, recognition, and mitigation* (Review in engineering geology, 7, pp. 139–155). Boulder: Geological Society of America.

Kochel, R. C. (1990). Humid fans of the Appalachian Mountains. In A. H. Rachocki & M. Church (Eds.), *Alluvial fans: A field approach* (pp. 109–129). New York: Wiley.

Kochel, R. C., & Johnson, R. A. (1984). Geomorphology, sedimentology, and depositional processes of humid- temperate alluvial fans in central Virginia. In E. Koster & R. Steel (Eds.), *Gravels and conglomerates* (Canadian Society of Petroleum Geologists, 10, pp. 109–122). Calgary: Canadian Society of Petroleum Geologists.

Latham, R. S., Wooten, R., Witt, A. C., & Brame, S. (2006). Investigation of the Peeks Creek debris flow of September 2004, Macon County, North Carolina. *14th annual David S. Snipes/ Clemson hydrogeology symposium field trip guidebook*. Clemson: Clemson University.

Latham, R. S., Wooten, R. M., Cattanach, B. L., Merschat, C. E., & Bozdog, G. N. (2009). Rock slope stability analysis along the North Carolina section of the Blue Ridge Parkway: Using a geographic information system (GIS) to integrate site data and digital geologic maps. *43rd US rock mechanics symposium and 4th US-Canadian rock mechanics symposium* (p. 12). American Rock Mechanics Association, 28 June–1 July 2009, 171.

Lee, L. G., Tanner, P. A., Horne, C. S., & Greer, S. (2011). The Qualla Boundary flash flood of 14 and 15 July 2011. http://www.erh.noaa.gov/er/gsp/localdat/cases/2011/14JulyQuallaBoundaryFF. Accessed 14 July 2014.

Leigh, D. (2009). Geomorphic assessment of the Ferrebee Memorial Park at Bridge 3 along the Nantahala River, US Highway 74, Swain County North Carolina, unpublished report. In T. Benyshek (Ed.), *Archeological deep testing and geomorphic study for the proposed replacement of bridge 3 across the Nantahala River*, TIP-4286, Swain County, North Carolina, Appendix 1:11.

Lewis, Y. W., Wayllace, A., Lu, N., Godt, J. W., & Wooten, R. M. (2013). Hydro-mechanical properties and stability analysis of four landslide-prone hillslopes. In: *Abstracts with programs*, Denver, 27–30 Oct 2013 (Vol. 45, p. 720). Boulder: Geological Society of America.

Lu, N., Godt, J. W., & Wu, D. T. (2010). A closed-form equation for effective stress in variably saturated soil. *Water Resources Research, 46*, 5.

Mazza, N. A., & Wieczorek, G. F. (1997). A volumetric sediment analysis for Kinsey Run debris flow, triggered during an extreme storm event, Madison, County, Virginia. In *Abstracts with programs* (p. 64). Geological Society of America, Northeast section, 29:1. Boulder: Geological Society of America.

Miller, D. K., Barros, A. P., Lee, L., Wilson, A. M., & Wooten, R. (2012). Weather conditions leading to the Gunter Fork Trail debris flows of 14–15 July. In: Abstract, 2011 Great Smoky Mountain Science Colloquium, 22 Mar 2011.

Mills, H. H. (1982). Long-term episodic deposition on mountain foot slopes in the Blue Ridge Province of North Carolina; evidence from relative-age dating. *Southeastern Geology, 23*, 123–128.

Mills, H. H. (1998). Surficial deposits and landforms on the south and west piedmont slopes of Roan Mountain, Mitchell County, North Carolina. In *Southeastern Friends of the Pleistocene 1998 Field Trip Guidebook* (pp. 9–34), Boone.

Mills, H. H., & Allison, J. B. (1995a). Weathering rinds and the evolution of Piedmont slopes in the southern Blue Ridge Mountains. *Journal of Geology, 103*, 379–394.

Mills, H. H., & Allison, J. B. (1995b). Controls on the variation of fan-surface age in the Blue Ridge Mountains of Haywood County, North Carolina. *Physical Geography, 15*, 465–480.

Mills, H. H. & Grainger, D. E. (2002). Cosmogenic isotope burial dating reveals 1.5 million-year-old fan deposit in Blue Ridge Mountains of North Carolina. In *Abstracts with programs* (pp. A–32), Geological Society of America, 34, Lexington. Boulder: Geological Society of America.

Mills, H. H., Brakenridge, R. G., Jacobson, R. B., Newell, W. L., Pavich, M. J., & Pomeroy, J. S. (1987). Appalachian mountains and plateaus. In W. L. Graf (Ed.), *Geomorphic systems of North America* (Centennial special, Vol. 2, pp. 5–50). Boulder: Geological Society of America.

Montgomery, D. R., & Dietrich, W. E. (1994). A physically based model for the topographic control on shallow landsliding. *Water Resources Research, 30*, 1153–1171.

Morgan, G. A., & Wieczorek, G. F. (1996). *Debris flows and landslides resulting from the June 27, 1995 storm in the North Fork Moormans River, Shennandoah National Park, Virginia*. US Geological Survey Open-File Report 96-503. 10p map scale 1:24000. Reston: US Geological Survey.

Morgan, B. A., Wieczorek, G. F., Campbell, R. H., & Gori, P. L. (1997). *Map showing debris-flow hazards in areas affected by the June 27, 1995 storm in Madison County, Virginia*. US Geological Survey Open File Report 97-438. map scale 1:24,000. Reston: US Geological Survey.

Morgan, B. A., Iovine, G., Chirico, P., & Wieczorek, G. F. (1999a). *Map Inventory of debris flows and floods in the Lovingston and Horseshoe Mountain, Virginia, 7.5-minute quadrangles, from the August 19–20, 1969, storm in Nelson County, Virginia*. US Geological Survey Open-File Report 99-0518. Reston: US Geological Survey.

Morgan, B. A., Wieczorek, G. F., & Campbell, R. H. (1999b). *Map of rainfall, debris flows, and flood effects of the June 27, 1995, storm in Madison County, Virginia*. US Geological Survey Investigations Series Map I-2623A. map scale 1:24,000. Reston: US Geological Survey.

Morrissey, M. M., Wieczorek, G. F., & Morgan, B. A. (2004). Transient hazard model using rada data for predicting debris flows in Madison County, Virginia. *Environmental and Engineering Geoscience, 5*, 285–296.

Myers, E. W. (1902). A study of the southern river floods of May and June 1901. *Engineering News and American Railway Journal, 48*, 102–104.

Neary, D. G., & Swift, L. W., Jr. (1987). Rainfall thresholds for triggering a debris avalanching event in the southern Appalachian Mountains. In J. E. Costa & G. F. Wieczorek (Eds.), *Debris flows, avalanches: Process, recognition, and mitigation* (Geological Society of America Review Engineering Geology, Vol. 7, pp. 81–92).

Neary, D. G., Swift, L. W., Jr., Manning, D. M., & Burns, R. G. (1986). Debris avalanching in the southern Appalachians: An influence on forest soil formation. *Soil Science Society of America Journal, 50*, 465–471.

O'Loughlin, C., & Watson, A. J. (1979). Root wood strength deterioration in Pinus radiata after clearfelling. *New Zealand Journal of Forestry Science, 9*, 284–293.

Otteman, R. (2001). *Using GIS to model debris flow susceptibility for the Bent Creek Experimental Forest near Asheville, North Carolina*. Thesis, Greenville: East Carolina University.

Ownby, S. E., Miller, C. F., Berquist, P. J., Carrigan, C. W., Wooden, J. L., & Fullagar, P. D. (2004). U-Pb geochronology and geochemistry of a portion of the Mars Hill terrane, North Carolina–Tennessee: Constraints on origin, history, and tectonic assembly. In R. P. Tollo, J. McLelland, & M. J. Bartholomew (Eds.), *Proterozoic tectonic evolution of the Grenville orogen in North*

America (Memoir (Geological Society of America), 197, pp. 609–632). Boulder: Geological Society of America.

Pack, R. T., Tarboton, D. G., & Goodwin, C. N. (1998). *Terrain stability mapping with SINMAP, technical description and users guide for version 1.00, 4114-0*. Terratech Consulting Ltd, Salmon Arm, British Columbia.

Pollen, N., & Simon, A. (2005). Estimating the mechanical effects of riparian vegetation on stream bank stability using a fiber bundle model. *Water Resources Research, 41*, W07025.

Pomeroy, J. S. (1980). Storm-induced debris avalanching and related phenomena in the Johnstown area, Pennsylvania, with references to other studies in the Appalachians. *US Geological Survey Professional Paper, 1191*, 24.

Pomeroy, J. S. (1984). Storm-induced slope movements at East Brady, northwestern Pennsylvania. *US Geological Survey Bulletin, 1618*, 16.

Pomeroy, J. S. (1991). Map showing late 1977 debris avalanches southwest of Asheville, western North Carolina. US Geological Survey open file report 91–334, map scale 1:24,000.

Pontrelli, M. D., Bryan, G., & Fritsch, J. M. (1999). The Madison County, Virginia, flash flood of 27 June 1995. *Weather and Forecast, 14*, 384–404.

Reneau, S. L., & Dietrich, W. E. (1987). The importance of hollows in debris flow studies; Examples from Marin County, California. In J. E. Costa, G. F. Wieczorek (Eds.), *Debris flows/avalanches: Process, recognition, and mitigation*: *Vol. 7. Reviews in engineering geology* (pp. 165–180). Boulder: Geological Society of America.

Roering, J. J., Schmidt, K. M., Stock, J. D., Dietrich, W. E., & Montgomery, D. R. (2003). Shallow landsliding, root reinforcement, and the spatial distribution of trees in the Oregon Coast Range. *Canadian Geotechnical Journal, 40*, 237–253.

Roth, D. M. (2015). Tropical Cyclone Rainfall in the Southeastern United States Tropical Rainfall Point Maxima. Available via http://www.hpc.ncep.noaa.gov/tropical/rain/tcmaxima.html, Hydrometeorological Prediction Center, United States National Oceanic and Atmospheric Administration, National Weather Service. Accessed 7 Feb 2015.

Sakals, M. E., & Sidle, R. C. (2004). A spatial and temporal model of root cohesion in forest soils. *Canadian Journal of Forest Research, 34*, 950–958.

Sas, R. J., & Eaton, L. S., Jr. (2008). Quartzite terrrains, geologic controls, and basin denudation by debris flows: Their role in long term landscape evolution in the central Appalachians. *Landslides, 5*, 97–106.

Schmidt, K., Roering, J. J., Stock, J. D., Dietrich, W. E., Montgomery, D. R., & Schaub, T. (2001). The variability of root cohesion as an influence on shallow landslide susceptibility in the Oregon Coast Range. *Canadian Geotechnical Journal, 38*, 995–1024.

Schneider, R. H. (1973). *Debris slides and related flood damage resulting from Hurricane Camille 19–20 August, and subsequent storm, 5–6 September, 1969 in the Spring Creek drainage basin, Greenbrier County, West Virginia*. Dissertation, University of Tennessee.

Schrader, F. E. (1945). *Notable local floods of 1939; Part 2. Flood of July 5, 1939 in eastern Kentucky*: *US Geological Survey water-supply paper no 967-B* (pp. 41–59). Washington DC: US Government Printing Office.

Schwarz, M., Cohen, D., & Or, D. (2010). Root-soil mechanical interactions during pullout and failure of root bundles. *Journal of Geophysical Research, 115*, F04035. doi:10.1029/2009JF001603.

Scott, R. C. Jr. (1972). *Geomorphic significance of debris avalanching in the Appalachian Blue Ridge Mountains*. Dissertation, University of Georgia.

Seiwa, K., Miwa, Y., Akasaka, S., Kanno, H., Tomita, M., Saitoh, T., Ueno, N., Kimura, M., Hasegawa, Y., Konno, M., & Masaka, K. (2013). Landslide-facilitated species diversity in a beech-dominant forest. *Ecological Research, 28*, 29–41.

Selby, M. J. (1993). *Hillslope materials and processes*. Oxford: Oxford University Press.

Sidle, R. C., & Ochiai, H. (2006). Landslides: Processes, prediction and land use. *American Geophysical Union, Water Resources Monograph, 18*, 312.

Sidle, R. C., Ziegler, A. D., Negishi, J. N., Nik, A. R., Siew, R., & Turkelboom, F. (2006). Erosion processes in steep terrain-Truths, myths, and uncertainties related to forest management in southeast Asia. *Forest Ecology and Management, 224*, 199–225.

Silver, T. (2003). *Mount Mitchell and the Black Mountains: An environmental history of the highest peaks in eastern America*. Chapel Hill: University of North Carolina Press.

Simpson, P. S., & Simpson, J. H. (1970). *Torn land*. Lynchburg: JP Bell Co.

Soller, D. R., & Mills, H. H. (1991). Surficial geology and geomorphology. In J. W. Horton & V. A. Zullo (Eds.), *The geology of the Carolinas* (pp. 290–308). Knoxville: University of Tennessee Press.

Southworth, S., Schultz, A., Aleinkoff, J. N., & Merschat, A. J. (2012). *Geologic map of the Great Smoky Mountains National Park region, Tennessee and North Carolina*: Scientific investigations map 2997, scale 1:100,000. Reston: US Geological Survey.

Tao, J., & Barros, A. P. (2014). Coupled prediction of flood response and debris flow initiation during warm- and cold-season events in the southern Appalachians, USA. *Hydrology Earth System Sciences, 18*, 367–388.

Tao, J., Wilson, A. M., Barros, A. P., Wooten, R. M. (2012). *Using a spatially dense, high elevation rain gauge network and a hydrologic model to enhance predictability of landslides in the southern Appalachians*: *Abstracts with programs (p. 420) 44:7*, Charlotte. Boulder: Geological Society of America.

Tennessee Valley Authority. (1940). *Floods of August 1940 in Tennessee River Basin*. Report no. 0-243-67S. Knoxville: TVA Water Control Planning Department, Hydraulic Data Division.

Tennessee Valley Authority. (1964). *Floods on the French Broad River, Davidson River, King Creek, Nicholson Creek in the vicinity of Brevard, NC*. Report no. 0–6373. Knoxville: TVA Water Control Planning Department, Hydraulic Data Division.

Turner, M. G., Pearson, S. M., Bolstad, P. V., & Wear, D. N. (2003). Effects of land-cover change on spatial pattern of forest communities in the southern Appalachian Mountains (USA). *Landscape Ecology, 18*, 449–464.

US Department of Agriculture. (1902). *Message from the President of the United States transmitting a report of the Secretary of Agriculture in relation to the forests, rivers, and mountains of the southern Appalachian region*. Washington, DC: Government Printing Office.

US Department of Agriculture – Forest Service. (2014). Assessment for the Nantahala and Pisgah National Forests. http://www.fs.usda.gov/Internet/FSE_DOCUMENTS/stelprd3793034.pdf. Accessed 14 July 2014.

US Geological Survey. (1949). *Floods of August 1940 in the southeastern states. US Geological Survey water supply paper 1066*. Washington, DC: US Government Printing Office

Waldron, L. J. (1977). The shear resistance of root-permeated homogeneous and stratified soil. *Journal of the Soil Science Society of America, 41*, 843–849.

White, P. S., Collins, B., & Wein, G. (2011). Natural disturbances and early successional habitats. In C. H. Greenberg, F. R. Thompson III, & B. S. Collins (Eds.), *Sustaining Young Forest Communities: Ecology and management of early successional habitats in the Central Hardwood Region, USA*. New York: Springer.

Whittecar, G. R., & Ryter, D. W. (1992). Boulder streams, debris fans, and Pleistocene climate change in the Blue Ridge Mountains of Central Virginia. *Journal of Geology, 100*, 487–494.

Wieczorek, G. F., Gori, P. L., Campbell, R. H., & Morgan, B. A. (1995). *Landslide and debris-flow hazards caused by the June 27, 1995, storm in Madison County, Virginia (includes discussion of mitigation options)*: *US Geological Survey open-file report 95-822*. Reston: US Geological Survey.

Wieczorek, G. F., Morgan, B. A., & Campbell, R. H. (2000). Debris flow hazards in the Blue Ridge of Central Virginia. *Environmental and Engineering Geoscience, VI*, 3–23.

Wieczorek, G. F., Mossa, G. S., & Morgan, B. A. (2004). Regional debris flow distribution and preliminary risk assessment from severe storm events in the Appalachian Blue Ridge Province, USA. *Landslides, 1*, 53–59.

Wieczorek, G. F., Eaton, L. S., Morgan, B. A., Wooten, R. M., & Morrissey, M. (2009). *An examination of selected historical rainfall-induced debris-flow events within the central and southern Appalachian Mountains of the eastern United States*: US Geological Survey open-file report 2009-1155. Reston: US Geological Survey.

Williams, G. P., & Guy, H. P. (1971). Debris-avalanches; A geomorphic hazard. In D. R. Coates (Ed.), *Environmental geomorphology, 1st annual geomorphology symposia series* (pp. 25–46). Binghamton State University of New York Publication in Geomorphology.

Williams, G. P., & Guy, H. P. (1973). *Erosional and depositional aspects of Hurricane Camille in Virginia, 1969*. US Geological Survey prof paper 804.

Winandy, J. E., & Rowell, R. M. (2005). Chemistry of wood strength. In R. M. Rowell (Ed.), *Handbook of wood chemistry and wood composites* (pp. 303–347). Boca Raton: CRC Press.

Witt, A. C. (2005). A brief history of debris flow occurrence in the French Broad River Watershed, western North Carolina. *The North Carolina Geographer, 13*, 58–82.

Witt, A. C., & Heller, M. J. (2012). Landslide hazard mapping in Page County, Virginia: GIS processes and preliminary findings. In *Abstracts with programs* (p. 467): Geological Society of America 44:7, Charlotte. Boulder: Geological Society of America.

Witt, A. C., & Heller, M. J. (2013). Development of a landslide hazard mapping project in Page County, Virginia: Preliminary results and methods. 13th annual technical forum on geohazards in transportation in the Appalachian Region, 31 July–1 August 2013, Harrisonburg.

Witt, A. C., & Heller, M. J. (2014). Landslide hazard mapping in Page County, Virginia: Processes, results and lessons learned. In *Abstract with programs* (p. 23). Geological Society of America 46, Blacksburg. Boulder: Geological Society of America.

Witt, A. C., Smith, M. S., Latham, R. S., Douglas, T. J., Gillon, K. A., Fuemmeler, S. J., Bauer, J. B., & Wooten, R. M. (2007a). Life, death and landslides: the August 13–14, 1940 storm event in Watauga County, North Carolina. In *Abstract with programs* (p. 76). Geological Society of America Southeast Section 39, Savannah. Boulder: Geological Society of America.

Witt, A. C., Wooten, R. M., Latham, R. S., Fuemmeler, S. J., Gillon, K. A., Douglas, T. J., & Bauer, J. B. (2007b). A comparative study of two GIS slope stability models for use in the North Carolina landslide hazard mapping program. In V. R. Schaefer, R. L. Schuster, A. K. Turner (Eds.), *Conference presentations 1st North America landslide conference*, Vail Colo, Association of Engneering and Environmental Geologist Special Publication, Vail.

Witt, A. C., Wooten, R. M., Gillon, K. A., Latham, R. S., Douglas, T. J., Fuemmeler, S. J., & Bauer, J. B. (2008). Using an observational database in the preliminary statistical analysis of debris flow run out and its application to inundation modeling in Watauga County, North Carolina, USA. 33rd international geological congress, 6–14 Aug 2008, Oslo.

Witt, A. C., Wooten, R. M., Gillon, K. A., Douglas, T. J., Fuemmeler, S. F., & Bauer, J. B. (2012). Determining potential debris flow inundation zones for an emergency response using LAHARZ: The Ghost Town debris flow, Maggie Valley, North Carolina, USA. In E. Eberhardt, C. Froes, A. K. Turner, S. Leroueil (Eds.), *Landslides and engineered slopes: Protecting society through improved understanding*, 11th international symposium on landslides and 2nd North America symposium on landslides, 2–8 June 2012, Banff.

Witt, A. C., Heller, M. J., & Eaton, L. S. (in press). *Slope movement hazard maps of Page County, Virginia*. Division of Geology and Mineral Resources, 3 sheets, map scale 1:48,000. Charlottesville: Virginia Department of Mines Minerals and Energy.

Wooten, R. M., Latham, R. S. (2004). *Report on the May 5–7, 2003 debris flows on slopes underlain by sulfidic bedrock of the Wehutty, Nantahala, and Copper Hill Formations, Swain County, North Carolina*. North Carolina Geological Survey Report of Investigations to North Carolina Division of Emergency Management. Raleigh: North Carolina Geological Survey.

Wooten, R. M., Latham, R. S. (2006). *August 31, 2006 Embankment Failure – Debris Flow at the Cascades Development, Haywood County, North Carolina*. North Carolina Geological Survey report of Investigation to Haywood County Erosion and Sedimentation Control Officer. Raleigh: North Carolina Geological Survey.

Wooten, R. M., Carter, M. W., & Merschat, C. E. (2003a). Geology of Gorges State Park. *North Carolina Geological Survey Information Circular, 31*, 60.

Wooten, R. M., Merschat, C. E., & Carter, M. W. (2003b). *Map of slope movements and related surficial deposits, Gorges State Park, Transylvania County, North Carolina*. North Carolina Geological Survey Geologic Map Series 10-B, map scale 1:12,000. Raleigh: North Carolina Geological Survey.

Wooten, R. M., Latham, R. S., Clinton, B. D., Reid, J. C., & Idol, T. (2004). The August 13, 1916 Lake Toxaway dam failure in Gorges State Park, Transylvania County, North Carolina: A retrospective from the geologic record. In Poster, 12th annual David W. Snipes Clemson University hydrogeology symposium, Clemson University.

Wooten, R. M., Reid, J. C., Latham, R. S., Medina, M. A., Bechtel, R., & Clark, T. W. (2005). An overview of the North Carolina Geological Survey's geologic hazards program – phase I: In Proceedings of 56th highway geological symposium, Wilmington, NC, 4–6 May 2005.

Wooten, R. M., Latham, R. S., Witt, A. C., Fuemmeler, S. J., Gillon, K. A., Douglas, T. J., & Bauer, J. B. (2006). *Slope movement hazard maps of Macon County, North Carolina*, North Carolina Geological Survey, Geologic Hazard Map Series 1, 3 sheets, map scale 1:48,000. Raleigh: North Carolina Geological Survey.

Wooten, R. M., Latham, R. S., Witt, A. C., Gillon, K. A., Douglas, T. J., Fuemmeler, S. J., Bauer, J. B., & Reid, J. C. (2007). Landslide hazards and landslide hazard mapping in North Carolina. In V. R. Schaefer, R. L. Schuster, A. K. Turner (Eds.), 1st North America landslide conference, Association of Engineering and Environmental Geologist Special Publication 23 (pp. 458–471), Vail.

Wooten, R. M., Gillon, K. A., Witt, A. C., Latham, R. S., Douglas, T. J., Bauer, J. B., Fuemmeler, S. J., & Lee, L. G. (2008a). Geologic, geomorphic, and meteorological aspects of debris flows triggered by Hurricanes Frances and Ivan during September 2004 in the southern Appalachian Mountains of Macon County, North Carolina (southeastern USA). *Landslides, 5*, 31–44.

Wooten, R. M., Witt, A. C., Gillon, K. A., Douglas, T. J., Latham, R. S., Fuemmeler, S. J., & Bauer, J. B. (2008b). *Slope movement hazard maps of Watauga County, North Carolina*. North Carolina Geological Survey Geologic Hazard Map series 3, 4 sheets, map scale 1:36,000. Raleigh: North Carolina Geological Survey.

Wooten, R. M., Gillon, K. A., Douglas, T. J., Witt, A. C., Bauer, J. B., & Fuemmeler, S. J. (2009a). *Report on the Bear Trail debris flow of January 7, 2009a*. North Carolina Geological Survey Report of investigation to Haywood County Erosion and Sedimentation Control Officer. Raleigh: North Carolina Geological Survey.

Wooten, R. M., Witt, A. C., Gillon, K. A., Douglas, T. J., Fuemmeler, S. J., Bauer, J. B., & Latham, R. S. (2009b). *Slope movement hazard maps of Buncombe County, North Carolina*. North Carolina Geological Survey Geologic Hazard Map Series 4, 3 sheets, map scale 1:52,000. Raleigh: North Carolina Geological Survey.

Wooten, R. M., Latham, R. S., Gillon, K. A., Bauer, J. B., Merschat, C. E., & Carter, M. W. (2010a). Stop 3- Lake Toxaway Dam. In J. Gill-Shaler (Ed.), From the Atlantic to the Appalachians and Back!, 2010, Annual meeting, fieldtrip #1 guidebook, Association of Environmental and Engineering Geologist, Charleston.

Wooten, R. M., Witt, A. C., Douglas, T. J., Gillon, K. A., Fuemmeler, S. J., Bauer, J. B., Latham, R. S., & Nickerson, J. G. (2010b). The North Carolina Landslide Program: 3864 km^2 mapped and 6,042 landslide features later – Findings and observations. In *Abstracts with programs*, 2010. Annual meeting, Association of Environmental and Engineering Geologists, Charleston. Denver: Association of Environmental and Engineering Geologists.

Wooten, R. M., Witt, A. C., Douglas, T. J., Fuemmeler, S. J., Bauer, J. B., Gillon, K. A., & Latham, R. S. (2011). *Digital data and maps of the slope movement hazards for Henderson County, North Carolina*. North Carolina Geological Survey Digital Data Series Geologic Hazard Map Series 5 (DDS-GHMS-5). Raleigh: North Carolina Geological Survey.

Wooten, R. M., Douglas, T. J., Bauer, J. B., Fuemmeler, S. J., Gillon, K. G., Witt, A. C., & Latham, R. S. (2012). Geologic, geomorphic and geotechnical aspects of debris flow initiation sites in the Blue Ridge Mountains of western North Carolina at study locations in Buncombe, Haywood, Henderson, Jackson, Macon, Transylvania and Watauga Counties. In *Abstracts with programs* (p. 147), Charlotte. Boulder: Geological Society of America 44.

Wooten, R. M., Cattanach, B. L., Bozdog, G. N., & Bullard, A. R. (2014). The North Carolina Geological Survey's response to landslide events in 2012 and 2013 in the Blue Ridge Mountains of western North Carolina: Using a geodatabase, LiDAR and ortho-imagery in a geographic information system to augment field investigations and map production. In *Abstract with programs* (p. 23): Geological Society of America Southeast Sect 46, Blacksburg. Boulder: Geological Society of America.

Young, C. H. (1973). Investigation of landslides resulting from severe storm. USDA Forest Service in-house report, 11 June 1973, Nat For in North Carolina, Asheville.

Zeedyk, W. D. (1973). *Storm and flood report – National Forests in North Carolina*. USDA Forest Service in-house report, 13 June 1976 Nat For in North Carolina, Asheville.

Chapter 10
Structural and Compositional Change in Great Smoky Mountains National Park since Protection, 1930s–2000s

Julie P. Tuttle and Peter S. White

Abstract When Great Smoky Mountains National Park (GSMNP) was placed under strict protection in 1934, about 20 % of the landscape was old-growth forest that had never been logged or farmed, and about 80 % was second growth recovering from logging and settlement. We might expect that the structure of GSMNP's old-growth forests today would capture the natural range of variation of these southern Appalachian forests, subject only to localized natural disturbances such as landslides, flooding, windthrow, ice storms, and fire. Despite protection, however, multiple indirect, diffuse anthropogenic disturbances including exotic pests, atmospheric deposition, changes in herbivory, and changes in fire regime have continued to affect both old-growth and successional GSMNP forests. Here, we employ a mid-1930s vegetation survey and a compilation of more recent vegetation datasets (1990s–2000s) to compare the historic and present-day range of variation in GSMNP forest structure and composition. Widespread changes in structure reflect succession from historical disturbance and the overlay of continued disturbance, including in formerly undisturbed areas. Species indicative of disturbance and reduced fire frequency are abundant across the landscape in patterns reflecting the legacy of historical disturbance types, continued disturbance, and the interaction of these disturbances with environment.

Keywords Disturbance range of variation • Forest structure and composition • Great Smoky Mountains • Natural and anthropogenic disturbance • Southern Appalachians

J.P. Tuttle (✉)
Curriculum for the Environment and Ecology, University of North Carolina, CB #3280, Chapel Hill, NC 27599-3280, USA
e-mail: jptuttle@live.unc.edu

P.S. White
Department of Biology, University of North Carolina, CB #3280, Chapel Hill, NC 27599-3280, USA
e-mail: peter.white@unc.edu

C.H. Greenberg, B.S. Collins (eds.), *Natural Disturbances and Historic Range of Variation*, Managing Forest Ecosystems 32,
DOI 10.1007/978-3-319-21527-3_10

10.1 Introduction

Spatial and temporal variability in environment and disturbance, interacting with species adaptations, creates structural and compositional variability across forested landscapes. This variability has been shown to be important for the persistence of populations (both plants and wildlife), ecosystem function, and ecosystem services (Landres et al. 1999). With increased, pervasive anthropogenic impacts on forests and the need for effective management across landscapes and regions, understanding and comparing historic and present-day range of variation in forest structure and composition serves important scientific, conservation, and management purposes. In eastern USA forests, previous studies of disturbance and forest structure have focused on single disturbance types (e.g., Clinton and Baker 2000), temporal sequences dating from specific disturbance events (e.g., Reilly et al. 2006), specific communities or site conditions (e.g., Jenkins et al. 2011), and restricted spatial scales (e.g., Elliott and Swank 1994). Research elsewhere has begun to focus on our critical need to understand the aggregate structural and compositional effects of multiple, widespread disturbances of different types across forested regions (Ohmann et al. 2007). To that end, in this chapter, we investigate the aggregate, landscape-level effects of multiple disturbances on forest structure and composition across Great Smoky Mountains National Park (GSMNP). Assessing the state of our study area's mix of old-growth and successional forests with respect to a range of natural and anthropogenic disturbance histories may provide a valuable reference for other southern Appalachian forests that are more intensively used and managed.

Great Smoky Mountains National Park is the largest protected area in the Central Hardwood Region (CHR), encompassing a wide range of elevation and site conditions and harboring nearly the full range of southern Appalachian forest communities. The rugged topography, breadth of forest types, and high biodiversity in this region inspired its designation as a national park in 1934. While much of the park, like other areas in the southern Appalachians, has a history of Native American land use (see Grissino-Mayer Chap. 6; Leigh Chap. 8; Greenberg et al. Chap. 12), European settlement, and logging, approximately 20 % of the park was free of direct human activities (i.e., logging, settlement, and herded livestock grazing) at the time of park creation. Since 1934, the entire park has been protected from further direct uses other than small areas used for recreation. Despite protection, however, multiple indirect, diffuse anthropogenic disturbances or changes in disturbance regime—including exotic pest introductions, reduced fire frequency, atmospheric deposition, and changes in herbivory—have continued to impact nearly all GSMNP forests. Fortunately, the National Park Service hired forester Frank Miller to conduct a park-wide vegetation survey in the 1930s; this detailed survey of forest structure and composition at the time of park formation provides a rare snapshot of landscape-level variation in forests both disturbed and undisturbed by humans. We expect the state of 'undisturbed' forests at that time to lie within the natural, historic range of variation for GSMNP forest structure and composition, thereby representing an important reference for comparison to present-day forests. Likewise, areas impacted by pre-park human activities represent the initial conditions for forests that were subsequently protected from direct human use. We expect a comparison of present-day forests to these 1930s 'bounding' reference conditions to provide insights on the stability of old-growth forests, the state of widely disturbed forests after approximately 75 years of succession, and the effects of ongoing disturbances overlaid on both historically undisturbed and disturbed forests.

While no subsequent studies in GSMNP (including Whittaker 1956) have been as comprehensive as Miller's survey (MacKenzie and White 1998), the combined wealth of middle to late twentieth century plot-level studies in GSMNP—many of them efforts to understand and monitor forest dynamics with respect to disturbance—spans a range of park regions and environments (White and Busing 1993). Here, we compare a compilation of recent forest plot datasets to the Miller survey plots to assess differences in forest structure and composition, within and between time periods, with respect to disturbance history and dominant environmental gradients. Specifically, we use maps of historic anthropogenic disturbance and terrain variables that represent temperature and moisture gradients to ask the following questions: (1) how do environment and disturbance history influence the distribution of basal area (BA), density, and relative species composition in each time period?; (2) what is the range of variation in BA, density, and average tree size, as measured by quadratic mean diameter (QMD) (Curtis and Marshall 2000), with respect to disturbance history in each time period?; and (3) how does the abundance of species functional groups, defined by species' responses to disturbance and environment, differ across disturbance categories and time periods?

10.2 Study Area

10.2.1 Physical Environment

Great Smoky Mountains National Park is a 212,000-ha (2,120 km^2) preserve that lies along the North Carolina-Tennessee border at approximately 35 °N and 83 °W (Fig. 10.1). The topography is steep and complex, with elevation ranging from 267 to 2025 m. From low to high elevation, annual precipitation increases from 140 to >200 cm, the mean annual July temperature decreases from 22 to 14 °C, and the mean January temperature decreases from 4 to −2 °C (Busing et al. 2005). Bedrock is primarily Precambrian sandstone, phyllite, schist, and shale, but one section is dominated by gneiss, granite, and schist, and there are small areas of limestone and related rocks. Soils are acidic and infertile and mostly consist of Ultisols and Inceptisols (USDA Natural Resources Conservation Service 2009).

10.2.2 Vegetation

The forest vegetation of GSMNP is a complex of broad-leaved deciduous hardwoods and needle-leaved evergreen conifers (Jenkins 2007) controlled by environmental gradients, soil conditions, and disturbance (Fig. 10.1; see Greenberg et al.

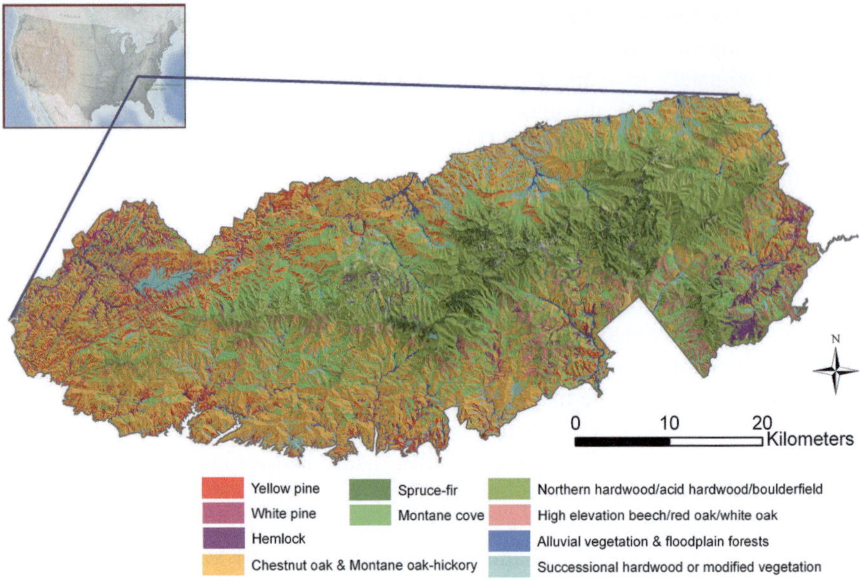

Fig. 10.1 Great Smoky Mountains National Park, North Carolina and Tennessee, USA, with topographic relief and major forest types. Vegetation classification created by the Center for Remote Sensing and Mapping Science (now the Center for Geospatial Research), University of Georgia, Athens (Madden et al. 2004)

Chap. 1). At low elevations, community composition varies along a moisture gradient controlled by topography and soil characteristics, from diverse, mesic cove hardwood forests (on relatively higher pH soils) and less diverse Eastern hemlock (*Tsuga canadensis*, acid cove) forests (on lower pH soils) through oak (*Quercus*)-dominated mid-slope positions to dry ridges dominated by yellow pine (*Pinus pungens, P. ridgida, P. virgniana, P. echinata*) and xerophytic hardwoods. The degree of pine dominance on the ridges at any one time is likely controlled by fire history and outbreaks of the southern pine beetle (*Dendroctonus frontalis*) (see Nowak et al. Chap. 4). With increasing elevation, mesic cove hardwood and acid cove forests grade into northern hardwood forests and then, above about 1500 m in the central and eastern part of the park, into forests dominated by the needle-leaved evergreens red spruce (*Picea rubens*) and Fraser fir (*Abies fraseri*) and the hardwood yellow birch (*Betula alleghaniensis*), with the occasional occurrence of high-elevation 'beech gap' (*Fagus grandifolia*) forests. At high elevations in the western part of the park, northern hardwoods dominate on north-facing and cooler landscape positions, with high-elevation oak forests often occupying south-facing and warmer landscape positions. Historically, American chestnut (*Castanea dentata*) was a frequent dominant or subdominant in all of these forests except for the highest elevation spruce-fir forests.

10.2.3 Natural Disturbance

Both before and after federal protection in 1934, the landscape of GSMNP experienced a wide variety of natural and anthropogenic disturbances, which collectively affected all parts of the landscape (Harmon et al. 1983). However, these disturbances varied widely in frequency, size, and magnitude. In this and the following section, we briefly review natural and anthropogenic disturbances, with special reference to disturbances that cause tree mortality, thus causing structural and compositional variation and initiating successional dynamics, and those disturbances that can be mapped or otherwise quantified at the landscape scale.

Natural disturbances that cause the mortality of dominant trees and produce abrupt and discrete change in GSMNP ecosystems include windstorms, fire, insect damage (most notably the southern pine beetle), ice storms, debris slides, and flood scour (see Oak et al. Chap. 3; Nowak et al. Chap. 4; Peterson et al. Chap. 5; Grissino-Mayer Chap. 6; Lafon Chap. 7; Leigh Chap. 8; Wooten et al. Chap. 9). The most common form of natural disturbance in park forests consists of small canopy gaps where individual trees have succumbed to wind, ice, insects, fungi, or drought (White et al. 1985; Runkle and Yetter 1987; Elliott and Swank 1994; see Hart Chap. 2; Oak et al. Chap. 3). Of these, we know most about the gap dynamics in mesic forests (specifically cove and spruce-fir forests). Although there are instances of disturbance from tornadoes and tropical depressions in the park (Peterson et al. Chap. 5), most windstorm damage consists of patches of relatively small blowdowns, ranging from several to perhaps 1–25 tree stems (Runkle 1985; Busing 1993; Busing and White 1993). These patches are scattered (at any one time) across the park's landscape and have occurred with a frequency of about 0.5–3 % of the canopy stems per year (Runkle 1985; Busing 1993). These disturbances have been studied at the stand scale but have not been mapped for GSMNP at the landscape scale.

The second most important type of natural disturbance in the park is lightning-ignited fire and its interaction with mortality to mature pine caused by the southern pine beetle, a native insect (Barden and Woods 1976; Harmon 1982; Cohen et al. 2007; Flatley et al. 2013). Evidence of pre-park fires can be derived from fire scars (Harmon 1982; Flatley et al. 2013), charcoal (Welch 1999), and the structural and compositional characteristics of current stands (Harrod et al. 2000). However, understanding the spatial and temporal patterns of historical lightning fire in the park is complicated by centuries of overlay with pre-park fires set both by Native Americans and later European settlers (Barden and Woods 1976; Fowler and Konopik 2007). Since park establishment, lightning-ignited fires have exhibited a pattern of increasing frequency from wetter to drier sites along topographic gradients and decreasing frequency with elevation, particularly above 1,220 m (1940–1979, Harmon 1981). These fires have tended to be warm-season fires (Harmon 1981; Cohen et al. 2007) and, unsuppressed, tend to burn over larger areas and for longer periods of time than suppressed fires (Cohen et al. 2007). Cohen et al. (2007) reported 140 lightning-ignited fires (128 suppressed, 12 unsuppressed) for GSMNP between 1940 and 2006, or approximately 2.1 fires per year. Mean (±SE) annual number of lightning-ignited fires in the surrounding National Forests of the Blue Ridge Mountains ecoregion ranges from 2.0 ± 0.3 to 4.2 ± 0.7 per 2,000 km^2 (1970–2013, Greenberg et al. Chap. 1, Table 1.6).

Other natural disturbances are sometimes intense but are spatially less extensive. Ice storms are a phenomenon of middle to high elevations, usually causing damage to tree crowns and occasionally resulting in treefall (Lafon 2006; Butler et al. 2014; Lafon Chap. 7). Like windstorms, these disturbances are not easily mapped. Intense rainstorms cause debris slides at high elevations and flood scour along stream channels. Debris slides occur where topography causes water to funnel into the steep headwaters of streams, and the sudden downslope fall of soils and vegetation leaves open bedrock behind (Clark 1987; Wooten Chap. 9). Downstream, stormwater and flood-borne debris often scour the sides of the rocky stream channels (see Leigh Chap. 8). Debris slides and flood scour are intense but spatially restricted in the park landscape, affecting a very small percentage of park vegetation. Debris slides are documented in the park's surficial geology map (USDA Natural Resources Conservation Service 2009), and effects of stream scour are documented indirectly in the GSMNP vegetation classification (Madden et al. 2004) by yellow birch-dominated forests along streams.

10.2.4 Anthropogenic Disturbance

Before European settlement of the southern Appalachians, Native American villages were located in major valleys and floodplains, and human activities on surrounding mountain slopes likely consisted of hunting and gathering on upper slopes and ridgetops as well as intentional use of fire in particular topographic settings (Delcourt and Delcourt 1997). Concentrated European settlement occurred between the late 1700s and 1930 and was generally restricted to low-elevation (below 750 m), productive valley-flat sites and adjacent forest areas (Pyle 1988). Diffuse settlement and related human activities, such as early-style logging, gathering, livestock grazing, and use of low-intensity fire, extended to the areas surrounding concentrated settlement and were pervasive in the western end of the park (Pyle 1985). Livestock grazing, especially for the hard mast in the late summer and fall, occurred in pastures and forests around settlements. Livestock was also often taken to high-elevation summer pasture on grassy balds (Pyle 1985).

In the early to mid-1800s, logging was not mechanized and targeted only the most valuable species and sizes of trees (not the largest, not the smallest) (Pyle 1985). Logging companies soon began buying larger tracts of land, however, and built large logging camps and railroad lines, so that between about 1880 and 1930, corporate logging was a major disturbance to the park landscape, sometimes affecting entire watersheds and usually less selective in terms of species and tree sizes. Slash fires sometimes followed logging, and on some sites, these fires combined with soil erosion after rainstorms, causing delayed or absent regeneration of forests (some sites remain without a continuous forest cover 90 years after logging).

Park establishment ended the human-set fires of settlement and logging. The fire regime since 1934 has been dominated by arson and accidental fire with some lightning fire as well (Harmon 1981). However, reduction of human-set fires related to settlement and suppression of both human- and lightning-ignited fires has likely reduced fire frequency (Harmon 1982; Flatley et al. 2013) and extent (Cohen et al. 2007) in historically fire-maintained areas of the park. Between 1940 and 1979, the mean (±SD) annual number of human-set fires in GSMNP was 13.3 ± 10.02 (Harmon 1981). These fires tended to originate in more mesic areas and at lower elevations than lightning-ignited fires during the same period, and human-set fire frequency peaked in April and November (Harmon 1981). Similarly, for 1930–2003, Flatley et al. (2011) found a mean (±SD) annual number of anthropogenic fires of 5.5 (4.9) per 1000 km^2 (approximately half the area of the park). Although these fires may be more likely to originate in more mesic areas, Flatley et al. (2011) found no significant relationship between slope position and percent of total area burned. Mean (±SE) annual number of human-set fires in the surrounding National Forests of the Blue Ridge ecoregion ranges from 16.1 ± 1.5 to 73.2 ± 7.8 per 2,000 km^2 (1970–2013, Greenberg et al. Chap. 1, Table 1.6). Wildland fire use policy and prescribed fires have been introduced to the park during the last two decades (Cohen et al. 2007).

About 20 % of the park area has no history of direct anthropogenic disturbance (Pyle 1985, 1988). However, during the twentieth century, the remnants of undisturbed stands have been impacted by a wide range of human-mediated disturbances that have caused major canopy disturbance, including exotic pest introductions and air pollution (acid and nitrogen deposition, ozone). In terms of dominant tree cover, the effects of four exotic pest species have been particularly dramatic: chestnut blight (*Cryphonectria parasitica*; 1920s to the present, with most dominant trees affected by the 1930s and dead by the 1950s) (Woods and Shanks 1959), balsam woolly adelgid (*Adelges piceae*; 1960s to the present, with most Fraser fir populations drastically impacted by 1990) (Eagar 1984), beech bark disease (an insect-fungus disease complex initiated by *Cryptococcus fagisuga*; late 1980s to the present, with high mortality of mature American beech trees) (Vandermast 2005; Taylor 2012), hemlock woolly adelgid (*Adelges tsugae*; early 2000s to the present) (Roberts et al. 2009; Onken and Reardon 2010; Krapfl et al. 2011), and European wild boar (*Sus scrofa*; rooting impacts concentrated in mesic forests and floodplains) (Bratton 1975). Air pollution mostly affects tree growth rates and not mortality (e.g., Johnson et al. 1992). Deposition of nitrogen and sulfur compounds through acid rain and fog increases with elevation and cloud exposure and therefore probably has greatest impact on forests of the highest elevations (Aneja et al. 1992). Air quality regulation greatly lowered sulfur deposition by the 1990s. Near-ground ozone exposure affects species susceptible to leaf damage, such as yellow buckeye (*Aesculus flava*) and black cherry (*Prunus serotina*) (Chappelka et al. 1999).

10.3 Data and Methods

10.3.1 The 1930s Forest Plots

From 1934 to 1938, Frank Miller led a park-wide vegetation survey to map and classify the park's forest types. As the central focus of this effort, vegetation was sampled in 1378 plots in a grid across the entire park, although some watersheds were sampled more or less intensively than others. Some areas in the southwestern part of the park were not sampled because they were outside the proposed park boundary at that time. In each rectangular plot of 1 × 2 chains (20.12 m × 40.23 m), all live and dead trees greater than 4 in. (10.16 cm) in diameter at breast height (dbh) were identified to species, recorded as live or dead, and tallied in four size classes (converted to metric here): 10.16 to <30.48 cm, 30.48 to <60.96 cm, 60.96 to <91.44 cm, and ≥91.44 cm. Site environmental variables were recorded in the field, including but not limited to topographic variables such as elevation, slope steepness, and slope aspect. Surveyors noted whether or not plots had been burned or cut over, recording the known or estimated date of last burn or logging where possible. Plot locations were described on the data sheets, marked with an X on USGS topographic quadrangle maps, and later verified by Miller and his surveyors. In the 1980s, personnel at the Uplands Field Research Laboratory (UFRL) of GSMNP digitized the plot data and transferred the plot locations by hand onto modern USGS topographic quadrangle maps. UFRL personnel annotated the digital dataset with calculations of live and dead BA (m^2 per ha, based on geometric mean diameter for each size class) and density (stems per ha) for each species in each plot (see MacKenzie and White 1998 for further description of the Miller plot data). In 2004, Tuttle digitized the plot locations and, more recently, matched these locations to the plot data and performed extensive data cleaning, including assessing the accuracy of plot locations using the field data and GIS.

The Miller plot data have been described or used in several previous studies. MacKenzie and White (1998) summarized the Miller data for all plots with BA ≥8 and ≤110 m^2 per ha (950 plots) and classified the data into community types using two-way indicator species analysis, a hierarchical divisive clustering method. Busing et al. (1993) summarized the composition and structure of the northern hardwood—spruce-fir ecotone in the 1930s using a subset of the Miller plots. Although the Miller plots were not permanently marked, several studies have compared subsets of the Miller data to more recent, comparable plot samples to investigate change over time for specific geographic regions, portions of environmental gradients, or forest types in GSMNP (Walker 1978; Harrod 1999; Knebel 1999; Vandermast 2005; Tuttle 2007). Our study is the first to attempt a park-wide comparison of the Miller data to more recent datasets. Because we were interested in variation across types of disturbance, we selected all plots matched to a location and with at least one recorded tree ≥10 cm dbh for our study (1,284 plots). We excluded two plots that were extreme outliers in our initial analyses, yielding 1,282 plots for comparison. The Miller plot data reflect the standing-dead American chestnut trees on the landscape at that time; chestnut blight was in progress, but live American chestnut trees still dominated the landscape. In this study, we treated all standing-dead American chestnut trees as live, to obtain a better estimate of the true dominance of this species in historic GSMNP forests. As a result, any early responses to chestnut blight that had begun may create some inflation of BA or density in our results. However, upon reviewing the history of the progression of chestnut blight in GSMNP (Woods and Shanks 1959), we believe that any such initial responses would have minimal effect on our results.

10.3.2 The 1990s–2000s Forest Plots

Using lists or databases of GSMNP forest datasets compiled by White and Busing (1993), Fridley (unpubl. database), and GSMNP personnel, we identified all plot-based datasets that included stem diameter and identification of species for individual trees. Of the stem data that we obtained, representing a variety of research and monitoring studies between the 1960s and 2010, we selected plots from the 1990s and 2000s to maximize time since the 1930s as well as coverage across the park. Most datasets include field environmental data collected for each plot, including topographic variables such as elevation, slope steepness, and slope aspect. After data cleaning, including assessing the accuracy of plot locations using field data and GIS, we developed a compilation of 490 plots for comparison to the Miller data. These plots occur in 17 out of 22 park watersheds, but the majority of plots are concentrated in three areas: the Abrams Creek watershed on the western end of the park, the Cataloochee Creek watershed on the southeastern end of the park, and the high-elevation spruce-fir zone in the central portion of the park. Most plot sizes in our compilation are 20 m × 50 m or 20 m × 40 m (similar in size to the Miller plots), with the exception of 20 m × 20 m plots in the spruce-fir zone and a few larger plots scattered elsewhere. We limited our compilation dataset to trees ≥10 cm dbh to match the Miller data, and we calculated BA (m^2 per ha) and density (stems per ha) for each tree species in each plot.

10.3.3 Disturbance History and Environment

We used several pre-existing, spatially referenced datasets of GSMNP land use and fire history to assign anthropogenic disturbance and fire history categories to each plot in our two datasets. Miller's final vegetation maps, which were recently digitized and georeferenced by GSMNP personnel, show areas that were cut or burned, as determined by synthesis of his extensive vegetation survey data as well as historical records of logging, land-use, and forest conditions. Pyle (1985, 1988) extensively researched, compiled, and mapped records of pre-park anthropogenic disturbance (mostly by Europeans) in several categories, including corporate logging, concentrated settlement, livestock grazing, diffuse disturbance (regions with small, scattered areas of settlement, cutting, grazing, or fire), and areas with no known or recorded anthropogenic disturbance (hereafter referred to as undisturbed, although these plots would have been subject to natural disturbance). Pyle (1985) also compiled and mapped records of pre-park fire, largely consisting of post-logging slash fires but also areas with records or evidence of diffuse, smaller fires. Pyle's compilation included evidence of cutting or fire recorded on the Miller plot data sheets and on Miller's final park vegetation maps. Miller's final vegetation maps show additional areas of cutting or burning not included in Pyle's generalization. A GIS layer of disturbance history similar to Pyle's maps exists in the GSMNP geospatial database, but it differs from Pyle's maps in important ways, and the layer is currently of unknown provenance. We scanned and georeferenced maps from Pyle (1985) and used both these maps and the digitized Miller maps as a guide to reclassify or modify polygons in the GIS disturbance history layer to match Pyle's maps and categories (Fig. 10.2). We combined several fire history datasets for GSMNP into a GIS layer of all known, mapped areas of burn before park establishment (Fig. 10.2), including evidence of prior burn from the 1930s plot data sheets, burned polygons from the digitized Miller vegetation map and the GSMNP fire history layer, and burned areas from Pyle's maps (Pyle 1985).

Whittaker (1956), in his classic work describing vegetation- and species-environment relationships in GSMNP, identified two dominant environmental gradients controlled by topography: elevation as a surrogate for temperature and moisture and topographic position, shape, and exposure as factors affecting moisture availability. Digital terrain variables derived from digital elevation models (DEMs) are now widely used to represent regional, watershed, and local topographic controls on microclimate and water drainage (Moore et al. 1991; Fridley

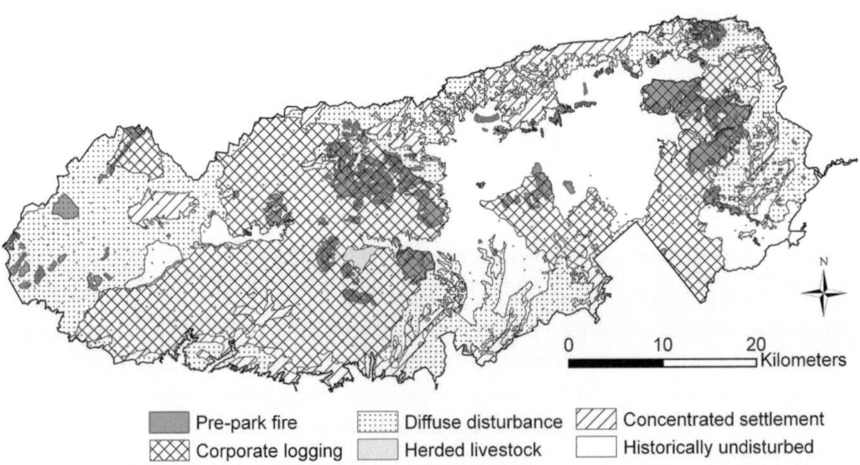

Fig. 10.2 Pre-park disturbance history and fire in GSMNP, based on Pyle's compilation (1985), the Miller survey, and the park's fire history database. Irregular dots, including in undisturbed areas, represent locations of Miller survey plots for which evidence of fire was recorded

2009), and many studies of southern Appalachian forests have quantified the relationship between terrain variables and forest structure and composition (e.g., Bolstad et al. 1998; Simon et al. 2005). These variables often overlap in their representations of microclimate and soil water balance and differ in the extent to which they isolate or integrate the effects of particular microclimatic or hydrologic processes; as a result, terrain variables are often correlated, and sets of multiple, sometimes interacting terrain variables often best represent the aggregate topographic effects on site conditions in vegetation models (e.g., Simon et al. 2005).

For this study, we assembled a set of terrain variables from a combination of field data and data layers developed by GSMNP personnel and derived originally from a hybrid LiDAR- (primarily) and National Elevation Dataset-based, 3-m-resolution DEM. Terrain variables considered in our analyses include elevation; slope aspect, steepness, and curvature; potential annual solar radiation (RAD) (in watt hours per m^2); topographic wetness index (TWI) (Beven and Kirkby 1979; Moore et al. 1991); topographic position index (TPI) (Weiss 2001); terrain shape index (TSI) (McNab 1989; Bolstad et al. 1998); and landform index (LFI) (McNab 1993). Topographic wetness index represents relative moisture drainage at a site as a function of upslope contributing area (drainage input) and local slope steepness (drainage output); higher values of TWI imply greater relative moisture balance. Topographic position index represents relative slope position as the difference between a site's elevation and the mean elevation of surrounding sites; negative values imply lower slope or valley-bottom positions, positive values imply upper slope or ridgetop positions, and values near zero imply flat or midslope positions. Terrain shape index represents local land surface shape (surface curvature gradient) as the mean slope gradient between the center of the plot and the plot boundary; low and high values correspond to concave and convex sites, respectively. Landform index is the average of the horizon angle in eight directions and ostensibly integrates several topographic factors that affect water and air drainage as well as exposure to (or shelter from) radiation and winds; negative or low values correspond to ridgetop positions that are not surrounded by higher ridges, and high values correspond to valley-bottom positions surrounded by steep slopes.

For slope steepness and aspect, we used values from field data where available. For missing field values and for all other terrain variables besides LFI, we extracted values for each plot location from 10-m (elevation) or 30-m (all others) spatial grids derived by GSMNP personnel from the original 3-m LiDAR DEM. We used the Whitebox Geospatial Analysis Tools software (Lindsay 2012) to derive LFI from the LiDAR-derived 30-m DEM. For several of these DEM-derived variables, values extracted from the 30-m layers performed better in our preliminary analyses than 10-m values (where available or where separately derived by us). In addition, we reasoned that values derived from the fine-scale LiDAR DEM but rescaled or resampled to 30-m resolution more accurately represented average site conditions for the plot sizes in our dataset, most of which measured 20 m × 40 m or 20 m × 50 m. For LFI, the longer computation time for derivation from a 3-m or 10-m DEM was prohibitive and would have provided little benefit over the 30-m version for this variable based on watershed-level topography.

10.3.4 Dataset Matching

Because the 1990s–2000s plots were concentrated in three areas of the park and originated from different site selection strategies, differences in the spatial and topographic distributions of the two datasets could bias our comparisons. To avoid spurious comparisons, we used Optmatch, an optimal matching approach developed for comparative observational studies (Hansen and Klopfer 2006), to match the 1930s plots ('controls') to 1990s–2000s plots ('treatment'). Optmatch finds an optimal matching solution for two datasets by minimizing multivariate distances between plots based on user-specified matching criteria. Previous vegetation studies in GSMNP and preliminary statistical modeling of our datasets indicated that anthropogenic disturbance history, elevation, LFI, TPI, RAD, and park region were important predictors of forest structure and composition. We categorized the park into geographic regions similar in climatic, geologic, and topographic setting, approximately representing northwest, southwest, northeast, and southeast quadrants, with the highest elevations in the central and eastern part of the park combined as a single region. We then limited valid matches to plots within the same region, disturbance history category, elevation class (<600 m, 600–900 m, 900–1,200 m, 1,200–1,600 m, and >1,600 m), and LFI class (<9.5, 9.5–16.5, and >16.5). We specified elevation (<100-m difference between plots), LFI, TPI, and RAD as matching criteria, selected Mahalanobis distance as the metric to optimize, and specified a matching ratio for number of 1930s plots matched to 1990s–2000s plots, to further exclude poor matches resulting from the large number of 1930s plots relative to the 1990s–2000s plots. A matching ratio of ≤3:1 produced the best combination of matching and sample size, yielding 529 plots from the 1930s and 460 plots from the 1990s–2000s for comparison. Box plots used to compare the distribution of the matched datasets on the continuous matching criteria (elevation, LFI, TPI, and RAD) indicated that the datasets were well matched within combinations of elevation and LFI classes (environmental/topographic 'zip codes', Dean Urban pers. comm.; Jobe 2006), with slight offsets of distribution for some variables in three infrequently sampled zip codes with low abundance on the landscape; low sample sizes did not permit further improvement of these matches.

10.3.5 Statistical Modeling of Structure and Composition

To evaluate differences in the drivers of forest structure between the two time periods, we modeled BA and density for the full, unmatched dataset in each time period as a function of disturbance history category, topography, and pre-park fire. Exploratory generalized additive models (GAM) of BA and density for each time period revealed that some predictors, particularly elevation, exhibited a nonlinear relationship with BA or density; however, once significant interactions between predictors were taken into account, relationships with individual predictors were (approximately) linear, and we subsequently used linear regression. For disturbance history categories, we chose to perform contrasts against the most disturbed category, which we hypothesized to be corporate logging, to highlight significant differences between disturbance categories that might not be apparent if all disturbance categories were contrasted against undisturbed. Model selection was guided by forward and backward stepwise selection by AIC. Because our models are not intended for prediction, we compared the final models in terms of overall variance explained as well as the identity, sign, and relative importance of significant predictor variables. Relative importance for predictors in each model was calculated using the Lindeman, Merenda, and Gold (LMG) variance decomposition method available in the relaimpo package for R statistical software (Groemping 2007; R Core Team 2014).

We used adonis in the R package vegan (Oksanen et al. 2013) to model multivariate, distance-based differences in species composition between plots in each time period, again using the full, unmatched dataset, as a function of disturbance history, fire, and topographic gradients. Adonis is a method for permutational multivariate analysis of variance using distance matrices that allows the use of semimetric distance matrices, such as those commonly used in ecological studies (Anderson 2001). To attempt to isolate tree species composition from structure, we used Bray-Curtis dissimilarity of plots based on species' *relative* BA within plots as our distance measure. Because adonis can identify significant relationships resulting from difference in either location (mean) or dispersion (variance), we used betadisper in the Vegan package (Oksanen et al. 2013) to check for multivariate homogeneity of multivariate dispersion among groups in the datasets (Anderson 2006). This procedure confirmed that the significant predictors in our models reflected differences in location rather than dispersion. Differences in the overall drivers of forest composition between time periods were evaluated by comparing the final adonis models in terms of the identity and importance of significant predictor variables.

10.3.6 Comparisons of Structure and Functional Composition by Disturbance History

To determine the overall effects of pre-park and ongoing anthropogenic disturbance on forest structure, we calculated mean and coefficient of variation (CV) for BA, density, and QMD in each time period for the following groups: all plots, undisturbed plots, disturbed plots, and plots in each disturbance history category, excluding categories with sample size ≤ 10. We compared these metrics for unmatched and matched datasets to assess the effects of matching on results and interpretation. However, further analyses were based on comparison of the matched datasets to minimize the influence of regional and topographic biases in the 1990s–2000s dataset on our results. Histograms of BA, density, and QMD for disturbed and undisturbed subsets of the matched datasets were used to compare the landscape distribution of structural characteristics in these samples. Finally, we evaluated the effects of disturbance on functional composition by first categorizing species into functional groups (Table 10.1) and then plotting mean and standard error of BA for each functional group in each disturbance history category. Species were assigned to functional groups based on disturbance-related functional characteristics described in the literature (USDA Forest Service 2014; Burns and Honkala 1990; Ninemets and Valladares 2006; see Wilfahrt et al. Chap 11), such as shade tolerance, drought tolerance, fire sensitivity or response, reproduction strategy, growth rate, and response to release.

Table 10.1 Functional groups used for the tree species of GSMNP. Species were assigned to groups based on disturbance-related functional characteristics: shade tolerance, drought tolerance, fire sensitivity or response, growth rate, and regeneration/recruitment strategy

Group	General characteristics and list of dominant species in group
S	Late-successional and extremely shade-tolerant species of mostly mesic sites, as well as understory gap responders
	Slow growth rates and abundant reproduction in shady environments
	Dominant species in group: *Abies fraseri, Acer pensylvanicum, A. saccharum, A. spicatum, Aesculus flava, Amelanchier sp., Fagus grandifolia,Ostrya virginiana, Picea rubens, Tilia sp., Tsuga canadensis*
V	Vigorous disturbance responders
	Shade intolerant (most species) or capable of rapid, abundant response to disturbance in colonization, sprouting, and growth rates
	Dominant species in group: *A. rubrum, Liquidambar styraciflua, Liriodendron tulipifera, Pinus strobus, Prunus pensylvanica, Robinia pseudoacacia, Sassafras albidum, Sorbus americana, Fraxinus americana*
M	Moderate disturbance responders
	Intermediate shade tolerance, abundant response to disturbance
	Dominant species in group: *Betula alleghaniensis, B. lenta, Halesia tetraptera, Magnolia acuminata, M. fraseri, P. serotina*
O	All oak species
	Drought tolerant, intermediate shade tolerance, slow dispersal and growth, sprouting after disturbance
	Dominant species in group: *Quercus alba, Q. coccinea, Q. montana, Q. rubra, Q. velutina*
D	Slow-growing, drought-tolerant species
	Intermediate shade tolerance
	Dominant species in group: *Carya cordiformis, C. glabra*, C. *tomentosa, Nyssa sylvatica,Oxydendrum arboreum*
Y	All yellow pine species
	Shade intolerant, extremely drought tolerant, strong responders to large canopy disturbance or intense fire if a seed source is nearby
	Dominant species in group: *P. echinata, P. pungens, P. rigida, P. virginiana*
C	*Castanea dentata*
	Similarities in niche and functional traits to oak species as well as disturbance responders such as *A. rubrum*
	Placed in its own group because of its former wide dominance on the landscape and the pervasive impact of its decimation by chestnut blight

10.4 Results

10.4.1 Influence of Disturbance History and Environment on Structure and Composition

Topography, anthropogenic disturbance history, and pre-park fire explain between 12.6 % and 24.8 % (adjusted R^2 11.5–24.3 %) of the variance in density and BA across the two time periods (Table 10.2). Different approaches for determining relative importance of regression predictors can yield different results, so the LMG variance decomposition results here should be interpreted with caution; however, our rankings according to LMG (Table 10.2) illustrate several interpretable changes in the variables that best explain how GSMNP forests are structured.

Disturbance category and elevation were the most important predictors of both BA and density in the 1930s (Table 10.2). The least disturbed areas (diffuse disturbance and undisturbed) have higher BA and density than the highly disturbed areas (corporate logging and settlement) with widespread clearing in various stages of

Table 10.2 Significant predictors, sign of relationship, and relative importance (LMG variance decomposition) of predictors for linear regression models of BA and density, for the unmatched datasets. For disturbance classes, contrasts were performed against corporate logging. For pre-park fire, burned plots (1) were contrasted with unburned plots (0)

	BA				Density			
	1930s		1990s–2000s		1930s		1990s–2000s	
Predictor	Sign	Decomp. R^2	Sign	Decomp. R^2	Sign	Decomp. R^2	Sign	Decomp. R^2
Disturbance history class		0.154		0.028		0.037		0.041
Diffuse disturbance	+		+		+		–	
Undisturbed	+		+		+		–	
Settlement			+					
Pre-park fire (1)	–	0.020			–	0.011		
Elevation (m)	+	0.067	–	0.013	+	0.037		
LFI	–	0.001	–	0.049	+	0.012	–	0.041
TPI	–	0.005	–	0.013			+	0.043
TWI			+	0.010	+	0.004		
TSI			+	0.015				
RAD					–	0.004		
Elevation x LFI			+	0.029				
Elevation x TPI			+	0.009				
LFI x TPI	+	0.002						
LFI x TWI					–	0.024		
Total R^2		0.248		0.168		0.128		0.126
Adjusted R^2		0.243		0.149		0.122		0.115

regrowth. Pre-park fire and topography are also important predictors of 1930s BA and density. In the 1990s–2000s, disturbance category remains important but ranks lower as a predictor of vegetation structure (Table 10.2). Basal area remains higher for the least disturbed areas, and settlement areas have accumulated significantly more BA than areas of corporate logging. Elevation and topography, particularly LFI, are relatively more important predictors of BA than disturbance for the 1990s–2000s. Density of the least disturbed areas also remains significantly different from corporate logging areas in the 1990s–2000s. However, the sign of this relationship has reversed from the 1930s, with highly disturbed areas exhibiting a greater abundance of stems. Elevation is no longer an important predictor of density for the 1990s–2000s, and the relationship of topography to density in the 1990s–2000s appears simple: overall, density is higher on ridgetops (high TPI) and lower in valleys and coves (high LFI).

Adonis models of compositional dissimilarity reveal differences between time periods similar to those for BA and density (Table 10.3). While disturbance categories and pre-park fire are significant predictors in both time periods, the relative importance of pre-park fire in structuring composition decreases in the 1990s–2000s. Not surprisingly, elevation, the dominant environmental gradient in the park, is the most important predictor of composition for both time periods, and elevation more strongly determines the composition in the 1990s–2000s than in the 1930s. While disturbance category and fire are more important than individual topographic variables in the 1930s model, topographic variables emerge as equally important predictors in the 1990s–2000s model. Overall, the variance explained for compositional dissimilarity nearly doubles from the 1930s to the 1990s–2000s.

10.4.2 *Effects of Disturbance History on Structural Range of Variation*

Interestingly, the results for the matched datasets (Table 10.4b) are strikingly similar to the results with no matching (Table 10.4a), and we limit our discussion to results for the matched datasets (Table 10.4b). Mean BA for undisturbed plots shows little difference between the two time periods. In all disturbance categories, mean BA has increased from the 1930s to the 1990s–2000s, approaching the value for undisturbed plots. The highly disturbed areas show the greatest increases in mean BA, nearly doubling for corporate logging and more than tripling for settlement areas. Similarly, mean density has increased for all disturbance categories, nearly doubling for diffuse disturbance and more than doubling for corporate logging and settlement areas. Mean density in all disturbance categories has surpassed undisturbed mean density for both 1930s and 1990s–2000s, by 40.7–66.8 % and 12.2–25 %, respectively. In contrast to BA, mean density for undisturbed plots has increased by approximately one-third. Mean QMD is similar across time periods for all disturbance categories, with the exception of an increase of 3.9 cm for settlement. For undisturbed plots, mean QMD shows a notable decrease of 5.6 cm.

Table 10.3 Adonis model results for the unmatched datasets using relative BA, for (a) the 1930s dataset and (b) the 1990s–2000s dataset

	Df	Sum of squares	Mean square	F	R^2	Sig.
(a)						
Elevation (m)	1	32.63	32.63	95.50	0.064	0.01**
Disturbance history class	6	13.73	2.29	6.70	0.027	0.01**
Pre-park fire	1	9.38	9.39	27.47	0.019	0.01**
LFI[a]	1	7.68	7.68	22.48	0.015	0.01**
TPI[b]	1	4.48	4.48	13.12	0.009	0.01**
RAD[c]	1	2.33	2.33	6.81	0.005	0.01**
LFI x TPI	1	1.86	1.86	5.44	0.004	0.01**
TPI x RAD	1	0.59	0.59	1.73	0.001	0.06.
Residuals	1,268	433.25	0.34		0.856	
Total	1,281	505.94			1	
(b)						
Elevation (m)	1	35.70	35.70	127.39	0.186	0.01**
Disturbance history class	6	8.30	1.38	4.94	0.043	0.01**
Pre-park fire	1	2.24	2.24	8.01	0.012	0.01**
LFI[a]	1	6.11	6.11	21.81	0.032	0.01**
TPI[b]	1	2.36	2.36	8.43	0.012	0.01**
RAD[c]	1	1.84	1.84	6.55	0.010	0.01**
TWI[d]	1	0.24	0.24	0.84	0.001	0.59
LFI×TPI	1	0.60	0.60	2.15	0.003	0.02*
LFI×RAD	1	0.74	0.74	2.63	0.004	0.01**
LFI×TWI	1	0.67	0.67	2.37	0.003	0.05*
Residuals	474	132.83	0.28		0.693	
Total	489	191.62			1	

[a]Landform index
[b]Topographic position index
[c]Potential annual solar radiation
[d]Topographic wetness index
Significance codes: ***0.001, **0.01, *0.05, . 0.1

Combining these structural characteristics yields four structural groups with, on average: (1) high BA, moderate density, and high QMD (1930s undisturbed); (2) low BA, low density, and low QMD (1930s disturbed); (3) high BA, high density, and moderate QMD (1990s–2000s undisturbed); and (4) moderate BA, high density, and low QMD (1990s–2000s disturbed).

Variation around mean BA and density, as represented by CV, shows marked differences across time periods and between disturbed and undisturbed categories (Table 10.4b). These differences are best understood by examining histograms of plot BA, density, and QMD for disturbed and undisturbed categories in each time period (Fig. 10.3). Basal area for 1930s undisturbed plots (Fig. 10.3a) exhibits a broad peak in the 10–60 m^2 per ha range with few plots <10 m^2 per ha and positive skew with a long tail of high-BA plots, up to nearly 190 m^2 per ha. The 1930s

Table 10.4 Comparison of BA, density, and QMD in each time period for (a) unmatched datasets and (b) matched datasets

Unmatched datasets	1930s						1990s–2000s					
Disturbance class	n	Mean BA (m^2/ha)	CV	Mean density (stems/ha)	CV	Mean QMD (cm)	n	Mean BA (m^2/ha)	CV	Mean density (stems/ha)	CV	Mean QMD (cm)
(a)												
All plots	1,282	21.4	0.89	239.6	0.71	32.6	490	34.3	0.47	489.0	0.52	31.6
Undisturbed – no fire, logging, or settlement	241	37.9	0.63	313.4	0.85	40.5	180	37.7	0.49	448.3	0.65	34.5
All disturbed	1,041	17.6	0.88	222.5	0.60	30.7	310	32.3	0.43	512.6	0.45	29.9
No logging or settlement, burned	54	24.8	0.53	240.9	0.45	35.9	1	—	—	—	—	—
Diffuse disturbance	428	21.0	0.82	240.4	0.50	32.4	192	32.5	0.46	479.5	0.25	31.2
Herded livestock	3	—	—	—	—	—	10	—	—	—	—	—
Corporate logging	483	14.4	0.92	206.5	0.69	29.2	76	30.9	0.38	556.1	0.46	27.6
Concentrated settlement	73	12.7	0.94	204.6	0.68	27.4	31	36.1	0.41	554.5	0.38	29.4
(b)												
Matched datasets	1930s						1990s–2000s					
Disturbance class	n	Mean BA (m^2/ha)	CV	Mean density (stems/ha)	CV	Mean QMD (cm)	n	Mean BA (m^2/ha)	CV	Mean density (stems/ha)	CV	Mean QMD (cm)
All plots	529	24.1	0.91	267.8	0.80	32.8	460	34.2	0.47	486.4	0.53	31.7

Undisturbed – no fire, logging, or settlement	132	39.3	0.69	338.2	0.98	40.1	167	37.8	0.50	451.3	0.66	34.5
All disturbed	397	19.0	0.88	244.4	0.61	30.4	293	32.2	0.43	506.5	0.46	30.1
No logging or settlement, burned	3	—	—	—	—	—	1	—	—	—	—	—
Diffuse disturbance	250	21.2	0.86	253.4	0.51	31.5	188	32.4	0.45	475.8	0.46	31.3
Herded livestock	1	—	—	—	—	—	4	—	—	—	—	—
Corporate logging	123	15.8	0.86	230.2	0.81	28.9	73	31.1	0.38	564.1	0.45	27.5
Concentrated settlement	20	11.1	0.75	214.9	0.66	24.8	27	34.0	0.39	555.2	0.41	28.7

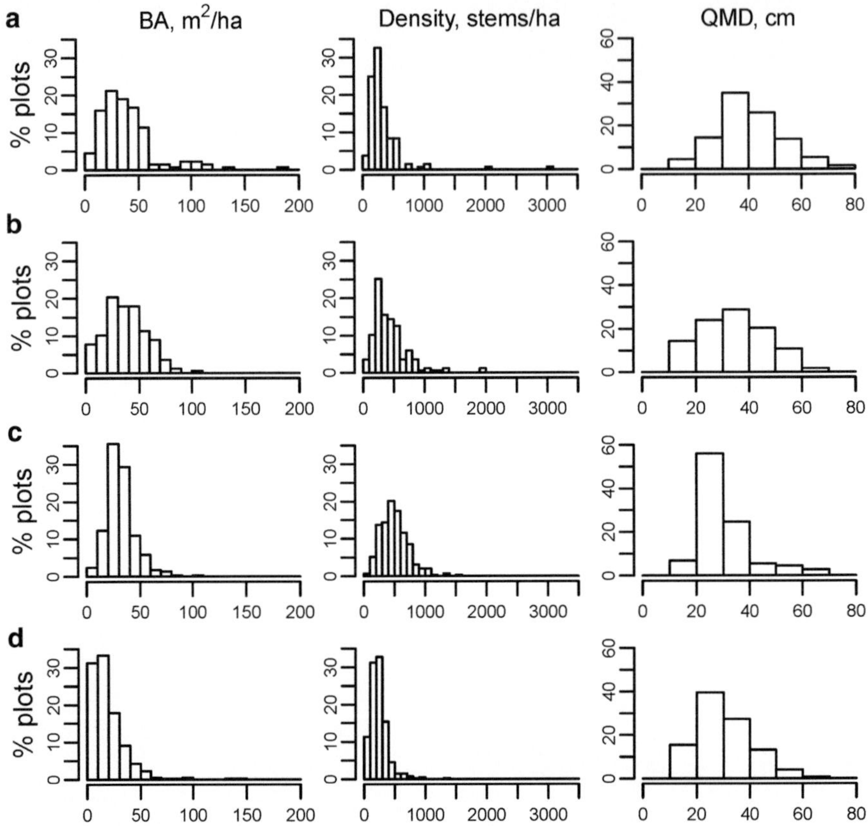

Fig. 10.3 Histograms (%) of plot BA, density, and QMD in the matched datasets for (**a**) 1930s undisturbed plots; (**b**) 1990s–2000s undisturbed plots; (**c**) 1990s–2000s disturbed plots; and (**d**) 1930s disturbed plots, arranged (top to bottom) from least anthropogenic disturbance to most anthropogenic disturbance and from longest time since disturbance to most recent disturbance

undisturbed density distribution (Fig. 10.3a) exhibits a narrow peak between 100 and 300 stems per ha with few plots <100 stems per ha and again positively skewed with scattered high-density plots up to nearly 3,100 stems per ha. The moderate and high CV for BA and density, respectively, reflects the scale, location, and positive skew of these distributions. Quadratic mean diameter for 1930s undisturbed plots (Fig. 10.3a) shows a unimodal, nearly symmetric distribution around a high mean QMD of 40.1 cm, with infrequent high QMD values up to 80 cm. However, while the 1990s–2000s undisturbed distributions are roughly similar in location (Fig. 10.3b), both BA and density distributions show changes in scale (broader peaks), less positive skew for BA with contraction of the higher BA plots, and some contraction of extreme high-density values but overall expansion of high-density plots. The corresponding QMD distribution is slightly positively skewed with a shift toward a lower mean QMD and a smaller range reflecting contraction of high-QMD plots.

For disturbed plots, the 1930s distributions exhibit narrow peaks of BA and density concentrated in the lowest classes and very few extreme high values; QMD is likewise positively skewed with a low mean (Fig. 10.3d). Of note, the 1930s disturbed density distribution is similar to the 1930s undisturbed distribution, although with a narrower peak that is more concentrated at lower densities; however, the strikingly different BA and QMD distributions paired with these similar density distributions sharply differentiate the two categories. In comparison to the 1930s disturbed plots, the 1990s–2000s disturbed BA distribution (Fig. 10.3c) exhibits a shift to a higher mean, and the BA peak remains narrow with positive skew. The density distribution has shifted to a broad peak at a higher mean that surpasses the 1930s undisturbed mean. The QMD distribution shows no change in mean from the 1930s and marked narrowing of the peak, with reductions in frequency of both low- and high-QMD plots.

10.4.3 Effects of Disturbance History on Functional Composition

Mean BA is low for all species groups across the disturbed categories and for several species groups at their elevational limits in the predominantly high-elevation undisturbed category (Fig. 10.4). American chestnut still dominated the 1930s

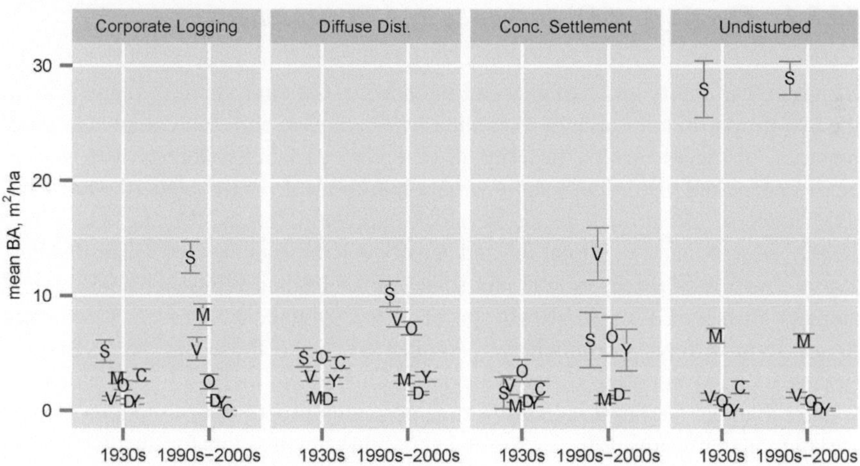

Fig. 10.4 Mean (±SE) BA for each species functional group in each time period, by disturbance history category. The herded livestock category is not shown because of low sample sizes. For species functional groups (see Table 10.1 for more detail), S = shade-tolerant and late-successional species, V = vigorous disturbance responders, M = moderate disturbance responders, O = all oak species, D = slow-growing, drought-tolerant species, Y = all yellow pine species, and C = American chestnut. American chestnut is absent in the 1990s–2000s for all categories except corporate logging

landscape (although much of it was already standing dead); mean BA for this single species was comparable to mean BA for functional groups made up of several species. In contrast, American chestnut is essentially absent from all categories in the 1990s–2000s dataset. In the undisturbed category, all species groups other than American chestnut exhibit no significant difference in mean BA.

The shade-tolerant species group increased significantly in mean BA for all disturbed categories, with the largest increase and dominance over other species groups in corporate logging areas as well as dominance in diffuse disturbance areas (Fig. 10.4). Vigorous disturbance responders increased dramatically in all disturbed categories and as a group are dominant or codominant in settlement and diffuse disturbance areas. Moderate disturbance responders increased and became a dominant species group in corporate logging areas. Average increases of this group in diffuse disturbance and settlement areas are moderate to low compared to increases for other species groups. Oaks as a group did not decrease in any category and increased significantly in diffuse disturbance and settlement areas. Slow-growing, drought-tolerant species dominated by hickory (*Carya*) species did increase in areas of diffuse disturbance and settlement but were of low mean BA and relatively unimportant for overall compositional change. As a group, yellow pine species did not decrease in mean BA in any category and exhibited a large, significant increase in settlement areas.

10.5 Discussion

10.5.1 Overlays of Disturbance and Environment

Sampling variability and variables not included in our analysis likely contribute to the low proportion of variance explained by the models and also complicate interpretation of model results. In terms of sampling variability, forests with similar environment and history will show variation at the scale of the plots (0.04–0.1 ha) used here. In old growth mesic hemlock-hardwood forests, Busing and White (1993) showed that CV of total BA and density was dependent on plot size, decreasing to 10–15 % of means only at plot scales of 0.5 ha, and Whittaker (1966) found similar values for CV in old growth spruce-fir forests at the 0.5 ha scale. Coefficient of variation in forests dominated by smaller trees and those that are even-aged (owing to stand initiating disturbances) would decrease to 10–15 % at plot sizes smaller than 0.5 ha, but it is likely that a substantial proportion of the unexplained variation in our results reflects expected variability among plots for the plot sizes in our samples.

Disturbance factors not available or excluded from our analysis include natural disturbance history at plot locations, time since pre-park anthropogenic disturbance, post-park fires, the discrete locations and types of disturbance in areas mapped as diffuse disturbance, and, for the 1990s–2000s models, unmapped post-park disturbances such as the decimation of American chestnut, American beech, and Fraser fir. In addition, the distribution of unmapped disturbances and unquantified severities is likely correlated with elevation and topography. For instance, the increase of BA and density with elevation in the 1930s may represent a range of factors, such as lower levels of diffuse disturbance at higher elevations or the correlation of most major disturbances with low elevation, rather than a true environmental gradient. Likewise, site topographic variables like LFI and TPI (and their interactions) as predictors of 1930s BA and density may represent unmapped variation in degree of disturbance from cove to ridge, such as preferential siting of agricultural fields and grazing areas or logging limited to accessible topographic positions or desirable forest types. Elevation and topography, particularly LFI, are relatively more important predictors of BA and density for the 1990s–2000s dataset compared to the 1930s dataset; once again, however, site topographic variables are likely correlated with such factors as fire suppression and the distribution and timeline of disturbance by exotic pests (via the niches of both the pests and host species). Nevertheless, the decreased importance of anthropogenic disturbance history (including human-set fires) combined with the increased importance of elevation and topography across the models for BA, density, and compositional dissimilarity indicates sorting of structure, species, and forest communities along environmental and disturbance gradients during 60–75 years of response to historic and recent disturbances.

In summary, a complicating factor in understanding the relative roles of disturbance and environment is that natural and anthropogenic disturbances are not independent of the environmental conditions that also affect species distributions, successional trajectories, and recovery times. Further, natural and historic anthropogenic disturbances overlap in spatial distribution in GSMNP but differ in timing, severity, and extent. For instance, catastrophic disturbances such as clear-cut logging in the high elevations (removal of the canopy over large areas), are inverse to the pattern of natural disturbance (low rates of canopy turnover through small gaps). Patterns of post-park anthropogenic disturbances further differ from historical disturbances. For instance, while decimation of American chestnut, Fraser fir, and American beech may follow environmental gradients related to niches of the host and pest species, these canopy disturbances have been widespread but diffuse (compared to widespread and intense logging) and have occurred over a relatively short time period (compared to natural-disturbance-related gap formation). The combined effects of widespread, major anthropogenic disturbances, particularly when multiple such disturbances occur on a relatively short time scale (within the same century or a few decades), can mask environmental gradients, including patterns along those gradients that were also shaped by natural disturbances.

10.5.2 Overlays of Disturbance Before and After Park Establishment

Our results for undisturbed forest in Miller's time represent the best estimate we have for structural variability in old growth, at least for forests above 900 m. Histograms for 1930s undisturbed plots appear to reflect a landscape dominated by natural disturbance and gap dynamics. In the absence of continued disturbance, we would expect the CV values and histograms for the 1990s–2000s undisturbed plots to be similar to results for the 1930s. However, while mean BA appears stable in these areas, the decrease in CVs, the increase in mean density, and the decrease in QMD likely reflect continued, widespread disturbance from several exotic pests that have removed the biggest trees and increased the number and size of canopy gaps on the landscape. These aggregate results for undisturbed areas should be interpreted with some caution, because some sampling biases and differences in environmental distribution that were not corrected by our plot matching method could be generating over- or underestimates of difference between our matched values. Specifically, the 1990s–2000s undisturbed plots were biased toward the highest elevations (≥1,600 m), while the 1930s plots were biased toward plots between 1,200 and 1,600 m elevation. Sampling bias in the design of several 1990s–2000s studies (toward plots containing Eastern hemlock), particularly in the 1,200–1,600 m range, likely biased our comparison toward higher 1990s–2000s mean BA; at the same time, the disproportionate number of 1990s–2000s plots ≥1,600 m likely biased our comparison to lower 1990s–2000s mean BA (from Fraser fir decline). Improved plot matching aimed at similar proportions of elevation and topographic settings across time periods would decrease the number of plots for comparison but could increase confidence in the aggregate differences between time periods.

We expected the 1990s–2000s disturbed distributions, in the absence of continued major disturbance, to show recovery over approximately 75 years of succession. While BA in historically disturbed areas has approached values in undisturbed areas, higher density and no change in QMD since the 1930s likely reflect the overlay of continued disturbance and changes in fire regime on ongoing recovery. Specifically, where disturbances resulted in even-aged stands, BA is distributed among more, but smaller stems; continued disturbances have removed large, canopy-dominant trees; and on drier sites, decreases in fire frequency and extent have allowed in-fill of stems (as also reported by Harrod et al. 1998). While we might expect that QMD would be slow to increase with the high stem density of early to middle succession, our results indicate that even as some historically disturbed plots may be thinning and shifting toward higher QMD, continued disturbance is reducing BA and increasing density, shifting plots back toward lower QMD.

While we might use the 1930s undisturbed histograms as a reference for the 1990s–2000s undisturbed plots, their use as a reference for the 1990s–2000s disturbed plot distributions may be limited because of the paucity of known undisturbed plots in the park's lower elevations, where disturbed plots are concentrated. Spatial and environmental matching of disturbed to undisturbed plots (where enough plots are available) would be necessary to make comparisons that account for differences in natural disturbance regimes (particularly fire) and species' characteristics along elevation and topographic gradients. Future examination of the diffuse disturbance plots alone might be instructive, in comparison to other disturbed and undisturbed plots (where enough matching plots are available). Even though the frequency and character of human-set fires might have differed from the natural fire regime and levels of forest clearing might not be characteristic of any natural disturbances in GSMNP, the diffuse nature of the disturbances appears to have generated an aggregate effect on structure intermediate to areas of more major disturbance (logging and settlement) and undisturbed areas. In addition, Pyle (1985) mapped areas of diffuse disturbance with 'big trees,' indicating even lower levels of diffuse disturbance in some areas. Such comparison could also help us better understand the effects of diffuse human activities on the landscape.

10.5.3 Dominance of Species Functional Groups: The Landscape Imprint of Disturbance and Environment

The clustering of all species functional groups at low mean BA for all 1930s disturbed categories reflects the widespread impacts of recent disturbance on all GSMNP tree species groups at the time of park formation. In contrast, species groups in the undisturbed category appear stable, despite the loss of American chestnut, decimation of Fraser fir, and decreases in American beech. The stability of mean BA for shade-tolerant species, which dominate the higher-elevationlandscapes, likely indicates that other shade-tolerant species increased in response to decreases caused by exotic pests. Tuttle (2007) found that red spruce, understory maples (*Acer pensylvanicum, A. spicatum*), and yellow birch all increased in the northern hardwood – spruce-fir ecotone in conjunction with decline of Fraser fir. However, yellow birch's species group (moderate disturbance responders) does not show increases in mean BA for the undisturbed category. In addition, both Fraser fir and American beech respond rapidly to disturbance in stands where they dominate, with abundant regeneration (Fraser fir) (Smith and Nicholas 2000) or vigorous sprouting (American beech) (Vandermast 2005). American mountain ash (*Sorbus americana*) and pin cherry (*P. pensylvanica*), both vigorous disturbance responders, increased in undisturbed areas in response to Fraser fir decline (Busing et al. 1988, Tuttle 2007), but these trees do not dominate stand BA, are short-lived, and may have already peaked in disturbance response. As for the structural comparisons, these aggregate compositional comparisons for undisturbed areas should be interpreted with some caution because of sampling and environmental biases between the matched plots. For instance, the possible bias of the 1990s–2000s plots toward more Eastern hemlock and more Fraser fir could have resulted in an overestimate of mean BA in the shade-tolerant species group.

Dramatic increases in vigorous disturbance responders in all disturbed categories reflect primarily dramatic increases in both red maple (*A. rubrum*) and yellow-poplar (*Liriodendron tulipifera*) as well as large increases in white pine (*P. strobus*). All three of these species are long-lived enough to persist in the landscape in the absence of further major disturbance, including fire. Red maple has a broad niche (Whittaker 1956), similar to American chestnut, and was already widely present in low abundance across the GSMNP landscape. The distinct dominance of the vigorous disturbance responder group in the settlement category supports observations that yellow-poplar attains high BA rapidly and dominates the canopy in many former settlement areas, suppressing canopy attainment by other species, particularly shade-tolerant species. The increase in the yellow pine group in settlement areas like reflects early-successional dominance of Virginia pine in some old fields. However, the similar mean BA of both oaks and shade-tolerant species in this category may indicate that the relatively short-lived Virginia pine stands are declining, and slower-growing species are increasing.

The clustering of different species groups in the diffuse disturbance category in the 1990s–2000s dataset may reflect the diffuse, patchy nature of the disturbances as well as the variety of topographic settings in this category that covers large portions of the GSMNP landscape at low and middle elevations. With a range of prepark disturbance types, sizes, and ages, species across all groups will respond somewhere, according to their adaptations for regeneration or recovery from disturbance. Increases in this category are also likely driven by both the decline of American chestnut and decrease in fire frequency and extent, particularly with fire suppression and the removal of frequent burning by humans in the western part of the park (Pyle 1985). Although mean BA did not decrease for the yellow pine group in this category, this result cannot distinguish between the effects of any post-park fires in maintaining yellow pine stands (less likely) or in-fill of other species where pines persist but fire is absent (more likely); clearly, the fire- and dry-site-adapted yellow pines are not abundant responders to landscape recovery from diffuse prepark disturbances, nor to continued disturbances such as chestnut blight.

The corporate logging category spans a wide range of elevation, topographic settings, and times since disturbance; as a result, it is not surprising that a variety of species groups, responding to different portions of the gradient, have significantly increased in mean BA as the logged landscape has become reforested. While the increase in the shade-tolerant species group may reflect recovery of conifers and mesic hardwoods in more mesic and higher-elevation logged areas, the secondary dominance of the two disturbance-responder groups in this category is indicative of response to widespread clear-cutting and, likely, to later disturbances such as chestnut blight, balsam woolly adelgid, and beech bark disease. Perhaps future analysis of co-occurring species in each time period as well as comparison of diffuse disturbance and corporate logging plots to undisturbed plots that overlap in elevation and topography (where enough matched plots are available) could help disentangle the mix of responses to different disturbances and their relationship with undisturbed reference conditions.

Differences in Eastern hemlock abundance between the 1930s and 1990s–2000s datasets are worthy of separate discussion. Much of the increase in the shade-tolerant species group across disturbed categories is accounted for by increases in Eastern hemlock mean BA (not shown). Eastern hemlock seems to have broadly responded with recovery and/or increases in response to disturbance, likely including increases from chestnut blight on more mesic sites (as observed in Woods and Shanks 1959) and some increases on slopes with reduced fire frequency. It is possible, too, that much Eastern hemlock was left on the landscape by settlers and some selective logging operations (Pyle 1985), fostering its increase after disturbance. Some of the apparent increase in this study could be an artifact of biased sampling in our 1990s–2000s compilation. However, Tuttle and White (2011) found that Eastern hemlock increased in relative frequency on the GSMNP landscape from 19.26 to 25.78 % between the 1930s and 1999. Although mean *relative* abundance did not exhibit an increase in this earlier study, the 1999 dataset used for comparison was collected along trails and therefore likely biased to ridgetop sites. Most, if not all, of the recent plots in this study were sampled prior to decline of Eastern hemlock in the park from the hemlock woolly adelgid. The observed massive decline of the widespread Eastern hemlock trees in GSMNP over the last several years represents yet another landscape-level set of canopy disturbances that will further distort the range of variation in these forests, even if recent treatment successes with Eastern hemlock continue.

10.6 Conclusions

Our results illustrate the ways that anthropogenic disturbances can both mask and interact with environmental gradients to structure CHR and temperate forests. After several decades of succession from widespread, major disturbance in all but approximately 20 % of GSMNP, historical anthropogenic disturbance has remained important as a predictor of forest structure and composition, although its importance has decreased relative to topography, and the level of importance differs with type of disturbance. The increased importance of topography implies sorting of structure and composition along environmental gradients. However, historical disturbances and more recent anthropogenic disturbances interact with topography and differ in pattern on the landscape, complicating attempts to understand how environmental gradients contribute to forest structure and composition across the region.

Re-sampled plots and consistent coverage of plots across the landscape would have improved our modeling and comparisons. Miller's survey plots were not permanently marked, making resampling and direct comparison impossible. More modern research projects have tended to be designed around immediate management problems limited to particular portions of the landscape and, as a result, have not provided consistent coverage of plots across the landscape. To disentangle the aggregate effects of anthropogenic disturbance, natural disturbance, and environment on these forests, a systematic and objective layout of permanent monitoring plots is needed (as has been recently initiated by GSMNP personnel). Fine-scale models of environmental variables over time would enable structural and compositional analysis that does not rely on topographic approximations of environment. We also need more specific, detailed maps (or models) of historic and recent disturbances, including the timeline, extent, and intensity of these events. Additional digitization and spatial referencing of historical disturbance records as well as increased field data collection and mapping with remote sensing tools, such as LiDAR, could improve our efforts going forward. These needs are even more pronounced on managed and private lands, where management activities and land use further complicate the picture.

In spite of the sampling limitations, our structural comparisons of matched plots reveal widespread, substantial increases in density in GSMNP forests, reflecting disturbance in formerly undisturbed areas, continued disturbance in undisturbed areas, and in-fill from the change in fire regime at lower elevations. While BA in disturbed areas may have begun to converge with pre-disturbance levels, we would expect a corresponding increase in QMD as canopy tree sizes increase through successional time. However, the lack of increase and possible decrease in QMD in disturbed and historically undisturbed areas, respectively, supports the interpretation that BA has been reduced or inhibited from recovery in all areas by the removal of large trees by exotic pests, which have targeted three long-lived, shade-tolerant species having large mature stem diameters and/or high BA in plots where they dominated (American chestnut, Fraser fir, and American beech). Vigorous and moderate disturbance responder species are abundant across all historical disturbance categories (with the exception of low moderate disturbance responder abundance in historically settled areas). The net effect of these structural and compositional changes is a higher number of disturbed, early- to mid-successional patches, a correspondingly lower number of old-growth patches, and lower structural diversity on the landscape than we would expect for forests subject only to natural disturbances. The widespread death of Eastern hemlock trees in GSMNP from the hemlock woolly adelgid is not included our 1990s–2000s dataset, so further changes in structure beyond those observed in our results have already occurred.

Structural diversity and disturbance are keys to maintaining the habitats on which biodiversity depends. Reintroduction of fire is important for maintaining structural diversity and persistence of fire-maintained species in landscapes historically structured by a natural fire regime. However, in forests not historically structured by fire, impeded succession and the ongoing, rapid loss of remaining old-growth forests historically structured by small canopy gaps—and the loss of the largest trees from these forests—are concerning. A long series of exotic pest species has altered structure and composition of eastern USA forests, and additional species are likely to invade, underscoring the importance of efforts to minimize further impacts and to develop management approaches that will allow the redevelopment of stand and landscape-level structural complexity consistent with old-growth forests (Franklin et al. 2002). While the pace of climate change in this region has been slow, future climate change will further affect GSMNP and other CHR forests. The broad extent and significant complexity of ongoing and impending disturbances highlight the need for region-wide assessment of the aggregate effects of multiple stressors across protected, managed, and private lands.

Acknowledgments We thank Mark Harmon and Scott Pearson for insightful review comments, Dean Urban and Jason Fridley for advice on datasets and analysis, and Cathryn Greenberg and Beverly Collins for the invitation to contribute our work to this book.

References

Anderson, M. J. (2001). A new method for non-parametric multivariate analysis of variance. *Austral Ecology, 26*, 32–46.

Anderson, M. J. (2006). Distance-based tests for homogeneity of multivariate dispersions. *Biometrics, 62*, 245–253.

Aneja, V. P., Robarge, W. P., Claiborne, C. S., Murthy, A., Soo-Kim, D., & Li, Z. (1992). Chemical climatology of high elevation spruce-fir forests in the southern Appalachian Mountains. *Environmental Pollution, 75*, 89–96.

Barden, L. S., & Woods, F. W. (1976). Effects of fire on pine and pine-hardwood forests in the southern Appalachians. *Forest Science, 22*, 399–403.

Beven, K. J., & Kirkby, M. J. (1979). A physically based, variable contributing area model of basin hydrology. *Hydrological Sciences Bulletin, 24*, 43–69.

Bolstad, P. V., Swank, W., & Vose, J. (1998). Predicting southern Appalachian overstory vegetation with digital terrain data. *Landscape Ecology, 13*, 271–283.

Bratton, S. P. (1975). The effect of the European wild boar, *Sus scrofa*, on gray beech forest in the Great Smoky Mountains. *Ecology, 56*, 1356–1366.

Burns, R. M., & Honkala, B. H. (tech cords) (1990). Silvics of North America: 1. Conifers; 2. Hardwoods. Agriculture handbook 654. Washington, DC: USDA Forest Service.

Busing, R. T. (1993). Three decades of change at Albright Grove, Tennessee. *Castanea, 58*, 231–242.

Busing, R. T., & White, P. S. (1993). Effects of area on old-growth forest attributes: Implications for the equilibrium landscape concept. *Landscape Ecology, 8*, 119–126.

Busing, R. T., Clebsch, E. E. C., Eagar, C. C., & Pauley, E. F. (1988). Two decades of change in a Great Smoky Mountains spruce-fir forest. *Bulletin of the Torrey Botanical Club, 115*, 25–31.

Busing, R. T., Clebsch, E. E. C., & White, P. S. (1993). Biomass and production of southern Appalachian cove forests reexamined. *Canadian Journal of Forest Research, 23*, 760–765.

Busing, R. T., Stephens, L. A., & Clebsch, E. E. C. (2005). *Climate data by elevation in the Great Smoky Mountains: A database and graphical displays for 1947–1950 with comparison to long-term data*. Data Series Report DS 115. Reston: USDI US Geological Survey.

Butler, S. M., White, A. S., Elliott, K. J., & Seymour, R. S. (2014). Disturbance history and stand dynamics in secondary and old-growth forests of the southern Appalachian Mountains, USA. *Journal of the Torrey Botanical Society, 141*, 189–204.

Chappelka, A., Somers, G., & Renfro, J. (1999). Visible ozone injury on forest trees in Great Smoky Mountains National Park, USA. *Water Air and Soil Pollution, 116*, 255–260.

Clark, G. M. (1987). Debris slide and debris flow historical events in the Appalachians south of the glacial border: Geol Soc Am. *Reviews in Engineering Geology, VII*, 125–137.

Clinton, B. D., & Baker, C. R. (2000). Catastrophic windthrow in the southern Appalachians: Characteristics of pits and mounds and initial vegetation responses. *Forest Ecology and Management, 126*, 51–60.

Cohen, D., Dellinger, B., Klein, R., & Buchanan, B. (2007). Patterns in lightning-caused fires at Great Smoky Mountains National Park. *Fire Ecology, 3*, 68–82.

Curtis, R. O., & Marshall, D. D. (2000). Why quadratic mean diameter? *Western Journal of Applied Forestry, 15*, 137–139.

Delcourt, H. R., & Delcourt, P. A. (1997). Pre-Columbian Native American use of fire on southern Appalachian landscapes. *Conservation Biology, 11*, 1010–1014.

Eagar, C. C. (1984). Review of the biology and ecology of the balsam woolly aphid in southern Appalachian spruce-fir forests. In P. S. White (Ed.), *The southern Appalachian spruce-fir ecosystem: Its threats and biology* (Research/resources management report SER-71, pp. 36–50). Atlanta: USDI National Park Service Southeast Regional Office.

Elliott, K. J., & Swank, W. T. (1994). Impacts of drought on tree mortality and growth in a mixed hardwood forest. *Journal of Vegetation Science, 5*, 229–236.

Flatley, W. T., Lafon, C. W., & Grissino-Mayer, H. D. (2011). Climatic and topographic controls on patterns of fire in the southern and central Appalachian Mountains, USA. *Landscape Ecology, 26*, 195–209.

Flatley, W. T., Lafon, C. W., Grissino-Mayer, H. D., & LaForest, L. B. (2013). Fire history related to climate and land use in three southern Appalachian landscapes in the eastern United States. *Ecological Applications, 23*, 1250–1266.

Fowler, C., & Konopik, E. (2007). The history of fire in the southern United States. *Human Ecology Review, 14*, 165–176.

Franklin, J. F., Spies, T. A., Van Pelt, R., Carey, A. B., Thornburgh, D. A., Berg, D. R., Lindenmayer, D. B., Harmon, M. E., Keeton, W. S., Shaw, D. C., Bible, K., & Chen, J. (2002). Disturbances and structural development of natural forest ecosystems with silvicultural implications, using Douglas-fir forests as an example. *Forest Ecology and Management, 155*, 399–423.

Fridley, J. (2009). Downscaling climate over complex terrain: High finescale (<1000 m) spatial variation of near-ground temperatures in a montane forested landscape (Great Smoky Mountains). *Journal of Applied Meteorology and Climatology, 48*, 1033–1049.

Groemping, U. (2007). Estimators of relative importance in linear regression based on variance decomposition. *American Statistician, 61*, 139–147.

Hansen, B. B., & Klopfer, S. O. (2006). Optimal full matching and related designs via network flows. *Journal of Computational and Graphical Statistics, 15*, 609–627.

Harmon, M. (1981). *Fire history of the Great Smoky Mountains National Park: 1940 to 1979.* NPS-SER Res/resource management report no 46. Gatlinburg: USDI National Park Service Southeast Region, Uplands Field Research Laboratory.

Harmon, M. (1982). Fire history of the westernmost portion of Great Smoky Mountains National Park. *Bulletin of the Torrey Botanical Club, 109*, 74–79.

Harmon, M. E., Bratton, S. P., & White, P. S. (1983). Disturbance and vegetation response in relation to environmental gradients in the Great Smoky Mountains. *Vegetatio, 55*, 129–139.

Harrod, J. C. (1999). *Disturbance history and ecological change in a southern Appalachian landscape: Western Great Smoky Mountains National Park 1936–1996.* Dissertation, University of North Carolina, Chapel Hill.

Harrod, J., White, P. S., & Harmon, M. E. (1998). Changes in xeric forests in western Great Smoky Mountains National Park, 1936–1995. *Castanea, 63*, 346–360.

Harrod, J. C., Harmon, M. E., & White, P. S. (2000). Post-fire succession and 20th century reduction in fire frequency on xeric cites on xeric southern Appalachian sites. *Journal of Vegetation Science, 11*, 465–472.

Jenkins, M. (2007). Vegetation communities of Great Smoky Mountains National Park. *Southeastern Naturalist, 6*, 35–56.

Jenkins, M. A., Klein, R. N., & McDaniel, V. L. (2011). Yellow pine regeneration as a function of fire severity and post-burn stand structure in the southern Appalachian Mountains. *Forest Ecology and Management, 262*, 681–691.

Jobe, R. T. (2006). *Biodiversity and scale: determinants of species richness in Great Smoky Mountains National Park.* Dissertation, University of North Carolina, Chapel Hill.

Johnson, A. H., McLaughlin, S. B., Adams, M. B., Cook, E. R., DeHayes, D. H., Eagar, C., Fernandez, I. J., Johnson, D. W., Kohut, R. J., Mohnen, V. A., Nicholas, N. S., Peart, D. R., Schier, G. A., & White, P. S. (1992). Why are red spruce declining at high elevations? A synthesis of epidemiological and mechanistic studies. In C. Eagar & M. B. Adams (Eds.), *The ecology and decline of red spruce in the eastern United States.* New York: Springer.

Knebel, L. G. (1999). *Changes in old growth hemlock forests of the Cataloochee watershed in the Great Smoky Mountains National Park: preparing for the arrival of the hemlock woolly adelgid (Adelges tsugae).* Thesis, Western Carolina University, Cullowhee.

Krapfl, K. J., Holzmueller, E. J., & Jenkins, M. A. (2011). Early impacts of hemlock woolly adelgid in *Tsuga canadensis* forest communities of the southern Appalachian Mountains. *The Journal of the Torrey Botanical Society, 138*, 93–106.

Lafon, C. W. (2006). Forest disturbance by ice storms in *Quercus* forests of the southern Appalachian Mountains, USA. *Ecoscience, 13*, 30–43.

Landres, P. B., Morgan, P., & Swanson, F. J. (1999). Overview of the use of natural variability concepts in managing ecological systems. *Ecological Applications, 9*, 1179–1188.

Lindsay, J. (2012). Whitebox Geospatial Analysis Tools (GAT) v 2.0. http://www.uoguelph.ca/~hydrogeo/Whitebox/. Accessed 11 Apr 2012.

MacKenzie, M. D., & White, P. S. (1998). The vegetation of Great Smoky Mountains National Park: 1935–1938. *Castanea, 63*, 323–336.

Madden, M., Welch, R., Jordan, T., Jackson, P., Seavey, R., & Seavey, J. (2004). *Digital vegetation maps for the Great Smoky Mountains National Park: Final report*. Center for Remote Sensing and Mapping Science. Athens: University of Georgia.

McNab, W. H. (1989). Terrain shape index: Quantifying effect of minor landforms on tree height. *Forest Science, 35*, 91–104.

McNab, W. H. (1993). A topographic index to quantify the effect of mesoscale landform on site productivity. *Canadian Journal of Forest Research, 23*, 1100–1107.

Moore, I. D., Grayson, R. B., & Ladson, A. R. (1991). Digital terrain modelling: A review of hydrological, geomorphological, and biological applications. *Hydrocarbon Processing, 5*, 3–30.

Ninemets, U., & Valladares, F. (2006). Tolerance to shade, drought, and waterlogging of temperate northern hemisphere trees and shrubs. *Ecological Monographs, 76*, 521–547.

Ohmann, J. L., Gregory, M. J., & Spies, T. A. (2007). Influence of environment, disturbance, and ownership on forest vegetation of coastal Oregon. *Ecological Applications, 17*, 18–33.

Oksanen, J., Blanchet, F. G., Kindt, R., Legendre, P., Minchin, P. R., O'Hara, R. B., Simpson, G. L., Solymos, P., Stevens, M. H. H., & Wagner, H. (2013). Vegan: Community ecology package. R package version 2.0-10. http://CRAN.R-project.org/package=vegan

Onken, B., & Reardon, R. (Eds.) (2010). *Fifth symposium on hemlock woolly adelgid east US*. USDA Forest service report FHTET-2010-07.

Pyle, C. (1985). *Vegetation disturbance history of Great Smoky Mountains National Park: An analysis of archival maps and records* (Research resource management report SER-77). Atlanta: USDI National Park Service Southeast Regional Office.

Pyle, C. (1988). The type and extent of anthropogenic vegetation disturbance in the Great Smoky Mountains before National Park Service acquisition. *Castanea, 53*, 183–196.

R Core Team. (2014). R: A language and environment for statistical computing, version 3.1.1. R Foundation for statistical computing, Vienna. http://www.R-project.org/. Accessed 10 July 2014.

Reilly, M. J., Wimberly, M. C., & Newell, C. (2006). Wildfire effects on plant species richness at multiple spatial scales in forest communities of the Southern Appalachians. *Journal of Ecology, 94*, 118–130.

Roberts, S., Tanksersley, R., & Orvis, K. (2009). Assessing the potential impacts to riparian ecosystems resulting from hemlock mortality in Great Smoky Mountains National Park. *Environmental Management, 44*, 335–345.

Runkle, J. R. (1985). Disturbance regimes in temperate forests. In S. T. A. Pickett & P. S. White (Eds.), *The ecology of natural disturbance and patch dynamics*. New York: Academic.

Runkle, J. R., & Yetter, T. C. (1987). Treefalls revisited: Gap dynamics in the southern Appalachians. *Ecology, 68*, 417–424.

Simon, S. A., Collins, T. K., Kauffman, G. L., McNab, W. H., & Ulrey, C. J. (2005). *Ecological zones in the southern Appalachians: First approximation* (Research paper SRS-41). Asheville: USDA Forest Service Southern Research Station.

Smith, G. F., & Nicholas, N. S. (2000). Size-and age-class distributions of Fraser fir following balsam wooly adelgid infestation. *Canadian Journal of Forest Research, 30*, 948–957.

Taylor, G. (2012). *Forest health monitoring summary, Great Smoky Mountains National Park*. Natural Resource Data Series NPS/APHN/NRDS-2012/378. Fort Collins: USDI National Park Service Natural Resource Stewardship and Science.

Tuttle, J. P. (2007). *Ecotonal change in high-elevation forests of the Great Smoky Mountains, 1930s–2004*. Thesis, University of Georgia, Athens.

Tuttle, J. P., & White, P. S. (2011). Understanding 65 years of change in the forests of Great Smoky Mountains National Park, USA. *Unpublished poster presentation, Biennial conference on international Biogeography Society*, Heraklion.

USDA Forest Service Rocky Mountain Research Station, Missoula Fire Sciences Laboratory (Producer). (2014). Fire effects info system. http://www.fs.fed.us/database/feis/plants/tree. Accessed 23 Nov 2014.

USDA Natural Resources Conservation Service. (2009). *Soil survey of Great Smoky Mountains National Park*. Tennessee and North Carolina.

Vandermast, D. B. (2005). *Disturbance and long-term vegetation change in the high-elevation deciduous forests of Great Smoky Mountains National Park*. Dissertation, University of North Carolina, Chapel Hill.

Walker, G. L. (1978). *A comparative vegetation analysis of cove hardwood communities in the Great Smoky Mountains National Park*. Thesis, Western Carolina University, Cullowhee.

Weiss, A. D. (2001). Topographic position and landforms analysis. *Poster presentation, 21st annual ESRI user conference, San Diego*. http://www.jennessent.com/downloads/tpi-poster-tnc_18x22.pdf. Accessed 14 Feb 2015.

Welch, N. J. (1999). The occurrence of fire in southern Appalachian yellow pine forests as indicated by macroscopic charcoal in soil. *Castanea, 64*, 310–317.

White, P. S., & Busing, R. T. (1993). *LTERM: Long-term monitoring and research in Great Smoky Mountains National Park*. Technical report NPS/SERGRSM/NRTR-93/10. Atlanta: USDI National Park Service Southeast Regional Office.

White, P. S., MacKenzie, M. D., & Busing, R. T. (1985). Natural disturbance and gap phase dynamics in southern Appalachian spruce-fir. *Canadian Journal of Forest Research, 15*, 233–240.

Whittaker, R. H. (1956). The vegetation of the Great Smoky Mountains. *Ecological Monographs, 26*, 1–80.

Whittaker, R. H. (1966). Forest dimensions and production in the Great Smoky Mountains. *Ecology, 47*, 103–112.

Woods, F. W., & Shanks, R. E. (1959). Natural replacement of chestnut by other species in the Great Smoky Mountains National Park. *Ecology, 40*, 349–361.

Chapter 11
Disturbance, Productivity, and Tree Characteristics in the Central Hardwoods Region

Peter A. Wilfahrt, Peter S. White, Beverly S. Collins, and Julie P. Tuttle

Abstract Community theory proposes that the taxonomic diversity and characteristics, or traits, of the trees found within a particular forest community is a function of both the productivity and disturbance history of that community. The theory also predicts that niche differentiation to the conditions caused by disturbance is strongest on productive sites and decreases along productivity gradients. Therefore, both taxonomic and trait diversity should be highest on productive sites at scales that encompass a range of disturbance histories, and lowest on poor quality sites where environmental conditions are strong 'filters' for plant traits. Using a large data set from the USDA Forest Service, we examine patterns of taxonomic and tree trait diversity with respect to recent disturbance events and productivity across the Central Hardwood Region. Our analyses reveal strong regional trends in diversity, less pronounced trends along the productivity gradient, and little effect of disturbance on tree diversity and characteristics.

Keywords Disturbance history • Productivity • Functional traits • Trait diversity • Tree diversity

---

P.A. Wilfahrt (✉)
Curriculum for the Environment and Ecology, University of North Carolina Chapel Hill, CB #3280, Chapel Hill, NC 27599, USA
e-mail: wilfahrt@live.unc.edu

P.S. White
Department of Biology, University of North Carolina, CB #3280, Chapel Hill, NC 27599-3280, USA
e-mail: peter.white@unc.edu

B.S. Collins
Department of Biology, Western Carolina University, NS 132, Cullowhee, NC 28734, USA
e-mail: collinsb@email.wcu.edu

J.P. Tuttle
Curriculum for the Environment and Ecology, University of North Carolina, CB #3280, Chapel Hill, NC 27599-3280, USA
e-mail: jptuttle@live.unc.edu

C.H. Greenberg, B.S. Collins (eds.), *Natural Disturbances and Historic Range of Variation*, Managing Forest Ecosystems 32,
DOI 10.1007/978-3-319-21527-3_11

11.1 Drivers of Forest Structure and Composition

Variability in climate, soil, and disturbance leads to differences in composition and structure of forests across landscapes and through time. In this chapter, we use a large data set from the USDA Forest Service's Continuous Forest Inventory (CFI) to extend earlier work (White et al. 2011; Wilfahrt et al. 2014) and explore environment-disturbance-vegetation relationships across the Central Hardwood Region (CHR). White et al. (2011) mapped variability of the kinds and frequency of disturbances across the CHR, while Wilfahrt et al. (2014) described the effect of physiographic region and stand age on characteristics, or 'functional trait' values, of dominant tree species. Here, we explore the influence of natural disturbances on forest composition and trait diversity across the CHR. Our work relies on two broad ecological frameworks: community theory that predicts diversity patterns as a function of disturbance and productivity, and a trait-based approach to examining mechanisms underlying variability in forest structure and composition.

11.1.1 Community Theory

Succession entails marked changes in forest composition and structure. Following a disturbance, short-lived but fast-growing colonizers that depend on low resource-competition and high resource levels are replaced by long-lived but slow growing species able to compete effectively at lower resource levels in the absence of disturbance. Species also differ in sensitivity or adaptation to particular disturbances; in other words, each type of disturbance filters community characteristics in the initial phases of succession. Applying the 'Intermediate Disturbance Hypothesis' (Connell 1978) to successional cycles over the landscape, one would predict that species diversity is maintained by a mix of patches that differ in time since disturbance and, thus, in composition and structure. Huston (1994) built on this idea to propose that disturbance and productivity (which is correlated with climatic, topographic, and other environmental gradients) interact to influence patterns of diversity. He predicted that the highest diversity should occur at intermediate site productivity and disturbance interval because higher productivity sites are characterized by faster growth rates, hastening the rate of succession and increasing the rate of competitive exclusion. In this chapter, we explore the effect of ecoregion (as a correlate of regional variation in climate, growth rates, and species distributions), site index (as a surrogate for local productivity), and disturbance history (focusing on recent natural disturbances) on forest diversity, structure, and composition.

11.1.2 Functional Traits

Our second framework is based on exploring these patterns from the perspective of tree characteristics, or 'functional traits,' rather than taxonomic, or species, diversity. Functional traits refers to characteristics that connect physiological characteristics of organisms to their environment and provide more nuanced understanding of their relative fitness in a given habitat (Webb et al. 2010) and connection to ecosystem properties (Flynn et al. 2011). Analyses of functional traits have allowed for general mechanistic understanding of forest community dynamics relative to successional processes (Wilfahrt et al. 2014), environmental variation (Coyle et al. 2014), biogeography (Swenson and Weiser 2010), and biotic interaction (Swenson and Weiser 2014) within eastern USA forests.

Succession in the CHR has been well-studied from a taxonomic point of view (e.g., Braun 1950), but it is less well characterized from a functional trait perspective. Evidence from tropical forests reveals that as succession proceeds, foliar traits shift from those favoring rapid resource acquisition and allocation to those that favor conservative resource use (Reich et al. 1995). More recent evidence from tropical forests found additional evidence for rapid resource acquisition traits, but also demonstrated how an abiotic filter (low water availability) can shift which traits are functionally important in early successional forests (Lohbeck et al. 2013). Wilfahrt et al. (2014) found that species with higher seed mass and wood density values increasingly made up forest communities as forest age increased; this pattern was most pronounced at higher latitudes in eastern deciduous forest, which may again suggest that abiotic filters variably affect trait selection. These studies demonstrate that disturbance-adapted species often tend toward rapid resource acquisition and allocation, fast growth, and low parental investment in individual seed production at the expense of leaf and individual longevity, but with the benefit of large seed crops; this is consistent with the earlier work of Grime (1977). Analysis of functional traits also might reveal the processes that affect forest structure and composition. It has been hypothesized that stressful conditions (low productivity due to environmental stresses) filter which traits are successful, leading to convergence and less variability in traits in stressed environments (Weiher and Keddy 1995). In contrast, high productivity sites should experience intense competition, leading to niche diversification and higher trait diversity. However, Coyle et al. (2014) found that this was not a general pattern in eastern USA forests along a water stress gradient, and it remains unclear how productivity affects functional trait patterns.

11.2 Ecosystem, Environment, and Disturbance Characteristics of the Central Hardwood Region

Across the CHR, generally from west to east, forests shift from pine (*Pinus*) and oak (*Quercus*) through oak-hickory (*Quercus-Carya*) and then to mixed-mesophytic and cove hardwood forests with increasing soil water holding capacity, depth, fertility, and the balance of precipitation to evapotranspiration (Braun 1950; Dragoni et al. 2011; see also Greenberg et al. Chap. 1). Tree species diversity also tends to increase, at least to a point, along this regional moisture gradient, with maximum diversity occurring where species dominant in intermediate moisture conditions coexist with more mesic understories (Chapman and McEwan 2012). These same shifts can occur at the local scale over differences in site quality (see Greenberg et al. Chap. 1). For example, along a moisture gradient in southern Indiana, there was a shift in species composition from dry-mesic dominated by white oak (*Q. alba*) and sugar maple (*Acer saccharum*) forests to drier forests dominated by species such as chestnut oak (*Q. montana*) (Jose and Gillespie 1998). The productivity of mesic forests dominated by beech (*Fagus grandifolia*) and sugar maple also decreased from more moist to drier sites. Species composition also has been shown to differ along a soil fertility (potential nitrogen mineralization) gradient, with oak forest on the least fertile sites and mixed mesophytic forests on the most fertile sites (Liu and Muller 1993). A modeling-based example found that a GIS-based integrated moisture index predicted that dry ridgetops in Ohio, which are characterized by low site index, were dominated by oak species, while mesic sites had increased yellow-poplar (*Liriodendron tulipifera*) and black cherry (*Prunus serotina*) abundances (Iverson et al. 1997).

Variation in site quality and species composition also correlates with the type and frequency of disturbances, especially fire. Broadly over CHR forests (see Greenberg et al. Chap. 1, Fig. 1.1), a gradient from pine and dry-site oaks (e.g., chestnut oak, scarlet oak (*Q. coccinea*)) to mixed mesophytic and cove hardwoods that vary widely in shade tolerance is associated with a decrease in 'ground based' disturbances such as fire and grazing, and increase in 'canopy' disturbances such as wind and ice (White et al. 2011; Greenberg et al. Chap. 1). For example, although tornadoes sometimes occur (see Peterson et al. Chap. 5), fire is the predominant disturbance to forests in the Ouachita Mountains, Arkansas Valley, and Boston Mountains ecoregions (also see Greenberg et al. Chap. 1, Table 1.6), while ice is the predominant disturbance in more northern and higher elevation ecoregions (White et al. 2011; Lafon Chap. 7). In effect, there may be an interaction between disturbance type and a gradient from poor quality sites with stress-tolerant species, and the traits associated with these conditions, to high quality sites with a mix of competitive species (sensu Grime 1977), and associated traits (see also Greenberg et al. Chap. 1).

Where one or more environmental variables are sufficiently extreme, open plant communities such as barrens, glades, savannahs, rock outcrops, and balds have developed in our study area. These communities often show evidence of recurring fire or other anthropogenic disturbances such as grazing, and often have a suite of endemic or disjunct plant species adapted to the environmental conditions and disturbance regime. Open communities are scattered throughout the CHR (see Wiser and White 1999), but here we restrict ourselves to the structural and compositional variability of forest ecosystems.

11.3 Data Set, Questions, and Methods

We used the two most recent surveys of Forest Inventory and Analysis (FIA) plots that are distributed across the CHR and had no record of logging to examine how productivity and disturbance relate to forest community types as well as taxonomic and trait diversity. More specifically, we addressed the following questions: (1) *do productivity and disturbance interact to influence species and trait diversity at the plot scale?*; (2) *how are patterns in functional traits correlated with forest community type, stand age, latitude, and diversity measures?*; and (3) *do disturbance and productivity interact to influence the mean trait values across communities and regions?*

From the 10,156 FIA plots selected, we defined disturbed plots as those exhibiting a net loss in basal area for the two most recent surveys, yielding 2,741 plots of mature trees, and 2,622 plots for saplings (saplings were not present on all plots). Using the 'site class' variable, we binned plots into two productivity categories: high ($\geq$8.4 m^3 per ha per year) and low (<8.4 m^3 per ha per year). Plots also were characterized by stand age, which we used as a proxy for successional age; community type, which represents the dominant genera or species of a forest plot; and geographic location (latitude and longitude), which we used to bin plots into groups of ecoregions representing similar climate, edaphic properties, and biogeographic histories. While only fuzzed geographic coordinates are provided by the USDA Forest Service, misclassification of fuzzed plots into regional groups could only occur along regional boundaries that are themselves fuzzy, and the large number of plots used here likely minimizes the impact of any misclassification on our analyses.

We characterized tree species using four functional traits and two tolerance indices: maximum height, seed mass, wood density, foliar nitrogen, and shade and drought tolerances. *Maximum height* of a species represents its potential for height, and therefore light capture, and capacity to move water against gravity. Being tall is advantageous for avoiding being shaded out, but requires considerable investment in structural wood and increases difficulty in maintaining hydraulic connectivity from roots to canopy leaves (Chave et al. 2009). This hydraulic effect may select for lower maximum heights in xeric environments due to decreased water potential. With respect to *seed mass*, heavier seeds are advantageous for mid- to late-successional species whose seeds are more likely to reach a given site under a closed canopy, as seedlings of heavy seeds are better competitors in low resource environments. In contrast, lighter seeds, which typically are wind-dispersed, maximize the chance of being in a location where a disturbance occurs (Leishman et al. 2000). Additionally, smaller seeds are more readily worked down into the soil profile, potentially insulating them from granivory and fire. Higher *wood density* provides greater structural support and reduced chance of breakage or cavitation (Chave et al. 2009), but results in lower hydraulic connectivity and slower water movement through the organism (Reich 2014). These species generally have slower growth, but greater longevity. Species with lower density woods can grow faster to occupy canopy gaps, but are in turn more susceptible to mortality from future disturbances. Slow growth may be selected for in low productivity sites as resource capture is lower and requires effective competition over longer time periods. *Leaf nitrogen* content correlates positively with photosynthetic capacity and net primary productivity (Reich 2014). High nitrogen content also correlates with increased susceptibility to herbivory and pathogens, as it provides high energetic reward and these leaves generally have little chemical or mechanical defense. Leaf nitrogen is independent of stand age (Reich et al. 1995; Wilfahrt et al. 2014), and is confounded by high responsiveness to soil fertility (Kazakou et al. 2014; Coyle et al. 2014); nevertheless, species with low leaf nitrogen tend to be slow-growing and highly defended while species with high leaf nitrogen tend toward rapid growth and low resource conservation strategies (Reich 2014). We used *shade* and *drought tolerance indices* provided by Russell et al. (2014; adapted from Niinemets and Valladares 2006) to provide a more holistic view of a species' ecological strategy, though it should be noted that such indices are ultimately controlled by physiologic properties. Shade intolerant species are more likely to be gap colonizers characterized by quick growth and short lifespans. Drought tolerant species are more likely to be characterized by slow growth and higher allocation to below-ground biomass, resulting in smaller-statured individuals (Russell et al. 2014).

We examined responses of both trees and saplings. FIA plots consist of four circular 7.32 m radius subplots separated by 36.6 m from center, with a circular 2.07 m radius microplot nested within each of the four subplots. Trees are defined as stems that are greater than or equal to 12.7 cm in diameter at breast height (DBH) and are tallied on all subplots. Saplings are defined as stems reaching breast height but less than 12.7 cm DBH, and are tallied on the nested microplots. We pooled all tree stems from the four subplots to represent plot level tree diversity, and all sapling stems from the four microplots to represent sapling diversity. Since 'disturbed plots' indicates a relatively recent disturbance, occurring anywhere from 1 to 12 years prior (median of 5 years) to the sampling time used, trees may be interpreted as post-disturbance residual individuals, while saplings may more accurately reflect post-disturbance recruitment. Tree and sapling stem counts for each plot were used with the species' mean trait values to calculate a community wide, abundance weighted mean for each of the six traits in each plot. For each plot, we calculated species richness (number of species), Shannon diversity (H'), and Rao's quadratic entropy (FDq) for the standardized values of each of the six traits. FDq provides an analog to species diversity (H') for trait diversity (Botta-Dukát 2005); it describes communities based on how different species are from each other in n-dimensional trait space, where n is the number of traits used. We used it as an abundance weighted measure and scaled all traits so they contributed equally to the metric, which resulted in values ranging from 0 (species richness of 1) to 0.55. We log-10 transformed species' seed mass values prior to analysis to normalize the data. Species were assigned a value from 1 to 5 on a continuous scale of shade and drought tolerance, where 1 indicates 'least tolerant' and 5 indicates 'most tolerant' for each respective index (Russell et al. 2014).

We used Principle Component Analysis (PCA) of the standardized abundance weighted mean trait values (i.e., mean = 0, and ± 1 standard deviation for each trait) over all 12,824 plots to examine trends in traits separately for both tree and saplings. Based on the PCA results and previous research (Wilfahrt et al. 2014) that showed geographic trends in mean trait values, we binned the 13 ecoregions of the CHR (see Greenberg et al. Chap. 1, Fig. 1.1) into five contiguous regional groups (Fig. 11.1). This reduced the specificity of our analysis, but allowed for broader

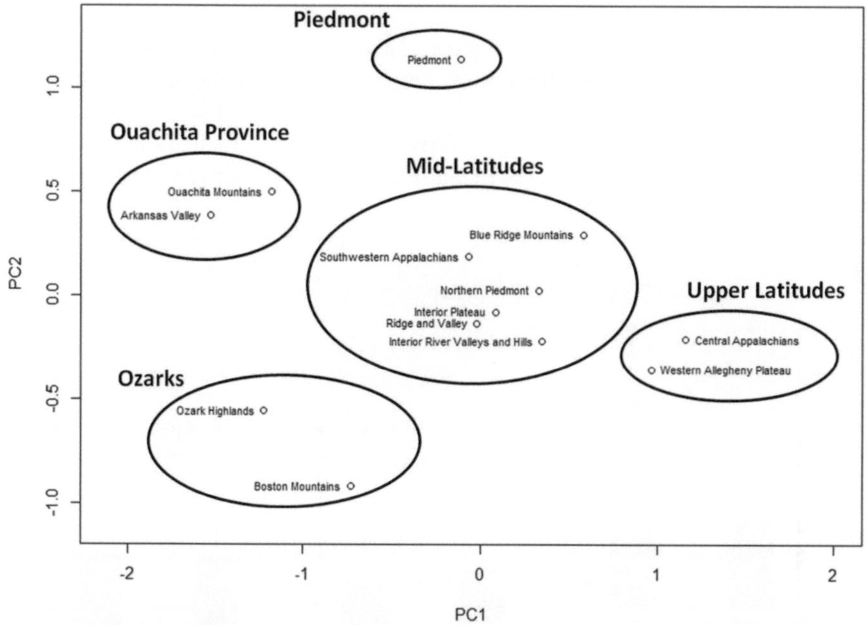

Fig. 11.1 Central Hardwood Region ecoregions merged into five regions based on principle component analysis

generalization of productivity and disturbance trends across the region, while still accounting for underlying climate and edaphic gradients. The five regional groups were: the Ouachita Province (encompassing the Ouachita Mountains and Arkansas Valley ecoregions); the Ozarks (encompassing the Ozark Highlands and Boston Mountains); the Mid-Latitudes (encompassing a central contiguous area of mountainous ecoregions including the Blue Ridge Mountains and Southwestern Appalachians, as well as the Northern Piedmont, Interior Plateau, Ridge and Valley, and Interior River Valley and Hills); the Piedmont; and the Upper Latitudes (encompassing the Central Appalachians and Western Allegheny Plateau).

11.4 Productivity and Disturbance Effects on Diversity and Trait Variability

Across the CHR, forests on less, or less recently, disturbed and more productive sites may be expected to have higher species richness and diversity than forests on more intensively disturbed, more recently disturbed, or less productive sites. However, the analyses showed that most variation in the diversity metrics occurred between the five regions (grouped ecoregions as defined above); within regions, productivity showed small changes and disturbance showed little variation

(Figs. 11.2, 11.3, and 11.4). Richness and Shannon-Weiner diversity generally increased from western to eastern regions. This likely reflects increasing regional productivity, though even the high productivity plots of the Ouachita Province and Ozarks regions do not approach the diversity levels of the eastern regions. Recruitment, however, may be less sensitive to underlying productivity as saplings showed little variation in species richness and diversity between disturbed and

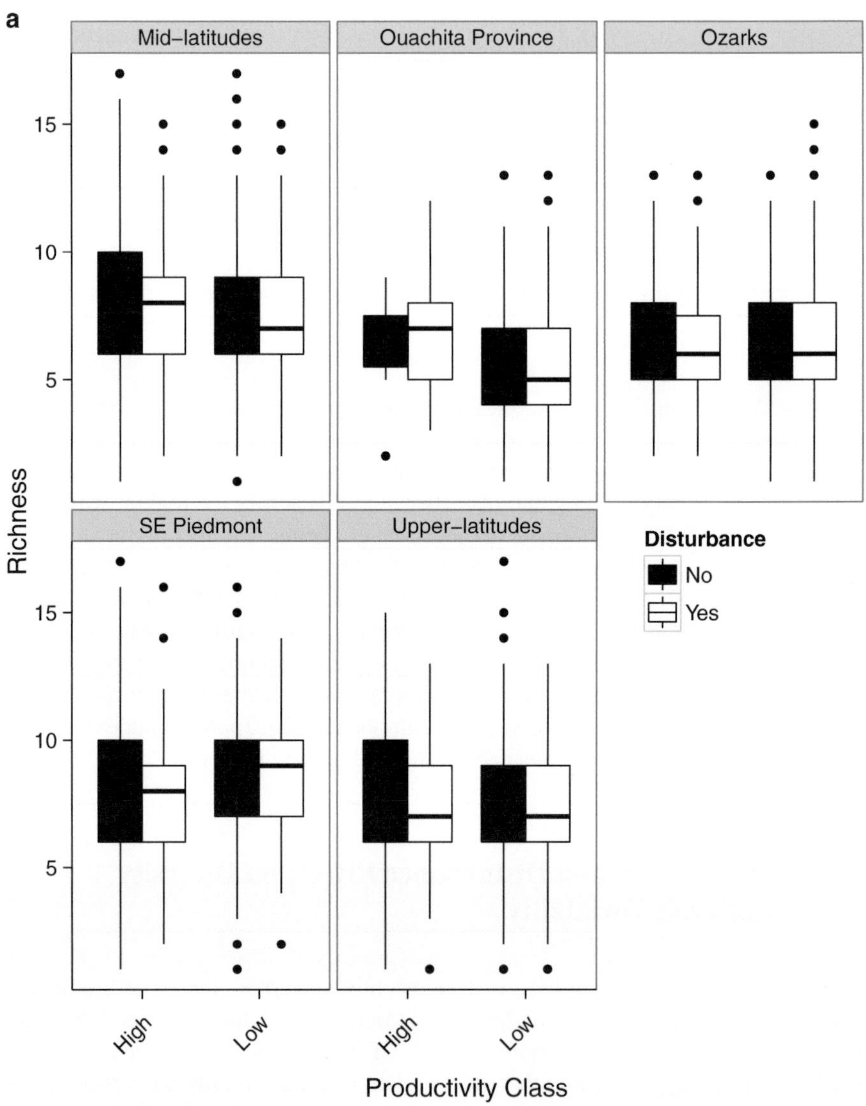

Fig. 11.2 Tree (**a**) and sapling (**b**) richness in high- to low-diversity plots within each of five regions (see Fig. 11.1) within the CHR

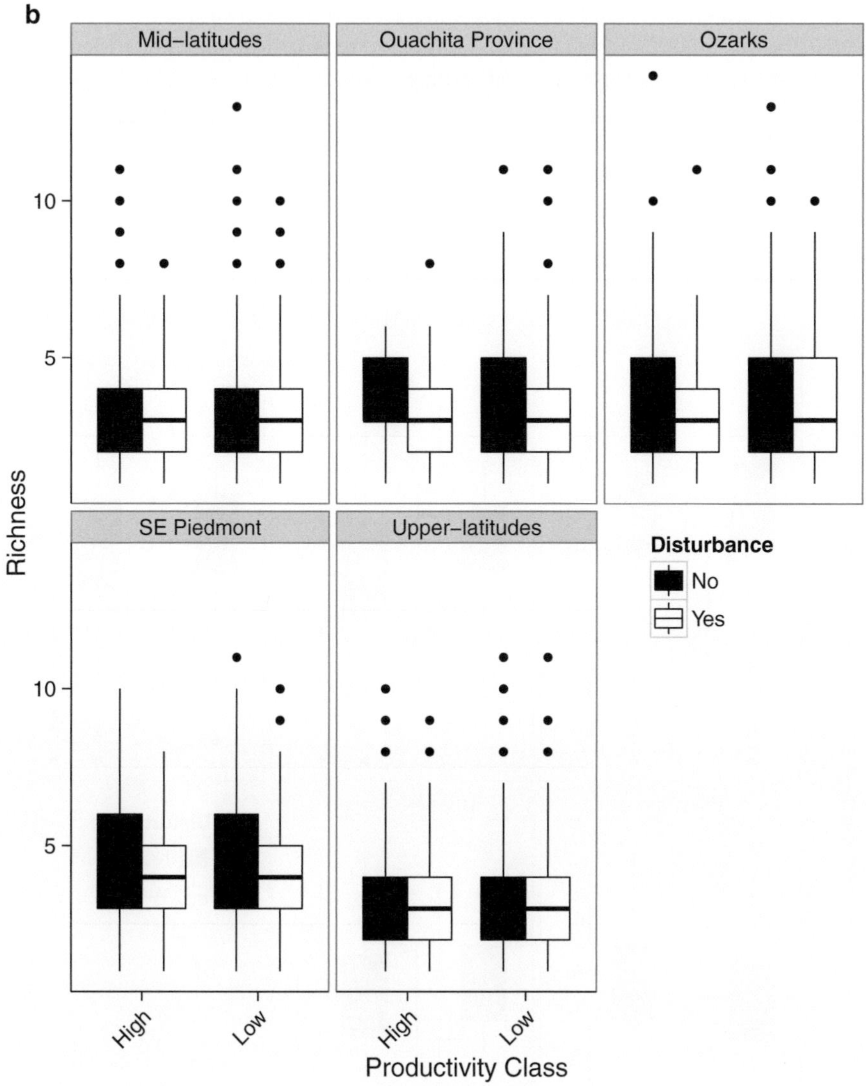

Fig. 11.2 (continued)

undisturbed plots or over productivity classes (Figs. 11.2 and 11.3). The decreased sapling diversity relative to trees is certainly due to the smaller sampling areas, but it is unclear if this masks patterns related to productivity or disturbance. In general, the patterns in the tree and sapling layers suggest recent disturbance has had little effect on, or potentially decreased species richness and diversity across CHR forests.

As indicated by Rao's FDq, diversity of tree traits was generally unaffected or only slightly affected by disturbance. However, in the Mid-Latitude, Ouachita

Province, and Upper-Latitudes regions, there was a trend toward lower trait diversity in the low productivity plots (Fig. 11.4a). Saplings again demonstrated decreased trait diversity (i.e., species with more similar traits) relative to trees, but this is more safely interpreted as a sampling artifact than any recruitment restriction, particularly due to the lack of difference between disturbed and undisturbed plots (Fig. 11.4b). Overall, variability in maximum height, seed mass, wood density, foliar nitrogen,

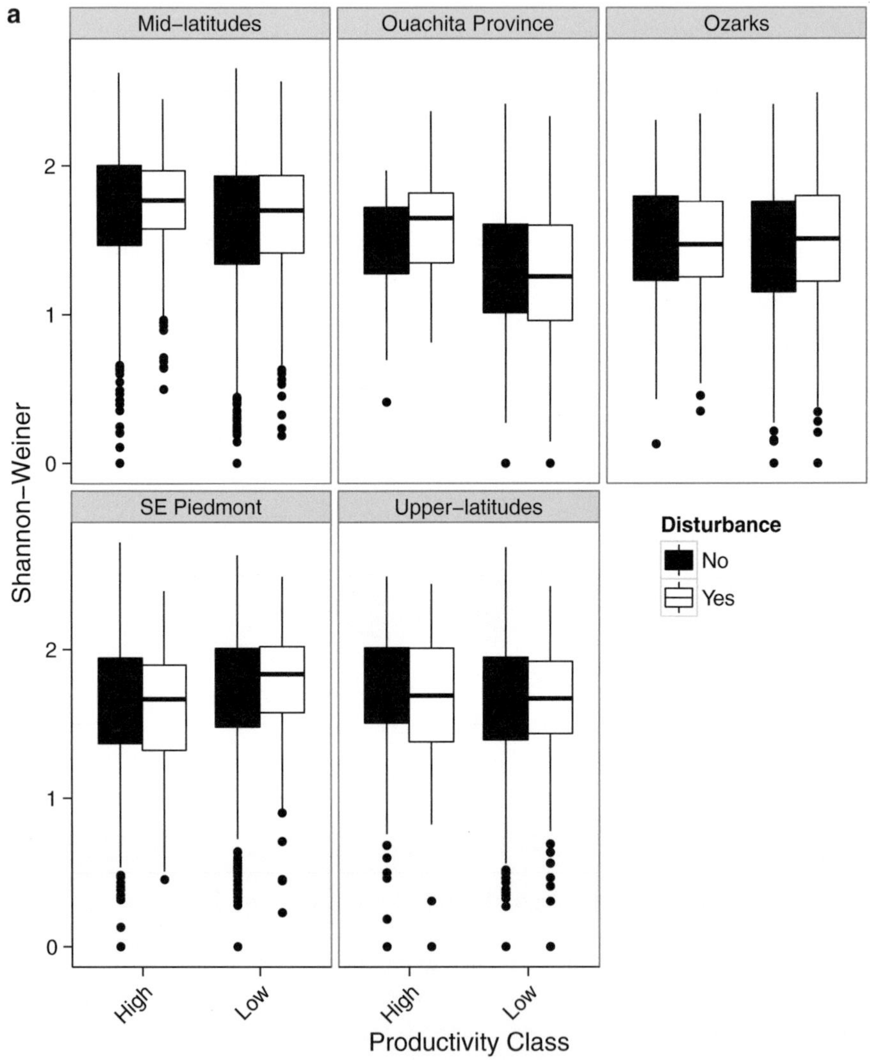

Fig. 11.3 Tree (**a**) and sapling (**b**) Shannon-Weiner diversity (H') in high- to low-diversity plots within each of five regions (see Fig. 11.1) within the CHR

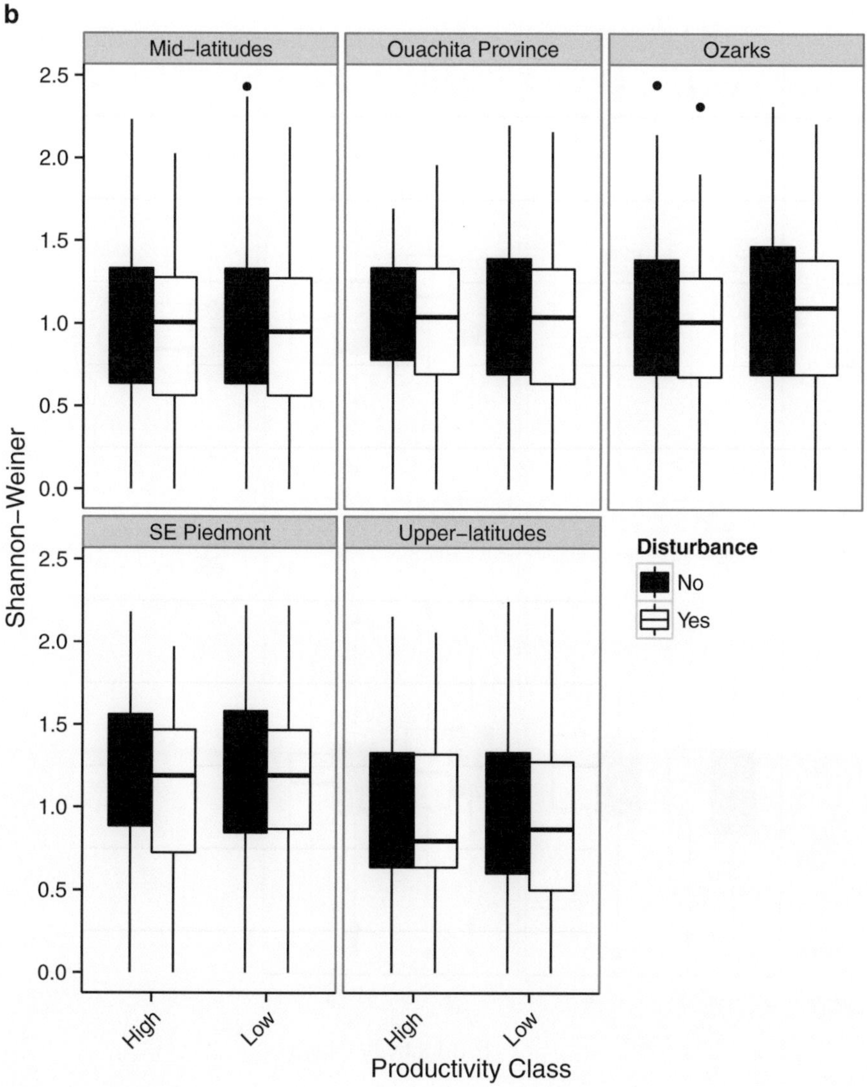

Fig. 11.3 (continued)

and shade and drought tolerances of saplings and trees shows no consistent pattern (e.g., no convergence or divergence) with disturbance or between high and low productivity classes in the CHR. It is important to emphasize, however, that FDq measures variability in trait values rather than any shift in mean trait values; below, we discuss patterns in functional traits over regions and stand ages, and how these are affected by disturbance or productivity.

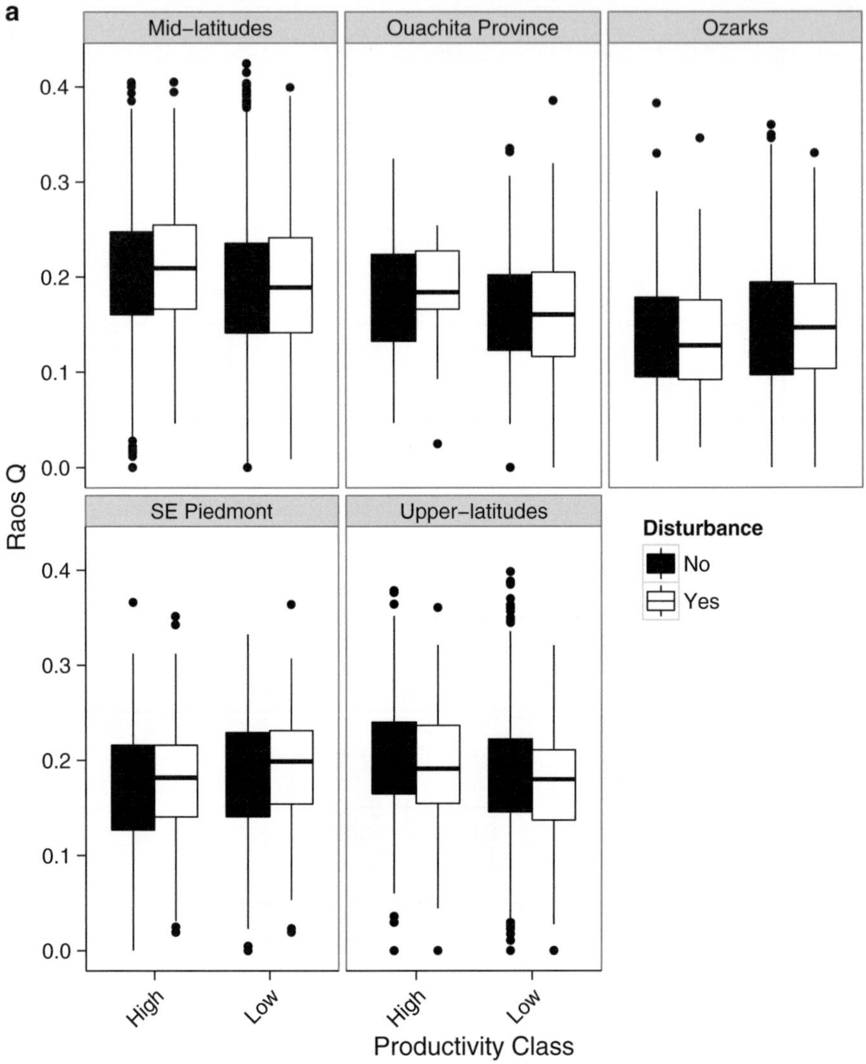

Fig. 11.4 Tree (**a**) and sapling (**b**) Rao's FDq index of trait diversity in high- to low-diversity plots within each of five regions (see Fig. 11.1) within the CHR

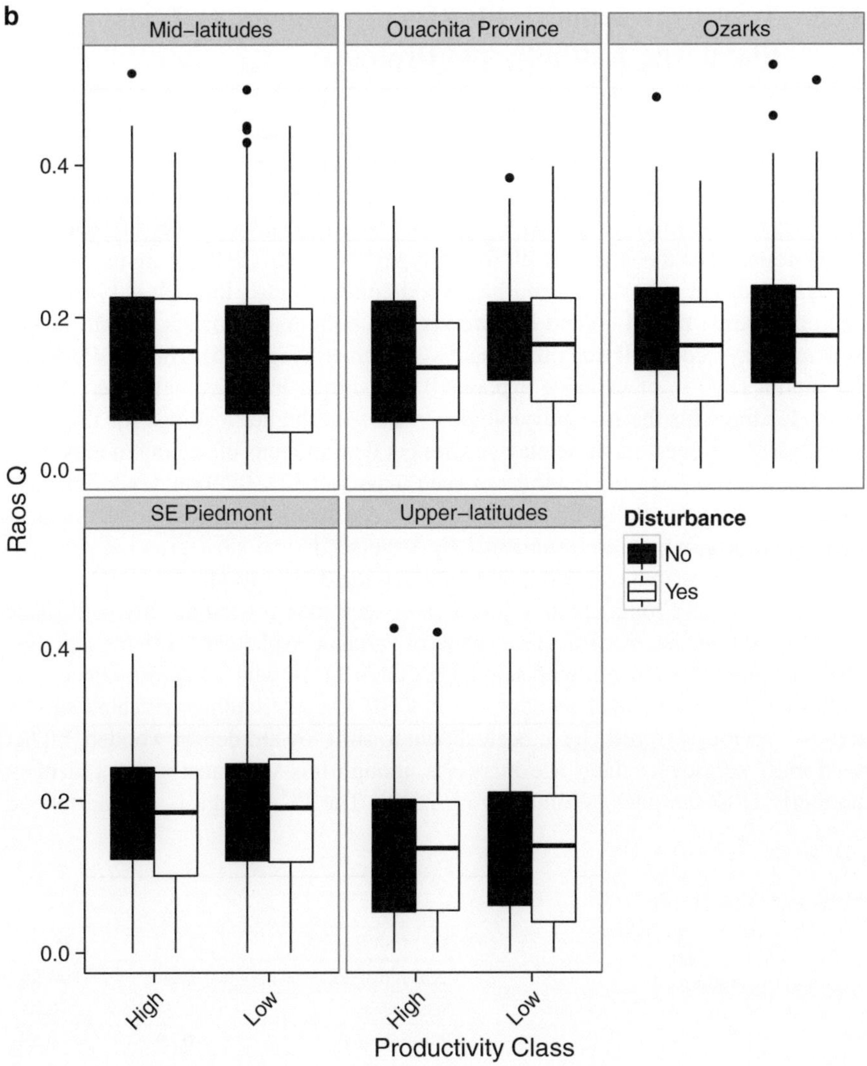

Fig. 11.4 (continued)

11.5 Trait Correlations with Region, Community Type, Stand Age, Latitude, and Diversity

The first PCA axis, which accounted for 36 % of the variance in traits of trees, was strongly and negatively correlated with drought tolerance (Table 11.1), and positively correlated with shade tolerance and maximum tree height. It was also correlated negatively, although less strongly, with seed mass and wood density. This axis likely represents the gradient of forest types over the CHR, from more open, drought-resistant pine and mixed hardwood-conifer forests through oak and oak-hickory forests, mixed upland hardwoods, and bottomland forests, to tall, mesic, and multi-layered maple and mixed hardwood forests (Fig. 11.5). The first PCA axis for saplings (30 % of variance explained) had similar loadings, but generally had lower loadings for the four physiological traits, and higher loadings for the tolerance indices, suggesting these may be stronger determinants of sapling membership in CHR forests. Negative loadings of seed mass (–0.32, –0.59) and wood density (–0.38, –0.58) on both the first and second PCA axes likely represent oak and oak-hickory forests, which are dominated by large seeds and slow-growing drought-tolerant species with hickories in particular having dense wood.

In addition to strong negative loadings of seed mass, wood density, and shade tolerance, the second tree PCA axis (29 % of variance explained; 24 % for saplings) has a negative relationship with stand age (Table 11.2), which suggests this axis is capturing the successional process within CHR forests. Saplings revealed similar trends. Previously, forests have been shown to shift toward denser wooded, higher seed mass species as stand age increases, though this was most evident at more northerly USA latitudes (Wilfahrt et al. 2014). The PCA suggests a trend toward

Table 11.1 Eigenvector values of trait loadings on the first two PCA axes for (a) trees and (b) saplings

(a)		
Tree trait loading	PCA1	PCA2
Leaf nitrogen	0.29	–0.24
Seed mass	–0.34	–0.58
Wood density	–0.39	–0.58
Maximum height	0.34	–0.03
Shade tolerance	0.43	–0.47
Drought tolerance	–0.59	0.23
(b)		
Sapling trait loading	PCA1	PCA2
Leaf nitrogen	–0.01	–0.24
Seed mass	–0.23	–0.59
Wood density	–0.33	–0.62
Maximum height	0.27	0.06
Shade tolerance	0.55	–0.43
Drought tolerance	–0.68	0.18

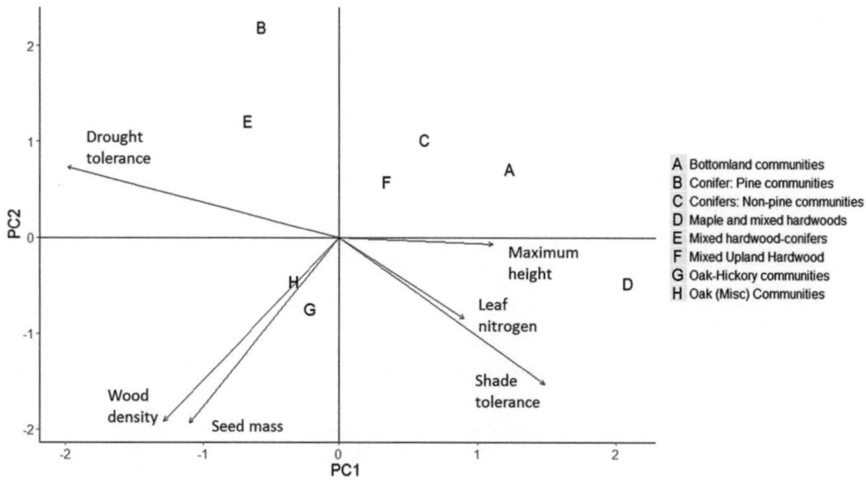

Fig. 11.5 Principle components analysis of trees, with vectors for six functional traits and distributions of community types in the CHR

Table 11.2 Correlations of plot level variables with two tree PCA axes

	PCA1	PCA2
Stand age	0.07	–0.39
Latitude	0.27	–0.28
Species richness	0.11	–0.03
Shannon-Weiner (H')	0.16	–0.16
Rao's Quadratic	0.27	0.11

denser wood and heavier seed mass as oak and oak-hickory forests of the CHR develop, and an increase in shade tolerance as maple and mixed hardwood forests age and develop dense, multi-layered canopies. The increase in trait diversity (Rao's Q) along PCA1 (Table 11.2), as shade tolerance increases, and drought tolerance decreases, also suggests that forests become more complex and support a wider range of functional types.

Clear geographic trends emerge when an ecoregion mean (for the 13 CHR ecoregions; see Greenberg et al. Chap. 1, Fig. 1.1) is calculated from their constituent plots' tree PCA values (Table 11.3). The first PCA axis has the lowest values in the western ecoregions, stretching from the Ozark Highlands to the Ouachita Mountains, and peak values in the Central Appalachian ecoregion. This is consistent with an increase in drought tolerant communities in the western ecoregions and the increased prevalence of mesic species (red maple (*A. rubrum*), beech) in the montane eastern ecoregions (Nowacki and Abrams 2008). The eastern ecoregions also have greater height and shade tolerance; this may result from higher productivity resulting in greater light competition which in turn selects for higher height potential in some species and greater shade tolerance in others. PCA axis 2 reveals variation in the western ecoregions of the CHR with Ouachita Province ecoregions having higher

Table 11.3 Mean plot PCA values for 13 ecoregions in CHR (see Greenberg et al. Chap. 1, Fig. 1.1 for ecoregions map)

Ecoregion	PCA1	PCA2
Arkansas Valley	−1.52773	0.384723
Blue Ridge Mountains	0.580282	0.293679
Boston Mountains	−0.73066	−0.9155
Central Appalachians	1.161306	−0.20883
Interior Plateau	0.087045	−0.07792
Interior River Valleys and Hills	0.350087	−0.21834
Northern Piedmont	0.330115	0.026002
Ouachita Mountains	−1.17797	0.499654
Ozark Highlands	−1.22079	−0.5557
Piedmont	−0.11341	1.137195
Ridge and Valley	−0.01801	−0.13165
Southwestern Appalachians	−0.06509	0.18845
Western Allegheny Plateau	0.965032	−0.35423

values than the Ozark regions. The Piedmont has the highest PCA2 values, indicating low seed mass, wood density, and shade tolerance, possibly a signal of high loblolly pine (*P. taeda*) presence.

11.6 Disturbance and Productivity Effects on Traits

Decreases in PCA axis 1 scores suggest trees increase in drought tolerance, wood density, and seed mass, and decrease in maximum height, shade tolerance, and leaf nitrogen from high to low productivity (Fig. 11.6). The increase in seed mass and wood density is further seen in PCA axis 2 scores, but these also reflect an increase, rather than a decrease, in shade tolerance and leaf nitrogen. That is, trees on low productivity sites tend to be shorter and have heavier seeds and denser wood, but vary in shade tolerance and leaf nitrogen. This may reflect greater abundance of slow-growing, stress tolerant oaks on low productivity sites. Saplings exhibited similar, albeit, weaker trends with PCA scores relative to productivity (Fig. 11.7). The increased loadings of shade tolerance on PCA axis 1 may indicate that saplings are less responsive to site conditions and more responsive to biotic plot conditions (e.g., shade from established trees). Further, generally equal or lower scores for saplings in disturbed compared to undisturbed plots suggests more shade intolerant recruitment into the understory following disturbance.

As noted above, PCA axis 2 appears to be capturing the successional process; higher scores for trees in disturbed plots compared to those in undisturbed plots in high productivity plots within all regions (Fig. 11.6) suggests composition of lighter seeds, less dense wood, and lower shade tolerance. Saplings, however, show no consistent patterns in disturbed compared to undisturbed plots on PCA2 (Fig. 11.7), which suggests no convergence on particular traits during recruitment into the understory.

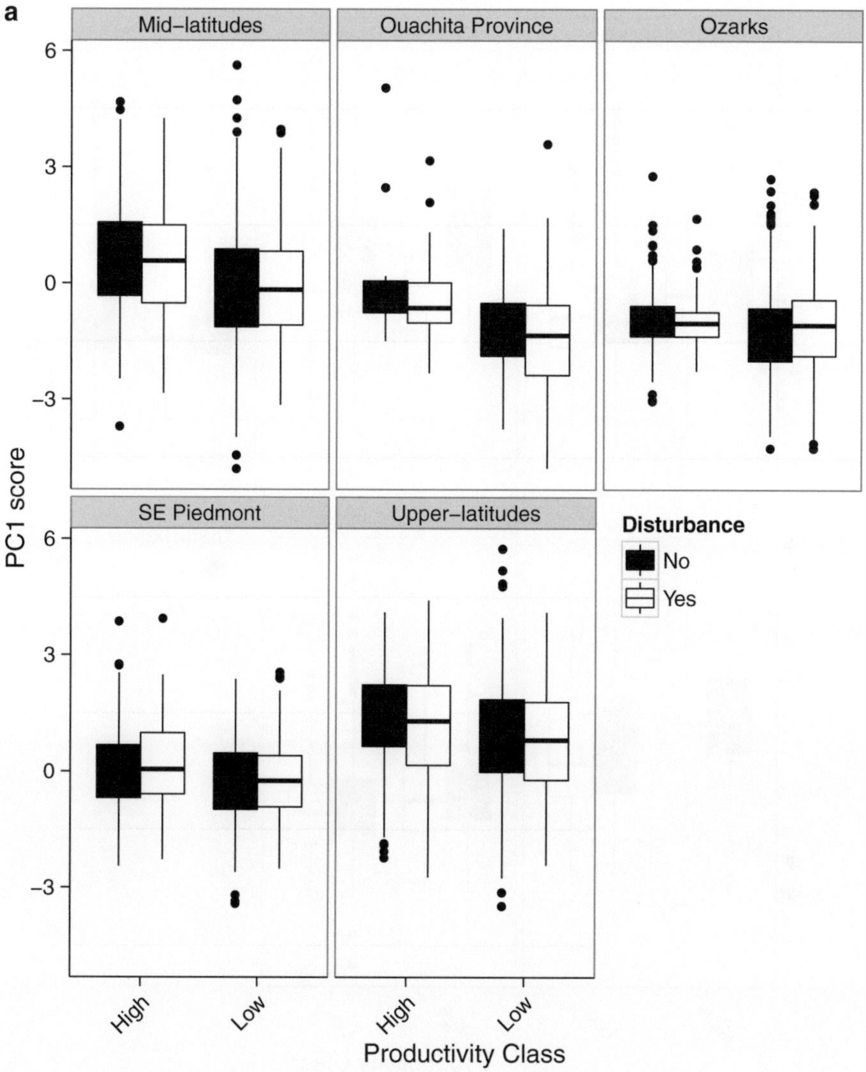

Fig. 11.6 Tree PCA1 (**a**) and PCA2 (**b**) axis scores for undisturbed and disturbed plots in high- to low-diversity plots within each of five regions (see Fig. 11.1) within the CHR

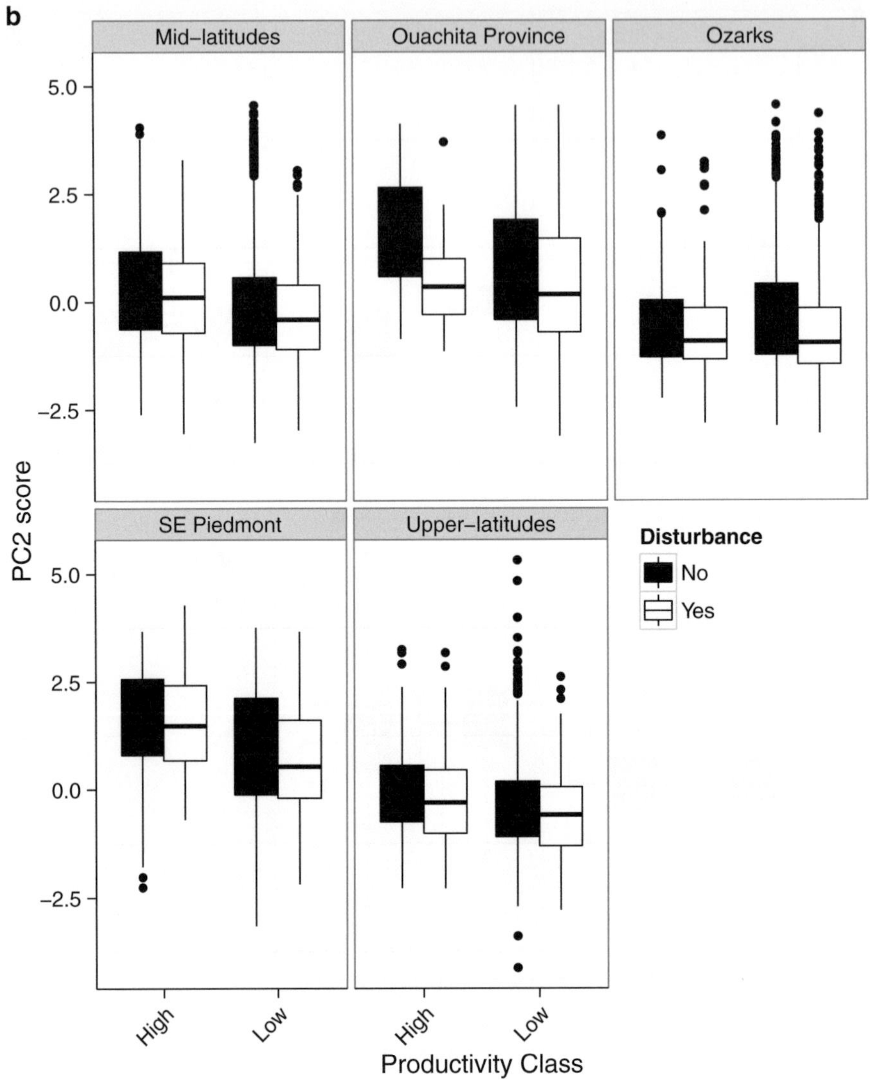

Fig. 11.6 (continued)

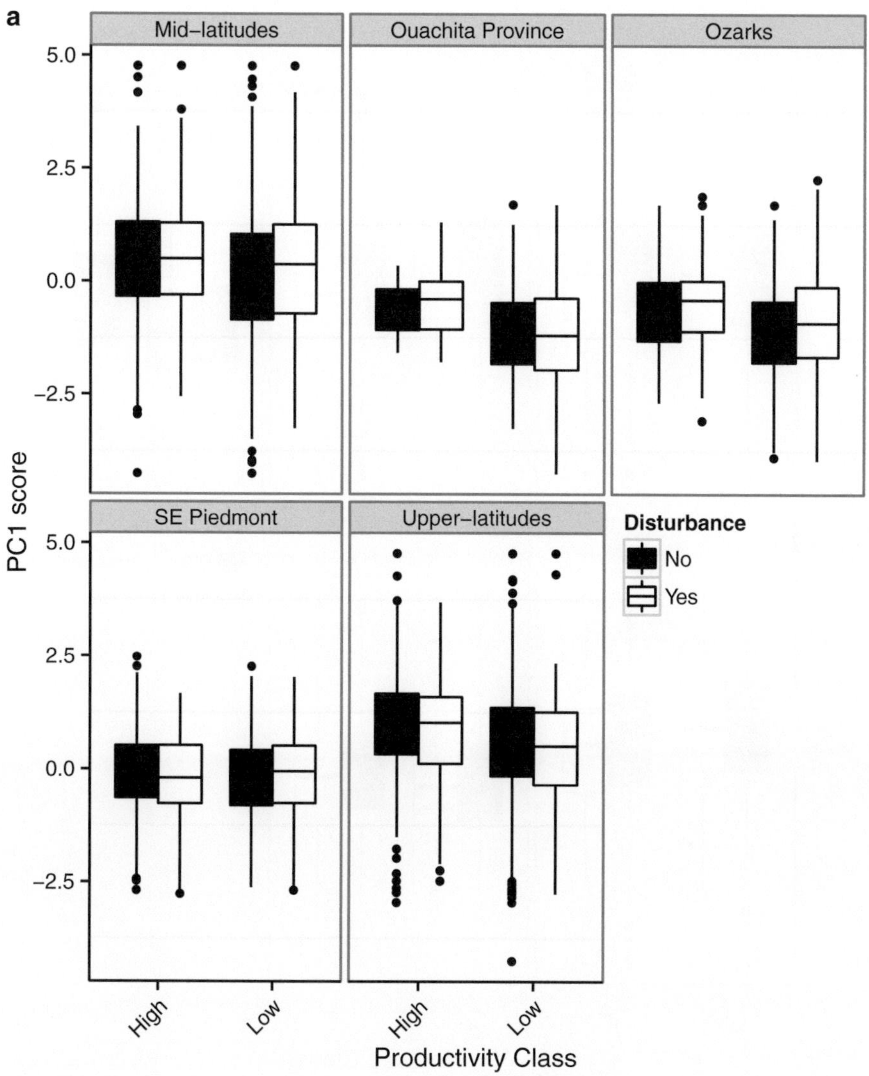

Fig. 11.7 Sapling PCA1 (**a**) and PCA2 (**b**) axis scores for saplings in disturbed and undisturbed plots in high- to low-diversity plots within each of five regions (see Fig. 11.1) within the CHR

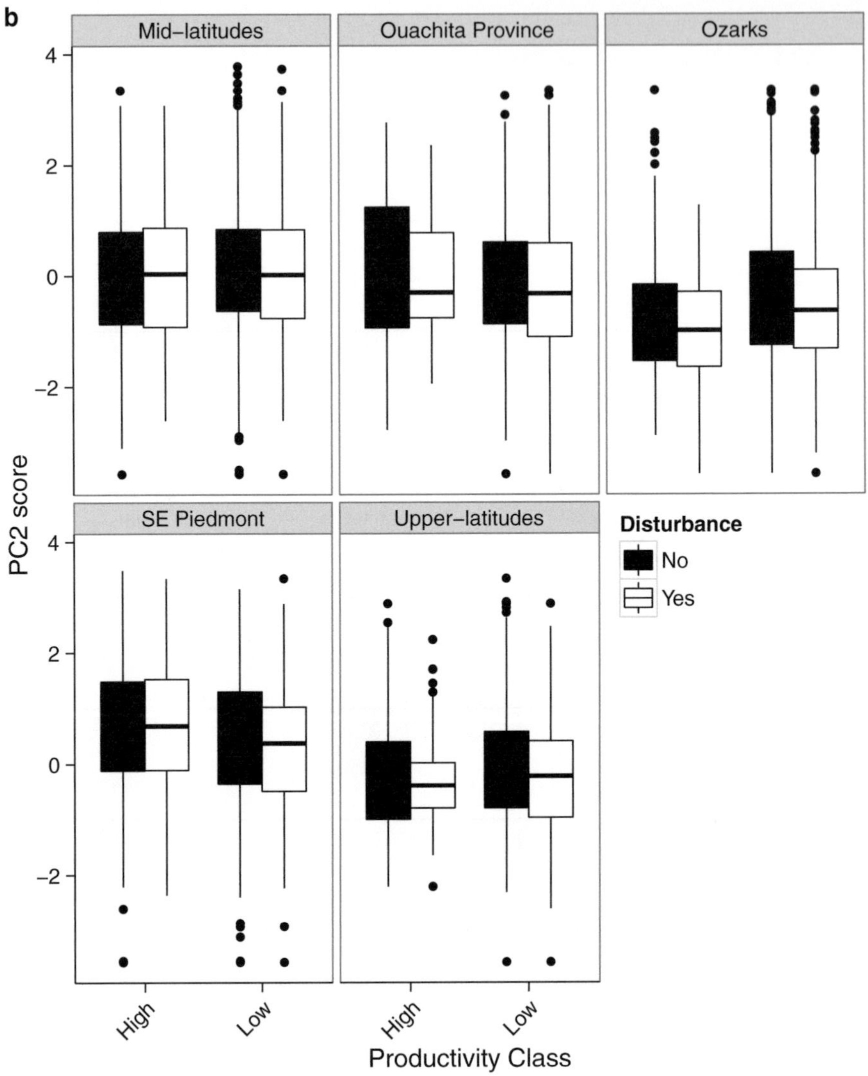

Fig. 11.7 (continued)

11.7 Conclusions

Our analyses confirm the gradient of forest types over the CHR, from more open, drought-tolerant communities in the western ecoregions through taller, structurally complex eastern forests with more abundant mesic and shade tolerant species such as red maple and beech (Nowacki and Abrams 2008). On more productive sites, these forests tend to have greater species richness and diversity. High productivity also is associated with taller forests, with greater shade tolerance and leaf nitrogen, but lower drought tolerance, wood density and seed mass. In agreement with Coyle et al. (2014), we found no convergence or less variability in tree traits with increasing stress; i.e., from high to low productivity, and diversity of tree traits was generally unaffected or only slightly affected by disturbance.

Our analyses also captured the successional process in CHR forests. The second PCA axis suggests a trend toward denser wood and heavier seed mass as oak and oak-hickory forests develop and an increase in shade tolerance as maple and mixed hardwood forests age and develop dense, multi-layered canopies. An increase in trait diversity as shade tolerance increases and drought tolerance decreases also suggests that forests become more complex and support a wider range of functional types as succession progresses. Productivity may influence the extent to which disturbance 'resets' forests to an earlier successional stage; in high productivity plots within all regions, higher PCA axis 2 scores for disturbed plots compared to undisturbed plots suggest composition of trees with lighter seeds, lower wood density, and lower shade tolerance. A lack of patterns in the sapling layer, however, suggests no convergence on particular traits during recruitment into the understory.

Acknowledgments We would like to thank Tara Keyser and Henry McNab for insightful commentary during the review process. Further, we thank members of the Plant Ecology Lab at UNC at Chapel Hill for helpful feedback in the early stages of this project. Finally, we would like to thank Cathryn Greenberg for inviting us to participate in the book and organized conference at the 2014 meeting of the Association of Southeastern Biologists.

References

Botta-Dukát, Z. (2005). Rao's quadratic entropy as a measure of functional diversity based on multiple traits. *Journal of Vegetation Science, 16*, 533–540.

Braun, L. E. (1950). *Deciduous forests of eastern North America*. New York: Hafner.

Chapman, J. I., & McEwan, R. W. (2012). Tree regeneration ecology of an old-growth central Appalachian forest : Diversity, temporal dynamics, and disturbance response. *The Journal of the Torrey Botanical Society, 139*, 194–205.

Chave, J., Coomes, D., Jansen, S., Lewis, S. L., Swenson, N. G., & Zanne, A. E. (2009). Towards a worldwide wood economics spectrum. *Ecology Letters, 12*, 351–66.

Connell, J. (1978). Diversity in tropical rain forests and coral reefs. *Science, 199*, 1302–1310.

Coyle, J. R., Halliday, F. W., Lopez, B. E., Palmquist, K. A., Wilfahrt, P. A., & Hurlbert, A. H. (2014). Using trait and phylogenetic diversity to evaluate the generality of the stress-dominance hypothesis in eastern North American tree communities. *Ecography, 37*, 814–826.

Dragoni, D., Schmid, H. P., Wayson, C. A., Potter, H., Grimmond, C. S. B., & Randolph, J. C. (2011). Evidence of increased net ecosystem productivity associated with a longer vegetated season in a deciduous forest in south-central Indiana, USA. *Global Change Biology, 17*, 886–897.

Flynn, D., Mirotchnick, N., Jain, M., Palmer, M., & Naeem, S. (2011). Functional and phylogenetic diversity as predictors of biodiversity-ecosystem-function relationships. *Ecology, 92*, 1573–1581.

Grime, J. (1977). Evidence for the existence of three primary strategies in plants and its relevance to ecological and evolutionary theory. *American Naturalist, 111*, 1169–1194.

Huston, M. A. (1994). *Biological diversity: The coexistence of species*. Cambridge: Cambridge University Press.

Iverson, L. R., Dale, M. E., Scott, C. T., & Prasad, A. (1997). A GIS-derived integrated moisture index to predict forest composition and productivity of Ohio forests (USA). *Landscape Ecology, 12*, 331–348.

Jose, S., & Gillespie, A. R. (1998). Compensatory mechanisms of central US hardwood forest communities in a changing environment. *Plant Ecology, 135*, 125–134.

Kazakou, E., Violle, C., Roumet, C., Navas, M.-L., Vile, D., Kattge, J., & Garnier, E. (2014). Are trait-based species rankings consistent across data sets and spatial scales? *Journal of Vegetation Science, 25*, 235–247.

Leishman, M. R., Wright, I., Moles, A. T., & Westoby, M. (2000). The evolutionary ecology of seed size. In M. Fenner (Ed.), *Seeds: The ecology of regeneration in plant communities* (2nd ed.). New York: CAB International.

Liu, Y., & Muller, R. N. (1993). Above-ground net primary productivity and nitrogen mineralization in a mixed mesophytic forest of eastern Kentucky. *Forest Ecology and Management, 59*, 53–62.

Lohbeck, M., Poorter, L., Lebrija-Trejos, E., Martinez-Ramos, M., Meave, J., Paz, H., Perez-Garcia, E., Romero-Peraz, I., Tauro, A., & Bongers, F. (2013). Successional changes in functional composition contrast for dry and wet tropical forest. *Ecology, 94*, 1211–1216.

Niinemets, Ü., & Valladares, F. (2006). Tolerance to shade, drought, and waterlogging of temperate northern hemisphere trees and shrubs. *Ecological Monographs, 76*, 521–547.

Nowacki, G., & Abrams, M. (2008). The demise of fire and "mesophication" of forests in the eastern United States. *BioScience, 58*, 123.

Reich, P. B. (2014). The world-wide "fast-slow" plant economics spectrum: A traits manifesto. *Journal of Ecology, 102*, 275–301.

Reich, P., Ellsworth, D., & Uhl, C. (1995). Leaf carbon and nutrient assimilation and conservation in species of differing successional status in an oligotrophic Amazonian forest. *Functional Ecology, 9*, 65–76.

Russell, M. B., Woodall, C. W., D'Amato, A. W., Domke, G. M., & Saatchi, S. S. (2014). Beyond mean functional traits: Influence of functional trait profiles on forest structure, production, and mortality across the eastern US. *Forest Ecology and Management, 328*, 1–9.

Swenson, N. G., & Weiser, M. D. (2010). Plant geography upon the basis of functional traits: An example from eastern North American trees. *Ecology, 91*, 2234–41.

Swenson, N. G., & Weiser, M. D. (2014). On the packing and filling of functional space in eastern North American tree assemblages. *Ecography, 37*, 001–007.

Webb, C. T., Hoeting, J. A., Ames, G. M., Pyne, M. I., & LeRoy, P. N. (2010). A structured and dynamic framework to advance traits-based theory and prediction in ecology. *Ecology Letters, 13*, 267–83.

Weiher, E., & Keddy, P. A. (1995). Assembly rules, null models, and trait dispersion: New questions from old patterns. *Oikos, 74*, 159–164.

White, P. S., Collins, B., & Wein, G. R. (2011). Natural disturbances and early successional habitats. In C. H. Greenberg, B. S. Collins, & F. R. Thompson III (Eds.), *Sustaining young forest communities: Ecology and management of early successional habitats in the Central Hardwood Region, USA*. New York: Springer.

Wilfahrt, P. A., Collins, B., & White, P. S. (2014). Shifts in functional traits among tree communities across succession in eastern deciduous forests. *Forest Ecology and Management, 324*, 179–185.

Wiser, S., & White, P. S. (1999). High elevation outcrops and barrens of the southern Appalachian mountains. In R. C. Anderson, J. A. Fralish, & J. M. Baskin (Eds.), *Savannas, barrens, and rock outcrop plant communities of North America*. Cambridge: Cambridge University Press.

Chapter 12
The Historic Role of Humans and Other Keystone Species in Shaping Central Hardwood Forests for Disturbance-Dependent Wildlife

Cathryn H. Greenberg, Kendrick Weeks, and Gordon S. Warburton

Abstract Multiple natural disturbance types historically created conditions that were suitable for many, but not all, disturbance-dependent wildlife species in the Central Hardwood Region (CHR). In addition, some wildlife species, such as beavers, passenger pigeons, elk, and bison, historically functioned as keystone species by creating or maintaining unique disturbed habitats that otherwise would be rare. For millennia, humans (Native Americans, and later European settlers) also created and maintained early successional habitat variants (estimated at 7–43 % of the CHR landscape in 1500 AD) including farmlands, old fields in different stages of succession, grasslands, and open woodlands by clearing for cultivation and settlements, frequent burning, and old field abandonment. In this chapter, we argue that humans were a keystone species in the CHR, having a major influence on the diversity, distribution, and abundance of many disturbance-dependent wildlife species by creating, maintaining, or greatly expanding specific, unique types of early successional habitats and some mature forest types dominated by shade-intolerant pioneer species, such as yellow pine. Determining the largely unknowable historic range of variation of natural disturbances, selecting an arbitrary moment on a temporally and spatially dynamic landscape as a reference, and subjectively deciding what should

---

C.H. Greenberg (✉)
USDA Forest Service, Southern Research Station, Bent Creek Experimental Forest,
1577 Brevard Rd, Asheville, NC 28806, USA
e-mail: kgreenberg@fs.fed.us

K. Weeks
North Carolina Wildlife Resources Commission,
346 South Mills River Rd, Mills River, NC 28759, USA
e-mail: Kendrick.Weeks@ncwildlife.org

G.S. Warburton
North Carolina Wildlife Resources Commission,
783 Deep Woods Drive, Marion, NC 28752, USA
e-mail: Gordon.Warburton@ncwildlife.org

C.H. Greenberg, B.S. Collins (eds.), *Natural Disturbances and Historic Range of Variation*, Managing Forest Ecosystems 32,
DOI 10.1007/978-3-319-21527-3_12

or should not be included as 'natural' may not serve as the most productive guide for conservation. Alternatively, forest and land use planning for diverse wildlife conservation might more logically start with clear objectives, and proceed with management activities targeted toward attaining them.

Keywords Wildlife • Keystone species • Humans • Ecosystem engineers • Disturbance-dependent birds

12.1 Introduction

Disturbance-dependent wildlife species require open structural conditions created immediately after forest disturbances or at some point early in the dynamic process of recovery and maturation. Historically, natural disturbances (e.g., Chaps. 1, 2, 3, 4, 5, 6, 7, 8, 9, this volume) provided habitats for many disturbance-dependent species by creating patches of young forest structure, from small canopy gaps to large swaths of partial or complete canopy removal, within a mature upland hardwood forest matrix of the Central Hardwood Region (CHR). In addition, some wildlife species, such as beavers (*Castor canadensis*), passenger pigeons (*Ectopistes migratorius*), elk (*Cervus canadensis*), and bison (*bison bison*), historically functioned as keystone species by creating or maintaining unique disturbed habitats such as wetlands or prairies that would otherwise be rare, thereby increasing the abundance, diversity, and distribution of wildlife species that required them. Hence, multiple natural disturbance types historically created conditions that were suitable for many, but not all, disturbance-dependent wildlife species. However, several breeding birds (Askins 2001) and other wildlife species of the CHR such as woodchucks (*Marmota monax*) and rabbits (*Sylvilagus floridanus*) require specific variants of disturbance-created habitats that were created, maintained, or expanded in large part by humans (*Homo sapiens*) through active land management by clearing, frequent burning, and land abandonment in and surrounding inhabited areas, for thousands of years (Delcourt and Delcourt 2004).

In this chapter, we argue that humans were a keystone species in the CHR, having a major influence on the diversity, distribution, and abundance of many disturbance-dependent wildlife species by creating, maintaining, or greatly expanding specific, unique types of early successional habitats and some mature forest types dominated by shade-intolerant pioneer species, such as yellow pine (*Pinus spp.*). Through land use and active land management by clearing, farming, and frequent burning Native Americans (and later European settlers) created settlements, gardens, farmlands, meadows and prairies, open woodlands, [river] canebreaks (*Arundinaria gigantea*), and old fields at varying stages of succession that included successional yellow pine forests (Delcourt and Delcourt 2004). We use results of studies on natural disturbances, paleoecology, archeology, and historical accounts by early explorers to illustrate and discuss likely dynamic scenarios of prehistoric (prior to European settlement) CHR landscapes, and availability of different early successional habitat variants required by disturbance-dependent wildlife. We focus our discussion on disturbance-created habitats rather than wildlife species per se, because habitat availability likely governed the distribution of many disturbance-dependent wildlife species historically, as it does today. We use disturbance-dependent breeding birds to illustrate possible scenarios of historic wildlife distribution because they are among the most thoroughly studied taxa, and many species are specialized in their associations with specific variants of early successional habitat (MacArthur and MacArthur 1961) that include disturbed young forests, but also other land uses and conditions commonly created by humans (Askins 2001).

12.2 Origin and Early History of Central Hardwood Forests: Climate, Megafauna, and Humans

Human arrival in the CHR coincided with retreating ice sheets and a warming climate more than 13,000 years before present (BP), as open tundra-boreal 'parklands' dominated by spruce (*Picea* sp.) and jack pine (*P. banksiana*) were slowly being replaced by temperate, deciduous forest migrating north from warmer more southerly refuges (Delcourt and Delcourt 2004). During the last ice age, glaciers in North America extended south as far as the Missouri and Ohio Rivers, and east to New England (Clark et al. 2009), and a colder, drier climate resulted in alpine tundra in the Blue Ridge Mountains above 1,450 m (Delcourt and Delcourt 2004). Between 18,000 years and 5,000 years BP, climate in the CHR shifted from arid-cool to the warm-humid climate of today (Carroll et al. 2002), interrupted by glacial re-advancing with associated cold and drought during the Younger Dryas period (12,800–11,500 years BP) (http://en.wikipedia.org/wiki/Younger_Dryas), and a rapid warming (more than 7° C in 50 years) around 11,500 years BP (the Pre-boreal transition phase) (http://en.wikipedia.org/wiki/Boreal_(age)). Warmer temperatures during the hypsithermal (7,500–5,000 years BP) were followed by a cooling trend (5,000–120 years BP) that included the Little Ice Age (600–120 years BP) (Carroll et al. 2002). Oak-chestnut-hickory (*Quercus-Castanea-Carya*) forests became widespread in the CHR by 3,000 years BP (Dickinson 2000; Delcourt and Delcourt 2004). Past climate change and associated disturbance regimes, even in recent millennia, has been a major 'background' natural disturbance in the CHR and resulted in major shifts in forest composition and habitats, as well as extinctions and reassembly of wildlife communities.

When humans first arrived, megafauna including ancient and modern bison (*Bison antiquus* and *B. bison*), elk, primitive horses (*Equus* spp.), wooly mammoths (*Mammuthus* sp.), mastodons (*Mammut* sp.), stag-moose (*Cervalces scotti),* and giant ground sloth (*Megalonyx* sp. and others), as well as extant modern wildlife species associated with boreal forest and tundra, such as muskox (*Ovibos muschatus*) and caribou (*Rangifer tarandus*) inhabited much of the CHR (Carroll et al. 2002; O'Gara and Dundas 2002; http://exhibits.museum.state.il.us/exhibits/larson/ice_age_animals.html; http://parks.ky.gov/parks/historicsites/big-bone-lick/history.

aspx). A mere 14,000 years ago or less, grazing, browsing, and trampling by keystone megafauna herbivores were important natural disturbances, shaping forests and maintaining open habitats (e.g., Svenning 2002) for many disturbance-dependent species.

Most of these megafauna are now extinct; the relative roles of an overall warming climate, associated shifts in vegetation composition and structure, and the arrival of big-game hunting humans (the Clovis culture) to their extinction is hotly debated (Burney and Flannery 2005; Koch and Barnosky 2006; Askins 2014). Theories that extinctions were precipitated through over-hunting by humans are supported by archaeological evidence at multiple sites, showing human arrival just prior to local extinctions of remaining megafauna (many populations collapsed from 14,800 to 13,700 years BP; Gill et al. 2009). Mosimann and Martin (1975) hypothesized and developed simulations illustrating how even a small founding population of humans could multiply quickly enough to extirpate the slow-reproducing megafauna under a steady hunting regime, with extinctions progressing in a front, beginning in Alaska where humans first entered North America, and moving slowly south and east (Burney 1993). The coincidental timing of extinctions of many megafauna species with the arrival humans is corroborated by a study of spores in lake sediments in upstate New York, Ohio, and Indiana (Gill et al. 2009). The study showed that the abundance of *Sporomiella*, a fungus that grows on the dung of herbivorous mammals, declined dramatically between 14,000 and 13,000 years BP, indicating that large mammals also declined during that period and coinciding with human arrival. This decline was followed by an increase in deciduous trees (possibly released by the absence of grazing and browsing by keystone megafauna), and more frequent fires (possibly set by humans and/or fueled by denser vegetation), as evidenced by increased charcoal particles in the sediments (Gill et al. 2009). Many of the mammals that became extinct at the end of the Pleistocene had survived similar glacial-interglacial transitions for hundreds of thousands of years, suggesting that humans may have played an important and perhaps complex role in their demise (Burney and Flannery 2005; Koch and Barnosky 2006).

Whether through their hypothesized role in extinction of megafauna or (and) their documented role in the more recent extinction or local extirpation of modern fauna, the predatory activities of humans have dramatically and directly influenced many wildlife populations in the CHR. In the last 250 years or less, European settlers were directly responsible for the extinction of several species including the Carolina parakeet (*Conuropsis carolinensis*) and passenger pigeon; the regional extirpation of wolves (*Canis lupus*), cougars (*Puma concolor*), bison, elk, and other species; and dramatic population declines of other species such as beavers through excessive and unregulated hunting and trapping. By removing keystone wildlife species such as beaver (wetland creators), or elk and bison (grazers) whose activities historically created or maintained disturbed, open habitats, humans also indirectly influenced the distributions and populations of many disturbance-dependent wildlife species.

Perhaps most importantly, humans themselves have historically functioned as a keystone species through their management and use of land on inhabited landscapes

Fig. 12.1 (**a**) Open woodland with grass-forb-woody understory (Similar to descriptions by early explorers in the Cumberland Plateau, Piedmont of the Carolinas and Georgia, and Appalachians on South- and West-facing aspects of the southern Appalachians) on private land in Sequatchie County, Tennessee in 2013. The woodland, initially closed canopy forest, was not mechanically thinned, and was burned every 2–3 years since the late 1970s (Craig Harper pers. comm; photo courtesy of Craig Harper); (**b**) Native prairie vegetation at Ft. Campbell in Kentucky and Tennessee managed with burning or mowing at 1–3 year intervals (Daniel Moss pers. comm.) (Photo courtesy of William Minser)

since before central hardwood forests existed as we know them today. By regularly creating, maintaining, or expanding early successional habitats, including many variants that might rarely be created by natural disturbances alone such as agricultural fields, old fields in different stages of succession, open woodlands (e.g., Fig. 12.1a), meadows or prairies (e.g., Fig. 12.1b), and successional yellow pine-dominated forests, humans historically were a major influence on abundance, distribution, and species diversity of disturbance-dependent fauna.

12.3 Early Successional Habitats: One Size Does Not Fit All

The term 'early successional habitat' is commonly, albeit erroneously, used generically to refer to any open, recently disturbed habitat that is transient unless maintained by recurring disturbances (Greenberg et al. 2011a). Although both young forest and truly early successional habitats share the features of openness in common, they differ considerably in many ways in the structure and composition of plants (Lorimer 2001; Greenberg et al. 2011a) and, because of that, the wildlife species that use them. In the CHR, high-severity natural disturbances such as large blowdowns, or anthropogenic disturbances such as regeneration harvests, create young forest with high woody stem density and thick cover for wildlife, but generally do not create bona fide successional conditions with high plant species turnover. Even after high-severity natural disturbances that substantially reduce canopy cover, plant species composition usually remains similar to the original mature forest, often with a transient addition of blackberry (*Rubus* spp.) or pokeweed (*Phytolacca americana*), as pre-existing shrubs and fallen or damaged trees resprout prolifically and tree seedlings grow from pre-established advance regeneration or seed (Lorimer 2001; Greenberg et al. 2011b). In the CHR open, young forest conditions typically last 8–15 years before canopy closure (Loftis et al. 2011; Thompson and Dessecker 1997).

In contrast, truly successional habitats occur when pioneer plants colonize treeless areas that are no longer disturbed. In the CHR, most successional habitats originate from anthropogenic land uses, such as abandoned pasture or farmlands, as colonizing shade-intolerant pioneer tree species, including yellow pines (e.g., pitch (*P. rigida*), shortleaf (*P. echinata*), or Virginia pine (*P. virginiana*)), yellow-poplar (*Liriodendron tulipifera*), or black locust (*Robinia pseudoacacia*) (Delcourt and Delcourt 2004), and shrubs create conditions suitable for other, later successional species in a relay sequence toward a mature hardwood forest (Lorimer 2001; Greenberg et al. 2011a).

Disturbance-dependent breeding birds are associated with open habitats created by disturbances, but many are relatively specialized, requiring specific but different variants of early successional habitats ranging from young forest to grasslands (Askins 2001; Hunter et al. 2001) (Table 12.1). In this chapter we use the term 'early successional habitat' as it is commonly used and understood in the wildlife literature to denote open conditions created by either natural or anthropogenic disturbances (Greenberg et al. 2011a). However, we emphasize that 'one size does not fit all' for disturbance-dependent wildlife species, and therefore different variants of early successional habitats, created by both natural and anthropogenic disturbances, and interactions between them, are required to maximize diversity and abundance of native, disturbance-dependent breeding birds.

Table 12.1 Associations of select disturbance-dependent breeding bird species of the CHR with different early successional habitat variants[a] created by natural or anthropogenic disturbances

Species	Early successional habitat variants[a]										
	GH	WM	OW	Sa	SS	SF	Pa	Ag	OF	Su	Ur
American goldfinch (*Carduelis tristis*)	X	X	X	X	X	X	X	X	X	X	–
American kestrel (*Falco sparverius*)	X	–	–	X	–	–	X	X	–	–	–
American robin (*Turdus migratorius*)	X	–	X	X	–	–	X	X	–	X	–
American woodcock (*Scolopax minor*)	–	–	X	–	X	X	–	–	–	–	–
Barn owl (*Tyto alba*)	X	–	–	X	–	–	X	X	–	–	–
Barn swallow (*Hirundo rustica*)	X	X	–	X	–	–	–	X	–	X	–
Bewick's wren (*Thryomanes bewickii*)	X	–	X	X	–	–	–	–	–	X	–
Blue grosbeak (*Passerina caerulea*)	–	–	X	–	X	–	–	–	X	–	–
Blue-winged warbler (*Vermivora pinus*)	–	–	X	X	X	–	–	–	X	–	–
Bobolink (*Dolichonyx oryzivorus*)	X	–	–	–	–	–	–	–	–	–	–
Brown thrasher (*Toxostoma rufum*)	–	–	X	X	X	X	–	–	–	X	–
Carolina wren (*Thyrothorus ludovicianus*)	X	–	X	X	–	–	–	–	–	X	–
Chestnut-sided warbler (*Setophaga pensylvanica*)	–	–	X	–	X	–	–	–	–	–	–
Chipping sparrow (*Spizella passerina*)	X	–	X	X	–	–	X	X	–	X	–
Common nighthawk (*Chordeiles minor*)	–	–	X	X	–	–	–	–	–	X	X
Common yellowthroat (*Geothlypis trichas*)	–	–	X	–	X	–	–	–	X	–	–
Dickcissel (*Spiza americana*)	X	–	–	X	–	–	X	X	–	–	–
Eastern bluebird (*Sialia sialis*)	X	–	X	X	–	–	X	X	–	X	–
Eastern kingbird (*Tyrannus tyrannus*)	X	–	–	X	–	–	–	–	–	–	–
Eastern meadowlark (*Sturnella magna*)	X	–	–	–	–	–	X	–	–	–	–
Eastern phoebe (*Sayomis phoebe*)	X	–	X	X	–	–	–	X	–	X	–
Eastern towhee (*Pipilo erythrophthalmus*)	–	–	X	X	X	X	–	–	–	X	–
Field sparrow (*Spizella pucilla*)	–	–	X	X	X	–	–	–	–	–	–
Golden-winged warbler (*Vermivora chrysoptera*)	–	X	X	–	X	–	–	–	–	–	–
Grasshopper sparrow (*Ammodramus savannarum*)	X	–	–	–	–	–	–	–	–	–	–
Gray catbird (*Dumetella carolinensis*)	–	–	X	–	X	X	–	–	–	X	–
Henslow's sparrow (*Ammodramus henslowii*)	X	X	–	–	–	–	X	–	–	–	–
Horned lark (*Eremophila alpestris*)	X	–	–	–	–	–	X	X	–	X	–
House wren (*Troglodytes aedon*)	X	–	X	X	–	–	–	–	–	X	–
Indigo bunting (*Passerina cyanea*)	–	–	X	X	X	–	–	–	–	X	–
Loggerhead shrike (*Lanius ludovicianus*)	X	–	X	X	–	–	X	X	–	–	–
Mourning dove (*Zenaida macroura*)	X	–	X	X	–	–	X	X	–	X	–
Northern bobwhite (*Colinus virginianus*)	X	–	X	X	–	–	–	–	–	–	–
Northern cardinal (*Cardinalis cardinalis*)	–	–	X	–	X	X	–	–	X	X	–
Northern mockingbird (*Mimus polyglotus*)	X	–	X	X	–	–	–	–	–	X	–

(continued)

Table 12.1 (continued)

Species	Early successional habitat variants[a]										
	GH	WM	OW	Sa	SS	SF	Pa	Ag	OF	Su	Ur
Northern rough-winged swallow (*Stelgidopteryx serripennis*)	X	X	–	–	–	–	–	X	–	–	–
Orchard oriole (*Icterus spurius*)	–	–	X	X	–	–	–	–	X	–	–
Prairie warbler (*Setophaga discolor*)	–	–	X	–	X	X	X	–	–	–	–
Purple martin (*Progne subis*)	X	X	–	X	–	–	X	X	X	–	–
Red-headed woodpecker (*Melanerpes erythrocephalus*)	–	X	X	X	–	–	–	–	–	–	–
Red-tailed hawk (*Buteo jamaicensis*)	X	–	X	X	–	–	X	X	X	X	X
Red-winged blackbird (*Agelaius phoeniceus*)	–	X	–	–	–	–	–	–	–	–	–
Ruffed grouse (*Bonasa umbellus*)	–	–	X	X	X	X	–	–	–	–	–
Savannah sparrow (*Passerculus sandwichensis*)	X	–	–	–	–	–	–	–	–	–	–
Song sparrow (*Melospiza melodia*)	X	–	X	X	–	–	–	–	–	X	–
Tree swallow (*Tachycineta bicolor*)	–	X	–	X	–	–	–	X	–	–	–
Vesper sparrow (*Pooecetes gramineus*)	X	–	–	–	–	–	–	–	–	–	–
White-eyed vireo (*Vireo griseus*)	–	–	X	–	X	X	–	–	–	–	–
Yellow-breasted chat (*Icteria virens*)	–	–	X	–	X	X	–	–	X	–	–

[a]*GH* grass-herbaceous, *WM* wet meadow, *OW* open woodland, *Sa* savanna, *SS* scrub-shrub, *SF* sapling forest, *Pa* pasture, *Ag* agricultural, *OF* old field, *Su* suburban, *Ur* urban

12.4 Natural Disturbances and Early Successional Habitats for Wildlife: Severe Weather, Weather-Related, and Biotic Agents

Historically, non-anthropogenic natural disturbances created variable age classes and structural conditions across small- to large areas, at different locations and times in a shifting mosaic of ephemeral patches, with young forest composing varying proportions of the vast CHR landscape at any given time. Mobile, disturbance-dependent wildlife species could use these transient young forest habitats opportunistically. However, in the hypothetical absence of human-caused disturbances, their populations would almost certainly have been variable, fluctuating spatially and temporally as conditions became available for brief periods before becoming unsuitable as forests recovered and matured.

12.4.1 Severe Weather

Based on records over the past 150 years or less, disturbance types, frequencies, and severities historically varied temporally and spatially within and among ecoregions, and locally across topography. For example, in the CHR hurricane-related winds are more frequent in eastern ecoregions, whereas tornados are more frequent in western ecoregions but also commonly occur in the Piedmont, Ridge and Valley, and Southwestern Appalachians ecoregions (see Peterson et al. Chap. 5). Damage from hurricane-related winds is variable. For example in the Blue Ridge Mountains, Hurricane Opal (whose track did not enter the ecoregion despite considerable damage there) caused single- to multiple-tree windthrows from downbursts of wind (McNab et al. 2004), whereas damage from Hurricane Hugo included tens of thousands of hectares rated as "total timberland damaged" (Dogett 1993). Ice storms (Lafon Chap. 7) and landslides (Wooten et al. Chap. 9) are most common on steep slopes in mountainous ecoregions such as the Blue Ridge Mountains. The impacts of severe weather disturbances on central hardwood forests ranged from small gaps (Hart Chap. 2) to large swaths of broken limbs and (or) fallen trees, creating a gradient of young forest conditions potentially suitable for gap, edge, scrub-shrub, or sapling-stage breeding bird species (Table 12.1).

12.4.2 Lightning-Ignited Fire

Lightning-ignited fires are rare in the deciduous forests of the CHR because fuels are generally too moist, discontinuous, or otherwise inadequate to allow spread except under severe drought conditions (e.g., Schroeder and Buck 1970; Barden 1997; Delcourt and Delcourt 1997; Flatley et al. 2013; also see Sect. 12.7.1). Schroeder and Buck (1970) estimated that about 1–5 lightning ignitions occur annually per 4,000 km^2 in the CHR. The wildfire reporting database for National Forests shows that within CHR National Forests, lightning ignites an average of 0.3–7.8 fires per 2,000 km^2 annually; in contrast humans, accidentally or purposefully, ignite an average of 4.8–93.9 wildfires (not including prescribed burns) per 2,000 km^2 annually (Greenberg et al. Chap. 1, Table 1.6). Guyette et al. (2006a) estimated fewer than one in 200 wildfires in the CHR were lightning ignited; the rest were due to arson, cigarettes, unattended campfires, or other anthropogenic causes. Historic fire frequencies positively corresponded with human population densities and far exceeded probable frequencies attributable to lightning ignition (Guyette et al. 2006a, b; Hart and Buchanan 2012; also see Grissino-Mayer Chap. 6; Leigh Chap. 8).

Studies in the Boston Mountains ecoregion suggest that topographic heterogeneity, or 'roughness' reduces fire frequency in general (Guyette et al. 2006b). However, lightning-ignited fires occur more frequently and with higher severity in some topographic positions, such as ridgetops and dry south-, southwest-, or west-facing slopes in the Blue Ridge Mountains ecoregion that tend to be low-quality, drier sites (Flatley et al. 2013). Not coincidentally, these topographic positions are also where Table Mountain pine (*P. rigida*), pitch pine, mountain laurel (*Kalmia latifolia*), and other plant species associated with dry, low-quality sites and occasional fire most commonly occur (Zobel 1969).

Under most conditions, wildfires in hardwood forests of the CHR are low-intensity, and changes to forest structure and breeding bird communities are relatively minor and transient (Greenberg et al. 2013). Exceptions may occur during infrequent, extreme drought conditions, or on steep topography and ridgetops with xeric, low site quality conditions that are more conducive to hot, high-severity fires that result in heavy tree mortality. A combination of low lightning ignition frequency and the relatively random location of lightning strikes across the vast CHR render it unlikely that lightning-ignited fires alone would have repeatedly burned the same landscapes with sufficient frequency needed for the development and maintenance of prairies, savannas, open woodlands, or yellow pine forests (see Sect. 12.4.3) with occasional exceptions. When they did occur, lightning-ignited high-severity fires likely created open, young forest conditions with abundant standing snags potentially suitable for yellow pine regeneration (Jenkins et al. 2011) and for gap-, edge, scrub-shrub, sapling-stage, or open woodland breeding bird species (Table 12.1) (e.g., Greenberg et al. 2013).

12.4.3 Southern Pine Beetle

Historically, Native Americans (and later European settlers) promoted the development and maintenance of pine forests over mature, climax upland hardwoods on inhabited landscapes by actively managing with frequent fire, and by abandoning farmlands that were often colonized by yellow pines such as shortleaf, Virginia, and pitch pine (Ashe and Ayers 1901; Mattoon 1915; Ashe 1922; Balch 1928; Delcourt and Delcourt 2004). Such yellow pine-dominated forests are disappearing due to a combination of (1) southern pine beetle (*Dendroctonus frontalis*) attacks on yellow pine forests that established when Native Americans or European settlers (through the mid-1900s) were actively clearing, burning, and abandoning lands (see Nowak et al. Chap. 4), and (2) a dramatic reduction in the frequency of anthropogenic fire ignitions and (or) suppression of wildfires when they do occur (see Sect. 12.7.1). Without active land management by humans, yellow pine-dominated forests would likely have been limited in distribution to low quality sites and topographic positions where hardwood competition is reduced and lightning-ignited fires are more frequent. On a hypothetical CHR landscape without humans, the impact of southern pine beetles may have been minor across much of the landscape, because yellow pine forests would have been much more restricted in their distribution.

12.5 Keystone Wildlife Species: Bison, Elk, and Beaver as Agents of Disturbance and Ecosystem Engineers

Many animal species respond to, rather than drive, forest structure. However, some species were themselves agents of disturbance, functioning as keystone species by helping to create and maintain open, early successional conditions and wetlands that enabled many other wildlife species to thrive. Elk, bison, beaver, and perhaps even passenger pigeons were especially notable 'ecosystem engineers' that historically had considerable impacts on the CHR landscape, often in close association with humans. In fact, a mutualism developed between Native Americans, and their large ungulate prey. Native Americans created prairies, open woodlands, and forest edge through frequent burning and clearing that enabled elk and bison to thrive; in turn, their grazing helped to arrest forest succession and maintain these grass-dominated habitats that ignited easily and carried fire across broad areas (Delcourt et al. 1998; McClafferty 2000; Delcourt and Delcourt 2004). This 'graze and burn' disturbance regime, co-managed by Native Americans and large ungulates, created structural conditions promoting higher densities and greater distributions of wildlife species requiring open, grassy habitats than would be possible in its absence (e.g., Table 12.1).

12.5.1 Passenger Pigeon

Passenger pigeons numbered 3–5 billion in eastern North America until the late 1800s when their population declined until their extinction in 1914 (Yeoman 2014). They ranged throughout most of eastern North America, wintering south of latitude 36° N and breeding in the northern part of the CHR including southern New York, west-southwest across Pennsylvania, northern West Virginia, and Kentucky as well as parts of the midwestern USA (Ellsworth and McComb 2003). Based on their massive numbers and collective mass, passenger pigeon flocks are thought to have been an important agent of low-intensity disturbance throughout much of the CHR, similar to that of ice storms or lower-intensity windstorms, by increasing light levels in forests through limb and tree breakage (Ellsworth and McComb 2003). They also covered the ground with several cm of feces at nesting and roosting sites, killing the understory vegetation and inhibiting plant growth for several years (see Ellsworth and McComb 2003), and potentially altering fuels structure by killing understory plants and creating coarse woody debris (Ellsworth and McComb 2003). Flocks, estimated at 105,000 pigeons per ha, commonly formed columns over 1 km wide and 400–450 km long (King 1866 as cited in Schorger 1955; Ellsworth and McComb 2003) and numbered in the millions (see Yeoman 2014). In 1871, naturalist A.W. Schorger estimated a communal nesting site covering 2,200 km^2 of sandy oak barrens in Wisconsin, with 136 million breeding adults (Yeoman 2014). Roosting and nesting areas, commonly 0.02–10 km^2 but sometimes as large as 130 km^2, were used by an estimated 27,000–36,000 kg per ha of pigeons (Ellsworth and McComb 2003).

Ellsworth and McComb (2003) estimated that historically, passenger pigeons affected approximately 0.5–0.8 % of their total winter range annually through their use of temporary and long-term roosts; breakage of smaller limbs and trees (lower-intensity disturbance) affected an estimated 8 % of their breeding area annually (Ellsworth and McComb 2003). Ellsworth and McComb (2003) suggest that low-intensity damage in nest areas likely resulted in a light environment favoring shade-tolerant tree species such as American beech (*Fagus grandiflora*), eastern hemlock (*Tsuga canadensis*), and sugar maple (*Acer saccharum*), and establishment of intermediate (moderately shade-tolerant) species such as oaks, in gaps. In contrast, severe physical damage in roost areas may have resembled damage from high winds, or even hot fires that top-kill most plants and add nutrients to the soil, resulting in high light levels and release of intermediate tree species such as oaks or eastern white pine (*P. strobus*) (Ellsworth and McComb 2003). The gradient of structural conditions created by passenger pigeons were likely suitable for gap-associated breeding birds such as cerulean warblers (*Setophaga cerulea*) (Perkins 2006) where damage was relatively light, to edge- and open area species such as indigo buntings (*Passerina cyanea*) where damage was more severe. Passenger pigeons also may have played an important role in seed and nut dispersal. Hence, prior to their extirpation by humans, passenger pigeons may have functioned as a keystone species by affecting forest composition as well as forest structure that in turn affected other wildlife species.

12.5.2 Beaver

Historically, beavers inundated riparian forests by damming slow-moving streams, creating large areas of boggy scrub-shrub wetlands with dead standing trees, or beaver meadows (see Greenberg et al. Chap. 1, Fig. 1.2 h) that supported wetland-dependent plants and animals. Beaver populations were estimated at between 60 and 400 million in pre-colonial North America (Seton 1929), and they occurred virtually anywhere with suitable water and food plant resources (Baker and Hill 2003). In his travels, Bartram noted that beaver 'abounded' north of Georgia (Van Doren 1928). Given the extremely high populations of beaver in pre-European times, the entire CHR surely included an extensive mosaic of beaver ponds and swamps of various sizes and configurations. Hey and Phillipi (1995) estimated that 40 million beavers in 1600 would have maintained 206,795 km^2 of water surface in the upper Mississippi and Missouri River basins; current beaver populations there may pond about 2,023 km^2. In the eastern USA, heavy beaver trapping for the fur trade began in the 1600s (Naiman et al. 1988). Between 1853 and 1877 the eastern USA Hudson Bay Company alone shipped three million pelts to England (http://www.ohiohistorycentral.org/w/American_Beaver?rec=1124). Beaver populations nearly vanished throughout North America by the 1900s due to excessive trapping for the fur trade (Naiman et al. 1988; Baker and Hill 2003).

Inundation of dammed waterways provides multiple benefits for wildlife and increases local landscape diversity. Beaver pond complexes provide standing water, edge, standing snags killed by flooding, plant diversity, and vertical stem diversity all in one area (Baker and Hill 2003). Wetland vegetation including sedges (*Carex* spp.), bulrush (*Scirpus* spp.), and cattails (*Typha* spp.) rapidly colonize newly created wetland complexes. Many wildlife species including waterfowl, wading birds, red-winged blackbirds (*Agelaius phoeniceus*), woodpeckers, migrating songbirds, raptors, aquatic furbearers such as mink (*Neovison vison*) and muskrat (*Ondatra zibethicus*), amphibians, and semi-aquatic reptiles such as bog turtles (*Clemmys muhlenbergii*) and water snakes colonize beaver ponds and wetlands, and use them for breeding and habitat (Baker and Hill 2003; Rosell et al. 2005). Historically, when beaver populations were high, at least some beaver ponds may have persisted for over 1,000 years (Ben Tanner unpubl. data). Abandoned beaver ponds eventually drain and are colonized by shrubs and grasses, sometimes lasting for years before eventually reverting to forest (Askins 2000). Historically, these beaver meadows and disturbed areas surrounding beaver ponds were likely extensive along floodplains, and provided habitats for shrub-scrub and some grassland bird species, rodents, lagomorphs, ungulates, and their predators (Askins 2000; Baker and Hill 2003).

Historically, frequent and continual creation and abandonment of beaver ponds across the CHR created diverse wetland habitats that facilitated a much higher local diversity, as well as abundance, landscape distribution and population connectivity for many more wildlife species than today. Some species with poor dispersal ability, such as bog turtles, have reached critically low population levels likely in part because of severely diminished beaver populations and the habitats they engineered (US Fish and Wildlife Service 2001); more than 90 % of mountain bog habitat has been lost (Walton 2006). Thus, prior to their near-extirpation by humans, beaver were historically perhaps one of the greatest animal agents of disturbance given the spatial extent and effects of their habitat alteration activities.

12.5.3 Elk

More than 10 million elk were estimated to have occurred in North America prior to the arrival of Europeans (Seton 1929). Modern elk have occupied the CHR for at least 20,000 years (McClafferty 2000; O'Gara and Dundas 2002), since (and for long after) boreal forest and tundra dominated the landscape (Delcourt and Delcourt 2004). There are many historical reports of large numbers of elk in the CHR. Place names such as Banner Elk, North Carolina (Blue Ridge Mountains ecoregion), Elk River in West Virginia (Central Appalachians and Allegheny Plateau ecoregions), and Elk Creek in Kentucky (Interior Plateau ecoregion) also suggest that elk were once widespread (VDGIF 2010). By the late 1800s or early 1900s they had been eliminated by over-hunting (O'Gara and Dundas 2002; Innes 2011).

Elk are grazers, primarily using grasslands or open prairies, but usually remaining within 400 m of mature forests which they use for cover (Peek 2003). They feed primarily on grasses, shrubs, and forbs depending on the season and location (Peek 2003). Although elk populations were likely controlled by hunting, predation by wolves, black bear (*Ursus americanus*), and cougars, and diseases, their numbers, as seen and reported, were clearly sufficiently high to impact vegetation structure and composition in and around the open areas where they occurred. At high densities elk can alter species composition and structure, reduce or eliminate shrubs or seedlings, decrease plant diversity, and create browse lines (McClafferty 2000).

Even at low to moderate densities, elk browsing can slow the rate of succession (Putnam 1996). Selective grazing can stimulate the growth of palatable grasses at moderate densities or favor undesirable forage species at higher densities (Woodward et al. 1994; Schreiner et al. 1996). Although elk play an important role in maintaining open grasslands by grazing and browsing, it is unlikely that they can create them from a starting point of mature hardwood forest. In all likelihood, Native Americans facilitated the presence of elk in the CHR through frequent burning that created and maintained meadows, prairies and open woodlands.

12.5.4 Bison

Modern bison have been present (until recently) in the CHR since at least the Pleistocene (http://exhibits.museum.state.il.us/exhibits/larson/ice_age_animals.html). As with elk, grasslands created by widespread clearing and frequent burning by Native Americans provided suitable conditions for bison to thrive (Rostlund 1960) for thousands of years (http://exhibits.museum.state.il.us/exhibits/larson/ice_age_animals.html), as the CHR landscape transitioned from boreal parkland to deciduous forest. Reports by early explorers, archeological finds, place names such as Buffalo Lick in the Piedmont ecoregion of northeastern Georgia (http://www.bartramtrail.org/pages/articles.html), and a buffalo wallow in central North Carolina (Simmons 1999), indicate that bison were widespread throughout much of the CHR. Buffalo traces were made and used during their seasonal migrations from the plains of central Illinois, through forests to the salt licks of Kentucky (Interior Plateau and Interior Valleys and Hills ecoregions). These traces were wide enough to be used as travel routes by Native Americans and later by European settlers (http://keithbobbitt.com/Log%20Cabins/NorthCarolinaRoute.htm).

Reports by early explorers indicate that bison were quite numerous, and grasslands and woodlands were sufficiently abundant to support them (Rostlund 1960). In 1722, Catesby noted "The buffalo. They range in droves feeding upon the open savannas morning and evening, and in the sultry time of the day they retire to shady rivulets and streams" in the Piedmont ecoregion (as cited in Rostlund 1960). Michaux (1805) reported seeing bison in groups of 150–200 in his travels through Kentucky in the early 1800s. Ramsey (1853, as cited in Moss 2001) described prairies around Nashville, Tennessee (Interior Plateau ecoregion) "luxuriant growth of native grasses, pastured over as far as the eye could see, with numerous herds of deer [*Odocoileus virginianus*], elk, and buffalo." Captain Timothe de Monbreun, a French hunter and trapper from Illinois, traveling down the Cumberland River near Palmyra, Tennessee (Montgomery County) reported seeing large herds of buffalo in 1777 (Kellogg 1939 as cited in Moss 2001). In 1780, buffalo were killed by Colonel John Donelson and his party near the Cumberland River along the Kentucky-Tennessee line (Williams 1928 as cited by Moss 2001). Bison disappeared from the southeast in the middle 1800s largely due to over-hunting by European settlers (Rostlund 1960). They were extirpated from North Carolina by 1765, Maryland by 1775, and Virginia by 1797 (Rostlund 1960). Michaux (1805) noted that bison were frequently shot by settlers for their tongues, with the remainder of the carcass wasted.

In the CHR, bison used fire-maintained prairies and shrub-grass woodlands for feeding (Tesky 1995). Bison are grazers, eating up to 14 kg of grass daily (Evans and Pobasco 1977), although they may switch to woody browse species if grasses are not available. Similar to elk, bison grazing pressure can lead to changes in plant composition and structure, impede forest succession (Reynolds et al. 1982) and help to maintain the grass-dominated communities they depend on. Bison can also affect local stands of timber by horning and thrashing during the rut (Reynolds et al. 1982). Trails on hillsides can drain upland areas and change vegetative composition (Reynolds et al. 1982), and wallows can become pools of water for other species to use (Tesky 1995). Prior to their regional extirpation by humans, both bison and elk were likely instrumental in retarding forest succession by grazing that, in conjunction with frequent burning by humans, helped to create and maintain open oak woodlands, prairies and savannas.

12.6 Humans as a Keystone Species and Ecosystem Engineer on the Historic Landscape

For millennia, *Homo sapiens* was a predominant keystone species and ecosystem engineer in the CHR that created and maintained many variants of open, early successional habitats and forest age classes for a wide variety of disturbance-dependent species in and surrounding their settlements (Carroll et al. 2002). Native Americans were active land managers, using fire to clear forests for settlements and agriculture, and to improve visibility, facilitate travel, increase native fruit production, and create edge and open, grassy habitats to attract game species (Van Lear and Waldrop 1989; Yarnell 1998; Carroll et al. 2002; Delcourt and Delcourt 2004). Social organization became more centralized and sophisticated over millennia, and by 1,000 BP concentrated settlement patterns and agriculture "generated a landscape patchwork of fragmented forests, cultivated land, and nutrient-depleted old fields abandoned as fallow land" (Delcourt and Delcourt 2004). Habitats created and maintained primarily by Native Americans included settlements, farmlands and gardens, abandoned fields of different age-classes, and forests of pioneer species colonizing abandoned fields including river cane (forming dense canebreaks) and yellow pines (Delcourt and Delcourt 2004). Through their land management activities across large landscapes adjacent to settlements, Native Americans – and later European settlers – created grassland, garden-residential 'suburb,' agricultural field, old-field, yellow pine forest, and open woodland habitats, thereby influencing species diversity, distribution, and abundance of many disturbance-dependent breeding birds with these specific habitat associations (Table 12.1).

12.6.1 *Historic Land Use and Forest Management by Native Americans in the Central Hardwood Region*

As early as 10,000 years BP Native Americans lived in long-term settlements in valleys and lowlands near rivers throughout much of the CHR. Archaeological evidence from the Shenandoah Valley of Virginia, the Little Tennessee River Valley of east Tennessee, eastern Kentucky, Watauga Valley of North Carolina, West Virginia, and the Valley and Ridge province indicate that such settlements were widespread. By the Late Archaic (4,500 years BP) Native Americans cultivated crops and managed forests surrounding and far beyond their settlements by burning frequently to facilitate travel and visibility, promote seed, berry, and nut production, and production of grasses and forage to attract elk, deer, bison, and other game species (Chapman et al. 1982; Williams 1989). These activities increased through the late Holocene, with a substantial increase in burning frequency by 3,000 years BP (Yarnell 1998; Delcourt and Delcourt 2004; Grissino-Mayer Chap. 6; Leigh Chap. 8). Delcourt and Delcourt (2004) suggest that by 3,000 years BP frequent burning by Native Americans promoted upland forests dominated by more fire tolerant oak, chestnut, hickory, and walnut (*Juglans* spp.), even as the prevailing climate would otherwise be expected to promote more fire-intolerant, mesophytic species. Frequent burning promoted the development and maintenance of savannas, prairies, open woodlands and pine forests that were once interspersed throughout the CHR (Flatley et al. 2013). By 1,000 BP Native American social organization was highly sophisticated in the CHR, with widespread dependence on agriculture (Delcourt and Delcourt 2004).

American Indians cleared land for villages and agriculture by girdling trees through peeling or burning away bark, and by burning. Older fields with declining soil fertility and productivity were abandoned to be reclaimed by forest, and new fields were cleared progressively and in rotation (Williams 1989), creating a mosaic of different-aged forests and forest structures in the areas surrounding settlements. Disturbances from agriculture, field abandonment, and frequent burning affected forest composition near settlements. Before agriculture became widespread, only 10 % of wood charcoal from archaeological sites was from pioneer species such as yellow-poplar, pine, red cedar (*Juniperus virginianus*), or river cane; subsequently (and prior to the arrival of Europeans) it rose steadily to 50 % (Chapman et al. 1982; Yarnell 1998).

Wood was used to build structures, and for fuel (Williams 1989). Assuming that fuel wood use by Native Americans was similar to that of European settlers in the Blue Ridge Mountains ecoregion, an average family used about 3.62 m^3 (15 cords) per year for cooking and warmth, which was likely salvaged from cleared land, second growth forest in abandoned fields, cull and undersized trees, or the surrounding forest (Nesbitt and Netboy 1946; Williams 1989). Model simulations for the Little Tennessee River Valley (Baden 1987 as cited in Delcourt and Delcourt 2004) indicate that the area required for growing enough maize for one person increased from 0.1 to nearly 0.4 ha between 1,000 and 250 years BP as dependency on maize increased. Soil fertility and harvest yield generally decline sharply within 10 years on maize fields, forcing abandonment of old and clearing of new fields.

Clearing, agriculture, and widespread burning by Native Americans influenced the forests and early successional habitats, but were almost certainly most pronounced near settlements along valley bottoms and surrounding uplands (Van Lear and Waldrop 1989; Milner and Chaplin 2010). By 3,000 years BP anthropogenic fire resulted in "a fine-grained patchwork of vegetation on upper hillslopes and ridgetops that include prehistoric garden plots, open patches with mixed crops of domesticated species, abandoned Indian old-fields reverting back into early-successional grassland barrens, thickets of shrubs, and even-aged stands of pitch pine or tulip poplar trees" on the Cumberland Plateau with mixed mesophytic forests in more sheltered topographic positions (Delcourt and Delcourt 2004). A schematic representation of land use by Native Americans illustrates the different variants of early successional habitats they created in and surrounding villages by clearing, burning, and agricultural activities (Fig. 12.2).

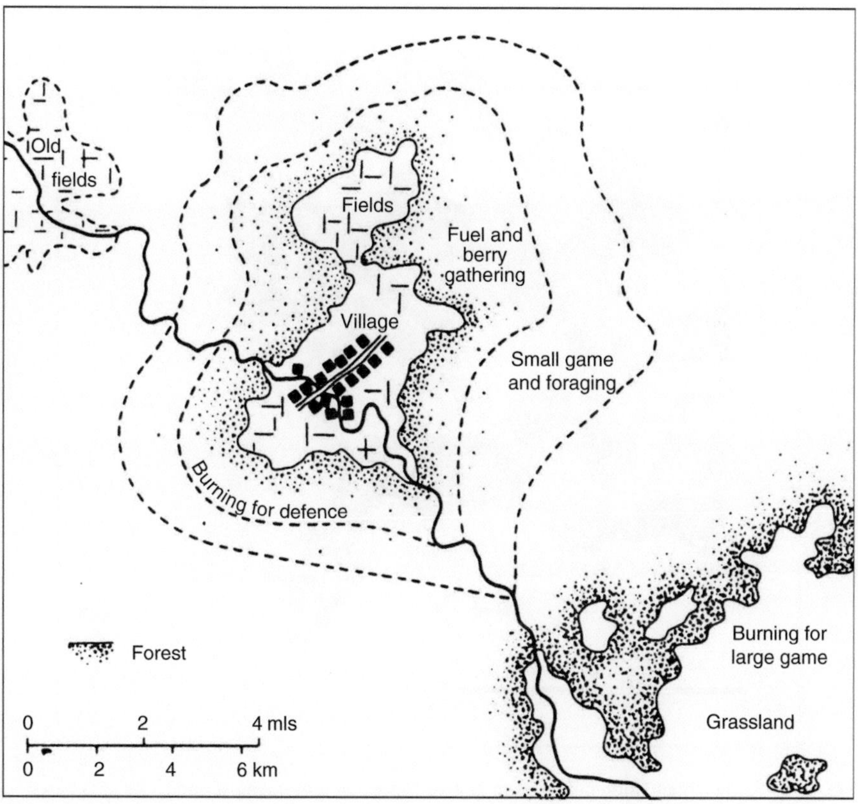

Fig. 12.2 Schematic of a Native American settlement and surrounding managed landscape (From Williams 1989)

Estimates of the Native American population ca. 1500 AD in the eastern USA range between 0.5 and 2.6 million, based on archaeological and historical information (Milner and Chaplin 2010). Their settlements and agricultural fields were located in spatially discrete, irregularly distributed aggregates across productive land, primarily along coastlines or (in the CHR) river valleys (Fig. 12.3) (Milner and Chaplin 2010). Nearly all Native Americans lived in villages of 50–1,000 people surrounded by fields (e.g., Fig. 12.2; Williams 1989). Some settlements were quite large; tens of thousands of people lived along a 100 km stretch along the Mississippi River and into the surrounding uplands in Cahokia, near east St. Louis (Ozark Highlands ecoregion), during the Mississippian period (800–1500 AD) (Denevan

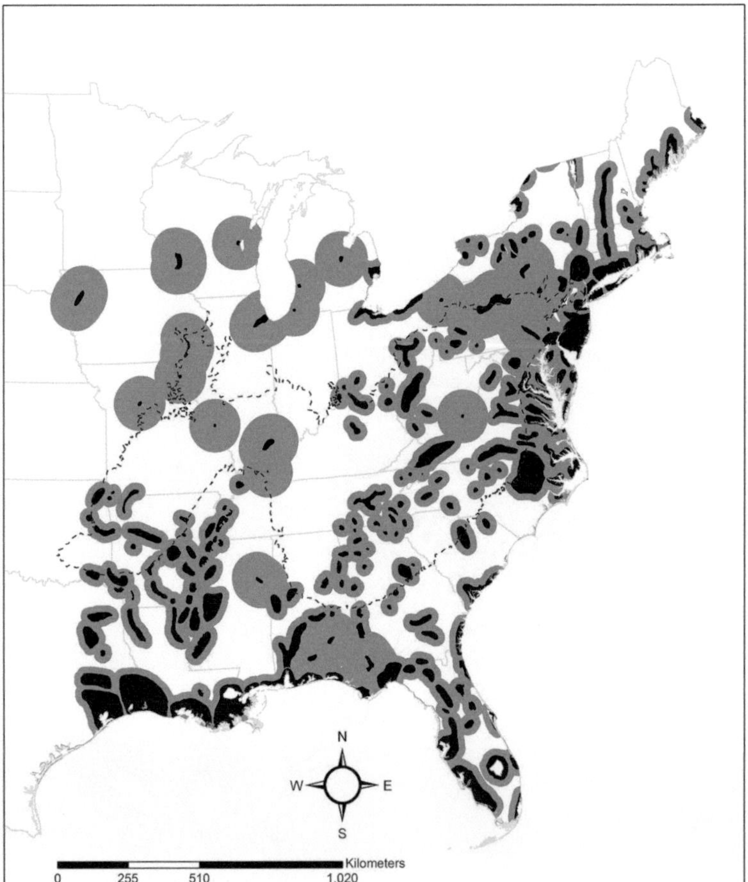

Fig. 12.3 Known population polygons of Native American settlements around 1500 AD based on archaeological and historical evidence (*black*), with buffers (*gray*) encompassing the potential zone of human influence, such as frequent fire, surrounding settlements (Adapted from Milner and Chaplin 2010)

1992; Milner and Chaplin 2010). Both population levels and the locations of settlements were dynamic over time. For example, a major depopulation occurred in the Midwest sometime between 1300 and 1500 AD, prior to European contact. Settlements sometimes relocated, likely as soil and fuel resource were depleted (Williams 1989).

Milner and Chaplin (2010) calculated the area of known Native American settlements ca. 1500 AD, and used nearest-neighbor statistics to calculate buffers surrounding them that incorporate the area of probable environmental impact. Based on their modeling, Native Americans impacted at least 7.1 % (settlements only), and up to 42.6 % (including buffers) of the landscape within the CHR ca. 1500 AD, prior to settlement by Europeans (Fig. 12.3). Landscapes heavily impacted by Native Americans likely expanded, contracted, and shifted with their populations and movements. Large tracts of mature or old-growth forest subjected primarily to non-anthropogenic natural disturbances likely created a matrix between populated areas and surrounding areas of environmental impact (Fritz 2000; Delcourt and Delcourt 2004). During this prehistoric moment, in a temporally and spatially dynamic landscape, suitable habitat was likely widely available for breeding bird species associated with mature- and young forests created by natural disturbances, as well as for species requiring different variants of land uses and early successional habitats that were created primarily by Native Americans.

12.6.2 Descriptions of Native American Land Use and Historic Landscapes by Early European Explorers

Accounts of pre-settlement landscapes by early naturalists and explorers are scant, and potentially biased, as most likely took routes most easily traversed and likely near or between Native American population centers. Nonetheless, several descriptions provide insight into landscape condition – thus the availability of different early successional wildlife habitats – at specific times and places, in different central hardwood ecoregions prior to or with minimal settlement by people of European descent.

In 1540, writings from Hernando DeSoto's expedition described the landscape of the Blue Ridge Mountains ecoregion inhabited by the Lamar and Qualla cultures as "including palisaded towns and large expanses of cultivated fields" ... "Ridges were well-wooded, and outside the cultivated valleys, the land was all forest" (as cited in Yarnell 1998). In Virginia, the Shenandoah Valley between the Blue Ridge Mountains and the Alleghenies was described in the mid-1700s as a vast grass prairie covering more than 2,590 km^2, which was burned annually by Native Americans (Van Lear and Waldrop 1989). In 1670 German explorer John Lederer described the Roanoke Valley along the Virginia-North Carolina border and along the border with West Virginia as forested, but "where it was inhabited by Indians, it lay in open in spacious plains," and "by the industry of these Indians as... very open and clear of wood" (John Lederer as cited in Williams 1989).

Prairie-like openings throughout the Piedmont ecoregion, some up to 40 km across, were described by several early explorers and traders (see Barden 1997). In 1540 Spanish and Portuguese narratives described "many fine fields… the forest was more open and there were very good fields along streams… They traveled a full league [5 km] through a garden-like land of fruit-bearing trees, among which a horse could be ridden without any trouble" near Camden, South Carolina. Further north along the Catawba River they describe several plains and "many fine fields of tilled lands" (Rostlund 1957 as cited in Barden 1997). In 1567 Spanish explorer Juan Pardo describes "very large and good plains… clear land… beautiful plains" including one near Charlotte North Carolina (Rostlund 1957 as cited in Barden 1997) during his travels through the Piedmont ecoregion. Other travelers (e.g., John Lederer in 1670; John Speed in 1676; John Lawson 1701; Catesby in 1720s (as cited in Barden 1997)) describe large savanna and prairie throughout the Piedmont ecoregion maintained by frequent burning. In winter 1701, John Lawson noted "the woods being newly burnt and on fire in many places," and in the 1720s Mark Catesby observed many fires started by Native Americans, observing that "in February and March the inhabitants have a custom of burning the woods, which causes such a continual smoke, that not knowing the cause, it might be imagined to proceed from the fog… an annual custom of the Indians in their huntings, of setting the woods on fire many miles in extent" (as cited in Barden 1997).

The Coosa chiefdom, including an area from the confluence of the French Broad and Tennessee Rivers to around Talladega Alabama in the Southwestern Appalachians, Ridge and Valley, and Blue Ridge Mountains ecoregions, was described by the DeSoto expedition (1540) as "thickly settled in numerous towns with fields between, extending from one to another" (US Congress as cited in Williams 1989). Bartram (Van Doren 1928) describes endless savannas along the Tennessee River to the west of the Appalachians in the Interior Plateau ecoregion in his 1775 travels.

The first white settlers in western Kentucky encountered the Big Barrens (Interior River Valleys and Hills, and Interior Plateau ecoregions), encompassing an estimated 12,950–15,540 km^2. They described it as vast grassland with only occasional stunted trees unsuitable for building material, fences, and fuel (Owen 1856 and Hussey 1876 as cited in McInteer 1946). Early writers thought that the open prairie vegetation of the Big Barrens was created and maintained through frequent burning by Native Americans to attract game, and later by the first European settlers to maintain pasturage for cattle (Michaux 1805; McInteer 1946). The prairie vegetation of the Big Barrens may be partly explained by its unique limestone geology, but a rapid shift to forest vegetation as well as cultivated fields and pasture by the early 1800s – soon after white settlement – indicates that frequent fire was instrumental in maintaining these open, prairie conditions (see McInteer 1946; Baskin et al. 1994).

Historic accounts of the Ouachita Mountains ecoregion indicate that forests were more open, with lower tree density and basal area and more shortleaf pine than today; extensive prairies occurred in the western Ouachita Mountains (Foti and Glenn 1990). Le Page du Pratz of Natchez wrote of his travels from the Natchez to the St. Francis in the late 1720s "during the summer, the grass is too high for travelling;

whereas in the month of September the meadows, the grass of which is then dry, are set on fire and the ground becomes smooth, and easy to walk on: and hence it is, that at this time, clouds of smoke are seen for several days together to extend over a long track [sic] of country; sometimes to the extent of between 20 [100 km] and 30 [150 km] leagues in length, by two or three leagues in breadth, more or less…" and after rains "the game spread themselves all over the meadows and delight to feed on the new grass..." (du Pratz 1774 as cited in Foti and Glenn 1990). He writes further "The lands we find in going up the Black [Ouachita] River… in general may be considered as one very extensive meadow, diversified with little groves, and cut only by the Black River and little brooks, bordered with wood up to their sources" (du Pratz 1774 as cited in Foti and Glenn 1990). Dunbar and Hunter described the hills near the Ouachita River in 1804 and 1805 as being sometimes barren, with oak-dominated and pine woods variously in the area (Rowland 1930 as cited in Foti and Glenn 1990). In 1819–1820, botanist Edwin James described the Ouachita Mountains as covered with small and scattered trees or nearly treeless with some denser forests along the bases of mountains east of Hot Springs (James 1823 as cited in Foti and Glenn 1990). In 1844, Featherstonhaugh wrote that Indian fires thinned the forests but did not destroy them and "now that Indians have abandoned the country, the undergrowth is rapidly occupying the ground again" (Featherstonhaugh 1844 as cited in Foti and Glenn 1990).

During their 1818–1819 travels through the Ozark Highlands ecoregion, Henry Schoolcraft and Levi Pettibone noted valley bottoms of dense, pristine deciduous forest, valley walls covered with oak, hickory and pine forests, and uplands covered by a mosaic of prairie, oak savanna, oak woods with open undergrowth, and open grassy glades or barrens. These were probably enlarged and maintained by the Osage Indians who set fire to prairies to drive game into the wooded areas where the animals could be more easily killed (Rafferty 1996). Brackenridge wrote of his voyage up the Missouri River in 1811 that "… notwithstanding the ravages of fire, the marks of which are everywhere to be seen, the woods, principally hickory, ash, and walnut formed a forest tolerably close" (Brackenridge 1816 as cited in Spetich et al. 2011). Between the early nineteenth and late twentieth century, tree density in the Boston Mountains ecoregion tripled, increasing from 133 to 378 trees per ha, and from 123 to 287 per ha in the Ozark Highlands ecoregion, likely due to changes in the cultural practice of intentional burning (Foti 2004).

Clearly, Native Americans throughout the CHR created abundant open conditions in and surrounding their settlements by clearing for settlements and cropland, and by their frequent, widespread use of fire to manage fields, woodlands, and grasslands. Through their land management activities, Native Americans functioned as a keystone species by creating specific variants of early successional habitats required by different disturbance-dependent species. Without human-created habitats, species strongly associated with grasslands, savannas or prairies (e.g., elk, bison, bobwhite quail (*Colinus virginianus*), eastern meadowlark (*Sturnella magna*), field sparrow (*Spizella pucilla*), grasshopper sparrow (*Ammodramus savannarum*)), abandoned fields (e.g., yellow-breasted chat (*Icteria virens*), blue grosbeak (*Passerina caerulea*), or gardens and suburbs (e.g., song sparrow (*Melospiza melodia*), northern mockingbird (*Mimus polyglotus*), or chipping sparrow (*Spizella passerina*)) would likely have been uncommon or highly restricted in their distribution (e.g., beaver meadows) in the CHR (Table 12.1).

12.6.3 European Settlement

DeSoto's explorations in 1540 mark the first Native American contact with Europeans, and the start of Native American depopulation from newly introduced diseases and warfare (Yarnell 1998). European settlement of the CHR began in the mid- to late 1700s (Williams 1989). By the early 1800s most Native American populations had been severely reduced and secondary forests began to overtake their abandoned fields and farmlands (Yarnell 1998).

European settlers in the CHR continued the Native American practice of burning, and perhaps increased the area and frequency. Human habitation was concentrated in the river valleys and lowlands, where agriculture and burning made their greatest mark on the landscape and surrounding slopes. As the post-Civil War population of settlers increased in the CHR, so did populations of free-ranging cattle, pigs, and other domestic animals – even at higher elevations of the southern Appalachian ecoregion. The frequency – often annual – of burning large landscapes also increased to expand the area of woodlands and grasslands as pasturage (Ashe and Ayers 1901; Yarnell 1998). In 1885, ornithologist William Brewster (1886) wrote of the Blue Ridge Mountains ecoregion "Much of the low country, especially those portions bordering or near the larger streams, is under cultivation… Extensive areas, however, are everywhere still clothed in forest, either of vigorous second-growth or fine old timber." Brewster (1886) also wrote "in many places… trees are scattered about in groups or singly at intervals of one or more hundred feet, with grassy openings in between, giving the country a park-like appearance." Yellow pine forests, open woodlands, and grasslands remained abundantly available surrounding settled areas during this period due to the land management activities of European settlers (Ashe and Ayers 1901; Mattoon 1915; Ashe 1922; Balch 1928).

12.7 Recent Changes in Land Use and Condition: The Past 120 Years

Many variants of early successional or heavily disturbed habitats were likely at their historical high for several decades in the early 1900s for several reasons: (1) much of the CHR was systematically and progressively logged, resulting in large areas of young forest, with new areas cut as others grew up and matured; (2) large wildfires were common, ignited both intentionally and by sparks from trains using railroads built to extract timber, fueled by recently cutover, dry forestland; (3) family-run farming practices commonly left weedy fencerows, fallow fields, and untilled patches; (4) the demise of American chestnut (*C. dentata*) created forests with large numbers of standing dead trees, followed by heavily perforated canopies lasting for many years as the 'mighty giants' fell and before their replacement in the canopy by oak and other forest tree species.

Conversely, both young forest and other early successional habitats may be at an historical low today because: (1) family-run farming operations have diminished since the 1960s, replaced by industrialized farming practices using equipment and herbicides that eliminate weed and brush cover; (2) forests on public lands have been allowed to mature for the past century, with dramatic reductions in regeneration harvest levels in recent decades (Shifley and Thompson 2011); (3) human population growth, land ownership patterns, urban sprawl, and second homes have fragmented forests and removed large areas from the wildland base.

Additionally, in the past century, humans have had substantial indirect impact on forest structure and tree species dominance by introducing non-native forest pests and pathogens that have effectively wiped out (or soon will) entire tree species within the CHR (Hicks 1998). In the early 1900s introduction of the non-native chestnut blight (*Cryphonectria parasitica*) gradually killed all mature American chestnut trees, then a dominant species throughout most of the CHR that composed up to 50 % of forest trees in some locations. Since then, gypsy moth (*Lymantria dispar*), balsam wooly adelgid (*Adelges piceae*), hemlock wooly adelgid (*Adelges tsugae*) and others have (or soon will) dramatically altered CHR by killing important tree species that are dominant in several forest types. Introductions of non-native wildlife species such as starlings (*Sturnus vulgaris*) and house sparrows (*Passer domesticus*), and increases in domestic and feral predators such as dogs (*Canis lupus familiaris*) and cats (*Felis catus*) also negatively affect wildlife populations and communities.

12.7.1 Reduced Fire Frequency: Suppression… or Changes in Cultural Burning Practices?

Many of the open woodlands, yellow pine forests, prairies, and other fire-maintained conditions in the CHR began to transition to closed canopy hardwood forests between the 1920s and 1940s, after the federal government began a campaign to curtail the use of fire across the landscape (Spetich et al. 2011). Fire suppression policies followed on the heels of several notable fires that burned vast areas of coniferous forest and killed people in the western USA or in northern states (e.g., upstate New York and Minnesota) (Spetich et al. 2011). Catastrophic crown fires are integral to the ecology of some coniferous ecosystems such as lodgepole pine (*P. contorta*) forests in the western USA, boreal forests in northern states (Schoennagel et al. 2004), and sand pine (*P. clausa*) scrub in Florida (Greenberg et al. 1995); in these ecosystems, mortality of mature pine trees is rapidly followed by release of their seed onto the bare, fire-'prepared' seedbed, with regenerating forests developing directly back to the original pine forest type. In contrast, crown fires are nearly unheard of in hardwood forests of the CHR. Wildfires in the CHR are typically surface fires that generally do not kill mature trees, and result in few long-term changes to either fuels or forest structure as shrubs rapidly resprout, and leaf litter is replenished as deciduous leaves drop the following autumn (Stottlemyer et al. 2006; Waldrop et al. 2007, 2010).

Historically, continuous, grassy fuels likely carried fire across frequently burned prairies, savannas and open woodlands that were locally interspersed with closed canopy forests throughout the CHR (Flatley et al. 2013) (e.g., Fig. 12.1). However, the relatively low frequency of lightning-ignited fire (e.g., Tuttle et al. Chap. 10; Greenberg et al. Chap. 1, Table 1.6), and the high fire frequency required to create and maintain a grassy ground cover, suggests that these habitats would have been rare in the absence of frequent, intentional burning by Native Americans (and subsequently by European settlers) (Lorimer 2001; Delcourt and Delcourt 2004; Guyette et al. 2006a; see Sect. 12.4.2). A hypothetical historical CHR landscape without forest management by Native Americans may have been dominated by primarily mature or old growth forest, interspersed with beaver-engineered wetlands and meadows along waterways, and subject to sporadic and varying natural disturbances (see Chaps. 1, 2, 3, 4, 5, 6, 7, 8, and 9, this volume) including occasional low-severity (e.g., low tree mortality) lightning-ignited fire; high-severity (e.g., heavy tree mortality) lightning-ignited fires were likely mainly limited to specific topographic positions and (or) under infrequent severe drought conditions. The decline of open, fire-maintained habitats in the CHR resulted from (1) eliminating the accepted cultural practice of frequently and repeatedly burning forests by Native American and (later) European settlers; and (2) suppressing wildfires that were predominantly ignited by humans, either intentionally or accidentally (e.g., Tuttle et al. Chap. 10; Greenberg et al. Chap. 1, Table 1.6 and Sect. 12.4.2), rather than suppression of natural (lightning-ignited) wildfires, or fire suppression policies per se.

12.8 Habitat Requirements of Disturbance-Dependent Wildlife: Were Natural Disturbances Enough?

Breeding bird species in the CHR differ in their associations with specific structural features (MacArthur and MacArthur 1961; Askins 2001), and across the gradient of conditions created by different natural and anthropogenic disturbances (Table 12.1). Many forest interior species, such as the wood thrush (*Hylocichla mustelina*) and ovenbird (*Seiurus aurocapillus*), are primarily associated with relatively large tracts of undisturbed closed-canopy forests but often use young, productive forests with abundant fruit and insect foods (Greenberg et al. 2011b) to forage during the post-fledgling period (Lanham and Whitehead 2011; Stoleson 2013). Others, such as the northern cardinal (*Cardinalis cardinalis*), eastern titmouse (*Baeolophus bicolor*),

Carolina chickadee (*Poecile carolinensis*), Carolina wren (*Thyrothorus ludovicianus*), eastern towhee (*Pipilo erythrophthalmus*), and American robin (*Turdus migratorius*) are generalists, able to thrive across a wide range of forest conditions and land uses.

Disturbance-dependent species are associated with open habitats created by disturbances, but may differ in their specific requirements. Some, such as indigo buntings, can thrive in most open habitats with adequate perch and nest sites, including small to extensive patches of young forest created by natural disturbances such as hurricane-related winds (Greenberg and Lanham 2001), tornadoes (Newbold 1996), or high-severity fire (e.g., Greenberg et al. 2013), or by anthropogenic disturbances such as recently harvested forest (e.g., Greenberg et al. 2014), shrubby pasturelands, or even utility rights-of-way (Lanham and Whitehead 2011). In contrast, requirements of many other CHR disturbance-dependent species are relatively specialized, and often associated with early successional habitats that are and historically were created primarily by humans (Table 12.1). On a hypothetical historical (Holocene) CHR landscape without humans, most of these more specialized species would likely have had lower population levels and narrower distributions. This suggests that many such species either co-evolved with Native American forest management, or were able to expand their ranges in response to land uses by Native Americans in the CHR.

12.8.1 Breeding Bird Response to Natural Disturbances

Our literature search indicates a paucity of studies on wildlife response to natural disturbances in the CHR. Studies of breeding birds in upland hardwood forests of the Ozark Mountains ecoregion of Arkansas (Prather and Smith 2003) and in the Ridge and Valley ecoregion of Tennessee (Newbold 1996) reported that the number and abundance of early successional species increased, and densities of most mature forest species remained high in tornado-damaged sites compared to mature forest. In the Blue Ridge Mountains ecoregion Greenberg and Lanham (2001) also reported higher species richness and relative abundance of breeding birds – including closed canopy-, edge-, and gap-associated species – in large (0.1–1.2 ha), 'incomplete' canopy gaps created by hurricane-related downbursts of wind that downed many, but not all trees. These studies of natural disturbances, together with studies of anthropogenically-disturbed (recently harvested) young forests in several different ecoregions indicate that the presence of a residual, partial canopy and dense shrubs or stump sprouts from damaged trees provide suitable habitat for forest canopy-, shrub-, and some disturbance-dependent species, resulting in higher species richness and density of breeding birds compared to mature forest (see Greenberg et al. 2014). However, many other disturbance-dependent breeding bird species of the CHR do not commonly occur in young forest, instead requiring more open ground, grass cover, or old field conditions (Table 12.1) that historically were created primarily by humans.

12.8.2 Breeding Bird Response to Anthropogenic Disturbances

Several breeding bird species in the CHR are uniquely associated with specific human-modified environments (Table 12.1). For example, eastern meadowlarks require open fields with tall, continuous grass cover. Bobwhite quail require continuous, tall grass and shrub cover with open- or no forest canopy. Golden-winged warblers (*Vermivora chrysoptera*) require open, grassy areas with some shrub and sapling cover in a forested matrix, at elevations greater than 850 m in the Blue Ridge Mountains ecoregion or lower elevations at higher latitudes (Rosenberg et al. in press). Chipping sparrows require open, mowed areas. Eastern bluebirds (*Sialia sialis*) require wide open fields where nest boxes are provided (or high-severity burns with standing snags (Greenberg et al. 2013), which are rare in the CHR). Field sparrows, yellow-breasted chats, and blue grosbeaks require abandoned pastures and old fields with mosaics of grass, shrubs, and saplings. Song sparrows and northern mockingbirds occur almost exclusively in garden habitats or suburban residential areas (Table 12.1).

By creating required habitat conditions for species that would otherwise be rare or limited in distribution, humans – first Native Americans and later European settlers – have functioned as a keystone species for thousands of years. Native Americans created and maintained savannas, prairies, and open woodlands by frequent, intentional burning; these habitats were used by elk and bison that helped to maintain them by grazing. Villages and agricultural fields were created by clearing and burning, and abandoned to revert to old fields, yellow pine forests, or other successional conditions. Historically, these anthropogenically-created habitats allowed many disturbance-dependent breeding bird species with specific requirements for differing variants of early successional habitats to occur and thrive within the CHR.

12.9 Historic Abundance and Shifting Distributions of Breeding Birds

Habitat availability affects both local and regional distributions of mobile species such as disturbance-dependent breeding birds that can opportunistically exploit ephemeral habitats. Historically, populations of breeding bird species dependent on anthropogenically-created habitats such as gardens, old fields, and grasslands likely tracked spatially and temporally dynamic Native American populations, settlement patterns, and increasing reliance on agricultural crops. Over the past several decades changes in land use and management practices have reduced the quantity and suitability of anthropogenically-disturbed habitats for many disturbance-dependent breeding bird species, with direct and indirect effects on their populations. The greater prairie chicken (*Tympanuchus cupido*), a grassland associate once common in the Big Barrens region of south-central Kentucky were extirpated largely due to the disappearance of vast, fire-maintained prairies within the region (Hunter et al. 2001). More than 70 % of eastern grassland-associated breeding bird species such as grasshopper sparrow, savannah sparrow (*Passerculus sandwichensis*), Henslow's sparrow (*Ammodramus henslowii*), Vesper sparrow (*Pooecetes gramineus*), bobolink (*Dolichonyx oryzivorus*), and eastern meadowlark have significantly declined over the past several decades (Askins 1993, 2000, 2001; Knopf 1994; Sauer et al. 2000) due to intensive management of pasture and haylands and loss or fragmentation of agricultural grasslands to development.

Similarly, nearly half of shrub-early successional breeding birds have significantly declined over the past several decades (Sauer et al. 2000). The disappearance of family farms, where low-intensity livestock grazing created and maintained a mosaic of grass, shrubs, and saplings, has resulted in the steep decline in golden-winged warbler populations in the Blue Ridge Mountains ecoregion where they were once common (Rosenberg et al. in press). Other bird species that depend on scrub/shrub or other early successional habitats, such as blue-winged warbler (*Vermivora pinus*), prairie warbler (*Setophaga discolor*), yellow-breasted chat, common yellowthroat (*Geothlypis trichas*), chestnut-sided warbler (*Setophaga pensylvanica*), American woodcock (*Scolopax minor*), field sparrow, indigo bunting, brown thrasher (*Toxostoma rufum*), and eastern towhee, have declined from 10 to 60 % in the eastern USA over the past 40 years (Partners in Flight 2013). Declines in ruffed grouse (*Bonasa umbellus*) populations are associated with a dearth of dense, sapling stage forest that develops 8–15 years after regeneration harvests (Thompson and Dessecker 1997). This is partly due to changes in federal land management policies that have greatly reduced timber harvests on National Forests. These strong correlations between land use change and populations of bird species associated with specific land uses or conditions suggests that their populations, and those of other wildlife species with similar habitat requirements, also likely increased or decreased historically with the shifting availability of suitable habitats created by weather-related natural disturbances, pests and pathogens, keystone wildlife species, and humans.

Historical shifts in the geographic distribution of some breeding bird species are documented, and are associated with changes in habitat availability corresponding with a dynamic climate and changing land uses or conditions. For example, red-cockaded woodpeckers (*Picoides borealis*) occurred until only a few decades ago at the northern extent of their range in isolated stands of shortleaf or Virginia pine or pine-oak in the Southwestern Appalachians and Blue Ridge Mountains ecoregions of Tennessee and Kentucky (Nicholson 1977). These small populations have disappeared as their fire-maintained yellow pine forest habitats declined due to southern pine beetle outbreaks (Nicholson 1977; see Nowak et al. Chap. 4), elimination of the cultural practice of intentional burning, and fire suppression. Historically, red-cockaded woodpeckers and other yellow pine forest associates such as brown-headed nuthatches (*Sitta pusila*) and pine warblers (*Setophaga pinus*) may have been more widely distributed in the CHR in yellow pine forests that commonly established on farmlands and old fields abandoned first by Native Americans and later by European settlers through the mid-twentieth century.

Similarly, the breeding distribution of golden-winged warblers has expanded and contracted within some CHR ecoregions over the past 150 years in response to regional changes in land use or condition, hence habitat availability (Rosenberg et al. in press). During the late nineteenth and early twentieth century widespread timber cutting and later abandonment of agricultural lands provided abundant early successional habitat for golden-winged warblers (Rosenberg et al. in press). In his visits to the Blue Ridge Mountains ecoregion Brewster (1886) wrote "Common in Jackson and Macon Counties, ranging 2,000–4,100 feet [610–1,250 meters], and haunting open oak woodlands, and second growth on hillsides. In many such tracts it seemed to be the most abundant and characteristic species…" Just a century later, golden-winged warblers only occur in isolated, higher elevation locations where habitat exists (Rosenberg et al. in press).

Brewster (1886) went on to describe abundant populations of other disturbance-dependent breeding bird species that today are relatively rare in the Blue Ridge Mountains ecoregion. Bobwhite quail were "Abundant everywhere, in grain fields in the valleys, oak woodland over the mountain sides, and throughout the balsam forests that cover the higher peaks and ridges..."; golden eagles [*Aquila chrysaetos*] were "frequently seen… said to breed on inaccessible cliffs and ledges of the higher mountains, whence they often descend into the valleys to prey on young geese, lambs, etc." Bartram noted many species associated with early successional habitats such as blue linnet (indigo bunting), yellow-breasted chats, and golden-winged warblers (Van Doren 1928).

Just as some disturbance-dependent species have declined or even disappeared from large areas of the CHR, they may persist in patches of suitable habitat, or colonize or recolonize areas if suitable habitats become available. For example, Henslow's sparrows were recorded on the Cumberland Plateau in Tennessee (Southwestern Appalachians ecoregion) for the first time after a large, local farm was left unmanaged for a few years, allowing suitable grassland conditions to develop (Lance 2014). American kestrels (*Falco sparverius*) and bobwhite quail are common in and near the large, open, grassy meadows at the Sandy Mush Game Lands in Buncombe County, but rare throughout most of western North Carolina. Prairie warblers (Greenberg et al. 2013) and red-headed woodpeckers (*Melanerpes erythrocephalus*) (Greenberg et al. unpubl. data) recently colonized tracts of upland hardwood forest in the Blue Ridge Mountains ecoregion, within several years after experimental high-severity burns killed most trees. Bobolinks, rare in the CHR, recently colonized hayfields managed with low-intensity mowing in Watauga County, North Carolina (K. Weeks, pers. obs.). These examples illustrate how 'if you build it they will come,' and suggest that managing for specific variants of early successional habitats could increase populations and distributions of targeted disturbance-dependent species.

12.10 Wildlife Conservation and Management within the Historic Range of Variation

Historically, humans were a keystone species, having a profound influence on the abundance, distribution, and diversity of disturbance dependent breeding birds and other wildlife species. Land management activities and land uses by Native

Americans such as clearing for settlements and agriculture, farmland abandonment, and frequent burning to create and maintain open grasslands and woodlands provided variants of early and later successional habitats for specialized species that would otherwise have been rare or absent in the CHR. On a hypothetical historical landscape without humans, mobile disturbance-dependent species that require edge and young forest conditions would likely have been transient in their distributions as natural disturbances created suitable conditions lasting only several years before canopy closure. However, except for those that could use once-common beaver meadows, breeding bird species requiring grassland, old field, garden-'suburb,' open woodland, and even yellow pine forest habitats (with some exceptions) would have been uncommon or absent within the CHR.

Forest management within the historic range of variation of natural disturbances first requires a subjective decision regarding whether prehistoric land management activities by humans should be regarded as a natural disturbance. If not, it becomes challenging to envision an historic CHR that approaches accuracy, because humans have occupied the CHR for more than 13,000 years; prior to human occupation, much of the CHR was in tundra-boreal forest with megafauna as primary agents of disturbance. Further, under that paradigm forest management is irrelevant; a laissez faire approach will allow non-anthropogenic disturbances to occur regardless, and any management would, by definition, be outside the historic range of variation. Both a 'no management' and a 'manage for a landscape as it might have been without humans' approach are confounded by a current forest condition that has been modified by humans for millennia and likely alters and influences the effects of natural disturbances. If so, then what moment in time should be selected to reference 'historic conditions' in a long and dynamic history of human population levels, settlement patterns, and land management practices – both Native American and European – on the landscape? Reference conditions including specific quantities, patch sizes, compositions, and structures simulating the gradient of conditions created by both anthropogenic and non-anthropogenic disturbances must be specifically defined.

Alternatively, forest and land use planning for diverse wildlife conservation might more logically start with clear objectives, and proceed with management activities targeted toward attaining them. Rather than debating an unknowable and dynamic historical condition, or what should or should not be considered a natural disturbance, a more pertinent question might be (1) do we want to maintain viable populations of diverse disturbance-dependent species? If yes, then (2) where, how much, and what methods should be used to attain targeted forest composition and structural conditions, and different variants of early successional habitats required by those species?

Clear conservation objectives and targeted management activities are confounded by the 'real world' of conservation planning, land management, and current forest condition (Foster and Aber 2004; Warburton et al. 2011; Zenner Chap. 14). The majority (90 %) of land within the CHR is in private ownership (see Greenberg et al. Chap. 1); its use and management is the decision of landowners, and may or may not be compatible with a landscape level conservation vision or strategy. Much of the land base that was historically mature or disturbed forest, or variants of early successional habitats managed by Native Americans is now urban, suburban, plantation, even-aged forest, or industrialized agriculture; tracts of land that were once continuous wildlands are now fragmented by development. On public lands, where large tracts provide the greatest opportunity for wildlife and wildlands conservation, policy, public opinion, and human values other than biodiversity must also be considered including (among others) aesthetics, recreation, endangered species, and forest products. Conservation management on a landscape scale will require a multi-pronged strategy by citizens, local, state, and tribal governments, non-governmental organizations, and the federal government to address deficiencies in the conservation of natural resources that Americans value. An important, currently deficient component of conservation delivery is management for disturbance-dependent wildlife and their habitats.

Acknowledgements We thank George R. Milner and George Chaplin for fruitful insights, and for allowing us to use their figure. Gary Wein provided GIS support. We thank Scott Stoleson, Robert Askins, and William (Billy) Minser for reviewing an earlier version of this chapter and helping to improve it. E. Daniel Moss and Craig Harper also provided useful information and insights.

References

Ashe, W. W. (1922). Forest types of the Appalachian and White Mountains. *Journal of the Elisha Mitchell Scientific Society, 37*, 183–198.

Ashe, W. W., & Ayers, H. B. (1901). Forests and forest conditions in the southern Appalachians. In *A report of the secretary of agriculture in relation to the forests, rivers, and mountains of the southern Appalachian region* (pp. 45–68). Washington, DC: US Gov Print Office.

Askins, R. A. (1993). Population trends in grassland, shrubland, and forest birds in eastern North America. *Current Ornithology, 11*, 1–34.

Askins, R. A. (2000). *Restoring North America's birds*. New Haven: Yale University Press.

Askins, R. A. (2001). Sustaining biological diversity in early successional communities: The challenge of managing unpopular habitats. *Wildlife Society Bulletin, 29*, 407–412.

Askins, R. A. (2014). *Saving the world's deciduous forest*. New Haven: Yale University Press.

Baker, B. W., & Hill, E. P. (2003). Beaver (*Castor canadensis*). In G. A. Feldhamer, B. C. Thompson, & J. A. Chapman (Eds.), *Wild mammals of North America: Biology, management and conservation* (2nd ed., pp. 288–310). Baltimore: The Johns Hopkins University Press.

Balch, R. E. (1928). *The influence of the southern pine beetle on forest composition in western North Carolina*. Thesis, New York State College of Forestry, Ithaca.

Barden, L. S. (1997). Historic prairies in the Piedmont of North and South Carolina, USA. *Natural Areas Journal, 17*, 149–152.

Baskin, J. M., Baskin, C. C., & Chester, E. W. (1994). The big barrens region of Kentucky and Tennessee: Further observations and considerations. *Castanea, 59*, 226–254.

Brewster, W. (1886). An ornithological reconnaissance in western North Carolina. *Auk, 3*(94–112), 173–179.

Burney, D. A. (1993). Recent animal extinctions: Recipes for disaster. *American Scientist, 81*, 531–541.

Burney, D. A., & Flannery, T. F. (2005). Fifty millennia of catastrophic extinctions after human contact. *Trends in Ecology and Evolution, 20*, 395–401.

Carroll, W. D., Kapeluck, P. R., Harper, R. A., & Van Lear, D. H. (2002). Background paper: Historical overview of the southern forest landscape and associated resources. In D. N. Wear &

J. G. Greis (Eds.), *Southern forest resource assessment* (General technical report SRS 53, pp. 583–605). Asheville: USDA Forest Service Southern Research Station.

Chapman, J., Delcourt, P. A., Cridlebaugh, P. A., Shea, A. B., & Delcourt, H. R. (1982). Man-land interaction: 10,000 years of American impact on native ecosystems in the lower Little Tennessee River Valley, eastern Tennessee. *Southeastern Archaeology, 1*, 115–121.

Clark, P. U., Dyke, A. S., Shakun, J. D., Carlson, A. E., Clark, J., Wohlfarth, B., Mitrovica, J. X., Hostetler, S. W., & McCabe, A. M. (2009). The last glacial maximum. *Science, 325*, 710–714.

Collins, S. L., & Uno, G. E. (1985). Seed predation, seed dispersal, disturbance in grasslands: A comment. *The American Naturalist, 125*, 866–872.

Delcourt, H. R., & Delcourt, P. A. (1997). Pre-Columbian Native American use of fire on southern Appalachian landscapes. *Conservation Biology, 11*, 1010–1014.

Delcourt, P. A., & Delcourt, H. R. (2004). *Prehistoric Native Americans and ecological change: Human ecosystems in eastern North America since the Pleistocene*. Cambridge: Cambridge University Press.

Delcourt, P. A., Delcourt, H. R., Ison, C. R., Sharp, W. E., & Gremillion, K. J. (1998). Prehistoric human use of fire, the eastern agricultural complex, and Appalachian oak-chestnut forests: Paleoecology of cliff palace pond, Kentucky. *American Antiquity, 63*, 263–278.

Denevan, W. M. (1992). The pristine myth: The landscape of the Americas in 1492. *Annals of the Association of American Geographers, 82*, 369–385.

Dickinson, W. R. (2000). Changing times: The Holocene legacy. *Environmental History, 5*, 483–502.

Dogett, C. A. (1993). A method to assess large-scale forest damage: A case study. *Southern Journal of Applied Forestry, 17*, 197–199.

Ellsworth, J. W., & McComb, B. C. (2003). Potential effects of passenger pigeon flocks on the structure and composition of presettlement forests of eastern North America. *Conservation Biology, 17*, 1548–1558.

Evans, K. E., & Pobasco, G. E. (1977). *Wildlife of the prairies and plains* (General technical report NC 29). St. Paul: USDA Forest Service North Central Forest Experiment Station.

Flatley, W. T., Lafon, C. W., & Grissino-Mayer, H. D. (2013). Fire history, related to climate and land use in three southern Appalachian landscapes in the eastern United States. *Ecological Applications, 23*, 1250–1266.

Foster, D. R., & Aber, J. D. (2004). *Forests in time: The environmental consequences of 1,000 years of change in New England*. New Haven: Yale University Press.

Foti, T. L. (2004). Upland hardwood forests and related communities of the Arkansas Ozarks in the early 19th century. In M. A. Spetich (Ed.), *Upland oak ecology symposium: History, current conditions, and sustainability* (General technical report SRS 73, pp. 21–29). Asheville: USDA Forest Service Southern Research Station.

Foti, T. L., & Glenn, S. M. (1990, September 19–20). The Ouachita Mountain landscape at the time of settlement. In *Conference on restoring old growth forests in the interior highlands of Arkansas and Oklahoma* (pp. 49–65). Morrilton: Winrock International.

Fritz, G. J. (2000). Levels of native biodiversity in eastern North America. In P. E. Minnis & W. J. Elisens (Eds.), *Biodiversity and Native America* (pp. 223–247). Norman: University of Oklahoma Press.

Gill, J. L., Williams, J. W., Jackson, S. T., Lininger, K. B., & Robinson, G. S. (2009). Pleistocene megafaunal collapse, novel plant communities, and enhanced fire regimes in North America. *Science, 326*, 1100–1103.

Greenberg, C. H., & Lanham, D. J. (2001). Breeding bird assemblages of hurricane-created gaps and adjacent closed canopy forest in the southern Appalachians. *Forest Ecology and Management, 153*, 251–260.

Greenberg, C. H., Neary, D. G., Harris, L. D., & Linda, S. (1995). Vegetation recovery following high-intensity wildfire and silvicultural treatments in sand pine scrub. *The American Midland Naturalist, 133*, 149–163.

Greenberg, C. H., Collins, B. S., Thompson, F. R., III, & McNab, H. R. (2011a). What are early successional habitats, how can they be sustained, and why are they important? In C. H. Greenberg, B. S. Collins, & F. R. Thompson III (Eds.), *Sustaining young forest communities: Ecology and management of early successional habitats in the Central Hardwood Region, USA* (pp. 1–10). New York: Springer.

Greenberg, C. H., Perry, R. W., Harper, C. A., Levey, D. J., & McCord, J. M. (2011b). The role of recently disturbed upland hardwood forest as high quality food patches. In C. H. Greenberg, B. S. Collins, & F. R. Thompson III (Eds.), *Sustaining young forest communities: Ecology and management of early successional habitats in the Central Hardwood Region, USA* (pp. 121–141). New York: Springer.

Greenberg, C. H., Waldrop, T. A., Tomcho, J., Phillips, R. J., & Simon, D. (2013). Bird response to fire severity and repeated burning in upland hardwood forest. *Forest Ecology and Management, 304*, 80–88.

Greenberg, C. H., Franzreb, K. E., Keyser, T. L., Zarnoch, S. J., Simon, D. M., & Warburton, G. S. (2014). Short-term response of breeding birds to oak regeneration treatments in upland hardwood forest. *Natural Areas Journal, 34*, 409–422.

Guyette, R. P., Dey, D. C., Stambaugh, M. C., & Muzika, R. (2006a). Fire scars reveal variability and dynamics of eastern fire regimes. In M. B. Dickinson (Ed.), *Proceedings of fire in eastern oak forests: Delivering science to land managers* (General technical report NRS-P 1, pp. 20–39). Newton Square: USDA Forest Service Northern Research Station.

Guyette, R. P., Spetich, M. A., & Stambaugh, M. C. (2006b). Historic fire regime dynamics in the Boston Mountains, Arkansas, USA. *Forest Ecology and Management, 234*, 293–304.

Hart, J. L., & Buchanan, M. L. (2012). *History of fire in eastern oak forests and implications for restoration. Proceedings of 4th fire in eastern oak forests conference* (General technical report NRS-P 102). Newtown Square: USDA Forest Service Northern Research Station.

Hey, D. L., & Phillipi, N. S. (1995). Flood reduction through wetland restoration: The Upper Mississippi River Basin as a case study. *Restoration Ecology, 3*, 4–17.

Hicks, R. R. (1998). *Ecology and management of central hardwood forests*. New York: Wiley and Sons.

Hunter, W. C., Buehler, D. A., Canterbury, R. A., Confer, J. L., & Hamel, P. B. (2001). Conservation of disturbance-dependent birds in eastern North America. *Wildlife Society Bulletin, 29*, 440–455.

Innes, R. J. (2011). *Cervus elaphus*. In *Fire effects information system, [Online]*. USDA Forest Service, Rocky Mountain Research Station, Fire Sciences Laboratory (Producer). http://www.fs.fed.us/database/feis/. Accessed 18 Nov 2014.

Jenkins, M. A., Klein, R. N., & McDaniel, V. L. (2011). Yellow pine regeneration as a function of fire severity and post-burn stand structure in the southern Appalachian Mountains. *Forest Ecology and Management, 262*, 681–691.

Knopf, F. L. (1994). Avian assemblages on altered grasslands. *Studies in Avian Biology, 15*, 247–257.

Koch, P. L., & Barnosky, A. D. (2006). Late quaternary extinctions: State of the debate. *Annual Review of Ecology, Evolution, and Systematics, 37*, 215–250.

Lance, T. (2014). *The Henslow's sparrow thrives at Fort Campbell* (pp. 16–19). Tennessee Conservation, May/June 2014.

Lanham, D. J., & Whitehead, M. A. (2011). Managing early successional habitats for wildlife in novel places. In C. H. Greenberg, B. S. Collins, & F. R. Thompson III (Eds.), *Sustaining young forest communities: Ecology and management of early successional habitats in the Central Hardwood Region, USA* (pp. 209–224). New York: Springer.

Loftis, D. L., Schweitzer, C. J., & Keyser, T. L. (2011). Structure and species composition of upland hardwood communities after regeneration treatments across environmental gradients. In C. H. Greenberg, B. S. Collins, & F. R. Thompson III (Eds.), *Sustaining young forest communities: Ecology and management of early successional habitats in the Central Hardwood Region, USA* (pp. 59–71). New York: Springer.

Lorimer, C. G. (2001). Historical and ecological roles of disturbance in eastern North American forests: 9,000 year of change. *Wildlife Society Bulletin, 29*, 425–439.

MacArthur, R. H., & MacArthur, J. W. (1961). On bird species diversity. *Ecology, 42*, 594–598.

Mattoon, W. R. (1915). *Life history of the shortleaf pine* (Bulletin of the U.S. Department of Agriculture 244). Washington, DC: U.S. Department of Agriculture.

McClafferty, J. A. (2000). *An assessment of the biological and socioeconomic feasibility of elk restoration in Virginia*. Thesis, Virginia Polytechnical Institute and State University, Blacksburg.

McInteer, B. B. (1946). A change from grassland to forest vegetation in the "Big Barrens" of Kentucky. *The American Midland Naturalist, 35*, 276–282.

McNab, W. H., Greenberg, C. H., & Berg, E. C. (2004). Landscape distribution of large hurricane-related canopy gaps in a southern Appalachian watershed. *Forest Ecology and Management, 196*, 435–447.

Michaux, F. A. (1805). *Travels to the westward of the Allegheny Mountains, in the states of the Ohio, Kentucky, and Tennessee, in the year 1802*. Whitefish: Kessinger Legacy Reprints.

Milner, G. R., & Chaplin, G. (2010). Eastern North American population at CA AD 1500. *American Antiquity, 75*, 707–726.

Mosimann, J., & Martin, P. S. (1975). Simulating overkill by Paleoindians. *American Scientist, 63*, 304–313.

Moss, E. D. (2001). *Distribution and reproductive success of native grassland birds in response to burning and field size at Fort Campbell Military reservation: Special focus on Henslow's and grasshopper sparrows*. Thesis, University of Tennessee, Knoxville.

Naiman, R. J., Johnston, C. A., & Kelley, J. C. (1988). Alteration of North American streams by beaver. *Bioscience, 38*, 753–762.

Nesbitt, W. A., & Netboy, A. (1946). The history of settlement and land use in the Bent Creek forest. *Agricultural History, 20*, 121–127.

Newbold, C. D. (1996). *The effects of tornado and clearcut disturbances on breeding birds in a Tennessee oak-hickory (Quercus-Carya spp.) forest*. Thesis, University of Tennessee, Knoxville.

Nicholson, C. P. (1977). The red-cockaded woodpecker in Tennessee. *The Migration, 48*, 53–62.

O'Gara, B. W., & Dundas, R. G. (2002). Distribution: Past and present. In D. E. Toweill & J. T. Ward (Eds.), *North American elk: Ecology and management* (pp. 67–119). Washington, DC: Smithsonian Institution Press.

Partners in Flight Science Committee. (2013). *Population estimates database, version 2013*. http://rmbo.org/pifpopestimates. Accessed 29 Apr 2014.

Peek, J. M. (2003). Wapiti, *Cervus elaphus*. In G. A. Feldhamer, B. C. Thompson, & J. A. Chapman (Eds.), *Wild mammals of North America: Biology, management, and conservation* (2nd ed., pp. 877–887). Baltimore: Johns Hopkins University.

Perkins, K. A. (2006). *Cerulean warbler selection of forest canopy gaps*. Thesis, West Virginia University, Morgantown.

Prather, J. W., & Smith, K. G. (2003). Effects of tornado damage on forest bird populations in the Arkansas Ozarks. *The Southwestern Naturalist, 48*, 292–297.

Putnam, R. J. (1996). *Competition and resource partitioning in temperate ungulate assemblies*. London: Chapman and Hall.

Rafferty, M. D. (1996). Introduction. In *Rude pursuits and rugged peaks: Schoolcraft's Ozark journal 1818–1819* (pp. 1–15). Fayetteville: The University of Arkansas Press.

Reynolds, H. W., Glaholt, R. D., & Hawley, A. W. L. (1982). Bison. In J. A. Chapman & G. A. Feldhamer (Eds.), *Wild mammals of North America: Biology, management and economics* (pp. 972–1007). Baltimore: The Johns Hopkins University Press.

Rosell, F., Bozsér, O., Collen, P., & Parker, H. (2005). Ecological impact of beavers *Castor fiber* and *Castor canadensis* and their ability to modify ecosystems. *Mammal Review, 35*, 248–276.

Rosenberg, K. V., Will, T., Buehler, D. A., Barker Swarthout, S., Thogmartin, W. E., & Chandler, R. (in press). *Historical and current distribution and population status of Golden-winged Warbler*. Studies in Avian Biology.

Rostlund, E. (1960). The geographic range of the historic bison in the southeast. *Annals of the Association of American Geographers, 50*, 395–407.

Sauer, J. R., Hines, J. E., Thomas, I., Fallon, J., & Gough, G. (2000). *The North American breeding bird survey, results and analysis 1966–1999, Version 98.1*. Laurel: USGS Patuxent Wildlife Research Center.

Schoennagel, T., Veblen, T. T., & Romme, W. H. (2004). The interaction of fire, fuels, and climate across Rocky Mountain forests. *BioScience, 54*, 661–676.

Schorger, A. W. (1955). *The passenger pigeon: Its natural history and extinction*. Madison: University of Wisconsin Press.

Schreiner, E. G., Krueger, K. A., Kirsten, A., Happe, P. J., & Houston, D. B. (1996). Understory patch dynamics and ungulate herbivory in old-growth forests of Olympic National Park, Washington. *Canadian Journal of Forest Research, 26*, 255–265.

Schroeder, M. J., & Buck, C. J. (1970). *Fire weather* (Agriculture handbook 360). Washington, DC: USDA Forest Service.

Seton, E. T. (1929). *Lives of game animals*. Garden City: Doubleday, Doran.

Shifley, S. R., & Thompson, F. R., III. (2011). Spatial and temporal patterns in the amount of young forests and implications for biodiversity. In C. H. Greenberg, B. S. Collins, & F. R. Thompson III (Eds.), *Sustaining young forest communities: Ecology and management of early successional habitats in the Central Hardwood Region, USA* (pp. 73–95). New York: Springer.

Simmons, G. (1999). *The lost landscape of the Piedmont*. Salisbury. http://www.learnnc.org/lp/pages/1683

Spetich, M. A., Perry, R. W., Harper, C. A., & Clark, S. L. (2011). Fire in eastern hardwood forests through 14,000 years. In C. H. Greenberg, B. S. Collins, & F. R. Thompson III (Eds.), *Sustaining young forest communities: Ecology and management of early successional habitats in the Central Hardwood Region, USA* (pp. 41–58). New York: Springer.

Stoleson, S. H. (2013). Condition varies with habitat choice in postbreeding forest birds. *Auk, 130*, 417–428.

Stottlemyer, A. D., Shelburne, V. B., Waldrop, T. A., Rideout-Hanzak, S., & Bridges, W. C. (2006). Preliminary fuel characterization of the Chauga Ridges region of the southern Appalachian Mountains. In K. F. Connor (Ed.), *Proceedings of 13th biennial southern silvicultural research conference* (General technical report SRS 92, pp. 510–513). Asheville: USDA Forest Service Southern Research Station.

Svenning, J. (2002). A review of natural vegetation openness in north-western Europe. *Biological Conservation, 104*, 133–148.

Tesky, J. L. (1995). *Bos bison*. In *Fire effects information system, [Online]*. USDA Forest Service, Rocky Mountain Research Station, Fire Sciences Laboratory (Producer). http://www.fs.fed.us/database/feis/. Accessed 18 Nov 2014.

Thompson, F. R., III, & Dessecker, D. R. (1997). *Management of early-successional communities in central hardwood forests with special emphasis on the ecology and management of oaks, ruffed grouse, and forest songbirds* (General technical report NC 195). St. Paul: USDA Forest Service North Central Research Station.

US Fish and Wildlife Service. (2001). *Bog turtle (Clemmys muhlenbergii), northern population recovery plan*. http://www.fws.gov/northeast/nyfo/es/bogturtle.pdf. Accessed 22 Jan 2015.

Van Doren, M. (1928). *Travels of William Bartram*. New York: Dover Publications Inc.

Van Lear, D. H., & Waldrop, T. A. (1989). *History, uses, and effects of fire in the Appalachians* (General technical report SE 54). Asheville: USDA Forest Service Southeastern Forest Experiment Station.

VDGIF. (2010). *Elk restoration and management options for southwest Virginia*. Virginia Department of Game and Inland Fisheries Publication. http://www.dgif.virginia.gov/wildlife/elk/management-plan/. Accessed 22 Jan 2015.

Waldrop, T. A., Brudnak, L., & Rideout-Hanzak, S. (2007). Fuels of disturbed and undisturbed sites in the southern Appalachian Mountains, USA. *Canadian Journal of Forest Research, 37*, 1134–1141.

Waldrop, T., Phillips, R. A., & Simon, D. A. (2010). Fuels and predicted fire behavior in the southern Appalachian Mountains after fire and fire surrogate treatments. *Forestry Sciences, 56*, 32–45.

Walton, E. M. (2006). *Using remote sensing and geographical information science to predict and delineate critical habitat for the bog turtle, Glyptemys muhlenbergii*. Thesis, University of North Carolina, Greensboro.

Warburton, G. S., Harper, C. A., & Weeks, K. (2011). Conservation of early successional habitats in the Appalachian Mountains: A manager's perspective. In C. H. Greenberg, B. S. Collins, & F. R. Thompson III (Eds.), *Sustaining young forest communities: Ecology and management of early successional habitats in the Central Hardwood Region, USA* (pp. 225–251). New York: Springer.

Williams, M. (1989). *Americans and their forests: An historical geography*. New York: Cambridge University Press.

Woodward, A., Schreiner, E. G., Houston, D. B., & Moorhead, B. B. (1994). Ungulate-forest relationships on the Olympic Peninsula: Retrospective exclosure studies. *Northwest Science, 68*, 97–110.

Yarnell, S. L. (1998). *The southern Appalachains: A history of the landscape* (General technical report SRS 18). Asheville: USDA Forest Service Southern Research Station.

Yeoman, B. (2014). *Why the passenger pigeon went extinct*. Audubon, May-June 2014.

Zobel, D. B. (1969). Factors affecting the distribution of *Pinus pungens*, an Appalachian endemic. *Ecological Monographs, 39*, 304–333.

Chapter 13
Climate Change and the Future of Natural Disturbances in the Central Hardwood Region

Virginia H. Dale, M. Joseph Hughes, and Daniel J. Hayes

Abstract Spatial patterns and ecological processes of the USA central hardwood forests reflect past climatic conditions and natural disturbance regimes. Changes in climate can lead to disturbances that exceed their natural range of variation, and the impacts of these changes depend on the vulnerability or resiliency of these ecosystems. Global Circulation Models generally project annual increases in temperature across the Central Hardwood Region over the coming decades, but changes in precipitation are less consistent. More unclear is how climate change affects severity and frequency of natural disturbances, such as severe storms, fires, droughts, floods, and insect outbreaks. We use a time-series satellite data record to map the spatial pattern and severity of broad classes of natural disturbances the southeast region. Between 5 % and 25 % of forest land was affected by disturbance each year since 1985 across the four regions. The time series reveals periodic droughts that are widespread and low-severity but associated with more localized, high-severity disturbances such as fire and insect outbreaks. The map also reveals extensive anthropogenic disturbance across the region in the form of forest conversion related to resource extraction and urban and residential development. Changes in climate

V.H. Dale (✉) • D.J. Hayes
Environmental Sciences Division, Oak Ridge National Laboratory,
Oak Ridge, TN 37831-3036, USA
e-mail: dalevh@ornl.gov; hayesdj@ornl.gov

M.J. Hughes
Department of Earth and Environment, Boston University, Boston, MA 02215, USA
e-mail: jhughes@utk.edu

C.H. Greenberg, B.S. Collins (eds.), *Natural Disturbances and Historic Range of Variation*, Managing Forest Ecosystems 32,
DOI 10.1007/978-3-319-21527-3_13

and disturbance regimes might affect these forests in the future via altering the exposure, sensitivity, and adaptive capacity of these ecosystems. Changes in climate are highly likely to expose forests to more frequent and severe disturbances, but ultimately how vulnerable or resilient forests are to these changes depends on their sensitivity and capacity to adapt to novel conditions.

Keywords Climate change • Natural disturbance • Resiliency • Vulnerability • Exposure

13.1 Pattern and Process of Central Hardwood Forests as a Function of Climate and Disturbance Regimes

The spatial patterns and ecological processes of southeastern upland hardwood forests are influenced by both climate and disturbance regimes (McNulty et al. 2013). Current forest cover is largely a result of climate, topography, the frequency and intensity of natural disturbances, and human activity. While forest cover in the southeastern USA declined in the 1800s as settlements and farming expanded, forest cover across the region has increased over the past century as agriculture moved to the midwest and other areas. The expansion of urban areas is the most prominent current and expected future driver of losses of southeastern forests (USDA Forest Service 2012; Wear et al. 2013; Zhao et al. 2013). Furthermore, air pollution is compromising the health of many trees and in some cases responsible for their death – such as occurs with red spruce (*Picea rubens*) in high elevations of the southern Appalachians (Johnson 1992). This chapter focuses on natural disturbances affecting hardwood trees in the southeastern Central Hardwood Region (CHR) of the USA but recognizes that human-instigated disturbances and changes to other types of vegetation cover affect the structure, function, and character of these ecosystems as a whole.

Forests in the CHR (as elsewhere) are impacted by a variety of disturbances including insects and disease, invasive species, wildfires, harvest, and extreme events such as droughts and storms (Dale et al. 2001; also see Chaps. 1, 2, 3, 4, 5, 6, 7, 8, 9, 12). These disturbances are driven by climate change, human activities, and natural causes (Fig. 13.1). While fires and insect and pathogen outbreaks are related to all three forces, ice storms, hurricanes and droughts are a function of both natural and climate change. Deforestation is solely a result of human activities such as development and urbanization. Based on satellite data analysis, Goetz et al. (2012) found that the distribution of major disturbance types in North America varies by topography, vegetation, weather patterns, climate gradients, and proximity to human settlement. Their analysis shows that storms have major impacts on the Gulf Coast region, insect damage is widespread, and forest harvest is a dominant disturbance in the southeastern USA.

Many natural disturbances are an integral part of forest dynamics across the CHR. While insect disturbance is more common than fires, it is often the leading

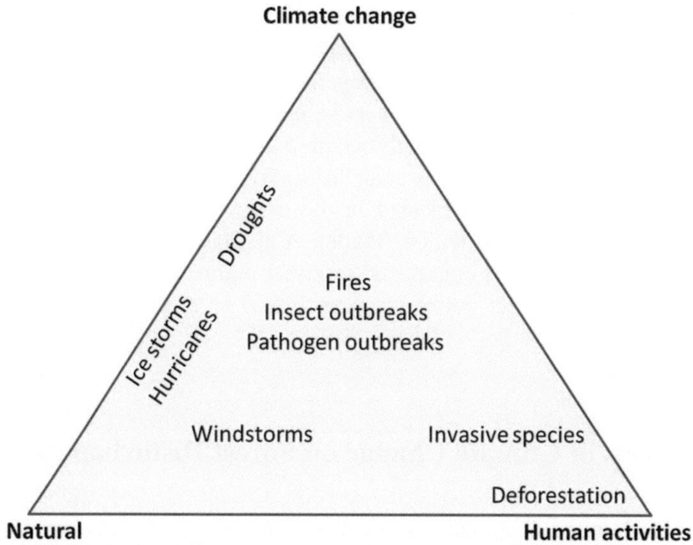

Fig. 13.1 Depiction of disturbances to central hardwood forests in relation to climate change, human activities, and natural causes. Fires and insect and pathogen outbreaks are related to all three forces. Deforestation is largely associated with human development and urbanization

edge to fires. For example, outbreaks of the native southern pine bark beetle (*Dendroctonus frontalis*) often cause the demise of large swaths of dense pine forest and create opportunities for succession of the hardwood understory (Nowak et al. Chap. 4). Disturbances are often interrelated; while insect disturbance is more common than wildfire in the region, it is often a precursor to fires. Such is the case when pine forests that have succumbed to beetles provide conditions ripe for subsequent fires (Xi et al. 2009). Furthermore, disturbance impacts can be amplified by general trends. For example, years of rising average temperatures can increase the susceptibility of forests to droughts (Jentsch et al. 2007). In addition, prevailing temperatures and precipitation patterns combine with competition to affect hardwood mortality (Yaussy et al. 2013; Oak et al. Chap. 3). The occurrence of more than one disturbance can induce even more extreme or long-lasting impacts than a single disturbance (Paine et al. 1998). Thus, identifying a single cause of tree mortality is rare because factors such as drought, warmer temperatures, pests and pathogens are often interrelated (Manion 1981; McDowell et al. 2008, 2011; Dukes et al. 2009; Allen et al. 2010). These combined factors are associated with recent large-scale die-off events of multiple tree species observed across the USA (Raffa et al. 2008; Allen et al. 2010; Oak et al. Chap. 3).

There is generally a strong relationship between fire and climate (Keane et al. 2008; Bowman et al. 2009), which suggests that, like other regions, central hardwood forests may be increasingly affected by large and intense fires. Yet fire occurrence in the southern Appalachian Mountains had little relationship to reconstructed annual drought conditions (Flatley et al. 2013). Aboriginal people used fire to clear forests for settlements and agricultural fields, and those fires likely affected even larger areas (Sauer 1950; Grissino-Mayer Chap. 6; Leigh Chap. 8; Greenberg et al. Chap. 12). Hence widespread and frequent anthropogenic ignitions during the late nineteenth and early twentieth centuries likely overwhelmed climatic influences as area-wide fires that burned across multiple stands occurred at 6–13-year intervals in the southern Appalachian Mountains (Flatley et al. 2013; Grissino-Mayer Chap. 6). However, fire suppression policies implemented in the early twentieth century have since reduced fire outbreaks in the region (Flatley et al. 2013). Even so, fires are more likely to follow protracted droughts and/or insect outbreaks. Seasonal dry periods, longer droughts, and insect outbreaks were linked to recent wildfires in Florida (Carter et al. 2014).

13.2 Effects of Climate Change on Forest Disturbances

Local, regional, and global changes in temperature and precipitation can influence the occurrence, timing, frequency, duration, extent, and intensity of disturbances (Baker 1995; Turner et al. 1998; Dale et al. 2001; Joyce et al. 2014). These impacts are manifest at multiple scales, from rapid and localized 'pulse' events such as wildfire and windstorms to more gradual and widespread 'press' disturbances such as drought and other stress-induced forest decline. Disturbances are a natural part of the ecological cycle driving central hardwood forest dynamics, but changes in climate can lead to disturbances that exceed their historic range of variation. Climate change will have both direct and indirect effects on forest disturbance because it is a key driver of ***vulnerability***. As with the impacts of climate change on any ecosystem (IPCC 2014), the vulnerability of forests to disturbance is a function of three fundamental components: exposure, sensitivity and adaptive capacity. Climate change thus influences forest vulnerability to disturbance in various ways depending on how it affects each of these components.

Climate change increases ***exposure*** if changes in temperature and precipitation drive a higher frequency or intensity of disturbance events. Extreme climate events such as temperature and precipitation anomalies, drought and wet periods, and storms can cause local- to widespread tree mortality depending on the severity, timing, and duration of the event. These extreme events can be a product of both natural variability and anthropogenic climate change, but a general warming trend and an increase in precipitation over the USA may be leading to an increase in these events (Karl et al. 1995; Easterling et al. 2000; National Climate Assessment 2014).

In the CHR, warmer temperatures and the associated acceleration of the hydrologic cycle could result in more frequent and intense storms (Emanuel 1987; Karl et al. 2009; USGCRP 2009). While the region has seen an increase in the number of rainfall events in recent decades (Karl and Knight 1998), warmer temperatures combined with longer periods between rainfall events has led to an increase in the area of this region experiencing moderate to severe drought (Karl et al. 2009). Such conditions can directly result in tree mortality, where warmer temperatures and drought conditions induce lower soil moisture and higher demand for evapotranspiration, thus leading to an overall water deficit in forests (Dietze and Moorcroft 2011). Drought conditions, tree mortality, and high-severity fire weather can create fuel conditions that lead to increased exposure to wildfire. Finally, climate change can influence the exposure of central hardwood forests to biotic disturbances in complex ways, as changes in temperature and precipitation control herbivore and pathogen survival, reproduction, dispersal, and distribution (Weed et al. 2013).

Climate change increases the ***sensitivity*** of forests to disturbance where variation in temperature and precipitation create environmental and forest health conditions that render forests more predisposed to disturbance (Adams et al. 2009, 2010). The mechanism proposed and experimentally verified by Adams et al. (2009) is that sensitivity of tree mortality to temperature depends on temperature-sensitive carbon starvation in response to protracted water stress and/or temperature-insensitive sudden hydraulic failure under extreme water stress (cavitation). For example, a significant shift in temperature sensitivity of a red spruce forest occurred after 1930s, when prior positive growth responses to warm temperatures shifted to negative responses coinciding with a period of clearcut harvesting (White et al. 2014). And a recent study examining a high-resolution, historical satellite data record found significant increasing trends in fire occurrence and area burned in areas of the western USA coinciding with increased drought severity (Dennison et al. 2014). Similarly, the combination of lower soil moisture and higher temperatures creates conditions that make central hardwood forests sensitive to larger and more intense fires than have typically occurred in the past (McNulty et al. 1996). Such conditions also lead to declines in tree health and vigor, which reduces their natural defenses causing central hardwood forests to become more susceptible to mortality during insect outbreaks (Williams and Liebhold 2002). Furthermore, changes in climate can set off a series of cascading effects that increases the sensitivity of forests to multiple disturbance agents interacting in complex ways (Ayres and Lombardero 2000). Storms and insect outbreaks result in tree mortality, and the increase in fuel loads coupled with drought conditions make forests susceptible to natural or human-caused fires. Alternatively, any stress or disturbance that damages but does not kill trees can result in forests becoming susceptible to pathogens and insect attacks. Overall, any increase in tree stress, damage or mortality causes forests to be more vulnerable to a greater extent, duration and severity of disturbances.

The composition, structure and function of forest systems are resilient to some disturbances, as their ***adaptive capacity*** has evolved to reflect their climatic conditions and natural disturbance regimes. Following disturbance, resilient forests recover to a similar state when the internal processes and interactions are self-reinforcing. In the southeast, for example, longleaf pine (*Pinus palustris*) forests are a fire-adapted ecosystem characterized historically by frequent, low-severity fires. This typical fire regime prevents the excessive buildup of fuels that can lead to more severe, high-intensity fires that result in large-scale changes in composition and structure. Fires are promoted in these ecosystems by readily flammable forest litter, and, in turn, fire removes competing understory vegetation and creates the site conditions needed for the regeneration of longleaf pine. The adaptive capacity of these and other resilient forest systems leads to the creation of persistent forest structure and functioning at specific scales in time and space unless there is a change in the prevailing disturbance regime. When novel or significantly altered disturbance regimes are introduced, these reinforcing feedbacks can weaken or be disrupted and thus render forests vulnerable to directional changes in structure and function that can lead to new system states.

13.3 Past Patterns of Climate, Disturbance, and Vegetation in the Central Hardwood Region

While this chapter focuses on future natural disturbances, it is grounded in past patterns of the central hardwood forests and how they responded to prevailing climate and natural disturbance regimes. The climate in this region is highly variable and influenced by latitude, elevation, topography and presence of large water bodies (Konrad et al. 2013). During the Little Ice Age – or period of Rapid Climate Change – in the sixteenth to nineteenth centuries, the CHR experienced cold periods separated by period of slight warming. Agricultural failures were common (Briffia 2000).

Disturbance regimes were quite different before Europeans settled in North America. Non-native pests and pathogens were not present on the historical landscape. Native Americans likely disturbed around 7 % of the forest by clearing settlement areas around 1500 AD and by as much as 43 % if the frequent burning associated with agriculture is included (Milner and Chaplin 2010; Greenberg et al. Chap. 12). Southern pine beetles activity on the landscape was largely related to frequent fires set by Native Americans and later by European settlers for land management purposes. Frequent fire maintained the pine-hardwood systems as open savannahs with widely spaced oaks and pines. But southern pine beetle kills more of the pines, adds fuel for future fires, and creates even-aged stands that are susceptible to future beetle outbreaks (Waldron et al. 2007). Also, some natural disturbances are now missing that historically were present, such as those resulting from high populations of beaver (*Castor Canadensis*), elk (*Cervus canadensis*), bison (*Bisonbison*), and passenger pigeons (*Ectopistes migratorius*) (Greenberg et al. Chap. 12). Similar to plant communities, animal communities have reassembled (Greenberg et al. 2014) as species responded in different ways to changes in climate and associated changes in vegetation patterns (McNab et al. 2014).

Prehistoric vegetation patterns in the CHR reflect prevailing climate and disturbance patterns and differ from current conditions. Based on the paleoecological evidence, Delcourt et al. (1981) developed a map for the eastern USA at 25,000 years ago that shows a mixed cool temperate forest belt across the southern Appalachians. Oak-hickory (*Quercus-Carya*) and southern pine forests occurred from South Carolina southwards to northern Florida replacing a xeric scrub that prevailed from about 28,000 to 25,000 year ago. From about 21,000 to 17,000 years ago, the southeastern USA was wooded or forested and more open than the forests there today. The cool climate pines were often interspersed by prairie herbs east of the Appalachians (e.g., Brook and Nickmann 1996), and the Atlantic coastal plain was dominated by an open jack pine (*P. banksiana*) forest (Wells 1992). The climate on the Gulf coastal plain experienced a more winter-wet and summer-dry pattern than today, and those conditions favored drought tolerant species of oaks and hickories rather than southern pines (Delcourt et al. 1981).

Delcourt (2002) documents how plant species migrated in response to climate warming or cooling (both of which have occurred about 20 times during the past two million years) and suggests that species survival depended on corridors or a chain of islands of suitable habitats. Delcourt and Delcourt (1998) point out that environmental changes in the southern Appalachian Mountains during the Quaternary period affected changes in biodiversity. Increased seasonality during the change from glacial to interglacial conditions in the late Pleistocene and Holocene resulted in landscape instability associated with a heterogeneous mosaic of habitats supporting a diverse biota. During the Holocene, ecotones between alpine, boreal, and temperate ecosystems changed latitude and longitude as a result of alteration in disturbance regimes and attainment of temperature thresholds (such as those that dictate the lower limit of alpine vegetation). Furthermore Delcourt (2002) poses that the continuity of habitats was disrupted by fires set by prehistoric Native Americans for game management (also see Greenberg et al. Chap. 12).

13.4 Mapping the Pattern and Severity of Disturbances over Recent Decades

To illuminate and understand the patterns of recent disturbances across central hardwood forests, we use Landsat imagery from 1984 to 2011 to map the spatial pattern and severity of forest disturbances for each year across the region. From this time-series, we explore the dynamics between disturbances with different severity classes and find that, in general, low-intensity disturbances such as drought and insect activity, which are predicted to increase in the southeast under climate change, interact to increase the frequency and scale of higher-impact disturbance events.

Landsat sensors image the earth every 16 days at 30 m resolution. For each year in the analysis, a single composite image is created by identifying clear-sky (i.e., absence of clouds and cloud shadow) pixels from across the multiple summertime Landsat images for that year and finding a weighted average of those pixels in such a way as to reduce the effects of phenology (Hughes and Hayes 2014). These composites are temporally interpolated in areas with no clear-sky views, and then a vegetation index is calculated from the spectral information. The time-series of the vegetation index for each pixel is then broken into segments, and a line is fit to the data within each segment in a process similar to the one described by Kennedy et al. (2010). Depending on the slope of the line, each segment represents a period of disturbance if negative, regeneration if positive, or stability for slopes near zero. For disturbance events, the magnitude of the slope represents its severity and the length of the line segment its duration.

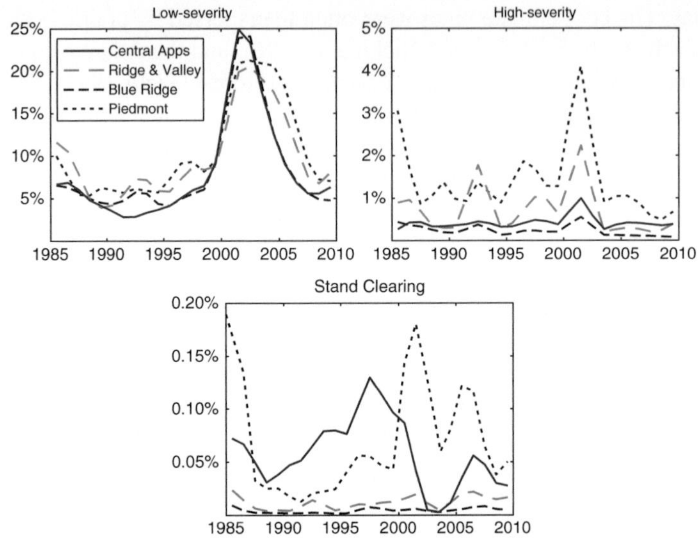

Fig. 13.2 Percent area of forests within four central hardwood ecoregions affected by low-severity (*top*), high-severity (*center*), and stand-clearing (*bottom*) disturbance events between 1985 and 2010

While it is not possible to distinguish between the various exact causal agents of disturbance, broad classes can be identified. We separate disturbances into three severity classes: (1) low-severity; (2) high-severity; and (3) stand-clearing. Low-severity disturbances have a small impact over a single year but can have significant aggregate effects when combined or when considered over multiple years. Common disturbances in this category include droughts and damage from hemlock wooly adelgid (*Adelges tsugae*), Asian gypsy moth (*Lymantria dispar*), or beech bark disease induced by beech scale insects (*Cryptococcus fagisuga*). High-severity disturbances are those that cause major biomass reduction while still leaving recognizable forest. These are disturbances such as extreme weather events, fires or severe pine beetle outbreaks, especially when combined with drought or other stresses. Stand-clearing disturbances are those that leave no forest. These are typically new development, mines, or forest clear-cuts and are almost always anthropogenic in cause.

In some parts of the CHR, between 5 % and 25 % of forest land has been affected by some sort of disturbance each year since 1985 (Fig. 13.2). The major drought that gripped the southeastern USA in the early 2000s is a key feature in the low-severity time series; less severe or localized droughts also appear as smaller humps in the 1990s and late 2000s. Importantly, these droughts are associated with increased high-severity disturbances in all regions during the same years, as water stress combined with insect disturbances, demonstrating the additive effects of multiple stresses. The history of coal mining in the Central Appalachians and logging in the Piedmont is evident in the graph of stand-clearing disturbances.

The most severe disturbance event in each pixel of the northern portion of the CHR is mapped in Fig. 13.3. Lexington, Kentucky, and the surrounding area is a

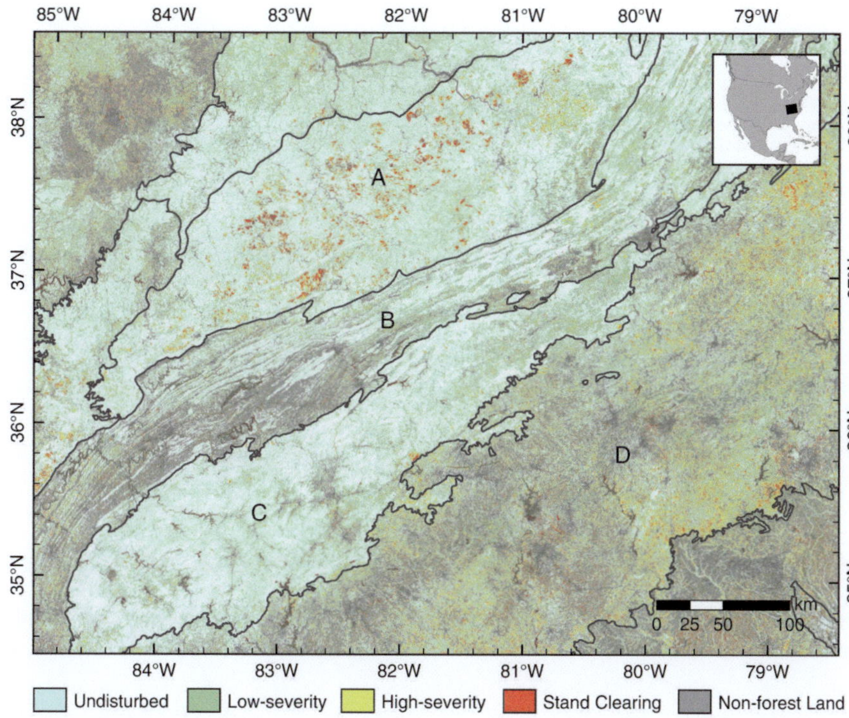

Fig. 13.3 Map of (*A*) a gradient of severities in disturbance in the Central Appalachians; (*B*) the Ridge and Valley; (*C*) the Blue Ridge Mountains; and (*D*) the Piedmont ecoregions of the CHR between 1984 and 2011

prominent non-forest feature in the northwest part of the map. The Central Appalachians are striking for their mixture of relatively undisturbed forest broken by strip mines and mountaintop removal sites. The Ridge and Valley region and the Blue Ridge Mountains, with their extensive network of protected areas, have been relatively undisturbed though do exhibit patches of stress from drought and insect disturbance. These patches are also associated with occasional fires, primarily resulting from standing dead pine that was previously disturbed by drought and southern pine beetle (Xi et al. 2009). The Piedmont has been heavily affected by southern pine beetles, fire, and the conversion of forest land to urban and residential use over the past 30 years.

13.5 Discussion: Trajectories of Change

A critical concern for managers and decision makers is how climate and environmental change will impact disturbance regimes in the future and how both ecological and social systems in the region will respond. The trajectory of forest disturbance

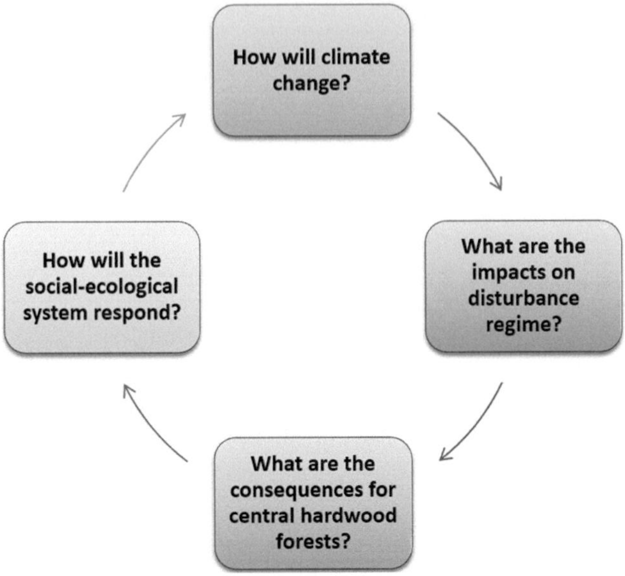

Fig. 13.4 A framework for understanding and predicting the future vulnerability of central hardwood forests to climate change and disturbance

in this CHR will be a function of the vulnerability or resiliency of the forests to future climate change (Fig. 13.4). As discussed in Sect. 13.2 above, the vulnerability of these forests will be determined by their exposure, sensitivity and adaptive capacity to future changes in disturbance. Changes in climate are highly likely to expose north central forests to novel conditions outside of the range of normal variability in the type, frequency and severity of disturbances to which they have adapted. But while exposure is likely to increase, ultimately how vulnerable or resilient north central forests are to these changes in disturbance will depend on their sensitivity and capacity to adapt to novel disturbance regimes. Furthermore, future changes in climate and disturbance will not be operating in a vacuum, and sensitivity and adaptive capacity will be strongly influenced by other factors such as forest management, land-use change and pollution.

Changes in future climate are anticipated to affect forest composition and tree physiology. Most Global Circulation Models project an increase in temperature for all months in the northern CHR, but changes in precipitation are less consistent (with some models projecting less precipitation and others projecting more) (Dale et al. 2009, 2010; Carter et al. 2014). Major shifts in forest species composition are projected for the southeast (Dale et al. 2010; Iverson et al. 2004, 2008). Nevertheless, xeric eastern forests may experience a loss of carbon even though they are adapted to periodic moisture deficits (Vose et al. 2012). In the eastern USA, elevated CO_2 and temperature may increase forest growth and potentially carbon storage, if sufficient water is available (Hanson and Weltzin 2000). Projections are for an increase in tree mortality as temperatures climb (Dietze and Moorcroft 2011). Regional projections of future greenhouse-gas-induced climatic warming indicate that the red spruce – Fraser fir (*Abies fraseri*) forests may become extinct in the southern Appalachians (McNab et al. 2014), and montane red spruce and balsam fir (*A. balsamea*) forests probably will occur between 44° and 49° N latitude (Delcourt and Delcourt 1998). The primary response of forests to drought will be reduction in net primary production and stand water use, as influenced by reductions in stomatal conductance (Hanson and Weltzin 2000), with the possibility of tree mortality as a result. Increased variation in summer drought and growing-season temperatures is another possible outcome of climate change. Sensitivity of regenerating forests to climatic variability is stand type-dependent and a function of species composition (Kardol et al. 2010). The direct effects of climatic variability on forest dynamics are likely to be small relative to successional processes or climate-related insect outbreaks (Kardol et al. 2010). The effects of changes in precipitation regimes on forest systems can be buffered by ecosystem processes such as deep rooting (Johnson et al. 2008). While deep rooted trees are less susceptible, severe or prolonged drought may cause even mature trees to be affected by insects or disease (Hanson and Weltzin 2000).

Observed changes in disturbance regimes might project into the future and with associated impacts on the patterns and processes in the north central forests. Disturbance regimes are expected to be further altered as a result of climate change. The warmer and drier conditions predicted by some climate models lead to a 30 % increase in the seasonal severity rating for fire hazard over the southeastern USA (Dale et al. 2001). Furthermore, an increase in the interannual variation in minimum winter temperatures is expected to favor more northerly outbreaks of southern pine beetles but could reduce more southerly outbreaks (Ungerer et al. 1999; Tran et al. 2007; Nowak et al. Chap. 4). Because insect populations typically migrate much faster than trees, many temperate tree species are likely to encounter nonnative insect herbivores that previously were restricted to other regions (Dale et al. 2001).

A prime force in reducing the area of forest in the southeastern USA in recent decades is urban development (USDA Forest Service 2012; Wear et al. 2013). Development pressures on forests are expected to continue into the future (Zhao et al. 2013). Hence economic incentives to keep land in forests are important to retain the ecological and social services provided by those systems (Oliver 2014).

Forest management can enhance resiliency to climate and disturbance by reducing competitive stress from climatic events as well as stresses caused by insects and pathogens (Yaussy et al. 2013). When tree species ranges shift with climate changes, migration is constrained by dispersal into newly suitable habitat (Iverson and McKenzie 2013). Addressing climate change effects on forests requires considering the interactions among land-use practices, energy options and disturbances (Dale et al. 2000, 2011). Because changes in climate and disturbance regimes are co-occurring with increases in landscape fragmentation, greater numbers of invasive species, changing social and economic conditions, and greater demands for ecosystem services from a growing population (Vose and Klepzig 2014), desired future conditions for north central USA forests are different from conditions that prevailed in the past. Assisted migration and protection may likely be necessary to protect some species from climate change and insect disturbances (Iverson et al. 2004; Greenberg et al. 2014). Biotic responses to late-Quaternary environmental changes indicate that a primary conservation emphasis should be to predict the effects of changes in landscape-level environmental processes on biodiversity (Delcourt and Delcourt 1998) and ecosystem interactions.

Acknowledgments This research was supported by the US Department of Energy (DOE). Oak Ridge National Laboratory is managed by UT-Battelle, LLC, for DOE under contract DE-AC05-00OR22725. We appreciate review comments provided by Forrest Hoffman, Cathryn Greenberg, and Louis Iverson.

References

Adams, H. D., Guardiola-Claramonte, M., Barron-Gafford, G. A., Villegas, J. C., Breshears, D. D., Zou, C. B., Troch, P. A., & Huxman, T. E. (2009). Temperature sensitivity of drought-induced tree mortality portends increased regional die-off under global change-type drought. *Proceedings of the National Academy of Science of the United States of America, 106*, 7063–7066.

Adams, H. D., Macalady, A. K., Breshears, D. D., Allen, C. D., Stephenson, N. L., Saleska, S. R., & Huxman, T. E. (2010). Climate induced tree mortality: Earth system consequences. *Eos, Transactions American Geophysical Union, 91*, 153–154.

Allen, C. D., Macalady, A. K., Chenchouni, H., Bachelet, D., McDowell, N., Vennetier, M., Kitzberger, T., Rigling, A., Breshears, D. D., Hogg, E. H., Gonzalez, P., Fensham, R., Zhang, Z., Castro, J., Demidova, N., Limp, J., Allard, G., Running, S. W., Semerci, A., & Cobb, N. (2010). A global overview of drought and head-induced tree mortality reveals emerging climate change risks for forests. *Forest Ecology and Management, 259*, 660–684.

Ayres, M. P., & Lombardero, M. J. (2000). Assessing the consequences of global change for forest disturbance from herbivores and pathogens. *Science of the Total Environment, 262*, 263–286.

Baker, W. (1995). Long-term response of disturbance landscapes to human intervention and global change. *Landscape Ecology, 10*, 143–159.

Bowman, D. M., Balch, J. K., Artaxo, P., Bond, W. J., Carlson, J. M., Cochrane, M. A., D'Antonio, C. M., DeFries, R. S., Doyle, J. C., Harrison, S. P., Johnston, F. H., Keeley, J. E., Krawchuk, M. A., Kull, C. A., Marston, J. B., Moritz, M. A., Prentice, I. C., Roos, C. I., Scott, A. C., Swetnam, T. W., van der Werf, G. R., & Pyne, S. J. (2009). Fire in the Earth system. *Science, 324*, 481–484.

Briffia, K. (2000). Annual climate variability in the Holocene: Interpreting the message of ancient trees. *Quaternary Science Reviews, 19*, 87–105.

Brook, G. A., & Nickmann, R. J. (1996). Evidence of late Quaternary environments in Northwestern Georgia from sediments preserved in Red Spider Cave. *Physical Geography, 17*, 465–484.

Carter, L. M., Jones, J. W., Berry, L., Burkett, V., Murley, J. F., Obeysekera, J., Schramm, P. J., Wear, D. (2014). Southeast and the Caribbean. In J. M. Melillo, T. C. Richmond, G. W. Yohe (Eds.), *Climate change impacts in the United States: The third national climate assessment* (pp. 396–417). US Global Change Research Program. doi:10.7930/J0Z60KZC.

Dale, V. H., Joyce, L. A., McNulty, S., & Neilson, R. P. (2000). The interplay between climate change, forests, and disturbances. *Science of the Total Environment, 262*, 201–204.

Dale, V. H., Joyce, L. A., McNulty, S., Neilson, R. P., Ayres, M. P., Flannigan, M. D., Hanson, P. J., Irland, L. C., Lugo, A. E., Peterson, C. J., Simberloff, D., Swanson, F. J., Stocks, B. J., & Wotton, B. M. (2001). Forest disturbances and climate change. *BioScience, 51*, 723–734.

Dale, V. H., Lannom, K. O., Tharp, M. L., Hodges, D. G., & Fogel, J. (2009). Effects of climate change, land-use change, and invasive species on the ecology of the Cumberland Forests. *Canadian Journal of Forest Research, 39*, 467–480.

Dale, V. H., Tharp, M. L., Lannon, K. O., & Hodges, D. G. (2010). Modeling transient response of forests to climate change. *Science of the Total Environment, 408*, 1888–1901.

Dale, V. H., Efroymson, R. A., & Kline, K. L. (2011). The land use – Climate change – Energy nexus. *Landscape Ecology, 26*, 755–773.

Delcourt, H. R. (2002). *Forests in peril: Tracking deciduous trees from ice-age refuges into the greenhouse world*. Granville: McDonald and Woodward Publ Co.

Delcourt, P. A., & Delcourt, H. R. (1998). Paleoecological insights on conservation of biodiversity: A focus on species, ecosystems, and landscapes. *Ecological Applications, 8*, 921–934.

Delcourt, H. R., West, D. C., & Delcourt, P. A. (1981). Forests of the southeastern United States: Quantitative maps for aboveground woody biomass, carbon, and dominance of major tree taxa. *Ecology, 62*, 879–887.

Dennison, P. E., Brewer, S. C., Arnold, J. D., & Moritz, M. A. (2014). Large wildfire trends in the western United States, 1984–2011. *Geophysical Research Letter, 41*, 2928–2933.

Dietze, M. C., & Moorcroft, P. R. (2011). Tree mortality in the eastern and central United States: Patterns and drivers. *Global Change Biology, 17*, 3312–3326.

Dukes, J. S., Pontius, J., Orwig, D., Garnas, J. R., Rodgers, V. L., Brazee, N., Cooke, B., Theoharides, K. A., Stange, E. E., Harrington, R., Ehrenfeld, J., Gurevitch, J., Lerdau, M., Stinson, K., Wick, R., & Ayres, M. (2009). Responses of insect pests, pathogens, and invasive species to climate changes in northeastern North America: What can we predict? *Canadian Journal of Forest Research, 39*, 231–248.

Easterling, D. R., Meehl, G. A., Parmesan, C., Changnon, S. A., Karl, T. R., & Mearns, L. O. (2000). Climate extremes: Observations, modeling, and impacts. *Science, 289*, 2068–2074.

Emanuel, K. A. (1987). The dependence of hurricane intensity on climate. *Nature, 326*, 483–485.

Flatley, W. T., Lafon, C. W., Grissino-Mayer, H. D., & LaForest, L. B. (2013). Fire history, related to climate and land use in three southern Appalachian landscapes in the eastern United States. *Ecological Applications, 23*, 1250–1266.

Goetz, S. J., Bond-Lamberty, B., Law, B. E., Hicke, J. A., Huang, C., Houghton, R. A., McNulty, S., O'Halloran, T., Harmon, M., Meddens, A. J. H., Pfeifer, E. M., Mildrexler, D., & Kasischke, E. S. (2012). Observations and assessment of forest carbon dynamics following disturbance in North America. *Journal of Geophysical Research, 117*(G2), G02022. http://dx.doi.org/10.1029/2011JG001733. Accessed 12 Jan 2015.

Greenberg, C. H., Perry, R. W., Franzreb, K. E., Loeb, S. C., Saenz, D., Rudolph, D. C., Winters, E., Fucik, E. M., Kwiatkowski, M. A., Parresol, B. R., Austin, J. D., & Tanner, G. W. (2014). Climate change and wildlife in the southern United States: Potential effects and management options. In J. M. Vose & K. D. Klepzig (Eds.), *Climate change adaptation and mitigation management options* (pp. 379–420). New York: CRC Press.

Hanson, P. J., & Weltzin, J. F. (2000). Drought disturbance from climate change: Response of United States forests. *Science of the Total Environment, 262*, 205–220.

Hughes, M. J., & Hayes, D. J. (2014). Automated detection of cloud and cloud-shadow in single-date Landsat imagery using neural networks and spatial post-processing. *Remote Sensing, 6*, 4907–4926.

IPCC. (2014). Climate change 2014: Impacts, adaptation, and vulnerability. In C. B. Field, V. R. Barros, D. J. Dokken, K. J. Mach, M. D. Mastrandrea, T. E. Bilir, M. Chatterjee, K. L. Ebi, Y. O. Estrada, R. C. Genova, B. Girma, E. S. Kissel, A. N. Levy, S. MacCracken, P. R. Mastrandrea, & L. L. White (Eds.), *Working Group II, contribution to the Fifth assessment report Intergovernmental Panel on Climate Change*. Cambridge: Cambridge University Press.

Iverson, L. R., & McKenzie, D. (2013). Tree-species range shifts in a changing climate: Detecting, modeling, assisting. *Landscape Ecology, 28*, 879–889.

Iverson, L. R., Schwartz, M. W., & Prasad, A. M. (2004). How fast and far might tree species migrate in the eastern United States due to climate change? *Global Ecology and Biogeography, 13*, 209–219.

Iverson, L., Prasad, A., & Matthews, S. (2008). Modeling potential climate change impacts on the trees of the northeastern United States. *Mitigation and Adaptation Strategies for Global Change, 13*, 487–516.

Jentsch, A., Kreyling, J., & Beierkuhnlein, C. (2007). A new generation of climate-change experiments: Events, not trends. *Frontiers in Ecology and the Environment, 5*, 365–374.

Johnson, A. H. (1992). The role of abiotic stresses in the decline of red spruce in high elevation forests of the eastern United States. *Annual Review of Phytopathology, 30*, 349–367.

Johnson, D. W., Todd, D. E., & Hanson, P. J. (2008). Effects of throughfall manipulation on soil nutrient status: Results of 12 years of sustained wet and dry treatments. *Global Change Biology, 14*, 1661–1675.

Joyce, L. A., Running, S. W., Breshears, D. D., Dale, V. H., Malmsheimer, R. W., Sampson, R. N., Sohngen, B., Woodall, C. W. (2014). Forests. In J. M. Melillo, T. C. Richmond, G. W. Yohe (Eds.), *Climate change impacts in the United States: The third national climate assessment* (pp. 175–194). US Global Change Research Program. doi:10.7930/J0Z60KZC.

Kardol, P., Todd, D. E., Hanson, P. J., & Mulholland, P. J. (2010). Long-term successional forest dynamics: Species and community responses to climatic variability. *Journal of Vegetation Science, 21*, 627–642.

Karl, T. R., & Knight, R. W. (1998). Secular trends of precipitation amount, frequency, and intensity in the United States. *Bulletin of the American Meteorological Society, 79*, 231–241.

Karl, T. R., Knight, R. W., & Plummer, N. (1995). Trends in high-frequency climate variability in the twentieth century. *Nature, 377*, 217–220.

Karl, T. R., Melillo, J. M., & Peterson, T. C. (Eds.). (2009). *Global climate change impacts in the United States*. Cambridge: Cambridge University Press.

Keane, R. E., Agee, J. K., Fule, P., Keeley, J. E., Key, C., Kitchen, S. C., Miller, R., & Schulte, L. A. (2008). Ecological effects of large fires on US landscapes: Benefit or catastrophe? *International Journal of Wildland Fire, 17*, 696–712.

Kennedy, R. E., Zhiqiang, Y., & Cohen, W. B. (2010). Detecting trends in forest disturbance and recovery using yearly Landsat time series: 1. LandTrendr—Temporal segmentation algorithms. *Remote Sensing of Environment, 114*, 2897–2910.

Konrad, C. E., Fuhrmann, C. M., Billiot, A., Keim, B. D., Kruk, M. C., Kunkel, K. E., Shafer, M., & Stevens, L. (2013). Climate of the southeast USA: Past, present and future. In K. T. Ingram, K. Dow, L. Carter, & J. Anderson (Eds.), *Climate of the southeast United States: Variability, change, impacts and vulnerability* (pp. 8–42). Washington, DC: Island Press.

Manion, P. (1981). *Tree disease concepts* (2nd ed.). Englewood Cliffs: Prentice Hall.

McDowell, N., Pockman, W. T., Allen, C., Breshears, D. D., Cobb, N., Kolb, T., Plaut, J., Sperry, J., West, A., Williams, D. G., & Yepez, E. A. (2008). Mechanisms of plant survival and mortality during drought: Why do some plants survive while others succumb to drought? *New Phytologist, 178*, 719–739.

McDowell, N. G., Beerling, D. J., Breshears, D. D., Fisher, R. A., Raffa, K. F., & Stitt, M. (2011). The interdependence of mechanisms underlying climate-driven vegetation mortality. *Trends in Ecology & Evolution, 26*, 523–532.

McNab, W. H., Spetich, M. A., Perry, R. W., Haywood, J. D., Laird, S. G., Clark, S. L., Hart, J. L., Torreano, S. J., & Buchanan, M. L. (2014). Climate-induced migration of native tree populations and consequences for forest composition. In J. M. Vose & K. D. Klepzig (Eds.), *Climate change adaptation and mitigation management options*. New York: CRC Press.

McNulty, S. G., Vose, J. M., & Swank, W. T. (1996). Potential climate change effects on loblolly pine forest productivity and drainage across the southern United States. *Ambio, 25*, 449–453.

McNulty, S., Caldwell, P., Doyle, T. W., Johnsen, K., Liu, Y., Mohan, J., Prestemon, J., & Sun, G. (2013). Forests and climate change in the southeast USA. In K. T. Ingram, K. Dow, L. Carter, & J. Anderson (Eds.), *Climate of the southeast United States: Variability, change, impacts and vulnerability* (pp. 165–189). Washington, DC: Island Press.

Milner, G. C., & Chaplin, G. (2010). Eastern North American population at A.D. 1500. *American Antiquity, 75*, 707–726.

National Climate Assessment. (2014). *US Global Change Research Program*. Washington, DC. http://nca2014.globalchange.gov/. Accessed 27 Jan 2015.

Oliver, C. D. (2014). Functional restoration of social-forestry systems across spatial and temporal scales. *Journal of Sustainable Forestry, 33*, S123–S148. doi:10.1080/10549811.2014.884003.

Paine, R. T., Tegner, M. J., & Johnson, E. (1998). Compounded perturbations yield ecological surprises. *Ecosystems, 1*, 535–545.

Raffa, K. F., Aukema, B. H., Bentz, B. J., Carroll, A. L., Hicke, J. A., Turner, M. G., & Romme, W. H. (2008). Cross-scale drivers of natural disturbances prone to anthropogenic amplification: The dynamics of bark beetle eruptions. *Bioscience, 58*, 501–517.

Sauer, C. O. (1950). Grassland climax, fire, and man. *Journal of Range Management, 3*, 16–21.

Tran, J. K., Ylioja, T., Billings, R. F., Régnière, J., & Ayres, M. P. (2007). Impact of minimum winter temperatures on the population dynamics of *Dendroctonus frontalis*. *Ecological Applications, 17*, 882–899.

Turner, M. G., Baker, W. L., Peterson, C. J., & Peet, R. J. (1998). Factors influencing succession: Lessons from large, infrequent disturbances. *Ecosystems, 1*, 511–523.

Ungerer, M. J., Ayres, M. P., & Lombardero, M. J. (1999). Climate and the northern distribution limits of *Dendroctonus frontalis* Zimmermann (Coleoptera: Scolytidae). *Journal of Biogeography, 26*, 1133–1145.

US Global Change Research Program CRP. (2009). Global climate change impacts in the United States. In T. R. Karl, J. M. Melillo, & T. C. Peterson (Eds.), *United States Global Change Research Program*. New York: Cambridge University Press.

USDA Forest Service. (2012). *Future of America's forests and rangelands, Resources Planning Act assessment* (General technical report WO, Vol. 87). Washington, DC: USDA Forest Service Washington Office.

Vose, J. M., & Klepzig, K. D. (2014). Introduction to climate change adaptation and mitigation management options. In J. M. Vose & K. D. Klepzig (Eds.), *Climate change adaptation and mitigation management options*. New York: CRC Press.

Vose, J. M., Peterson, D. L., & Patel-Weynand, T. (Eds.). (2012). *Effects of climatic variability and change on forest ecosystems: A comprehensive science synthesis for the US forest sector* (General technical report PNW-GTR, Vol. 870). Portland: USDA Forest Service Pacific Northwest Research Station.

Waldron, J. D., Lafon, C. W., Coulson, R. N., Cairns, D. M., Tchakerian, M. D., Birt, A., & Klepzig, K. D. (2007). Simulating the impacts of southern pine beetle and fire on the dynamics of xerophytic pine landscapes in the southern Appalachians. *Applied Vegetation Science, 10*, 53–64.

Wear, D. N., Huggett, R., Ruhong, R., Perryman, B., & Shan, S. (2013). *Forecasts of forest conditions in regions of the United States under future scenarios: A technical document supporting the Forest Service 2012 Resources Planning Act Assessment* (General technical report SRS-GTR, Vol. 170). Asheville: USDA Forest Service Southern Research Station.

Weed, A. S., Ayres, M. P., & Hicke, J. A. (2013). Consequences of climate change for biotic disturbances in North American forests. *Ecological Monographs, 83*, 441–470.

Wells, G. L. (1992). *The aeolian landscape of North America from the late Pleistocene*. Dissertation, University of Oxford, UK.

White, P. B., Soule, P., & van de Gevel, S. (2014). Impacts of human disturbance on the temporal stability of climate-growth relationships in a red spruce forest, southern Appalachian Mountains, USA. *Dendrochronologia, 32*, 71–77.

Williams, D. W., & Liebhold, A. M. (2002). Climate change and the outbreak ranges of two North American bark beetles. *Agricultural and Forest Entomology, 4*, 87–99.

Xi, W., Waldron, J. D., Lafon, C. W., Cairns, D. M., Birt, A. G., Tchakerian, M. D., & Klepzig, K. D. (2009). Modeling long-term effects of altered fire regimes following southern pine beetle outbreaks (North Carolina). *Ecological Restoration, 27*, 24–2.

Yaussy, D. A., Iverson, L. R., & Matthews, S. N. (2013). Competition and climate affects US hardwood-forest tree mortality. *Forest Science, 59*, 416–430.

Zhao, S. Q., Liu, S. G., Sohl, T., Young, C., & Werner, J. (2013). Land use and carbon dynamics in the southeastern United States from 1992 to 2050. *Environmental Research Letters, 8*, 044022.

Chapter 14
Managing Central Hardwood Forests Within the Context of the Historic Range of Variability (HRV): Challenges and Opportunities

Eric K. Zenner

Abstract The historical range of variability (HRV) has been an especially effective tool for guiding the implementation of ecosystem management in the western forests for which it was developed. HRV analyses typically focus on ecosystem components such as composition (species, cover types, lifeforms), structure (tree size classes, age classes, structural stages), and processes/ functions (regeneration, growth, mortality), and the spatio-temporal interactions of these with each other, the disturbance regime, and climate. In central hardwood forests, the most important contribution of HRV has been an increased appreciation for the importance of natural disturbances in maintaining the developmental processes responsible for critical habitat structures. HRV analyses can improve our understanding of the causes and consequences of historic variability in ecosystem components. HRV time series are particularly useful as a reference condition to assess, plan, evaluate, design, and implement ecosystem-restoration treatments, but are not intended to dictate or justify narrow management targets. Application of HRV in this region in the future will be challenged by the complexity of the small-scale gap dynamics disturbance regime, an onslaught of exotic species invasions, and uncertainty pertaining to climate change. However, as a framework for approaching these challenges, the HRV approach provides opportunities for understanding the processes that will be responsible for shaping forests into the future.

Keywords Climate change • Disturbance regime • Ecosystem management • Landscape structure • Resilience

E.K. Zenner (✉)
Department of Ecosystem Science and Management, The Pennsylvania State University, University Park, PA 16802, USA
e-mail: eric.zenner@psu.edu

C.H. Greenberg, B.S. Collins (eds.), *Natural Disturbances and Historic Range of Variation*, Managing Forest Ecosystems 32,
DOI 10.1007/978-3-319-21527-3_14

14.1 Historic Range of Variation (HRV) in Context

Forestry, like any good science, is adaptive. It was born of necessity and has been selectively modified over time to address the needs of the future through the application of logic to solve current problems. At the turn of the twentieth century, overexploitation had completely degraded our forests. Intensive management enabled their perpetuation, so sustained yield became the paradigm of the future. This was no accident or consequence of industrial collusion; it was a logical conclusion based on the knowledge of the day. Having *no* paradigm was rendering the overexploited forests into cutover moonscapes, and setting a minimum requirement of timber yield ensured that at least some forest integrity survived. Thus, adopting a sustained yield paradigm was the logical savior of the forest. In recent decades, forestry on public lands has evolved again. When sustained yield failed to provide the full range of desired products and services while pristine ecosystems perpetuated biodiversity, ecosystem management became the paradigm. This modern savior of the forest aims to ensure healthy, sustainable ecosystems that can maintain their structure and organization through time (Whitford et al. 1999; USDA Forest Service 2012), with the hope that they will continue to provide high-quality ecosystem services and products.

With the evolution to ecosystem management has come the need for a new orientation of silviculture, which has hitherto been defined as "the art and science of **controlling** the establishment, growth, composition, health, and quality of forests and woodlands to meet the diverse needs and values of landowners and society on a sustainable basis" (Society of American Foresters 2014). Under sustained yield, it was at least theoretically possible to control developmental and successional pathways when disturbance (harvest) was predictable, and variation—considered "something that clouded the average or normal conditions" (Wiens et al. 2012)—was limited. At the stand scale, such control typically was achieved by simplifying species composition and stand structure. This simplification led to concentration on a few desirable timber species, as well as reduction of horizontal variation in stand density, vertical complexity in the forest canopy, and key habitats such as dead trees (snags) and downed logs (coarse woody debris) (Franklin et al. 2007). At larger scales (e.g., ownership, landscape, district or National Forest), planning ensured a controlled, even flow of timber by determining the allowable annual cut and regulating timber harvests. Over a few rotations it was possible to achieve reliably balanced or even age class distributions (e.g., Johnson et al. 2009), with variation among, but not necessarily within, stands. The landscape was thus comprised of balanced age classes that promised predictability and efficiency and imparted the illusion of control.

Over time, the simplified species compositions and stand structures produced by management for sustained yield were found to be inadequate to ensure ecosystem integrity (Noss 1999). Further, variation or heterogeneity was not only found to be the norm in ecosystems, but apparently is required by the species adapted to them (Poisot et al. 2011). Consequently, the management approaches and forestry practices required to sustain ecosystems are necessarily more comprehensive and complex than those that seek to sustain timber yields. Benchmarking the average or 'normal' conditions that track so predictably through the orderly progression from harvest disturbance to steady growth is no longer sufficient to capture the variation associated with the multiple post-disturbance pathways and non-equilibrium processes of terrestrial ecosystems (Mori 2011; Romme et al. 2012). We find ourselves in need of new metrics to monitor management actions, and new standards by which we can evaluate our success.

The concept of the historical range of variability (HRV) has been promoted to fill this void. Although many definitions of HRV have been introduced into the literature (e.g., Kaufmann et al. 1994; Landres et al. 1999; Stephenson 1999; Hann and Bunnell 2001; Baker 2009; Binkley and Duncan 2009; Keane et al. 2009), they all characterize HRV as the range of some condition of the ecosystem (e.g., stand structure, patch size distribution, diversity) or some process (e.g., disturbance regime) that has occurred within an ecosystem across several spatial scales and over some specified time period. The central idea of the HRV concept is that historical conditions are assumed to represent natural or 'properly functioning' ecosystems (i.e., reference conditions). Retaining key ranges of variability in many ecosystem components or processes is crucial to ensure continued health, self-organization, and resilience (i.e., the ability to recover functionality following disturbance) of ecosystems and landscapes across spatio-temporal scales (Holling 1992; Cissel et al. 1994; Holling and Meffe 1996; Laughlin et al. 2004). The basic premise is that the ecological conditions most likely to preserve native species or conserve natural processes are those that sustained them in the past, when ecosystems were less affected by people (Manley et al. 1995; Egan and Howell 2001). Because fully functional contemporary reference ecosystems are difficult to find due to pervasive anthropogenic modifications that have converted many landscapes into human-dominated ecosystems (Noble and Dirzo 1997), reference states must often be defined from historical conditions (Safford et al. 2012). Thus, HRV analyses attempt to quantify the limits of the ranges of variability of different attributes of ecosystem structure and function over space and (historical) time. These limits can then be compared against the range of conditions produced in managed contemporary ecosystems (Aplet and Keeton 1999), and perhaps represent a reasonable set of bounds within which contemporary ecosystems should be managed (cf. Safford et al. 2012).

The HRV concept has been developed most successfully in coniferous forests of western North America. Historically, these forests were dependent on periodic wildfire, yet they were subjected to mandated fire suppression for nearly a century. This fire suppression policy was the first nationwide managerial effort to be codified (Omi 2005) and was also the result of a logical deduction: calamitous fires around the start of the twentieth century (albeit mostly in overexploited eastern and central forests along railways) devastated timber holdings and communities alike. These disasters could be prevented by assiduous and early fire control; thus, fire suppression became an important tool to protect the forest resource (Pyne 1982). In the western forests adapted to fire, however, its elimination gradually resulted in outcomes that proved to be undesirable, ranging from altered species compositions (regeneration failures of desired tree species) to the buildup of immense fuel loads that exploded into catastrophic fire when suppression efforts failed (Keane et al. 2002).

Fortunately, it has proved possible to mitigate many of the effects of fire suppression by mimicking the most obvious outcomes of the natural disturbance regime of the western USA—sometimes using fire itself as a restoration tool (e.g., Covington et al. 1997). Stand reconstructions of older forests using dendrochronology revealed the frequency and intensity of the wildfires in which these ecosystems had evolved (e.g., Fulé et al. 1997), documenting the HRV and establishing it as a framework for future restoration efforts. In general, in forest ecosystems dominated by only a few tree species (such as many western coniferous forests), the primary drivers of variability are exogenous disturbances, such as wind or fire. Accordingly, using HRV to guide management has become very important in the western region. Perhaps the greatest contribution of the HRV concept for the central hardwoods has been the resulting emphasis in forest ecology on the vital contribution of disturbance and on our awareness of the complexity of natural disturbances.

14.2 The Role of Disturbance

Employing HRV to manage landscapes that sustain ecosystems and the services they provide to society requires understanding the causes that impart variability, and being able to predict the consequences of variability—or the lack thereof (Keane 2009). This emphasis on disturbance comes at a time when our concept of disturbance is itself evolving (Mori 2011). Our model of disturbance has assumed the existence of a normal progression that can be disrupted, whether taken from academia ("Disturbances are relatively discrete events in time that **disrupt** the ecosystem, community, or population structure and bring about a change in resources, substrate availability, or the physical environment" (White and Pickett 1985)), NGOs ("A disturbance is any event that causes a **disruption** to the current state of an ecosystem" (National Wildlife Federation 2014)), or the informed public ("In ecology, a disturbance is a temporary change in average environmental conditions that causes a pronounced change in an ecosystem" (Wikipedia 2014)). Concepts of 'balance of nature' or 'equilibrium' are now being replaced by 'dynamics' and 'non-equilibrium' for predicting complexity, instability, and inevitable changes in ecosystems (White and Pickett 1985; Levin 1999; Phillips 2004; Mori 2011). Naturally recurring disturbances profoundly regulate the function, structure, and dynamics of ecosystems (Perry 1998; White et al. 1999). In non-equilibrating dynamic systems, disturbances play inherent and central roles (Phillips 2004; Moore et al. 2009); they drive most ecosystem processes and greatly influence landscape- and stand-level pattern, structure, and function, ultimately setting the bounds of the HRV (Keane 2013). Because disturbances are the primary agents of vegetation dynamics (Sprugel 1991; Laska 2001), natural disturbance regimes are among the most significant ecological processes that should be conserved or restored to maintain biodiversity (Yaffee 1999). It is thus no coincidence that disturbance regimes play a central role in the HRV concept, and that incorporating disturbances into management is a central tenet of HRV (Morgan et al. 1994; Landres et al. 1999).

Disturbance regimes express the cumulative effects of multiple disturbance events over space and time (Keane 2013). As such, they set the stage for multiple successional pathways of post-disturbance ecosystem development, which are highly variable, multi-directional, and influenced by the quality, quantity, and spatial arrangement of physical and biological legacies left at disturbed sites (Foster et al. 1998; Lindenmeyer and Franklin 2002; Fischer et al. 2006; Lindenmeyer and Noss 2006). These legacies can lead to diverging stand structures and development processes in the early phases of succession (Kashian et al. 2005). Consequently, disturbances not only impart immediate heterogeneity by providing early successional phases, but also imprint diversity during the long-term dynamics of ecosystems by providing differing initial conditions for ecosystem development. Disturbances create shifting mosaics of diverse plant communities and habitats across a landscape (Watt 1947), and the spatial and temporal variation of these communities is a foundation of biodiversity (Naveh 1994).

Keane (2013) provided an excellent overview of disturbance characteristics that comprehensively describe terrestrial disturbance regimes. Disturbance 'agent' refers to the entity that causes the disturbance (e.g., wind, fire, beetles), while the 'source' is the origin of the agent (e.g., lightning, snow load). 'Size' is the spatial extent of the area affected by a disturbance agent, and 'duration' is the length of time a disturbance event lasts, ranging from seconds (wind) and minutes (avalanches) to days (fire) and years (insects). 'Seasonality' refers to the time of year the disturbance event takes place, with differential spatial effects due to phenological differences in plant communities throughout the seasons. 'Frequency' is the number of disturbance events experienced over time, which can be expressed as a disturbance return interval or occurrence probability at a given point on a landscape (Baker and Ehle 2001) or as the number of years required to disturb an area of a given size, expressed as disturbance rotation or cycle (van Wagner 1978; Reed et al. 1998). 'Intensity' is a property of the disturbance agent and refers to the magnitude with which it affects an area (e.g., wind speed, heat output), whereas 'severity' is a property of the affected area and refers to the level of impact or amount of change sustained by the biophysical environment (e.g., level of mortality, change in fuel load, change in soil properties). 'Pattern' refers to the spatial variability (i.e., size, shape, location, intermingling) of areas on the landscape that are differentially affected by the disturbance, quantifying the spatio-temporal patchwork created by the disturbance regime. Finally, 'variability' refers to both the spatial and temporal variability of each of the aforementioned disturbance characteristics, setting the stage for interactions between previously created disturbance outcomes and future disturbance events (Mori 2011; Keane 2013) (Table 14.1).

The multidimensional nature of each subsequent disturbance event, coupled with the spatial and temporal variability of disturbance regimes, makes disturbances some of the most powerful and complex processes governing successional trajectories, plant community composition, age class distribution, landscape dynamics, and, ultimately, energy flow and biodiversity (Wiens et al. 1985; Hansen et al. 1991;

Table 14.1 Important characteristics to describe disturbance regimes (After Keane 2013)

Disturbance characteristic	Description	Example
Agent	Entity causing the disturbance	Fire, wind, insects, diseases
Source	Origin of the agent	Weather condition, lightning, hurricane, tornado
Size	Spatial extent of the affected area	Single tree, patch, stand, watershed, landscape
Duration	Length of the disturbance event	Minutes (tornado) to years (insect outbreak)
Seasonality	Time of year when disturbance takes place	Climatic and weather patterns throughout the year in relation to phenology of the vegetation
Frequency	Number of disturbance events over a time period	Time since last disturbance or insect/disease outbreak
Intensity	Magnitude of the disturbance	Wind speed, heat output, population level of insect
Severity	Level of impact or amount of change due to disturbance	Proportion of trees killed or tree crowns lost
Pattern	Spatial heterogeneity of disturbance effects	Spatial patchwork of severity
Variability	Spatial and temporal	Patchwork in space and time
Interaction	Positive or negative feedback among different disturbances and between disturbance and climate, vegetation, and other landscape characteristics	Previous fire may have reduced fuel in some location resulting in lower fire intensity in case of a repeated fire event for some time

Frelich and Lorimer 1991; Turner and Romme 1994; Turner et al. 1998; Keane 2013). In fact, the hierarchical context of natural disturbances at multiple spatiotemporal scales, combined with the effects of disturbance legacies and the possibility of multiple post-disturbance pathways, strongly affects ecosystem productivity (Turner et al. 2004), carbon dynamics (Litton et al. 2004), and biological diversity (Spies and Turner 1999; Lindenmeyer and Noss 2006), and gives rise to high levels of ecological heterogeneity at multiple scales (White and Jentsch 2001; Schoennagel et al. 2008; Williams and Bradstock 2008; Mori and Lertzman 2011).

Such complexity makes HRV inherently difficult to implement (North and Keeton 2008). First the ecological characteristics that are meaningful in relation to management objectives (Keane et al. 2002; Romme et al. 2012) must be identified. Each characteristic must then be quantified across the appropriate spatial scale/domain (Turner et al. 1993) and to a satisfactory spatial extent. However, this must also be done across an 'appropriate' time scale (i.e., to a sufficient temporal depth) to ensure that the historical ecosystem dynamics can be reasonably well tied to the underlying biophysical and ecological processes as well as anthropogenic influences that have created them (Keane et al. 2009). The choice of time scale is a tradeoff between a more recent time for which more data sources, including written and photographic records are available, and an earlier time when ecosystems were 'less' affected by people, both Native American and increasingly, over the last five centuries, Euro-Americans.

Although records demonstrate the pervasive and increasing influence of a burgeoning Euro-American population on the landscape over the last 200 years, historical Native American influences were a key factor shaping many ecosystems across the USA in the previous 12–14 millennia (Hicks 1997; Dey and Schweitzer 2014; Chaps. 1 and 12). For example, Native American burning of vast areas of land has long supplemented lightning-caused fires, defining historical fire regimes and affecting the ecological conditions of landscapes and forest communities (Barrett and Arno 1982; Kay 1995; Chap. 6). It has become increasingly clear that indigenous peoples did not live in harmony with nature (Krech 1999), and that societies lacking the technology for intensive agriculture have nonetheless caused profound environmental changes through extensive land use (Steffen et al. 2007; Ellis et al. 2013; also see Chap. 8). Distinguishing Native American influences, such as prescribed or accidental burning, from natural influences, such as the lightning-ignited fire regime in the historical data set is difficult (Slocum et al. 2007).

Quantifying the effect of Native Americans on ecosystem conditions is further complicated by steeply declining native populations after contact with European explorers, who brought with them lethal diseases. This population drop may have led to a temporary recovery of forest ecosystems from human influence, and the 'primeval forests' of legend (Hicks 1997). Thus, the pre-European settlement period selected by some as a benchmark for HRV evaluations may represent an anomaly, as ecosystems were responding to changes in disturbance regimes with more limited human influence (Dey and Schweitzer 2014). The corollary that historical ecosystems with 'minimal' human influence were more stable or in equilibrium (Perry 2002), self-regulating (Middleton 1999), and healthy (Swetnam et al. 1999) may thus be more romantic notion than historical fact. In fact, reference periods without human influence do not exist within the past 12–14 millennia, i.e., since boreal forests and steppe conditions extended into the southern USA (Dey and Schweitzer 2014; also see Chaps. 1 and 12). Any more recent HRV reconstructions increasingly reflect human influences.

To ensure a sufficient spatial context, the size of an evaluation area must be large enough to enable variation of the selected landscape attribute to be quantified relative to the spatial dynamics of the ecosystem processes (Turner et al. 1993), but not so large as to blur bioclimatic/biophysical boundaries. The optimal size of the evaluation area therefore not only depends on the dynamics of the major disturbance regimes, but also varies with ecosystem attributes (Tang and Gustafson 1997). While estimates of the HRV of some ecosystem attributes such as deadwood, coarse woody debris, tree density or standing volume can be assessed at the stand, landscape, and regional spatial scales, landscape composition and patch structure by necessity requires larger scales (Keane 2013). In the central hardwoods, for example, the exceptionally few forest stands old enough to permit at least partial historic temporal reconstructions are spatially disjunct and subject to widely varying anthropogenic influences, rendering reliable extrapolation virtually impossible.

Furthermore, our ability to reproduce only a simplified subset of disturbance outcomes may prove futile, if not dangerous, in light of such complexity; the likelihood of fundamental shifts in ecosystems (e.g., shifting from woodland to grassland) may increase if humans reduce resilience when altering disturbance regimes (Folke et al. 2004).

Realistically, the complexity of central hardwood forests can only be reproduced in computer simulations, if at all. The proper focus of HRV is not to reproduce disturbance regimes or even individual disturbances, but rather to reproduce the *outcomes* of natural historical disturbances regimes—the broad envelope of possible ecosystem conditions such as stand structure, patch size distribution, and diversity—to use as a reference condition to guide land management (Aplet and Keeton 1999). Management thus would use controlled anthropogenic disturbance (e.g., harvest) to mimic the outcomes of natural disturbance. This does not occur, however, within a vacuum: the outcomes to be mimicked are those that are consistent with currently desired ecosystem services. This is confounded by the fact that, in some systems, anthropogenic effects have interacted with natural factors for so long that they cannot be completely separated from natural variability (Keane 2013). The consequences of multiple subsequent, or even simultaneous, disturbances (i.e., combinations of planned human-induced change and unplanned natural processes or disturbances) can transform ecosystems. Dynamic ecosystems maintain complexity and are self-sustaining only within a range of conditions, beyond which they transition to other (perhaps undesirable) ecological states (Carpenter and Gunderson 2001; Egan and Howell 2001; Wu et al. 2006). Ecosystem resilience should most often be the first and foremost desired outcome of HRV, to ensure that whatever anthropogenic or natural change comes next, the forest (or at least **a** forest) will remain.

14.3 HRV in the Central Hardwood Forests

Using HRV to promote ecosystem resilience in central hardwood forests is unlikely to be as straightforward as in the western systems in which the concept was developed, due to differences in natural disturbance regimes, a lack of reference forests, a much higher tree species diversity, longer historic anthropogenic influence, and unprecedented future agents of change. The most dominant natural disturbance agent in many western coniferous ecosystems is the exogenous physical disturbance of fire, which occasionally selectively eliminates species or age classes of trees (e.g., based on bark thickness) but is frequently generic and widespread (Weaver 1974). Many of the obvious structural outcomes of such a disturbance, as measured by HRV, are cost-effectively reproducible using forest management practices that promote fewer but larger trees, more snags, more course woody debris, or open growing space, and often on relatively large scales. In central hardwood forests, primary current natural disturbances range from catastrophic disturbances associated with tropical storms and hurricanes to smaller wind events (Peterson et al. Chap. 5), ice (Lafon Chap. 7), and surface fires that typically operate on sub-stand scales (e.g., Nowacki and Abrams 2008; White et al. 2011; see also Grissino-Mayer Chap. 6). Given the enduring incidence of ice and wind storms or hurricanes, it is neither necessary, nor prudent, to reproduce catastrophic disturbances. In contrast, the structural outcomes of infrequent large-scale wind storms may be reproducible through overstory removal or clearcutting with reserves. Replicating the vastly more complex outcomes of far more frequent small-scale gap dynamics (Hart Chap. 2), however, requires detailed knowledge about HRV in gap dynamics that is largely lacking or biased toward atypical old-growth remnants (but see Hart Chap. 2).

The historic causes of variation in many western ecosystems have left tell-tale signals in the wood of long-lived species, enabling reconstruction of disturbance events often spanning many centuries. Such reconstructions are rare in eastern and central forests due largely to a lack of reference forests; few old trees or stumps remain, and those that do are often rotted or span scant centuries (e.g., Hart et al. 2012). Further, the duration of anthropogenic influence is longer in this region, such that the entire period for which historic reconstructions is possible is likely confounded by human influence. The oldest oak and pine forests may have been influenced by Native American burning, which persisted into the nineteenth century (Ruffner and Abrams 2002), and much of the 80–120 year old oak-dominated forest of today arose following clearcutting those oak and pine forests in the nineteenth century. These oak forests were also strongly influenced by the loss of American chestnut (*Castanea dentata*) following the introduction of chestnut blight (*Cryphonectria parasitica*) and by rising herbivore (e.g., deer (*Odocoileus virginianus*)) populations following expanded land clearing and the extirpation of predators. Although poorly documented, domestic grazing and subsistence wood use likely also shaped the structure of the central hardwood forests available for assessment under HRV.

Even assuming that HRV could be estimated and was able to capture ecosystem responses to changing processes (Veblen 2003), or that it could serve as a proxy for ecosystem health (Swetnam et al. 1999), the relevance of historic conditions is now challenged. Simply put, it is no longer possible to recreate the conditions that persisted centuries or even decades ago. In addition to the effects of anthropogenic fire, historic harvesting and then farm abandonment, and the introduction of chestnut blight, recent decades have seen the forests of the mid-Atlantic states exposed to soil acidification, nutrient impoverishment, and aluminum toxicity from acid deposition (Bailey et al. 2005). Millions of hectares of forest have been lost to development (Hicks 1998) and many more have become fragmented (Parker 1993). The human footprint in central hardwood ecosystems is tremendous and has persisted for as long as our records of HRV.

All these challenges pale, however, in comparison with the wave of unprecedented change that is now being ushered in by the combination of exotic species and climate change (see Chap. 13), resulting in emergent ecosystems with no historical analogs, even in the paleorecord (Gill et al. 2009; Hobbs et al. 2009; Hobbs 2013). In the past, the forest disturbance regime was based on physical exogenous forces that may not have been controllable, but were sufficiently predictable to enable regional adaptation. As a consequence, when these disturbances occurred, the result was typically greater landscape heterogeneity as new successional stages and developmental pathways were initiated. In the future, the disturbance regime will be based on a combination of physical forces and biological forces as ever-more exotic species are introduced to North America. Over 400 alien phytophagus insect species have already become established in North American forests and nearly every native woody plant genus in eastern North American forests has at least one alien insect and/or pathogen associated with it (Mattson et al. 1994). As temperatures rise and precipitation becomes more variable, central hardwood forests will be subject both to increasing storm events (Peterson 2000) and to stressors (e.g., crown damage, drought, flooding, winter respiration, disruptions to dormancy, loss of soil symbionts, etc.) that render individual trees more susceptible to pests and disease (Allen et al. 2010). Simultaneously, these pests, pathogenic symbionts, and diseases will themselves gain influences as disturbance agents as moister conditions promote fungi and droughts and warmer winters favor insects (Dukes et al. 2009).

Through the selective removal of ecosystem components and reduced biodiversity (Ghandi and Herms 2010), many disturbance agents have already begun to threaten the successful regeneration of individual tree species, in some cases leading to extirpation. For many decades, central and eastern forests have been further weakened by Dutch elm disease (*Ophiostoma novo-ulmi*) and gypsy moth (*Lymantria dispar*) (Davidson et al. 1999; Lovett et al. 2006). The smaller European elm bark beetle (*Scolytus multistriatus*), the most important vector of Dutch elm disease (Hanula and Berisford 1984), has decimated American elm (*Ulmus americana*) throughout the north (Barnes 1976; Karnosky 1979); hemlock woolly adelgid (*Adelges tsugae*) is causing wide-spread mortality of eastern (*Tsuga canadensis*) and Carolina (*Tsuga caroliniana*) hemlock throughout riparian areas in the Appalachian forests (Ellison et al. 2005); balsam woolly adelgid (*A. piceae*) has devastated Fraser fir (*Abies fraseri*) in high altitude forests of the southern Appalachian Mountains (Hollingsworth and Hain 1991); the emerald ash borer (*Agrilus planipennis*) threatens the future of ash (*Fraxinus* spp.) trees (Cappaert et al. 2005; Poland and McCullough 2006); and the Asian longhorned beetle (*Anoplophora glabripennis*) endangers the genera of maple (*Acer* spp.), poplar (*Populus* spp.), willow (*Salix* spp.), and elm (Townsend Peterson and Sachetti-Pereira 2004).

Projected climate change, which is likely to increase weather variability and lead to more frequent or severe extreme events (Meehl et al. 2000)—especially drought (Allen et al. 2010) and severe windstorms—may result in rapid directional changes that can push ecosystems into alternative states (Suding et al. 2004; Chap. 13). Predictable changes in climate and unpredictable stochastic natural disturbances resulting from changes in climate will push ecosystems beyond thresholds of acceptable structure and function (Chen et al. 2011). Further, changes in climatic means may create conditions that are no longer conducive for their growth (e.g., Lindner et al. 2010; Vose et al. 2012). Even though current forest ecosystems are a response to the past continually changing climate (Millar and Woolfenden 1999), future climates may be no-analog climates (Williams and Jackson 2007; Williams et al. 2007) in which species within a forest ecosystem will respond individually to expected radical shifts in local climate regimes (Aitken et al. 2008). Resulting novel ecosystems will be comprised of species assemblages without current analogs (Harris et al. 2006), indicating that past compositions/structures may not be resilient under future climate scenarios.

In fact, it is quite uncertain whether future disturbances will necessarily be associated with increased heterogeneity. It is still unclear how the life cycle processes of reproduction and mortality will interact with new climate conditions and hence shape species' responses to climate change (Keane et al. 2001; Ibanez et al. 2007). With the potential for changed biodiversity assemblages (Notaro et al. 2007) comes the potential for a resulting simplification of composition and structure. If the magnitude and rate of environmental change is sufficiently extreme, not only may previous HRV efforts prove inapplicable, but there may simply not be enough time to implement the approach in the future (Wiens et al. 2012; Higgs et al. 2014). When the fundamental assumption of HRV, climatic stationarity (i.e., a stable climate in the chosen reference period), is shown to be flawed, no historic period or range of historic conditions can be an appropriate reference for defining 'natural variability' (i.e., the spatio-temporal ecosystem variability driven by disturbances) (Landres et al. 1999, Mori 2011) or developing desired future condition targets (Millar 2014). Because the forests of the future will be exposed to entirely new combinations of disturbance agents, well beyond the evolutionary experience of the central hardwoods, it will not be possible to ensure resilience by recreating outcomes based on historic disturbances. Nonetheless, although the HRV concept is only marginally applicable in central hardwood forests, it has expanded our understanding of how species and systems respond to perturbations by creating a framework that emphasizes disturbance, and that perspective will be the key to managing for resilience and ultimately achieving our desired ecosystem services.

The goal of understanding HRV should not be to restore contemporary ecosystems to conditions that existed in the past. HRV should be less a model for narrowly defining desired future conditions (e.g., frequency distributions of different natural disturbances, vegetation cover type distributions, age class distributions, and patch size distribution) that future management is to recreate, than a tool for understanding processes such as tree colonization, adaptation, or extirpation in response to disturbances and past climates. By knowing what kinds of changes have occurred in the past and how ecosystems have responded, we can learn how these ecosystems functioned, under what ecological conditions they and the native species associated with them were sustained, deduce the mechanisms that underlie ecosystem dynamics and resilience—and then use those mechanisms to our advantage. Management decisions about everything from vertical and horizontal stand structures and age distributions to species composition and regeneration methods can be based on "knowledge of natural disturbance regimes (intensities, size distributions) and ecological processes" (Seymour and Hunter 1999). In the future, anthropogenic management will complement natural (and quasi-natural in the case of exotic species) disturbances to achieve desired ecosystem services by building resilience to avoid forest degradation and undesirable future conditions, such as the loss of forest extent, health and vitality, biodiversity, or protective and productive functions.

14.4 Role of History and Management in Central Hardwood Forests

Although the value of historical references in setting restoration goals appears to diminish (Millar et al. 2007), and with it the utility of the HRV approach (Wiens et al. 2012; Millar 2014), historical information and knowledge clearly plays a critical role in improving our understanding of how ecosystems have functioned in the past, identifying multiple ecological trajectories, relating ecological processes to ecosystem structure and composition, and establishing a baseline for gauging change, thus providing invaluable reference information for managers (Higgs et al. 2014). Contemporary ecosystems and the landscapes they comprise are shaped by an unlimited number of planned and unplanned, replicated and unreplicated experiments with shifting interactions (Diamond and Robinson 2010). The influence of the past results in detectable signatures (i.e., legacies) on contemporary ecosystem composition, structure, and function (Rhemtulla et al. 2009) from which managers can draw valuable lessons about the characteristics of a place (Higgs et al. 2014). Legacies link past events with the range of potential future ecosystem states, depending on decisions made by managers as well as events beyond our control (Suding et al. 2004).

In the Central Hardwood Region, the last 200 years in particular have seen a predominance of human influences shaping the present forest: settlement by subsistence farmers using European agricultural methods; abandonment of marginal agricultural lands that have since reverted back to forest; exploitative logging practices including clearcutting, burning, grazing, fire control, wildlife management; and pest introductions followed by a re-growing forest that is maturing (Hicks 1997; also see Chaps. 1 and 12). The planners and managers of today are challenged with promoting healthy ecosystems, maintaining areas of high conservation, enhancing resistance to climate change, enabling ecosystems to respond to change, and meeting society's demand for resources—all in landscapes that are sustainable, ecologically viable, and acceptable to society and facing unparalleled uncertainty. They will need to establish a clear set of realistic goals and priorities. Although some of these goals may be motivated partly by a sense of responsibility to ameliorate past damages (MacDougall et al. 2004) or the desire to recover and conserve native species or traditional ecological management practices (Higgs 2005), the ultimate goal is to shape future ecosystems that will be ecologically viable and enhance human well-being. Some of the activities required to achieve this outcome might include: (1) repairing ecosystem functions or other desired attributes (Ciccarese et al. 2012); (2) enhancing or enlarging specific ecosystems and habitat for species of concern (Thorpe and Stanley 2011); (3) restoring historic disturbance regimes by reintroducing disturbance agents such as fire in fire dependent communities (Arthur et al. 2012; Phillips et al. 2012; Brose et al. 2013); (4) manipulating forest structure by harvesting (Mitchell et al. 2002, Seymour et al. 2002); (5) enhancing understory composition by planting (Seymour et al. 2002); (6) regaining ecological integrity by restoring ecological processes that operate at larger landscape-level scales (forest landscape restoration) (Maginnis and Jackson 2007); (7) restoring interactions between function and structure ('functional restoration;' Phillips et al., 2012); and possibly (8) assisting species migration (Williams and Dumroese 2013). All these activities will succeed best if they are embedded in a framework of goals, strategies, and approaches to forest management.

By necessity, management strategies will be guided by a multidisciplinary, multiscale planning approach, and implemented by the application of a variety of silvicultural treatments within and across stands. Planners may have to re-think the scheduling of harvest and regeneration operations in space and time (e.g., rotation times, harvest unit sizes, harvest intensities), and become familiar with the suite of possibilities to enhance adaptation through stand- and landscape level management (e.g., Johnston 2009; Spies et al. 2010; FAO 2012). Some examples include (Spies et al. 2010): (1) altering the landscape structure to facilitate migration (i.e., corridors); (2) increasing the proportion of the landscape devoted to providing critical habitats and resilient ecosystem types to diffuse the risk that a single disturbance may have catastrophic effects on habitat; and (3) maintaining mature stands and older, well-established individual trees that are more resistant and resilient to disturbances and climate change where possible. Managers may have to modify silviculture to better match disturbance regimes, which may require thinking outside of traditional silvicultural systems and practicing a pragmatic, unemotional silviculture that does not lose sight of the site potential and the tree silvical requirements. Elements that should be considered include (1) application of a gradient of harvest intensities (from gaps to clearcuts with reserves); (2) retention of legacy trees and deadwood; (3) reintroduction of disturbances, such as fire in fire-dependent forest systems; (4) leaving some areas following natural disturbances to go through succession and stand development stages at their own pace (including refraining from salvage operations); and (5) application of variable density thinning regimes in dense young stands to provide more resources to promote resilience and vigor of individual tress and to promote diversity of species and stand structures.

14.5 Conclusions

The greatest potential for achieving desired future conditions by basing contemporary forest management on natural disturbance regimes and their HRV might be in ecosystems where these regimes have been altered primarily by past management, such as fire-dominated coniferous ecosystems. The combination of a natural disturbance regime that is feasible to emulate, historic management activities that can be easily modified in the future (ranging from simply ending fire suppression to implementing prescribed fires), a single, public landowner coordinating management over large expanses of forest, and areas with limited human infrastructure should facilitate meeting desired future conditions at sufficiently large spatial scales. In the eastern USA, a multitude of disturbance agents, a long history of forest conversion and poor management, and the loss of important tree species such as American chestnut will combine with large population centers, more human infrastructure, and primarily privately owned forest land to make achieving meaningful desired future conditions based on historic precedents of variability challenging, if not impossible, over large spatial scales.

That is not to say the HRV does not have any utility for forest managers in the eastern USA. The HRV framework is a new way of thinking about management challenges that places more emphasis on the relationships among disturbance regimes, ecological patterns, processes, and functions than previous management paradigms. Its biggest utility is in uniting different disciplines to develop management strategies to successfully regenerate desired species; wildlife biologists will advise on the structural characteristics needed for different species, landscape ecologists and forest planners will design landscapes that provide the necessary age diversity and connectivity to sustain wildlife populations, silviculturists will implement these strategies and plans, and human dimensions specialists will communicate these ideas to a public distrustful of 'centralized planning.' Considering the multitude of future challenges that await forest managers, rather than setting goals for narrow, and perhaps unattainable, desirable future conditions, the most effective approach may be to come together to define the undesirable future conditions we seek to avoid. In light of challenges posed by global and climate change, the most important contribution of HRV to forest management may be the recognition of the importance of H (history) on contemporary forests, rather than locking future generations into the RV of some past reference condition.

Acknowledgments I would like to thank Jeri Peck, Beverly Collins, Katie Greenberg, and several anonymous reviewers for their many helpful comments on earlier drafts.

References

Aitken, S. N., Yeaman, S., Holliday, J. A., Wang, T., & Curtis-McLane, S. (2008). Adaptation, migration or extirpation: Climate change outcomes for tree populations. *Evolutionary Applications, 1*, 95–111.

Allen, C. G., Macalady, A. K., Chenchouni, H., Bachelet, D., McDowell, N., Vennetier, M., Kitzberger, T., Rigling, A., Breshears, D. D., Hogg, E. H., Gonzalez, P., Fensham, R., Zhang, Z., Castro, J., Demidova, N., Lim, J.-H., Allard, G., Running, S. W., Semerci, A., & Cobb, N. (2010). A global overview of drought and heat-induced tree mortality reveals emerging climate change risks for forests. *Forest Ecology and Management, 259*, 660–684.

Aplet, G., & Keeton, W. (1999). Application of historical range of variability concepts to biodiversity conservation. In R. Baydack, H. Campa, & J. Haufler (Eds.), *Practical approaches to the conservation of biological diversity*. Washington, DC: Island Press.

Arthur, M., Alexander, H., Dey, D., Schweitzer, C., & Loftis, D. (2012). Refining the oak–fire hypothesis for management of oak-dominated forests of the Eastern United States. *Journal of Forestry, 110*, 257–266.

Bailey, S. W., Horsley, S. B., & Long, R. P. (2005). Thirty years of change in forest soils of the Allegheny Plateau, Pennsylvania. *Soil Science Society of America Journal, 69*, 681–690.

Baker, W. L. (2009). *Fire ecology in Rocky Mountain Landscapes*. Washington, DC: Island Press.

Baker, W. L., & Ehle, D. (2001). Uncertainty in surface-fire history: The case of ponderosa pine forests in the western United States. *Canadian Journal of Forest Research, 31*, 1205–1226.

Barnes, B. V. (1976). Succession in deciduous swamp communities of southern Michigan formerly dominated by American elm. *Canadian Journal of Botany, 54*, 19–24.

Barrett, S. W., & Arno, S. F. (1982). Indian fires as an ecological influence in the northern Rockies. *Journal of Forestry, 80*, 647–651.

Binkley, D., & Duncan, S. L. (2009). The past and future of Colorado's forests: Connecting people and ecology. *Ecology and Society, 14*, article 9.

Brose, P., Dey, D., Phillips, R., & Waldrop, T. (2013). A meta-analysis of the fire-oak hypothesis: Does prescribed burning promote oak reproduction in eastern North America? *Forest Science, 59*, 322–334.

Cappaert, D., McCullough, D. G., Poland, T. M., & Siegert, N. W. (2005). Emerald ash borer in North America: A research and regulatory challenge. *American Entomologist, 51*, 152–165.

Carpenter, S. R., & Gunderson, L. H. (2001). Coping with collapse: Ecological and social dynamics in ecosystem management. *BioScience, 51*, 451–457.

Chen, I.-C., Hill, J. K., Ohlemüller, R., Roy, D. B., & Thomas, C. D. (2011). Rapid range shifts of species associated with high levels of climate warming. *Science, 333*, 1024–1026.

Ciccarese, L., Mattsson, A., & Pettenella, D. (2012). Ecosystem services from forest restoration: Thinking ahead. *New Forest, 43*, 543–560.

Cissel, J. H., Swanson, F. J., McKee, W. A., & Burditt, A. L. (1994). Using the past to plan the future in the Pacific Northwest. *Journal of Forestry, 92*, 30–31.

Covington, W., Fulé, P., Moore, M., Hart, S., Klob, T., Mast, J., Sackett, S., & Wagner, M. (1997). Restoring ecosystem health in ponderosa pine forests of the southwest. *Journal of Forestry, 95*, 23–29.

Davidson, C. B., Gottschalk, K. W., & Johnson, J. E. (1999). Tree mortality following defoliation by the European gypsy moth (*Lymantria dispar*) in the United States: A review. *Forest Science, 45*, 74–84.

Dey, D. C., & Schweitzer, C. J. (2014). Restoration for the future: Endpoints, targets, and indicators of progress. *Journal of Sustainable Forestry, 33*(Sup 1), S43–S65.

Diamond, J., & Robinson, J. A. (2010). *Natural experiments of history*. Cambridge: Harvard University Press.

Dukes, J. S., Pontius, J., Orwig, D., Garnas, J. R., et al. (2009). Responses of insect pests, pathogens, and invasive plant species to climate change in the forests of northeastern North America: What can we predict? *Canadian Journal of Forest Research, 39*, 231–248.

Egan, D., & Howell, E. A. (Eds.). (2001). *The historical ecology handbook*. Washington, DC: Island Press.

Ellis, E. C., Kaplan, J. O., Fuller, D. Q., Vavrus, S., Klein Goldewijk, K., & Verburg, P. H. (2013). Used planet: A global history. *Proceedings of the National Academy of Sciences, 110*, 7978–7985.

Ellison, A. M., Bank, M. S., Clinton, B. D., Colburn, E. A., Elliott, K., Ford, C. R., Foster, D. R., Kloeppel, B. D., Knoepp, J. D., Lovett, G. M., Mohan, J., Orwig, D. A., Rodenhouse, N. L., Sobczak, W. V., Stinson, K. A., Stone, J. K., Swan, C. M., Thompson, J., Von Holle, B., & Webster, J. R. (2005). Loss of foundation species: Consequences for the structure and dynamics of forested ecosystems. *Frontiers in Ecology and the Environment, 3*, 479–486.

FAO. (2012). *Forest management and climate change: A literature review*. Rome: United Nations Food and Agriculture Organization.

Fischer, J., Lindenmayer, D. B., & Manning, A. D. (2006). Biodiversity, ecosystem function, and resilience: Ten guiding principles for commodity production landscapes. *Frontiers in Ecology and the Environment, 4*, 80–86.

Folke, C., Carpenter, S., Walker, B., Scheffer, M., Elmqvist, T., Gunderson, L., & Holling, C. S. (2004). Regime shifts, resilience, and biodiversity in ecosystem management. *Annual Review of Ecology, Evolution, and Systematics, 35*, 557–581.

Foster, D. R., Knight, D. H., & Franklin, J. F. (1998). Landscape patterns and legacies resulting from large, infrequent forest disturbances. *Ecosystems, 1*, 497–510.

Franklin, J. F., Mitchell, R. J., & Palik, B. J. (2007). *Natural disturbance and stand development principles for ecological forestry* (General technical report NRS, Vol. 19). Newtown Square: USDA Forest Service Northern Research Station.

Frelich, L. E., & Lorimer, C. G. (1991). Natural disturbance regimes in hemlock-hardwood forests of the Upper Great Lakes region. *Ecological Monographs, 61*, 145–164.

Fulé, P. Z., Covington, W. W., & Moore, M. M. (1997). Determining reference conditions for ecosystem management of southwestern ponderosa pine forests. *Ecological Applications, 7*, 895–908.

Ghandi, K. J. K., & Herms, D. A. (2010). Direct and indirect effects of alien insect herbivores on ecological processes and interactions in forests of eastern North America. *Biological Invasions, 12*, 389–305.

Gill, J. L., Williams, J. W., Jackson, S. T., Lininger, K. B., & Robinson, G. S. (2009). Pleistocene megafaunal collapse, novel plant communities, and enhanced fire regimes in North America. *Science, 326*, 1100–1103.

Hann, W. J., & Bunnell, D. L. (2001). Fire and land management planning and implementation across multiple scales. *International Journal of Wildland Fire, 10*, 389–403.

Hansen, A. J., Spies, T. A., Swanson, F. J., & Ohmann, J. L. (1991). Conserving biodiversity in managed forests. *BioScience, 41*, 382–392.

Hanula, J. L., & Berisford, C. W. (1984). Seasonal flight activity of the smaller European elm bark beetle, *Scolytus multistriatus* (Coleoptera: Scolytidae), and associates in Georgia. *Canadian Entomologist, 116*, 1251–1258.

Harris, J. A., Hobbs, R. J., Higgs, E., & Aronson, J. (2006). Ecological restoration and global climate change. *Restoration Ecology, 14*, 170–176.

Hart, J. L., Clark, S. L., Torreano, S. J., & Buchanan, M. L. (2012). Composition, structure, and dendroecology of an old-growth *Quercus* forest on the tablelands on the Cumberland Plateau, USA. *Forest Ecology and Management, 266*, 11–24.

Hicks, R. R. (1997). Humans and fire: A history of the Central Hardwoods. In *Proceedings: Workshop on fire, people, and the Central Hardwoods Landscape* (pp. 3–18), General technical report NE-274. Newtown Square: USDA Forest Service Northeastern Research Station.

Hicks, R. R. (1998). *Ecology and management of central hardwood forests*. New York: Wiley.

Higgs, E. S. (2005). The two culture problem: Ecological restoration and the integration of knowledge. *Restoration Ecology, 13*, 1–6.

Higgs, E., Falk, D. A., Guerrini, A., Hall, M., Harris, J., Hobbs, R. J., Jackson, S. T., Rhemtulla, J., & Throop, W. (2014). The changing role of history in restoration ecology. *Frontiers in Ecology and the Environment, 12*, 499–506.

Hobbs, R. J. (2013). Grieving for the past and hoping for the future: Balancing polarizing perspectives in conservation and restoration. *Restoration Ecology, 21*, 145–148.

Hobbs, R. J., Higgs, E., & Harris, J. A. (2009). Novel ecosystems: Implications for conservation and restoration. *Trends in Ecology & Evolution, 24*, 599–605.

Holling, C. S. (1992). Cross-scale morphology, geometry, and dynamics of ecosystems. *Ecological Monographs, 62*, 447–502.

Holling, C. S., & Meffe, G. K. (1996). Command and control and the pathology of natural resource management. *Conservation Biology, 10*, 328–337.

Hollingsworth, R. G., & Hain, F. P. (1991). Balsam woolly adelgid (Homoptera: Adelgidae) and spruce-fir decline in the southern Appalachians: Assessing pest relevance in a damaged ecosystem. *Florida Entomologist, 74*, 179–187.

Ibanez, I., Clark, J. S., LaDeau, S., & Lambers, J. H. R. (2007). Exploiting temporal variability to understand tree recruitment response to climate change. *Ecological Monographs, 77*, 163–177.

Johnson, P. S., Shifley, S. R., & Roberts, R. (2009). Chapter 7: Even-aged silvicultural methods. In P. S. Johnson, S. R. Shifley, & R. Roberts (Eds.), *The ecology and silviculture of oaks* (2nd ed., pp. 280–378). Cambridge: CAB International.

Johnston, M. (2009). *Vulnerability of Canada's tree species to climate change and management options for adaption: An overview for policy makers and practitioners*. Ottawa: Canadian Council of Forest Ministers.

Karnosky, D. F. (1979). Dutch elm disease: A review of the history, environmental implications, control, and research needs. *Environmental Conservation, 6*, 311–322.

Kashian, D. M., Turner, M. G., Romme, W. H., & Lorimer, C. G. (2005). Variability and convergence in stand structural development on a fire-dominated subalpine landscape. *Ecology, 86*, 643–654.

Kaufmann, M. R., Graham, R. T., Boyce, D. A., et al. (1994). *An ecological basis for ecosystem management* (General technical report RMRS, Vol. 246). Ft Collins: USDA Forest Service Rocky Mountain Research Station.

Kay, C. E. (1995). Aboriginal overkill and native burning: Implications for modern ecosystem management. *West Journal of Applied Forestry, 10*, 121–126.

Keane, R. E. (2013). Disturbance regimes and the historical range of variation in terrestrial ecosystems. *Encyclopedia of Biodiversity, 2*, 568–581.

Keane, R. E., Austin, M., Dalman, R., Field, C., Huth, A., Lexer, M., Peters, D., Solomon, A., & Wycoff, P. (2001). Tree mortality in gap models: Application to climate change. *Climatic Change, 51*, 509–540.

Keane, R. E., Ryan, K. C., Veblen, T. T., Allen, C. D., Logan, J., & Hawkes, B. (2002). *Cascading effects of fire exclusion in the Rocky Mountain ecosystems: A literature review* (General technical report RMRS, Vol. 91). Ft Collins: USDA Forest Service Rocky Mountain Research Station.

Keane, R. E., Hessburg, P. F., Landres, P. B., & Swanson, F. J. (2009). A review of the use of historical range of variation (HRV) in landscape management. *Forest Ecology and Management, 258*, 1025–1037.

Krech, S., III. (1999). *The ecological Indian myth and history*. New York: WW Norton.

Landres, P. B., Morgan, P., & Swanson, F. J. (1999). Overview of the use of natural variability concepts in managing ecological systems. *Ecological Applications, 9*, 1179–1188.

Laska, G. (2001). The disturbance and vegetation dynamics: A review and an alternative framework. *Plant Ecology, 157*, 77–99.

Laughlin, D. C., Bakker, J. D., Stoddard, M. T., Daniels, M. L., Springer, J. D., Gildar, C. N., Green, A. M., & Covington, W. W. (2004). Toward reference conditions: Wildfire effects on flora in an old-growth ponderosa pine forest. *Forest Ecology and Management, 199*, 137–152.

Levin, S. A. (1999). Towards a science of ecological management. *Conservation Ecology, 3*, 6.

Lindenmeyer, D. B., & Franklin, J. F. (2002). *Conserving forest biodiversity: A comprehensive multiscaled approach*. Washington, DC: Island Press.

Lindenmeyer, D. B., & Noss, R. E. (2006). Salvage logging, ecosystem processes, and biodiversity conservation. *Conservation Biology, 20*, 949–958.

Lindner, M., Maroschek, M., Netherer, S., Kremer, A., Barbati, A., Garcia-Gonzalo, J., & Marchetti, M. (2010). Climate change impacts, adaptive capacity, and vulnerability of European forest ecosystems. *Forest Ecology and Management, 259*, 698–709.

Litton, C. M., Ryan, M. G., & Knight, D. H. (2004). Effects of tree density and stand age on carbon allocation patterns in postfire lodgepole pine. *Ecological Applications, 14*, 460–475.

Lovett, G. M., Canham, C. D., Arthur, M. A., Weathers, K. C., & Fitzhugh, R. D. (2006). Forest ecosystem responses to exotic pests and pathogens in eastern North America. *BioScience, 56*, 395–405.

MacDougall, A. S., Beckwith, B. R., & Maslovat, C. Y. (2004). Defining conservation strategies with historical perspectives: A case study from a degraded oak grassland ecosystem. *Conservation Biology, 18*, 455–465.

Maginnis, S., & Jackson, W. (2007). What is FLR and how does it differ from current approaches? In J. Rietbergen-McCracken, S. Maginnis, & A. Sarre (Eds.), *The forest landscape restoration handbook*. London: Earthscan.

Manley, P. N., Brogan, G. E., Cook, C., Flores, M. E., Fullmer, D. G., Husare, S., Jimerson, T. M., Lux, L. M., McCain, M. E., Rose, J. A., Schmitt, G., Schuyler, J. C., Skinner, M. J. (1995). *Sustaining ecosystems: A conceptual framework*. Technical Publication R5-EM-TP-001. San Francisco: USDA Forest Service Pacific Southwest Region.

Mattson, W. J., Niemelä, P., Millers, I., & Inguanzo, Y. (1994). *Immigrant phytophagous insects on woody plants in the United States and Canada: An annotated list* (USDA General technical report NC, Vol. 169). St Paul: USDA Forest Service North Central Station.

Meehl, G. A., Zwiers, F., Evans, J., Knutson, T., Mearns, L., & Whetton, P. (2000). Trends in extreme weather and climate events: Issues related to modeling extremes in projections of future climate change. *Bulletin of the American Meteorological Society, 81*, 427–436.

Middleton, B. A. (1999). *Wetland restoration, flood pulsing, and disturbance dynamics*. New York: Wiley.

Millar, C. I. (2014). Historic variability: Informing restoration strategies, not prescribing targets. *Journal of Sustainable Forestry, 33*(Sup 1), S28–S42.

Millar, C. I., & Woolfenden, W. B. (1999). The role of climate change in interpreting historical variability. *Ecological Applications, 9*, 1207–1216.

Millar, C. I., Stephenson, N. L., & Stephens, S. L. (2007). Climate change and forests of the future: Managing in the face of uncertainty. *Ecological Applications, 17*, 2145–2151.

Mitchell, R., Palik, B., & Hunter, M. (2002). Natural disturbance as a guide to silviculture. *Forest Ecology and Management, 155*, 315–317.

Moore, S. A., Wallington, T. J., Hobbs, R. J., et al. (2009). Diversity in current ecological thinking: Implications for environmental management. *Environmental Management, 43*, 17–27.

Morgan, P., Aplet, G. H., Haufler, J. B., et al. (1994). Historical range of variability: A useful tool for evaluating ecosystem change. *Journal of Sustainable Forestry, 2*, 87–111.

Mori, A. S. (2011). Ecosystem management based on natural disturbances: Hierarchical context and non-equilibrium paradigm. *Journal of Applied Ecology, 48*, 280–292.

Mori, A. S., & Lertzman, K. P. (2011). Historic variability in fire-generated landscape heterogeneity of subalpine forests in the Canadian Rockies. *Journal of Vegetation Science, 22*, 45–58.

National Wildlife Federation. (2014). Disturbance. http://www.nwf.org/Wildlife/Wildlife-Conservation/Disturbance.aspx. Accessed 27 July 2014.

Naveh, Z. (1994). From biodiversity to ecodiversity: A landscape-ecology approach to conservation and restoration. *Restoration Ecology, 2*, 180–189.

Noble, I. R., & Dirzo, R. (1997). Forests as human-dominated ecosystems. *Science, 277*, 522–525.

North, M. P., & Keeton, W. S. (2008). Emulating natural disturbance regimes: An emerging approach for sustainable forest management. In R. Lafortezza, J. Chen, G. Sanesi, & T. R. Crow (Eds.), *Patterns and processes in forest landscapes* (pp. 341–372). Berlin: Springer Science + Business Media.

Noss, R. F. (1999). Assessing and monitoring forest biodiversity: A suggested framework and indicators. *Forest Ecology and Management, 115*, 135–146.

Notaro, M., Vavrus, S., & Liu, Z. (2007). Global vegetation and climate change due to future increases in CO_2 as projected by a fully coupled model with dynamic vegetation. *Journal of Climate, 20*, 70–88.

Nowacki, G. J., & Abrams, M. D. (2008). The demise of fire and "mesophication" of forests in the eastern United States. *BioScience, 58*(2), 123–138.

Omi, P. (2005). *Forest fires: A reference handbook*. Santa Barbara: ABC Clio.

Parker, G. R. (1993). Limitations on ecosystem management in the Central Hardwood Region In G. H. Aplet, N. Johnson, J. T. Olson, V. A. Sample (Eds.), *Defining sustainable forestry* (pp. 202–216). Washington, D.C.: Island Press.

Perry, D. A. (1998). The scientific basis of forestry. *Annual Review of Ecology and Systematics, 29*, 435–466.

Perry, G. L. (2002). Landscapes, space and equilibrium: Shifting viewpoints. *Progress in Physical Geography, 26*, 339–359.

Peterson, C. J. (2000). Catastrophic wind damage to North American forests and the potential impact of climate change. *Science of the Total Environment, 262*(3), 287–311.

Phillips, J. D. (2004). Divergence, sensitivity, and nonequilibrium in ecosystems. *Geographical Analysis, 36*, 369–383.

Phillips, R. J., Waldrop, T. A., Brose, P. H., & Wang, G. G. (2012). Restoring fire adapted forests in eastern North America for biodiversity conservation and hazardous fuels reduction. In J. A. Stanturf, P. Madsen, & D. Lamb (Eds.), *A goal-oriented approach to forest landscape restoration*. New York: Springer.

Poisot, T., Bever, J. D., Nemri, A., Thrall, P. H., & Hochberg, M. E. (2011). A conceptual framework for the evolution of ecological specialization. *Ecology Letters, 14*, 841–851.

Poland, T., & McCullough, D. (2006). Emerald ash borer: Invasion of the urban forest and the threat to North America's ash resource. *Journal of Forestry, 104*, 118–124.

Pyne, S. J. (1982). *Fire in America. A cultural history of wildland and rural fire*. Princeton: Princeton University Press.

Reed, W. J., Larson, C. P. S., Johnson, E. A., & Macdonald, G. M. (1998). Estimation of temporal variations in historical fire frequency from time-since-fire map. *Forest Science, 44*, 465–574.

Rhemtulla, J. M., Mladenoff, D. J., & Clayton, M. K. (2009). Legacies of historical land use on regional forest composition and structure in Wisconsin, USA (mid 1800s to 1930s to 2000s). *Ecological Applications, 19*, 1061–1078.

Romme, W. H., Wiens, J. A., & Safford, H. D. (2012). Setting the stage: Theoretical and conceptual background of historical range of variation. In J. A. Wiens, G. D. Hayward, H. D. Safford, & C. Goffen (Eds.), *Historical environmental variation in conservation and natural resource management* (pp. 3–18). Chichester: Wiley-Blackwell.

Ruffner, C. M., & Abrams, M. D. (2002). Dendrochronological investigation of disturbance history for a native American site in northwestern Pennsylvania. *Journal of the Torrey Botanical Society, 129*, 251–260.

Safford, H. D., Wiens, J. A., & Hayward, G. D. (2012). The growing importance of the past in managing ecosystems of the future. In J. A. Wiens, G. D. Hayward, & H. D. Safford (Eds.), *Historical ecological variation in conservation and natural resource management*. Oxford: Wiley-Blackwell.

Schoennagel, T., Smithwick, E. A. H., & Turner, M. G. (2008). Landscape heterogeneity following large fires: Insights from Yellowstone National Park, USA. *International Journal of Wildland Fire, 17*, 742–753.

Seymour, R., & Hunter, M. (1999). Principles of ecological forestry. In M. Hunter (Ed.), *Maintaining biodiversity in forest ecosystems* (pp. 22–61). Cambridge: Cambridge University Press.

Seymour, R. S., White, A. S., & deMaynadier, P. G. (2002). Natural disturbance regimes in northeastern North America—Evaluating silvicultural systems using natural scales and frequencies. *Forest Ecology and Management, 155*, 357–367.

Slocum, M. G., Platt, W. J., Beckage, B., Panko, B., & Lushine, J. B. (2007). Decoupling natural and anthropocentric fire regimes: A case study in Everglades National Park, Florida. *Natural Areas Journal, 27*, 41–55.

Society of American Foresters (SAF). (2014). Dictionary of forestry. http://dictionaryofforestry.org/dict/term/silviculture. Accessed 27 July 2014.

Spies, T. A., & Turner, M. G. (1999). Dynamic forest mosaics. In M. L. Hunter (Ed.), *Maintaining biodiversity in forest ecosystems* (pp. 95–160). Cambridge: Cambridge University Press.

Spies, T. A., Giesen, T. W., Swanson, F. J., Franklin, J. F., Lach, D., & Johnson, K. N. (2010). Climate change adaptation strategies for federal forests of the Pacific Northwest, USA: Ecological, policy, and socio-economic perspectives. *Landscape Ecology, 25*, 1185–1199.

Sprugel, D. G. (1991). Disturbance, equilibrium, and environmental variability: What is 'natural vegetation' in a changing environment? *Biological Conservation, 58*, 1–18.

Steffen, W., Crutzen, P. J., & McNeill, J. R. (2007). The Anthropocene: Are humans now overwhelming the great forces of nature. *Ambio, 36*, 614–621.

Stephenson, N. L. (1999). Reference conditions for giant sequoia forest restoration: Structure, process, and precision. *Ecological Applications, 9*, 1253–1265.

Suding, K. N., Gross, K. L., & Houseman, G. R. (2004). Alternative states and positive feedbacks in restoration ecology. *Trends in Ecology & Evolution, 19*, 46–53.

Swetnam, T. W., Allen, C. D., & Betancourt, J. L. (1999). Applied historical ecology: Using the past to manage the future. *Ecological Applications, 9*, 1189–1206.

Tang, S. M., & Gustafson, E. J. (1997). Perception of scale in forest management planning: Challenges and implications. *Landscape and Urban Planning, 39*, 1–9.

Thorpe, A. S., & Stanley, A. G. (2011). Determining appropriate goals for restoration of imperiled communities and species. *Journal of Applied Ecology, 48*, 275–279.

Townsend Peterson, A., & Sachetti-Pereira, R. (2004). Potential geographic distribution of *Anoplophora glabripennis* (Coleoptera: Cerambycidae) in North America. *American Midland Naturalist, 151*, 170–178.

Turner, M. G., & Romme, W. H. (1994). Landscape dynamics in crown fire ecosystems. *Landscape Ecology, 9*(1), 59–77.

Turner, M. G., Romme, W. H., Gardner, R. H., O'Neill, R. V., & Kratz, T. K. (1993). A revised concept of landscape equilibrium: Disturbance and stability on scaled landscapes. *Landscape Ecology, 8*, 213–227.

Turner, M. G., Baker, W. L., Peterson, C. J., & Peet, R. K. (1998). Factors influencing succession: Lessons from large, infrequent natural disturbances. *Ecosystems, 1*, 511–523.

Turner, M. G., Tinker, D. B., Romme, W. H., Kashian, D. M., & Litton, C. M. (2004). Landscape patterns of sapling density, leaf area, and aboveground net primary production in postfire lodgepole pine forests, Yellowstone National Park (USA). *Ecosystems, 7*, 751–775.

USDA Forest Service. (2012). National forest system land management planning. *Federal Register, 77*, 21162–212276.

van Wagner, C. E. (1978). Age-class distribution and the forest fire cycle. *Canadian Journal of Forest Research, 8*, 220–227.

Veblen, T. T. (2003). Historic range of variability of mountain forest ecosystems: Concepts and applications. *Forestry Chronicle, 79*, 223–226.

Vose, J. M., Peterson, D. L., & Patel-Weynand, T. (2012). *Effects of climatic variability and change on forest ecosystems: A comprehensive science synthesis for the US forest sector* (General technical report PNW, Vol. 870). Portland: USDA Forest Service Pacific Northwest Research Station.

Watt, A. S. (1947). Pattern and process in the plant community. *Journal of Ecology, 35*, 1–22.

Weaver, H. (1974). Effects of fire on temperate forests: Western United States. In T. T. Kozlowski (Ed.), *Fire and ecosystems* (pp. 279–319). New York: Academic Press.

White, P. S., & Jentsch, A. (2001). The search for generality in studies of disturbance and ecosystem dynamics. *Progress in Botany, 62*, 399–450.

White, P. S., & Pickett, S. T. A. (1985). Natural disturbance and patch dynamics: An introduction. In S. T. A. Pickett & P. S. White (Eds.), *The ecology of natural disturbance and patch dynamics* (pp. 3–13). New York: Academic Press.

White, P. S., Harrod, J., Romme, W. H., & Betancourt, J. (1999). Disturbance and temporal dynamics. In J. C. Johnson, A. J. Malk, W. T. Sexton, & R. Szaro (Eds.), *Ecological stewardship: A common reference for ecosystem management* (pp. 281–305). Oxford: Oxford University Press.

White, P. S., Collins, B., & Wein, G. R. (2011). Natural disturbances and early successional habitats. In C. Greenberg, B. Collins, & F. Thompson III (Eds.), *Sustaining young forest communities* (pp. 27–40). Dordrecht: Springer Netherlands.

Whitford, W. G., Rapport, D. J., & deSoyza, A. G. (1999). Using resistance and resilience measurements for "fitness" tests in ecosystem health. *Journal of Environmental Management, 57*(1), 21–29.

Wiens, J. A., Crawford, C. S., & Gosz, J. R. (1985). Boundary dynamics: A conceptual framework for studying landscape ecosystems. *Oikos, 45*, 421–427.

Wiens, J. A., Hayward, G. D., Safford, H. D., & Giffen, C. (2012). The growing importance of the past in managing ecosystems of the future. In J. A. Wiens, G. D. Hayward, & H. D. Safford (Eds.), *Historical ecological variation in conservation and natural resource management*. Oxford: Wiley-Blackwell.

Wikipedia. (2014). Disturbance (ecology). http://en.wikipedia.org/wiki/Disturbance(ecology). Accessed 27 July 2014.

Williams, R. J., & Bradstock, R. A. (2008). Large fires and their ecological consequences: Introduction to the species issue. *International Journal of Wildland Fire, 17*, 685–687.

Williams, M. I., & Dumroese, R. K. (2013). Preparing for climate change: Forestry and assisted migration. *Journal of Forestry, 114*, 287–297.

Williams, J. W., & Jackson, S. T. (2007). Novel climates, no-analog communities, and ecological surprises. *Frontiers in Ecology and the Environment, 5*, 475–482.

Williams, J. W., Jackson, S. T., & Kutzbach, J. E. (2007). Projected distributions of novel and disappearing climates by 2100 AD. *Proceedings of the National Academy of Sciences, 104*, 5738–5742.

Wu, J., Jones, B., Li, H., & Loucks, O. L. (2006). *Scaling and uncertainty analysis in ecology*. Dordrecht: Springer.

Yaffee, S. L. (1999). Three faces of ecosystem management. *Conservation Biology, 13*, 713–725.

Index

C.H. Greenberg, B.S. Collins (eds.), *Natural Disturbances and Historic Range of Variation*, Managing Forest Ecosystems 32,
DOI 10.1007/978-3-319-21527-3

MIX
Papier aus verantwortungsvollen Quellen
Paper from responsible sources
FSC® C105338

If you have any concerns about our products,
you can contact us on
ProductSafety@springernature.com

In case Publisher is established outside the EU,
the EU authorized representative is:
Springer Nature Customer Service Center GmbH
Europaplatz 3, 69115 Heidelberg, Germany

Printed by Libri Plureos GmbH
in Hamburg, Germany